Advances in
Heterocyclic
Chemistry

Volume 15

Editorial Advisory Board

A. Albert
A. T. Balaban
J. Gut
J. M. Lagowski
J. H. Ridd
Yu. N. Sheinker
H. A. Staab
M. Tišler

Advances in

HETEROCYCLIC CHEMISTRY

Edited by

A. R. KATRITZKY

A. J. BOULTON

School of Chemical Sciences
University of East Anglia
Norwich, England

Volume 15

Academic Press · New York and London · 1973

A Subsidiary of Harcourt Brace Jovanovich, Publishers

COPYRIGHT © 1973, BY ACADEMIC PRESS, INC.
ALL RIGHTS RESERVED.
NO PART OF THIS PUBLICATION MAY BE REPRODUCED OR
TRANSMITTED IN ANY FORM OR BY ANY MEANS, ELECTRONIC
OR MECHANICAL, INCLUDING PHOTOCOPY, RECORDING, OR ANY
INFORMATION STORAGE AND RETRIEVAL SYSTEM, WITHOUT
PERMISSION IN WRITING FROM THE PUBLISHER.

ACADEMIC PRESS, INC.
111 Fifth Avenue, New York, New York 10003

United Kingdom Edition published by
ACADEMIC PRESS, INC. (LONDON) LTD.
24/28 Oval Road, London NW1

LIBRARY OF CONGRESS CATALOG CARD NUMBER: 62-13037

PRINTED IN THE UNITED STATES OF AMERICA

Contents

CONTRIBUTORS vii

PREFACE ix

DEDICATION—ADRIEN ALBERT xi

Heterocyclic Oligomers
ADRIEN ALBERT AND HIROSHI YAMAMOTO

I. Introduction 1
II. Initiation and Termination of Oligomerization 2
III. Classification and Discussion of Oligomers 3
IV. Conclusion 64

The Oxidation of Monocyclic Pyrroles
G. P. GARDINI

I. Introduction 67
II. Autoxidation 68
III. Photo-oxidation 72
IV. Oxidation by Chemical Reagents 79
V. Pyrrole Blacks 95
VI. Conclusions 96
Note Added in Proof 97

The Chemistry of 4-Oxy- and 4-Keto-1,2,3,4-tetrahydroisoquinolines
J. M. BOBBITT

I. Introduction 99
II. Synthesis of 4-Oxy-1,2,3,4-tetrahydroisoquinolines . . 103
III. Reactions of 4-Oxy-1,2,3,4-tetrahydroisoquinolines . . 116
IV. Naturally Occurring 4-Oxy-1,2,3,4-tetrahydroisoquinolines . 122
V. The State of the Art of Isoquinoline Synthesis . . . 123
VI. Tables of Specific Compounds 128

Isotopic Hydrogen Labeling of Heterocyclic Compounds by One-Step Methods
G. E. CALF AND J. L. GARNETT

I. Introduction 137
II. Radiation-Induced Methods 138
III. Heterogeneous Metal-Catalyzed Exchange 149
IV. Current Trends in Labeling 174

The Chemistry of 1-Pyrindines
FILLMORE FREEMAN

I.	Introduction	187
II.	Preparation and Reactions	189
III.	Biomedical Applications	229

3-Oxo-2,3-dihydrobenz[d]isothiazole-1,1-dioxide (Saccharin) and Derivatives
H. HETTLER

I.	Introduction	234
II.	Nomenclature	234
III.	Structures and Physical Properties	235
IV.	The 3-Oxo-2,3-dihydrobenz[d]isothiazole-1,1-dioxide System	239
V.	The 3-Thioxo-2,3-dihydrobenz[d]isothiazole-1,1-dioxide System	260
VI.	3-Alkoxy and 3-Aryloxybenz[d]isothiazole-1,1-dioxides	262
VII.	3-Aminobenz[d]isothiazole-1,1-dioxide and Its Derivatives	265
VIII.	3-Imino-2,3-dihydrobenz[d]isothiazole-1,1-dioxide and Its Derivatives	269
IX.	Miscellaneous Compounds	270
X.	(Benz[d]isothiazolyl-3-oximino) alkane-1,1-dioxides	271
XI.	3-Cholorobenz[d]isothiazole-1,1-dioxide	273
XII.	Elimination Reactions with 3-Substituted Benz[d]isothiazole-1,1-dioxides	274

Applications of Nuclear Magnetic Resonance Spectroscopy to Heterocyclic Chemistry: Indole and Its Derivatives
SHIVAYOGI P. HIREMATH AND RAMACHANDRA S. HOSMANE

I.	Introduction	278
II.	The Spectrum of Indole	279
III.	Elucidation of α and β Substitutions	282
IV.	Solvent and Concentration Dependence of the 7-Proton Resonance	284
V.	Detection of α, β, and N–H Signals	285
VI.	Benzenoid Signals	286
VII.	Substituted Indoles	291
VIII.	Other Applications	318
IX.	Conclusions	322

AUTHOR INDEX 325

CUMULATIVE INDEX OF TITLES 348

Contributors

Numbers in parentheses indicate the pages on which the authors' contributions begin.

ADRIEN ALBERT, *Department of Medical Chemistry, John Curtin School of Medical Research, Canberra, Australia and Fachbereich Chemie, Universität Konstanz, West Germany* (1)

J. M. BOBBITT, *Department of Chemistry, University of Connecticut, Storrs, Connecticut* (99)

G. E. CALF, *Chemistry School, The University of New South Wales, Kensington N.S.W., Australia* (137)

FILLMORE FREEMAN, *Department of Chemistry, California State College, Long Beach, California* (187)

G. P. GARDINI, *Istituto Policattedra di Chimica Organica, Università di Parma, Parma, Italy* (67)

J. L. GARNETT, *Chemistry School, The University of New South Wales, Kensington N.S.W., Australia* (137)

H. HETTLER,* *Department of Chemistry, Max-Planck-Institute for Experimental Medicine, Göttingen, Germany* (233)

SHIVAYOGI P. HIREMATH,† *Department of Chemistry, Karnatak University, Dharwar-3, India* (277)

RAMACHANDRA S. HOSMANE, *Department of Chemistry, Karnatak University, Dharwar-3, India* (277)

HIROSHI YAMAMOTO, *Department of Medical Chemistry, John Curtin School of Medical Research, Canberra, Australia and Fachbereich Chemie, Universität Konstanz, West Germany* (1)

---

*Present address: 7858 Weil/Rhein, Hauptstr., 163, West Germany.

†*Present address:* Department of Chemistry, Karnatak University Postgraduate Centre, Gulbarga, India.

Preface

The present volume contains seven chapters, five of which deal with aspects of specific ring systems: 4-oxy- and 4-keto-1,2,3,4-tetrahydroisoquinolines (J. M. Bobbitt), the oxidation products of pyrroles (G. P. Gardini), the pyrindines (4- and 5-azaindenes) (F. Freeman), the NMR spectra of indoles (S. P. Hiremath and R. S. Hosmane), and the 1,2-benzoisothiazole-S-dioxides—(saccharin and its derivatives) (H. Hettler). In the general field of heterocyclic chemistry, hydrogen exchange reactions, mainly on metal surfaces, are dealt with (G. E. Calf and J. L. Garnett), and the remaining chapter is a systematic review of heterocyclic oligomers (A. Albert and H. Yamamoto). The volume is dedicated to Professor Adrien Albert, and this Preface is followed by a brief biography and bibliography.

A. R. KATRITZKY
A. J. BOULTON

Adrien Albert

Dedication—Professor Adrien Albert

This volume is dedicated to Adrien Albert to mark the occasion of his retirement as Head of the Department of Medical Chemistry of the John Curtin School of Medical Research, Australian National University, Canberra.

Adrien Albert is known by reputation to every heterocyclic chemist, and, happily, by many in person. His papers and books have had a decisive influence on the subject, and the marks of his teaching and research are deep-rooted. He has been a particular friend of these Advances having served on the Advisory Board since its inception. All heterocyclic chemists will wish him well in the new phase of his career now beginning.

Biographical Note (kindly supplied by Dr. Desmond J. Brown).

Adrien Albert was born in Sydney on November 19, 1907; he received his early education at the Scots College, Rose Bay, New South Wales; and then went on to the University of Sydney from which he graduated B.Sc. with First Class Honours and the University Medal in 1932. He proceeded to the faculty of medicine in the University of London where he obtained his Ph.D. in 1937. He was awarded the degree of D.Sc. from the University of London in 1947.

During the years 1938–1947 he taught at the University of Sydney where his research was supported by the Australian National Health and Medical Research Council. He became a Research Fellow at the Wellcome Research Institution in London from 1947 to 1948 and he also acted as advisor on medical chemistry to the Medical Directorate of the Australian Army from 1942 to 1947.

In 1948 he took up the position of Professor and Head of Department of Medical Chemistry of the John Curtin School of Medical Research, Australian National University and he was elected a Fellow of the Australian Academy of Science in 1958.

He retired at the end of 1972 from his Chair at Canberra and now occupies a Visiting Fellowship in the Research School of Chemistry of the Australian National University. His recreations include music, photography and the Australian flora.

List of Publications

ORIGINAL PAPERS

1. Chemotherapeutic Studies in the Acridine Series. Part I. 2,6- and 2,8-Diaminoacridines.
    (With W. H. Linnell.)
    J. Chem. Soc., 88-93 (1936).

2. Chemotherapeutic Studies in the Acridine Series. Part II. 2-Amino-, 2,5-, 2,7- and 2,9-Diaminoacridines.
(With W. H. Linnell.)
J. Chem. Soc., 1614-1619 (1936).

3. Chemotherapeutic Studies in the Acridine Series. Part III. 4-Amino-, 1,3-, 1,7- and 3,6-Diaminoacridines.
(With W. H. Linnell.)
J. Chem. Soc., 22-26 (1938).

4. Chemotherapeutic Studies in the Acridine Series. The Relation between Chemical Constitution and Biological Action in Simple Aminoacridines.
(With A. E. Francis, L. P. Garrod and W. H. Linnell.)
Brit. J. Exp. Pathol. **19**, 41-52 (1938).

5. The Effect of Soaps in Increasing the Water-Solubility of Essential Oils.
J. Soc. Chem. Ind. **58**, 196-199 (1939).

6. Acridine, Syntheses and Reactions. Part I. Synthesis of Proflavine from m-Phenylenediamine and its Derivatives.
J. Chem. Soc., 121-125 (1941).

7. Acridine Syntheses and Reactions. Part II. Synthesis of Proflavine from m-Phenylenediamine and its Derivatives (continued).
J. Chem. Soc. 484-487 (1941).

8. Improved Syntheses of Aminoacridines. Part I. The Five Isomeric Monoaminoacridines.
(With B. Ritchie.)
J. Soc. Chem. Ind. **50**, 120-123 (1941).

9. Improved Syntheses of Aminoacridines. Part II. 2,5-Diamino-7-ethoxyacridine, the Base of Rivanol.
(With W. Gledhill.)
J. Soc. Chem. Ind. **61**, 159-160 (1942).

10. The Influence of Chemical Constitution on Antiseptic Activity. Part I. A Study of the Mono-Aminoacridines.
(With S. Rubbo and M. Maxwell.)
Brit. J. Exp. Pathol. **23**, 69-83 (1942).

11. The Nature of the Amino-group in Aminoacridines. Part I. Evidence from Electrometric Studies.
(With R. Goldacre.)
J. Chem. Soc., 454-458 (1943).

12. The Nature of the Amino-group in Aminoacridines. Part II. Evidence from Chemical Reactions.
J. Chem. Soc., 458-462 (1943).

13. Aminoacridines. Some Partition and Surface Phenomena.
(With R. Goldacre and E. Heymann.)
J. Chem. Soc., 651-654 (1943).

14. Systematic Investigation of Dyeing Properties of Aminoacridines. The Five Monoaminoacridines.
(With C. L. Bird.)
J. Soc. Dyers Colour. **59**, 74-76 (1943).

15. Basicities of the Aminoquinolines: Comparison with the Aminoacridines and Aminopyridines.
    (With R. Goldacre.)
    Nature (London) **153**, 467-468 (1944).
16. The Antibacterial Action of Arsenic.
    (With J. Falk and S. Rubbo.)
    Nature (London) **153**, 712 (1944).
17. The Influence of Chemical Constitution on Antibacterial Activity. Part II. A general Survey of the Acridine Series.
    (With S. Rubbo, R. Goldacre, M. Davey and J. Stone.)
    Brit. J. Exp. Pathol. **26**, 160-192 (1945).
18. Absorption Spectra of Aminoacridines. Part III. The Hydroxyacridines.
    (With L. Short.)
    J. Chem. Soc., 760-763 (1945).
19. Improved Syntheses of Aminoacridines. Part III. 1,9-Dimethylproflavine.
    (With D. Magrath.)
    J. Soc. Chem. Ind., London **64**, 30-31 (1945).
20. Improved Syntheses of Aminoacridines. Part IV. Substituted 5-Aminoacridines.
    (With W. Gledhill.)
    J. Soc. Chem. Ind., London **64**, 169-172 (1945).
21. The Ionization of Acridine Bases.
    (With R. Goldacre.)
    J. Chem. Soc., 706-713 (1946).
22. Improved Syntheses of Acridines. Part V. The Dechlorination of 5-chloroacridines and a new Synthesis of Acridine.
    (With J. B. Willis.)
    J. Soc. Chem. Ind., London **65**, 26 (1946).
23. Acridine Syntheses and Reactions. Part III. Synthesis of Aminoacridines from Formic Acid and Amines.
    J. Chem. Soc., 244-250 (1947).
24. Improved Syntheses of Phenazines. Part I. 1-Aminophenazine and 1,3-Diaminophenazine.
    (With H. Duewell.)
    J. Soc. Chem. Ind., London **66**, 11-13 (1947).
25. The Influence of Chemical Constitution on Antibacterial Activity. Part III. A Study of 8-Hydroxyquinoline (oxine) and related Compounds.
    (With S. Rubbo, R. Goldacre and B. Balfour.)
    Brit. J. Exp. Pathol. **28**, 69-87 (1947).
26. The Choice of a Chelating Agent for inactivating Trace Metals. Part 1. A Survey of commercially available Chelating Agents.
    (With W. Gledhill.)
    Biochem. J. **41**, 529-533 (1947).
27. The Choice of a Chelating Agent for inactivating Trace Metals. Part 2. Derivatives of Oxine (8-Hydroxyquinoline).
    (With D. Magrath.)
    Biochem. J. **41**, 534-545 (1947).
28. Benzylamine and analogues of chemotherapeutic Diamidines.
    (With J. Mills and R. P. Royer.)
    J. Chem. Soc., 1452-1455 (1947).

29. Some Syntheses in the Benzoquinoline, Benzacridine and Phenanthridine Series.
   (With D. J. Brown and H. Duewell.)
   J. Chem. Soc., 1284-1295 (1948).
30. Acridine Syntheses and Reactions. Part IV. A New Synthesis of Aminoacridines from Formic Acid and Diarylamines.
   J. Chem. Soc., 1225-1230 (1948).
31. Ionization Constants of Heterocyclic Substances. Part I. The Strength of Heterocyclic Bases.
   (With R. Goldacre and J. Phillips.)
   J. Chem. Soc., 2240-2249 (1948).
32. Hydrogen Carriers produced by Hæmatin-catalysed Peroxidations.
   (With J. E. Falk.)
   Biochem. J. **44**, 129-136 (1949).
33. Acridine Syntheses and Reactions. Part V. A New Dehalogenation of 5-Chloroacridine and its Derivatives.
   (With R. Royer.)
   J. Chem. Soc., 1148-1151 (1949).
34. The Influence of Chemical Constitution on Antibacterial Activity. Part IV. A Survey of Heterocyclic Bases with special reference to Benzoquinolines, Phenanthridines, Benzacridines, Quinolines and Pyridines.
   (With S. Rubbo and M. Burvill.)
   Brit. J. Exp. Pathol. **30**, 159-175 (1949).
35. Therapeutic Interference: Some Interpretations.
   Symp. Soc. Exp. Biol. **3**, 318-326 (1949).
36. Quantitative Studies of the Avidity of naturally-occurring Substances for Trace Metals. Part 1. Amino-acids having only two Ionizing Groups.
   Biochem. J. **47**, 531-538 (1950).
37. The Influence of Chemical Constitution on Antibacterial Activity. Part V. The Antibacterial Action of 8-Hydroxyquinoline (Oxine).
   (With S. Rubbo and M. Gibson.)
   Brit. J. Exp. Pathol. **31**, 425-441 (1950).
38. 4-Aminopyridine.
   J. Chem. Soc., 1376 (1951).
39. Pteridine Studies. Part I. Pteridine, and 2- and 4-Amino- and 2- and 4-Hydroxy-pteridine.
   (With D. J. Brown and G. W. H. Cheeseman.)
   J. Chem. Soc., 474-485 (1951).
40. Pteridine Studies. Part II. 6- and 7-Hydroxypteridine and their Derivatives.
   (With D. J. Brown and G. W. H. Cheeseman.)
   J. Chem. Soc., 1620-1630 (1952).
41. Pteridine Studies. Part III. The Solubility and Stability to Hydrolysis of Pteridines.
   (With D. J. Brown and G. W. H. Cheeseman.)
   J. Chem. Soc., 4219-4232 (1952).
42. Analogues of 8-Hydroxyquinoline having additional cyclic Nitrogen Atoms. Part 1.
   (With A. Hampton.)
   J. Chem. Soc., 4985-4993 (1952).

43. Pteridine Syntheses. Part I. Leucopterin and Xanthopterin.
(With H. C. Wood.)
J. Appl. Chem. **2**, 591-592 (1952).

44. Quantitative Studies of the Avidity of naturally-occurring Substances for Trace Metals. Part 2. Amino-acids having three Ionizing Groups.
Biochem. J. **50**, 690-698 (1952).

45. Pteridine Syntheses. Part II. Isoxanthopterin.
(With H. C. Wood.)
J. Appl. Chem. **3**, 521-523 (1953).

46. Quantitative Studies of the Avidity of naturally-occurring Substances for Trace Metals. Part 3. Riboflavine, the Purines and Pteridines.
Biochem. J. **54**, 646-654 (1953).

47. The Influence of Chemical Constitution on Antibacterial Activity. Part VI. The Bactericidal Action of 8-Hydroxyquinoline (Oxine).
(With S. Rubbo and M. Gibson.)
Brit. J. Exp. Pathol. **34**, 119-130 (1953).

48. Pteridine Studies. Part IV. 4,6- and 4,7-Dihydroxypteridine.
(With D. J. Brown.)
J. Chem. Soc., 74-79 (1953).

49. Analogues of 8-Hydroxyquinoline having additional cyclic Nitrogen Atoms. Part II.
(With A. Hampton.)
J. Chem. Soc., 505-513 (1954).

50. Purine Studies. Part I. Stability to Acid and Alkali; Solubility; Ionization.
(With D. J. Brown.)
J. Chem. Soc., 2060-2071 (1954).

51. Pteridine Studies. Part V. The Monosubstituted Pteridines.
(With D. J. Brown and H. C. Wood.)
J. Chem. Soc., 3832-3839 (1954).

52. The Influence of Chemical Constitution on Antibacterial Activity. Part VII. The Site of Action of 8-Hydroxyquinoline.
(With A. Hampton, F. Selbie and R. Simon.)
Brit. J. Exp. Pathol. **35**, 75-84 (1954).

53. The Destruction of *Iso*nicotinic Acid Hydrazide in the Presence of Hæmin.
(With C. W. Rees.)
Biochem. J. **61**, 128-131 (1955).

54. The Solubility of 8-Hydroxymethylpurine.
Chem. Ind. (London) 202 (1955).

55. Pteridine Studies. Part VII. The Degradation of 4-, 6-, and 7-Hydroxypteridine.
J. Chem. Soc., 2690-2699 (1955).

56. Pteridine Studies. Part VIII. Methylation of Hydroxypteridines and Degradation of the Products.
(With D. J. Brown and H. C. Wood.)
J. Chem. Soc., 2066-2075 (1956).

57. Pteridine Studies. Part X. Pteridines with more than one Hydroxy- or Amino-Group.
(With J. Lister and C. Pedersen.)
J. Chem. Soc., 4621-4628 (1956).

58. The Transformation of Purines into Pteridines, and its biological Significance.
    Biochem J. **65**, 124-127 (1956).
59. Photo-reduction of Pteridines.
    Nature (London) **178**, 1077 (1956).
60. The Lability of 1,4,6-Triazanaphthalene.
    (With C. Pedersen.)
    J. Chem. Soc., 4683-4684 (1956).
61. The Solubility of Quinoline and the Hydroxyquinolines.
    Chem. Ind. (London) 252 (1956).
62. Ionization Constants of Heterocyclic Substances. Part II. Hydroxy-derivatives of six-membered rings.
    (With J. Phillips.)
    J. Chem. Soc., 1294-1304 (1956).
63. The Influence of Chemical Constitution on Antibacterial Activity. Part VIII. 2-Mercaptopyridine-N-oxide.
    (With C. W. Rees and A. J. Tomlinson.)
    Brit. J. Exp. Pathol. **37**, 500-511 (1956).
64. The mode of Action of Isoniazid.
    Nature (London) **177**, 525 (1956).
65. The Avidity of Tetracycline for Metallic Cations.
    (With C. W. Rees.)
    Nature (London) **177**, 433 (1956).
66. Why some Metal-binding Substances are antibacterial.
    (With C. W. Rees and A. J. Tomlinson.)
    Rec. Trav. Chim. Pays-Bas **75**, 819-824 (1956).
67. The Spectrum of Thiazole.
    Chem. Ind. (London), 1271 (1957).
68. Addition to Double-bonds in N-heteroaromatic six-membered Rings.
    Current Trends Heterocyc. Chem. Proc. Symp., 1957, 20-29 (1958).
69. Ionization Constants of Heterocyclic Substances. Part III. Mercapto-derivatives of six-membered Rings.
    (With G. B. Barlin.)
    J. Chem. Soc., 2384-2396 (1959).
70. Ionization Constants of Heterocyclic Substances. Part IV. The Effect of a tautomerizable *alpha*-Substituent on the Ionization of a Second Substituent.
    J. Chem. Soc., 1020-1023 (1960).
71. Pteridine Studies. Part XI. The Decomposition of 2-Hydroxypteridine by Acid and Alkali.
    (With F. Reich.)
    J. Chem. Soc., 1370-1373 (1960).
72. The Naphthyridines: Ionization Constants and Spectra.
    J. Chem. Soc., 1790-1793 (1960).
73. The Azaindoles: Ionization Constants and Spectra.
    (With T. K. Adler.)
    J. Chem. Soc., 1794-1797 (1960).

74. Quantitative Studies of the Avidity of naturally-occurring Substances for Trace Metals. Part 4. A potentiometric Method for highly insoluble Complexes, and its application to the Complexes of Adenine Anion.
(With E. P. Serjeant.)
Biochem. J. **76**, 621-624 (1960).

75. Pteridine Studies. Part XIII. Addition to 6-Hydroxypteridines.
(With F. Reich.)
J. Chem. Soc., 127-135 (1961).

76. Pteridine Studies. Part XV. The reduction of 2-Hydroxypteridine.
(With S. Matsuura.)
J. Chem. Soc., 5131-5137 (1961).

77. Quinazolines. Part I. Cations of Quinazoline.
(With W. Armarego and E. Spinner.)
J. Chem. Soc., 2689-2696 (1961).

78. Quinazolines. Part III. The Structure of the hydrated Quinazoline Cation.
(With W. Armarego and E. Spinner.)
J. Chem. Soc., 5267-5270 (1961).

79. Pteridine Studies. Part XVII. Addition to 2-Hydroxypteridines.
(With C. Howell.)
J. Chem. Soc., 1591-1596 (1962).

80. Pteridine Studies. Part XVIII. Reductions of Hydroxypteridines.
(With S. Matsuura.)
J. Chem. Soc., 2162-2171 (1962).

81. Pteridine Studies. Part XIX. Covalent Hydration of 2-Aminopteridines.
(With C. Howell and E. Spinner.)
J. Chem. Soc., 2595-2600 (1962).

82. Ionization Constants of Heterocyclic Substances. Part V. Mercapto-derivatives of Heterocycles with Two Nitrogen Atoms in the same ring.
(With G. B. Barlin.)
J. Chem. Soc., 3129-3141 (1962).

83. Naphthyridines. Part II. Covalent Hydration, Electron-deficiency, and Resonance Stabilization in 1,6-Naphthyridines.
(With W. L. F. Armarego.)
J. Chem. Soc., 4237-4242 (1963).

84. The Biological and Physical Properties of the Azaindoles.
(With Terrine K. Adler.)
J. Med. Chem. **6**, 480-483 (1963).

85. Resolution of a Racemic Pteridine during Paper Chromatography.
(With E. P. Serjeant.)
Nature (London) **199**, 1098-1099 (1963).

86. Triazanaphthalenes. Part II. Covalent Hydration in 1,4,6-Triazanaphthalenes.
(With G. B. Barlin.)
J. Chem. Soc., 5156-5166 (1963).

87. Pteridine Studies. Part XXIV. Competitive Covalent Hydration of 2,6-Dihydroxypteridine: Kinetics and Equilibria.
(With Y. Inoue and D. D. Perrin.)
J. Chem. Soc., 5151-5156 (1963).

88. Triazanaphthalenes. Part IV. Covalent Hydration in 1,4,5-Triazanaphthalenes.
(With G. B. Barlin.)
J. Chem. Soc., 5737-5741 (1963).

89. Pteridine Studies. Part XXV. Preparation, Hydration and Degradation of some Chloropteridines.
(With J. Clark.)
J. Chem. Soc., 1666-1673 (1964).

90. Pteridine Studies. Part XXVI. Acid Catalysis of Michael-type Reactions. Resolution of a Racemic Dipteridinylmethane during Paper Chromatography.
(With E. P. Serjeant.)
J. Chem. Soc., 3357-3365 (1964).

91. Diazaindenes ("Azaindoles"). Part II. Methyl Derivatives of 1,7-Diazaindene.
(With R. Willette.)
J. Chem. Soc., 4063-4065 (1964).

92. Pteridine Studies. Part XXVIII. Some 6- and 7-substituted Pteridines.
(With J. Clark.)
J. Chem. Soc., 27-32 (1965).

93. Acridine Syntheses and Reactions. Part VI. A new dehalogenation of 9-chloroacridine and its derivatives. Further acridine ionization constants and ultraviolet spectra.
J. Chem. Soc., 4653-4657 (1965).

94. Acridine Syntheses and Reactions. Part VII. Some binuclear acridines.
(With G. Catterall.)
J. Chem. Soc., 4657-4658 (1965).

95. Pteridine Studies. Part XXX. Some Michael-type Addition Reactions of 7-Hydroxypteridine.
(With J. McCormack.)
J. Chem. Soc., 6930-6934 (1965).

96. Penta-azaindenes (8-Azapurines). Part I. Covalent Hydration. Syntheses and Reactions.
J. Chem. Soc., B, 427-433 (1966).

97. Structures of the Anhydro-polymers of 2-Aminobenzaldehyde.
(With H. Yamamoto.)
J. Chem. Soc. B, 956-963 (1966).

98. Pteridine Studies. Part XXXII. Nucleophilic Addition Reactions of 2-Aminopteridine.
(With J. McCormack.)
J. Chem. Soc. C, 1117-1120 (1966).

99. Pteridine Studies. Part XXXIII. Equilibria between 3,4-Hydrated and 5,6,7,8-Dihydrated Cations.
(With T. Batterham and J. McCormack.)
J. Chem. Soc. B, 1105-1109 (1966).

100. Purine Studies. Part IV. A Search for Covalent Hydration in 8-Substituted Purines.
J. Chem. Soc. B, 438-441 (1966).

101. Oxidative Replacement of the Hydrazine-Group by Hydrogen and Deuterium in Azanaphthalenes.
   (With G. Catterall.)
   J. Chem. Soc. C, 1533-1541 (1967).
102. Pteridine Studies. Part XXXIV. Nucleophilic Addition Reactions of Pteridine-2-thiol.
   (With J. McCormack.)
   J. Chem. Soc. C, 63-68 (1968).
103. Pteridine Studies. Part XXXV. The Structure of the Hydrated Dimer Formed by the Action of Dilute Acid on 4-Methylpteridine.
   (With H. Yamamoto.)
   J. Chem. Soc. C, 1181-1187 (1968).
104. Pteridine Studies. Part XXXVI. The Action of Acid and Alkali on Pteridine.
   (With H. Yamamoto.)
   J. Chem. Soc. C, 2289-2292 (1968).
105. Pteridine Studies. Part XXXVII. Covalent Hydration in 4,6,7-Trimethyl- and 2,4,6,7-Tetramethyl-pteridine.
   (With H. Yamamoto.)
   J. Chem. Soc. C, 2292-2294 (1968).
106. Quinazoline Studies. Part XII. Action of Acid and Alkali on Quinazoline.
   (With H. Yamamoto.)
   J. Chem. Soc. C, 1944-1949 (1968).
107. Penta-azaindenes (8-Azapurines). Part III. A New Route to the 7-Methyl-8-azapurines.
   (With K. Tratt.)
   J. Chem. Soc. C, 344-347 (1968).
108. Penta-azaindenes (8-Azapurines). Part IV. A New Route to the 8-Methyl-8-azapurines through the 2-Methyl-1,2,3-triazoles.
   J. Chem. Soc. C, 2076-2083 (1968).
109. Penta-azaindenes (8-Azapurines). Part V. A Comparison of 1,2,3-Triazoles and Pyrimidines as Intermediates for the Preparation of 9-Substituted 8-Azapurines. Rearrangement of 6-Mercapto-8-azapurines and of 4-Aminotriazoles.
   J. Chem. Soc. C, 152-160 (1969).
110. Penta-azaindenes (8-Azapurines). Part VI. Methylation of 8-Azapurine.
   (With W. Pfleiderer and D. Thacker.)
   J. Chem. Soc. C, 1084-1085 (1969).
111. Penta-azaindenes (8-Azapurines). Part VII. Degradation by Acid of the 6-Methylthio-derivatives of 8-Azupurine and Purines to Thiol Esters such as 4-Amino-5-(methylthio)-carbonyl-1,2,3-triazole.
   J. Chem. Soc. C, 2379-2385 (1969).
112. Pteridine Studies. Part XXXIX. Pteridines unsubstituted in the 4-position. A New Synthesis from Pyrazines, *via* 3,4-dihydropteridines.
   (With K. Ohta.)
   J. Chem Soc. C, 1540-1547 (1970).
113. The Dimroth Rearrangement. Part XII. Transformation by Alkali of 4-Amino-3-benzyl-1,2,3-triazole and its 5-substituted derivatives into the corresponding 4-benzylamino-isomers.
   J. Chem. Soc. C, 230-235 (1970).

114. Pteridine Studies. Part XL. Synthesis of 4-Unsubstituted Pteridines from 3-Aminopyrazine-2-carbaldehyde.
(With K. Ohta.)
J. Chem. Soc. C, 2357-2362 (1971).

115. v-Triazolo[4,5-*d*]pyrimidines (8-Azapurines). Part VIII. Synthesis, from 1,2,3-Triazoles, of 7- and 8-Methyl-derivatives of 2,6-Disubstituted 8-Azapurines.
(With H. Taguchi.)
J. Chem Soc. Perkin Trans. 1, 449-456 (1972).

116. v-Triazolo[4,5-*d*]pyrimidines (8-Azapurines). Part IX. Nucleophilic Addition Reactions of Some 8-Azapurines.
(With W. Pendergast.)
J. Chem. Soc. Perkin Trans. 1, 457-461 (1972).

117. v-Triazolo[4,5-*d*]pyrimidines (8-Azapurines). Part X. New Routes to 8-Azapurines 5-Cyano-4-dimethylaminomethyleneamino-1-(also 2- and 3-)methyl-1,2,3-triazole.
J. Chem. Soc. Perkin Trans, *1*, 461-467 (1972).

118. v-Triazolo[4,5-*d*]pyrimidines (8-Azapurines). Part XI. The Preparation of 8-Benzyl-8-azapurine.
(With D. Thacker.)
J. Chem. Soc. Perkin Trans. 1, 468-469 (1972).

119. Pteridine Studies. Part XLI. New Routes to 4-Aminopteridines *via* 3-Dimethylaminomethyleneaminopyrazine-2-carbonitrile and related compounds.
(With K. Ohta.)
J. Chem. Soc. C, 3727-3730 (1971).

120. Pteridine Studies. Part XLII. Additive Reactions between Alcohols and Pteridines.
(With H. Mizuno.)
J. Chem. Soc. B, 2423-2427 (1971).

121. The Physical Basis for the Selectivity of Drugs.
(Rubbo Memorial Lecture.)
Med. J. Aust. **1**, 1013 (1972).

122. v-Triazolo[4,5-*d*]pyrimidines (8-Azapurines). Part XII. Cleavage by Some Active Methylene Reagents of the Pyrimidine Ring in 8-Azapurines and Eventual Closure to Give v-Triazolo[4,5-*b*]pyridines.
(With W. Pendergast.)
J. Chem. Soc. Perkin Trans. 1.

123. 1,2,3-Triazoles. Part 1. Some 4-Aminotriazole-5-carbaldehydes.
(With H. Taguchi.)
J. Chem. Soc. Perkin Trans. 1.

124. Cleavage of Pyrimidines and Fused Pyrimidines by Active Methylene Reagents with Eventual Closure to Give Pyridine Derivatives.
(With W. Pendergast.)
J. Chem. Soc. Perkin Trans. 1.

125. Pteridine Studies. Part XLIII. Reactions of Pteridine with Some Michael Reagents.
(With H. Mizuno.)
J. Chem. Soc. Perkin Trans. 1.

126. v-Triazolo[4,5-*d*]pyrimidines (8-Azapurines). Part XIII. Reaction of 8-Azapurines with Bifunctional Nucleophiles.
(With W. Pendergast.)
J. Chem. Soc. Perkin Trans. 1.

MAJOR REVIEWS

1. Ionization, pH, and Biological Activity.
   Pharmacological Reviews **4**, 136-167 (1952).
2. The Pteridines.
   Quarterly Reviews **6**, 197-237 (1952).
3. The Naturally Occurring Pteridines.
   Fortschritte der Chemie Organischer Naturstoffe **11**, 350-403 (1954).
4. The Relationship between Structure and Biological Activity.
   Ergebnisse der Physiologie **49**, 425-461 (1957).
5. The Activation of Metals by Chelation.
   The Strategy of Chemotherapy, Cambridge, 112-138 (1958).
6. The Design of Chelating Agents for Selective Biological Activity.
   Federation Proceedings **20**, (II), 137-147 (1961).
7. Ionization Constants.
   Physical Methods in Heterocyclic Chemistry **1**, 2-108 (1963). New York: Academic Press.
8. Correlations between Microbiological Morphology and the Action of Biocides.
   Advances in Applied Microbiology **5**, 1-50 (1963). New York: Academic Press.
9. Covalent Hydration in Nitrogen Heteroaromatic Compounds.
   (With W. Armarego.)
   Advances in Heterocyclic Chemistry **4**, 1-42 (1965). New York: Academic Press.
10. Hydration of C=N Bonds in Heteroaromatic Substances.
    Angewandte Chemie (Int. Ed. Engl.) **6**, 919-928 (1967).
11. (a) Ionization Constants of Pyrimidines and Purines. Ultraviolet Spectra of Pyrimidines and Purines.
    Synthetic Procedures in Nucleic Acid Chemistry **2** (1971) (in press).
12. Ionization Constants (Recent Advances).
    Physical Methods in Heterocyclic Chemistry, **3**, 1-26 (1971). New York: Academic Press.
13. Relations between Molecular Structure and Biological Activity: Stages in the Evolution of Current Concepts.
    Annual Review of Pharmacology **11**, 1-36 (1971).

BOOKS

1. The Acridines, their Preparation, Physical, Chemical and Biological Properties and Uses. London: Edward Arnold & Co., 1951. Second Edition (1966). Pp. xii + 604.
2. Selective Toxicity with Special Reference to Chemotherapy. London: Methuen & Co., 1951. Pp. ix + 228. Also Russian (1953) and Japanese (1956) editions. Also fourth edition (1968). Pp. xvi + 531. German and Russian editions in preparation.
3. Heterocyclic Chemistry. London: The Athlone Press, 1959. Also German edition (1963), Japanese edition (1964). Also second English edition (1968). Pp. xii + 547.
4. Ionization Constants. London: Methuen, 1962. (With E. P. Serjeant.) Also Russian (1965) and Japanese (1963) editions. Second English edition in press.)

Heterocyclic Oligomers

ADRIEN ALBERT AND HIROSHI YAMAMOTO

Department of Medical Chemistry, John Curtin School of Medical Research, Australian National University, Canberra, Australia

and

Fachbereich Chemie, Universität Konstanz, West Germany

I. Introduction	1
II. Initiation and Termination of Oligomerization	2
III. Classification and Discussion of Oligomers	3
A. Formation of Mononuclear Oligomers That Have No Repeating Heterocyclic Unit in the Molecule	4
B. Formation of Oligomers That Have Two (or More) Heterocyclic Nuclei Connected by a Side Chain	22
C. Formation of Oligomers That Have Two (or More) Heterocyclic Nuclei Joined to One Another by a Single Bond	40
D. Formation of Oligomers That Have Two (or More) Heterocyclic Nuclei Joined by a Double Bond	52
E. Formation of Oligomers That Have Two (or More) Heterocyclic Nuclei Joined by Two (or More) (Separate) Bonds	57
F. Formation of Miscellaneous Oligomers, Including Spiro Compounds	62
IV. Conclusion	64

I. Introduction

With the sole exception of photodimers,[1] heterocyclic oligomers, that is to say heterocycles formed from two to five units of a monomer, have not previously been reviewed. This neglect is surprising, because heterocyclic oligomers are both common and interesting. They arise frequently as by-products in the synthesis of heterocycles and sometimes they form the main product of the reaction. Very often, they are crystalline products, easily purified and characterized. In recent years, the constitution of many examples has been elucidated by taking advantage of modern physical methods.

[1] See, e.g., R. O. Kan, "Organic Photochemistry." McGraw-Hill, New York, 1966; E. Fahr, *Angew. Chem. Int. Ed. Engl.* **8**, 578 (1969); S. T. Reid, *Advan. Heterocycl. Chem.* **11**, 1 (1970).

The present review discusses conditions for both the initiation and the termination of the polymeric process, both from a theoretical standpoint (based largely on current applications of quantum mechanics to chemistry[2]) and from the practical point of view, namely, the promotion or avoidance of oligomerization. This discussion is followed by the classification of known oligomers into six main families according to the way in which the heterocyclic nuclei are united. Each family is subdivided according to the conditions which initiated the oligomerization. The advantage of the classification is that it brings together similarly constituted (and similarly formed) examples for comparison with one another and, particularly, with newly isolated substances whose structure is under consideration.

In our experience, determination of the constitution is most conveniently tackled in the following order. As soon as chromatography indicates that only a single substance is present, elemental analysis is obtained, and then the molecular weight. If the latter is sought through mass spectrography, the possibility of depolymerization in the ion chamber must be kept in mind. Ionization constants can be determined, if the oligomer has acidic or basic properties. The ultraviolet spectrum (of each molecular species) is then examined to indicate the extent of conjugation. Finally infrared and PMR spectra are used to assign particular groups.

This chapter is concerned mainly with substances whose heteroatoms are nitrogen, oxygen, or sulfur. Many heterocyclic copolymers of low molecular weight are included.

II. Initiation and Termination of Oligomerization

It is a well-established characteristic of polymerization reactions that a monomer changes to a polymer via an activated intermediate.[3] Chemical activation of a monomer requires: (1) a redistribution of electron densities in the bonds of the monomeric molecule (the intramolecular effect) and/or (2) a change of properties of existing (loose) intermolecular bonds between the reacting sites (intermolecular effect; an alteration of mutual orientation may suffice). For practical purposes,

---

[2] See, e.g., (a) R. B. Woodward and R. Hoffmann, "The Conservation of Orbital Symmetry." Academic Press, New York, 1970; (b) E. Heilbronner and H. Bock, "Das HMO-Modell und seine Anwendung." Verlag Chemie, Weinheim, 1968.

[3] See, e.g., C. S. Marvel, "An Introduction to the Organic Chemistry of High Polymers." Wiley, New York, 1959.

the reaction may start when the monomeric molecules have functional groups which are already reactive enough to undergo spontaneous intermolecular condensation, but, more often, one at least of the condensing groups must be activated by conversion to a new ionic species (by addition of acid or alkali) or to a free radical (e.g., by supplying energy in the form of heat or light). Whereas a monomer with only one functional group tends to give only a dimer, those with two such groups can give oligomers (linear or cyclic) of all sizes, and even linear polymers. When the monomer has more than three functional groups, the tendency to form polymers exceeds that to stop at oligomers.

High concentrations favor oligomerization, whereas at a lower concentration, an intramolecular reaction or depolymerization may compete with polymerization.[4] In fact, many cases are known where a monomer and its oligomers exist in equilibrium under various conditions.

In general, a reactive monomer does not polymerize indefinitely, but gives an oligomer, often in good yield provided that the product is stabilized (i.e., deactivated) by any of the following circumstances: (a) steric hindrance is imposed by the geometry of the product; (b) a cyclic product is formed; a metal cation is sometimes used to assist cyclization; (c) the pK_a (or the electronic state) is changed by an earlier stage of the reaction; and (d) the oligomeric product separates as a precipitate due to low solubility.

Further discussion of the termination of oligomerization, especially for monomers having two (or more) functional groups, will be found in the next section.

III. Classification and Discussion of Oligomers

The following six classes of oligomers are distinguished in this review:

A. Mononuclear oligomers that have no repeating heterocyclic unit in the molecule

B. Oligomers that have two (or more) heterocyclic nuclei connected by a side chain (nonheterocyclic rings, e.g., benzene, are regarded as side chains in this classification)

C. Oligomers that have two (or more) heterocyclic nuclei joined to one another by a single bond

D. Oligomers that have two (or more) heterocyclic nuclei joined by a double bond

---

[4] See, e.g., H. R. Allcock, "Heteroatom Ring Systems and Polymers." Academic Press, New York, 1967.

E. Oligomers that have two (or more) heterocyclic nuclei joined by two or more (separate) bonds

F. Miscellaneous oligomers, including spiro compounds

When an oligomer has the structure which belongs to more than one class, the *Prinzip der spätesten Systemstelle* is applied. The discussion of each class will be developed in the following order, based on the nature of the monomer: (1) heterocyclic oligomers from nonheterocyclic monomers, (2) heterocyclic oligomers from heterocyclic monomers, (3) heterocyclic co-oligomers from more than one kind of monomer, and (4) naturally occurring heterocyclic oligomers (including their synthesis).

Except for subclass (4), in which examples will be presented in the order used in *Beilstein's Handbuch*, further subclassification will follow the type of initiation of oligomerization, of which the following five are distinguished: (a) *acid-catalyzed oligomerizations*, (b) *base-catalyzed oligomerizations*, (c) *thermal oligomerizations*, including spontaneous polymerization under neutral conditions, (d) *photooligomerizations*, and (e) *miscellaneous* methods of initiation.

Examples of oligomers will now be presented in their classes and subclasses.

A. Formation of Mononuclear Oligomers That Have No Repeating Heterocyclic Unit in the Molecule

Subclass 1. Heterocyclic Oligomers from Nonheterocyclic Monomers

a. *Acid-Catalyzed Oligomerization.* In most cases, this is an intermolecular dehydration to give more geometrically and thermodynamically stable self-condensation products. A simple example is the first practical synthesis of 1,4-dioxane, made by Favorsky in 1906; ethylene glycol was distilled in the presence of zinc chloride or sulfuric acid[5] (for a better method see Section III,A,2).

1,3,5-Trioxane (**1a**)[6] was obtained somewhat similarly by distilling aqueous formaldehyde solution, slightly acidified with sulfuric acid. On an industrial scale trioxane is prepared in 96–98% yield by heating formaldehyde for 30–60 minutes in the presence of 30–50% H_3PO_4, $H_4P_2O_7$, or polyphosphoric acid while simultaneously removing trioxane as vapor.[7] Paraldehyde (**1b**)[6] is formed analogously by catalytic poly-

---

[5] For reviews, see W. Stumpf, "Chemie und Anwendungen des 1,4-Dioxans." Verlag Chemie, Weinheim, 1956; C. B. Kremer and L. K. Rochen, *in* "Heterocyclic Compounds" (R. C. Elderfield, ed.), Vol. 6, p. 1. Wiley, New York, 1956.

[6] For a review, see J. C. Bevington, *Quart. Rev.* **6**, 141 (1952).

[7] H. Fuchs and H. Sperber, German Patent 1,814,197; *Chem. Abstr.* **75**, 77482 (1971).

merization of acetaldehyde (e.g., with sulfur dioxide). These oligomers exist in equilibrium with the aldehydes, and are partly depolymerized on simple distillation.[8]

(1a) R = H
(1b) R = Me

(2)

(3)

Under other conditions,[9] aldehydes give dimers and even tetramers. The tetramer of acetaldehyde, made by the action of a trace of acid at about 0°, has been shown by X-ray diffraction analysis to be tetramethyltetraoxacyclooctane (2).[10] The acid-catalyzed dimerization of mercaptoacetaldehyde acetal $HSCH_2CH(OEt)_2$ gives 2,5-diethoxy-1,4-dithian (3), which can be converted to 1,4-dithiadiene by heating at 300° over alumina.[11]

2-Aminobenzaldehyde methylimide, on standing at room temperature, or more quickly on treatment with 10 N hydrochloric acid, yields the dimer (4a),[12] and the ethylimide gives a similar product (4b).[12] These are thought to be formed from a simpler eight-membered cyclic dimer, of which an alkylamino group is eliminated with simultaneous formation of an internal imino bridge.[12] Another example of the same ring system is the diformylanhydrotrimer of 2-aminobenzaldehyde (4c) formed as a by-product (20% yield) in the formylation of 2-aminobenzaldehyde with acetic formic anhydride.[13] Compound 4c is also formed in good yield from the anhydrotrimer of 2-aminobenzaldehyde (5a) (see below) by partial degradation with cold acetic formic anhydride.[13] When 2-methylaminobenzaldehyde is dissolved in 1.0 N hydrochloric acid at room temperature, an anhydro dimer separates,[14] to which structure 4d has been assigned.[12]

---

[8] W. H. Hatcher and B. Brodie, *Can. J. Res.* **4**, 574 (1931).
[9] M. Barón, *Nature (London)* **192**, 258 (1961); M. Barón, O. B. de Mandirola, and J. F. Westerkamp, *Can. J. Chem.* **41**, 1893 (1963).
[10] L. Pauling and D. C. Carpenter, *J. Amer. Chem. Soc.* **58**, 1274 (1936).
[11] W. E. Parham, H. Wynberg, and F. L. Ramp, *J. Amer. Chem. Soc.* **75**, 2065 (1953).
[12] A. Albert and H. Yamamoto, *J. Chem. Soc. C*, 1944 (1968).
[13] A. Albert and H. Yamamoto, *J. Chem. Soc. B*, 956 (1966).
[14] E. Bamberger, *Ber.* **37**, 966 (1904).

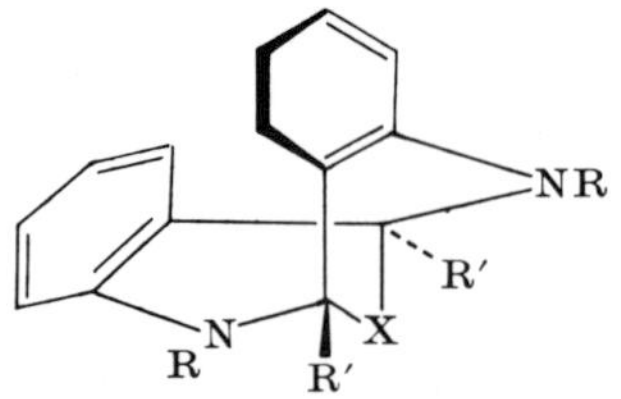

(4a) R = R' = H, X = NMe
(4b) R = R' = H, X = NEt
(4c) R = CHO, R' = H, X = NC₆H₄CHO(o)
(4d) R = Me, R' = H, X = O
(4e) R = H, R' = COOH, X = O

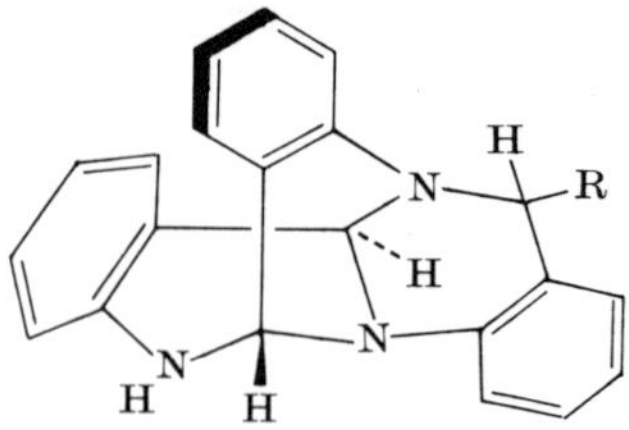

(5a) R = OH
(5b) R = NHC₆H₄CHO(o)

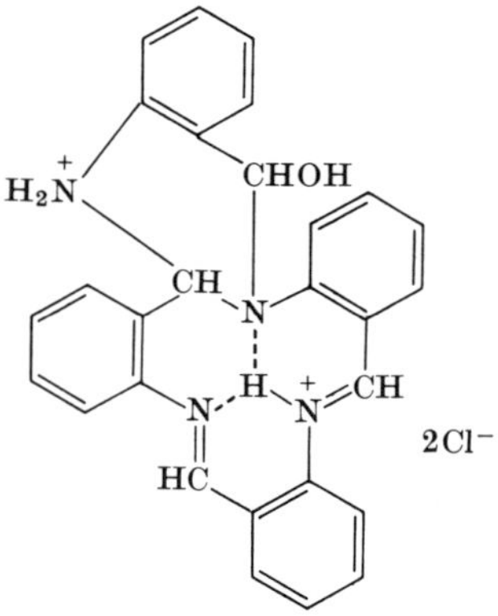

(6)

The anhydrotrimer of 2-aminobenzaldehyde, formed from the monomer on standing, or more quickly by the action of dilute acid,[15] has proved to have structure **5a**.[13,16] The anhydro tetramer (**5b**)[13,16] is obtained when the monomer is dissolved in 5 N hydrochloric acid and the precipitated red anhydro tetramer dihydrochloride,[15] for which structure **6** has been proposed,[13] is made basic with aqueous pyridine. At low concentrations (e.g., 10^{-5} M), equilibrium favors depolymerization to the monomer, but at higher concentrations (e.g., 10^{-2} M), **5a** and **6** are formed in proportions depending on the strength of acid. The reaction mechanism for the formation of these oligomers has been discussed.[13] The trimer (**5a**) and the tetramer (**5b**) consist of three tightly bound tetrahydroquinazoline rings and **6** has a macrocyclic structure. It is

---

[15] (a) F. Seidel, *Ber.* **59**, 1894 (1926); (b) F. Seidel and W. Dick, *ibid.* **60**, 2018 (1927); (c) E. Bamberger, *ibid.* **60**, 314 (1927).
[16] S. G. McGeachin, *Can. J. Chem.* **44**, 2323 (1966).

often observed that the incorporation of all active groups into small, compact structures, such as these, causes early termination of polymerization reactions, especially if the products separate from the reaction mixture as precipitates.

When 2-aminobenzaldehyde was heated in absolute ethanol in the presence of nickel nitrate, the tridentate and tetradentate macrocyclic complexes (**7** and **8**, respectively) were isolated.[17] The presence of copper(II) nitrate produced only the latter (**8**).[17] Although the monomer has two functional groups and the oligomers have reactive azomethine groups, the formation of macrocyclic ligands stabilized by a metal halts further polymerization.

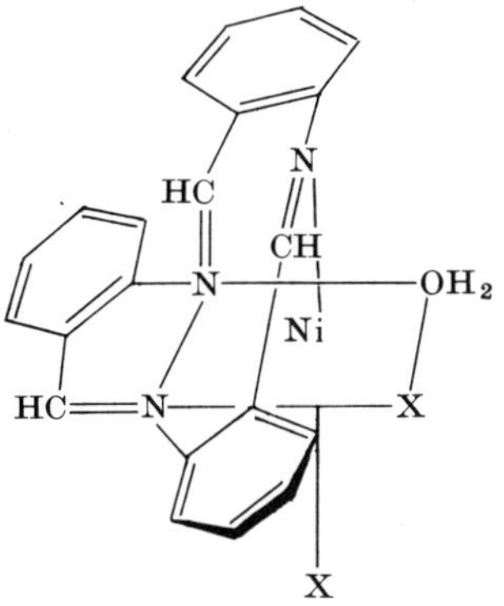

(7) 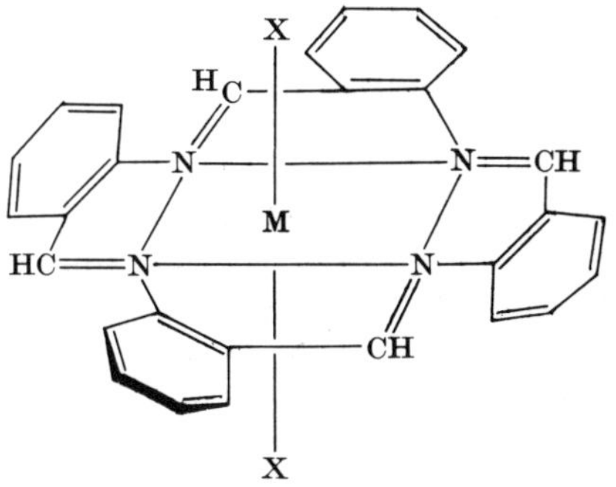   (8)

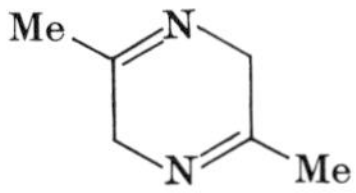

(9) 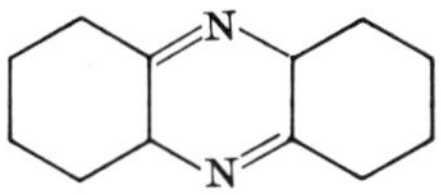   (10)

Many other examples are known of acid-catalyzed intermolecular Schiff reactions[18] to give cyclic dimers. Thus, condensation of two molecules of α-aminoketones gives 2,5-dihydropyrazines (e.g., **9** from 1-aminopropan-2-one), most of which can be converted into pyrazines by oxidation with mercuric chloride (i.e., the Gutknecht pyrazine synthesis[19]). The same type of self-condensation occurs on heating

---

[17] G. A. Melson and D. H. Busch, *Proc. Chem. Soc.*, 223 (1963); *J. Amer. Chem. Soc.* **86**, 4830, 4834 (1964); **87**, 1706 (1965).
[18] See e.g., L. F. Fieser and M. Fieser, "Advanced Organic Chemistry." pp. 497, 506, 816, 817, 1080. Reinhold, New York, 1961.
[19] For a survey of such reactions, see G. R. Ramage and J. K. Landquist, in "Chemistry of Carbon Compounds" (E. H. Rodd, ed.) Vol. 4B, p. 1319. Elsevier, Amsterdam, 1959.

2-aminocyclohexanone in 10% hydrochloric acid to give the dimer decahydrophenazine (**10**).[20] When a monomer has two functional groups, placed 1:3 to one another, an eight-membered cyclic dimer can be produced. Thus, the dibenzodiazocine (**11**) was prepared by Sternbach

(**11**)

(**12a**) R = H
(**12b**) R = Me

et al. by dimerizing 2-aminobenzophenone in an inert solvent catalyzed by a Lewis acid.[21] The bulky phenyl rings near the azomethine groups of **11** prevented further addition of the monomer to these C=N bonds, in contrast to the anhydrotrimer (**5a**) of 2-aminobenzaldehyde which is formed by the addition of another molecule of the monomer to a more exposed C=N bond.[13] An elaboration of the Schiff reaction is the formation of 2,5-dicyanopyrazines (known as the Gastaldi synthesis[19]) by cyclization of two molecules of an aminocyanomethyl ketone, $R \cdot CO \cdot CH(CN)\text{-}NH \cdot SO_3K$ (produced by treatment of an isonitrosomethyl ketone $R \cdot CO \cdot CH\text{=}NOH$ bisulfite compound with potassium cyanide), on heating in hydrogen chloride, followed by oxidation. A convenient one-step synthesis of adenine (**12a**) (in 40–50% yield) from five molecules of formamide by heating in a sealed tube in the presence of phosphoryl chloride has been effected.[22]

A compound having a carboxylic acid and an amino (or imino) group within the molecule can yield cyclic oligomers by intermolecular twofold amide formation. For example, oxanilic acid has been dimerized[23] in the presence of thionyl chloride to give the tetraketopiperazine (**13**). Self-condensation[24] of anthranilic acid on refluxing in xylene

[20] G. H. Alt and W. S. Knowles, *J. Org. Chem.* **25**, 2047 (1960).
[21] W. Metlesics and L. H. Sternbach, *J. Amer. Chem. Soc.* **88**, 1077 (1966).
[22] K. Morita, M. Ochiai, and R. Marumoto, *Chem. Ind. (London)*, 1117 (1968); M. Ochiai, R. Marumoto, S. Kobayashi, H. Shimazu, and K. Morita, *Tetrahedron* **24**, 5731 (1968).
[23] D. Buckley and H. B. Henbest, *J. Chem. Soc.*, 1888 (1956).
[24] A. Chatterjee and M. Ganguly, *J. Org. Chem.* **33**, 3358 (1968).

in the presence of phosphorus pentoxide gave mainly the trimer (**14**) together with a small amount of the tetramer (**15**). Similar cyclic anhydro oligomers can be produced from hydroxycarboxylic acids. For example, *o*-thymotic acid (3-isopropyl-6-methylsalicylic acid) gives a mixture of the anhydro dimer (**16**), a 12-membered cyclic anhydro trimer, and a 24-membered cyclic anhydro hexamer when heated in xylene in the presence of phosphorus pentoxide.[25] Salicylic acid gives a mixture of di-, tri-, tetra-, and hexasalicylide in yields varying with dehydrating agent and experimental conditions.[26]

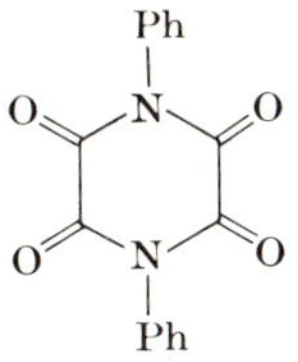

(**13**) 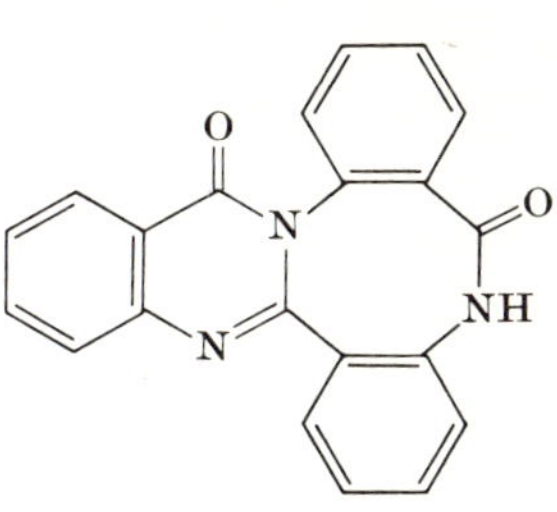   (**14**)

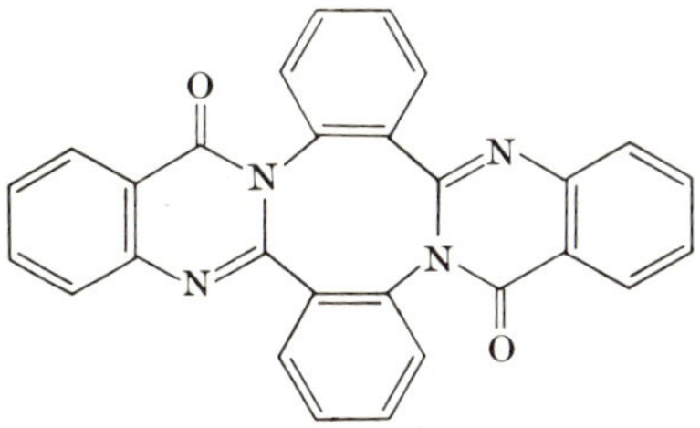

(**15**) 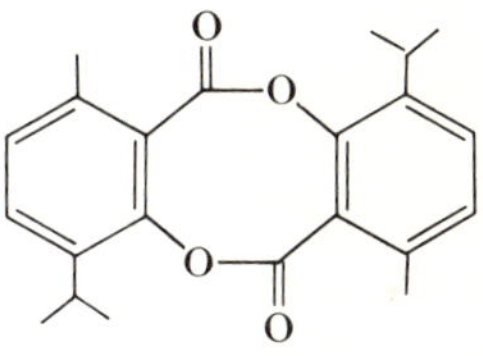   (**16**)

An example of acid-catalyzed oligomerization by intermolecular dehydration between an amino and a hydroxy group in a monomer is the usual industrial preparation of piperazine by heating ethanolamine and ammonium chloride at 250° in glass-lined vessels (for other methods of preparing piperazine, see Section III,A,3).

Nitriles may be trimerized with various acids, bases, or other catalysts.[27] The mechanism for the formation of *s*-triazines catalyzed by hydrochloric acid has been clarified.[28]

[25] W. Baker, *Ind. Chim. Belge* **17**, 633 (1952); *Chem. Abstr.* **46**, 10115 (1952).
[26] W. Baker, W. D. Ollis, and T. S. Zealley, *J. Chem. Soc.*, 201 (1951).
[27] See E. M. Smolin and L. Rapoport, "*s*-Triazines and Derivatives," pp. 52, 62, and 149. Wiley (Interscience), New York, 1959.
[28] C. Grundmann, G. Weisse, and S. Seide, *Ann.* **577**, 77 (1952); C. Grundmann, *Chem. Ber.* **97**, 3262 (1964).

b. *Base-Catalyzed Oligomerization.* Nitriles containing an active hydrogen atom become dimerized in the presence of sodium ethoxide (the Thorpe reaction[29]). Thus, oligomerization of an appropriate dinitrile forms a heterocyclic compound, although this may be, in some cases, competitive with intramolecular condensation to give a 2-cyanoenamine (i.e., Thorpe–Ziegler reaction[30]). Adipinonitrile, $NC(CH_2)_4CN$, gives a mixture of the dimer (**17**; 48% yield) and the trimer (**66**; 33%) in the presence of sodium;[31] the trimer is classified as a type B1 oligomer.

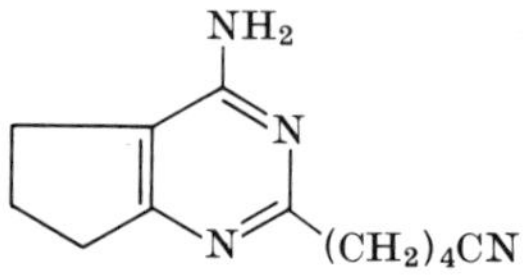

(**17**) 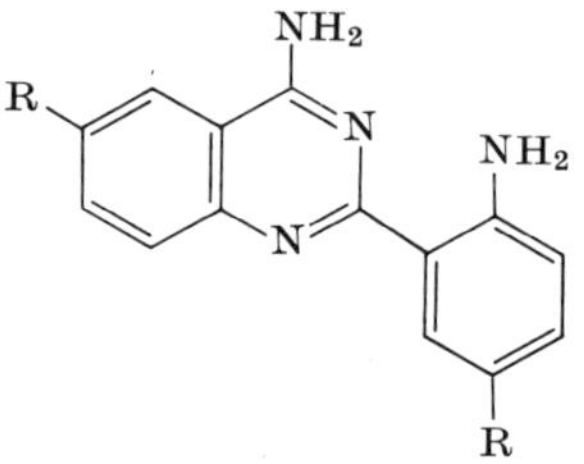   (**18**) 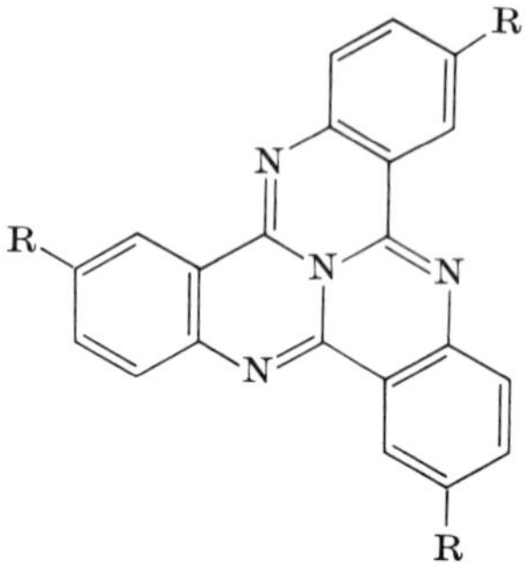   (**19**)

It is well known that the addition of ammonia or amines to nitriles leads to formation of various amidines;[32] this type of oligomerization can be applied to the synthesis of heterocycles. On heating in the presence of sodium amide (or benzenesulfonic acid in some cases), *o*-cyanoaniline produces a mixture of the quinazoline (**18**) by dimerization and the tricycloquinazoline (**19**) by trimerization; the trimer is obtained when the dimer is heated with another molecule of the monomer and sodium amide at 300°. The mechanism of this condensation has been reviewed.[33]

Alkali fusion of 2-aminoanthraquinone, alone or in inert solvents, produces a dimeric compound (Indanthrene) (**20**)[34] used for the vat dyeing of cotton.

[29] P. G. Stecher (ed.), "The Merck Index," 8th ed., p. 1220, and references therein. Merck, Rahway, New Jersey, 1968.
[30] For a review, see J. P. Schaefer and J. J. Bloomfield, *Org. Reactions* **15**, 1 (1967).
[31] K. Yamamoto, Japan Patent 70-27,979; *Chem Abstr.* **74**, 3653 (1971).
[32] See e.g., J. March, "Advanced Organic Chemistry: Reactions, Mechanisms, and Structure," p. 673. McGraw-Hill, New York, 1968.
[33] E. C. Taylor and A. McKillop, "The Chemistry of Cyclic Enaminonitriles and *o*-Aminonitriles," p. 233. Wiley (Interscience), New York, 1970.
[34] See Stecher,[29] p. 563, and references therein.

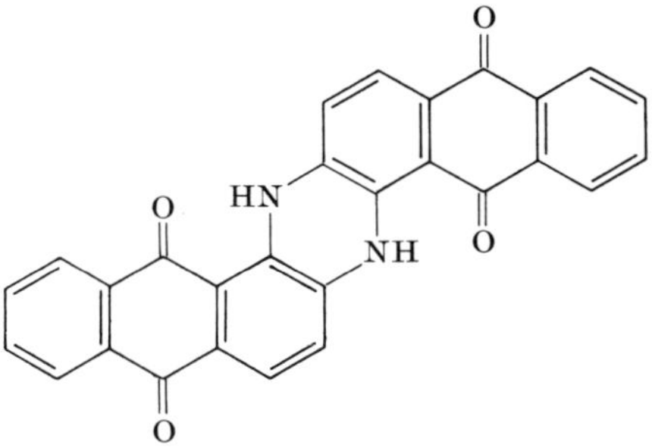

(20)

c. *Thermal Oligomerization.* Many nonheterocyclic compounds that possess bifunctional groups have been found to become condensed when merely heated; an equilibrium mixture of heterocyclic oligomers often results. Ketene, the best known example, quickly dimerizes; the C=C bond adds to the C=O bond to give the unsaturated β-lactone (21). Other ketenes (e.g., RR′C=C=O) dimerize to cyclobutanedione derivatives.[35] Lactide is formed by heating lactic acid at 180°–220° *in vacuo*,[36] and cyamelide (22) by heating cyanic acid (gaseous or liquid).[37] Thioaldehydes, such as thioformaldehyde, thioacetaldehyde, and thiobenzaldehyde, are known to exist as the cyclic trimers (23);[38] an equilibrium also exists between sulfur trioxide and the cyclic trimer (24).[39,40]

2,5-Piperazinedione, usually called 2,5-diketopiperazine, can be prepared by heating either glycine (e.g., in glycerol) or glycine ethyl ester hydrochloride.[41] An equilibrium between *N*-methylenemethylamine (MeN=CH$_2$) and its cyclic trimer (25a) is known to favor the latter

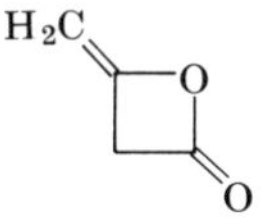

(21) 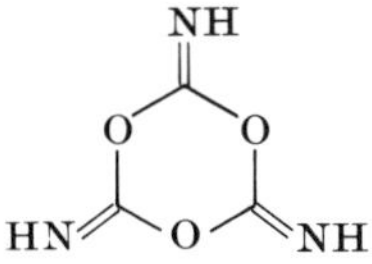   (22) 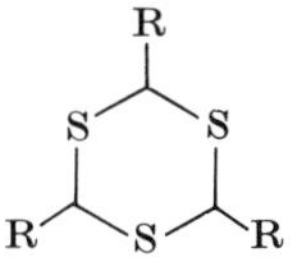   (23) 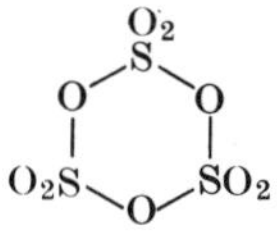   (24)

[35] See March,[32] pp. 636, 723.
[36] See Stecher,[29] p. 605.
[37] See Stecher,[29] p. 306.
[38] A. Schönberg and M. Z. Barakat, *J. Chem. Soc.*, 693 (1947), and references therein.
[39] H. Gerding, W. J. Nijveld, and G. J. Muller, *Z. Phys. Chem. Abt. B* **35**, 193 (1937).
[40] H. Gerding and W. J. Nijveld, *Rec. Trav. Chim. Pays Bas* **59**, 1206 (1940); Stecher,[29] p. 1005.
[41] E. Fischer, *Ber.* **39**, 2893 (1906).

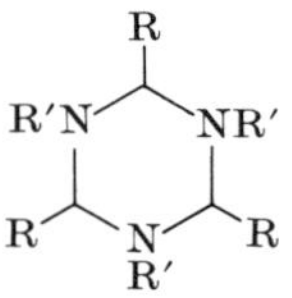

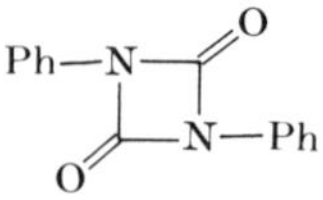

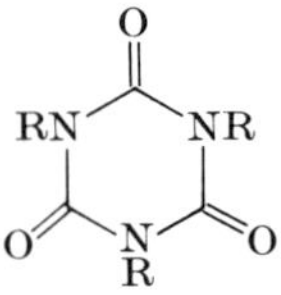

(25a) R = H, R' = Me   (26)   (27a) R = Ph
(25b) R = H, R' = COEt   (27b) R = H

at 100° *in vacuo*.[42] On heating phenyl isocyanate in the presence of sodium benzoate, an equilibrium mixture of the dimer (26) and the trimer (27a) is obtained.[43, 44] Cyanuric acid (27b) is formed spontaneously at 0° by polymerization of cyanic acid,[45] and it can be prepared by heating[46] urea at 120°. Melamine (triaminotriazine), which is condensed with urea, commercially, to form synthetic resins, is often prepared by heating dicyandiamide $H_2NC(=NH)NHCN$ under pressure.[27, 47]

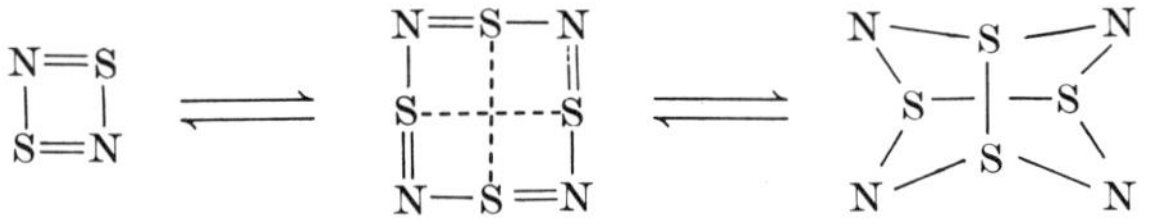

$$\begin{matrix} N=S \\ | \quad | \\ S=N \end{matrix} \rightleftharpoons \begin{matrix} N=S-N \\ | \quad \vdots \quad \| \\ S---|---S \\ \| \quad \vdots \quad | \\ N-S=N \end{matrix} \rightleftharpoons \begin{matrix} N \diagdown \quad S \diagup N \\ S-|-S \\ \diagup \quad S \diagdown \\ N \qquad \qquad N \end{matrix} \qquad (1)$$

(28)   (29a)   (29b)

At high temperatures, sulfur nitride (S≡N) gives an equilibrium mixture of the dimer (28) and the tetramer (29a); the latter is thought to exist partly in the bridged ring form (29b)[48] [Eq. (1)].

d. *Photooligomerization.* Photooligomerization of nonheterocyclic monomers to give a heterocyclic oligomer is rather rare.

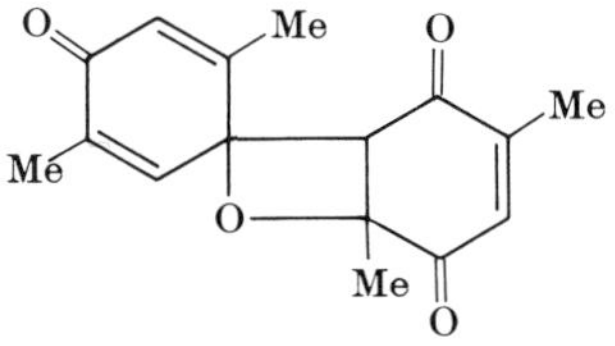

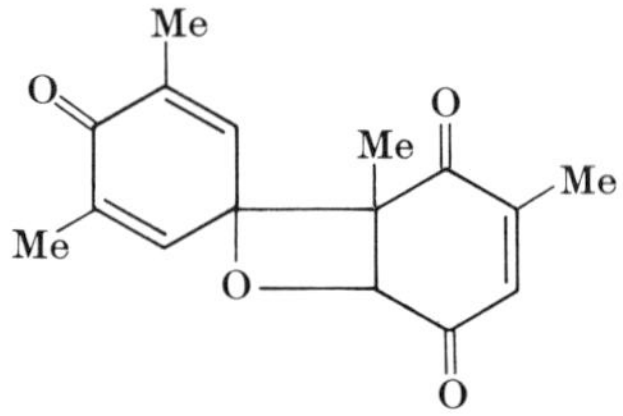

(30)   (31)

[42] B. G. Gowenlock and K. E. Thomas, *J. Chem. Soc. B*, 409 (1966).
[43] T. Yamabe, A. Nagasawa, H. Kitano, and K. Fukui, *Kogyo Kagaku Zasshi* **66**, 821 (1963).
[44] A. J. Bloodworth and A. G. Davies, *Chem. Commun.*, 24 (1965); S. Herbstman, *J. Org. Chem.* **30**, 1259 (1965).
[45] P. Klason, *J. Prakt. Chem.* [2] **33**, 129 (1886).
[46] F. Wohler, *Ann. Phys. Chem.* [2] **15**, 622 (1829).
[47] See Stecher,[29] p. 649.
[48] M. Becke-Goehring, *Quart. Rev.* **10**, 437 (1956).

2,5- and 2,6-Dimethylbenzoquinone dimerize by a C=C bond adding to an exocyclic C=O bond to give the products **30** and **31**, respectively.[49] Photodimerization of benzaldehyde–cyclohexylimide in ethanol has been reported to give the 1,3-diazetidine (**32**).[50] These compounds are all formed by 1,2-cycloaddition; a theoretical survey of photodimerization is given in Section III,E.

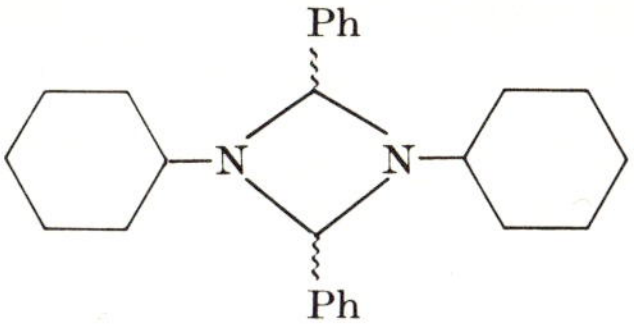

(**32**)

e. *Miscellaneous Oligomerization.* The dimer (**33**) is obtained when 3,5-di-*t*-butylanthranilic acid is treated with a strong dehydrating reagent (e.g., dicyclohexylcarbodiimide).[50a] The steric repulsion between the *t*-butyl groups on the phenyl rings of **33** may prevent further condensation with the monomer.

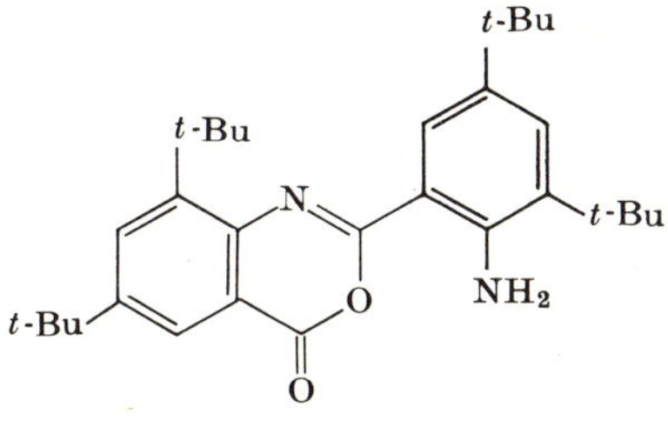

(**33**)

Subclass 2. Heterocyclic Oligomers from Heterocyclic Monomers

Mononuclear heterocycles formed by oligomerization of heterocyclic monomers are fewer than those from nonheterocycles. Ring opening of the monomer usually precedes condensation.

a. *Acid-Catalyzed Oligomerization.* Three-membered heterocycles often yield six-membered compounds; dioxane can be prepared industrially by passing oxiran (ethylene oxide) over a heated acidic catalyst (e.g.,

[49] R. C. Cookson, J. J. Frankel, and J. Hudec, *Chem. Commun.*, 16 (1965).
[50] R. O. Kan and R. L. Furey, *J. Amer. Chem. Soc.* **90**, 1666 (1968).
[50a] R. W. Franck and E. G. Leser, *J. Org. Chem.* **35**, 3932 (1970).

sodium hydrogen sulfate).[5] Cold 10 N hydrochloric acid and thiiran (ethylene sulfide) give a mixture of 1,4-dithian, with $Cl(CH_2)_2SH$ and $Cl(CH_2)_2S(CH_2)_2SH$.[51] The ring of N,N-dialkylaziridinium salts is opened by aqueous hydrochloric acid, yielding an equilibrium mixture containing dialkylaminoethyl chlorides, which easily dimerize to bis quaternary salts of piperazine.[52]

b. *Base-Catalyzed Oligomerization.* On refluxing isatin (indole-2,3-dione) with aqueous potassium hydroxide, followed by acidification of the product, the dimer (**4e**) was formed,[53] obviously through the ring-opened intermediate o-$H_2N \cdot C_6H_4COCO_2H$ and by a mechanism[12] similar to that responsible for the formation of the analog (**4d**).

c. *Thermal Oligomerization.* Goldman *et al.*[54] have reported that 3,4-dihydro-5H-benz[c]azepine (**34**) dimerizes on standing to the macrocyclic dimer (**35**); the latter is thermally stable, but depolymerizes to the monomer when a chloroform solution is allowed to stand.

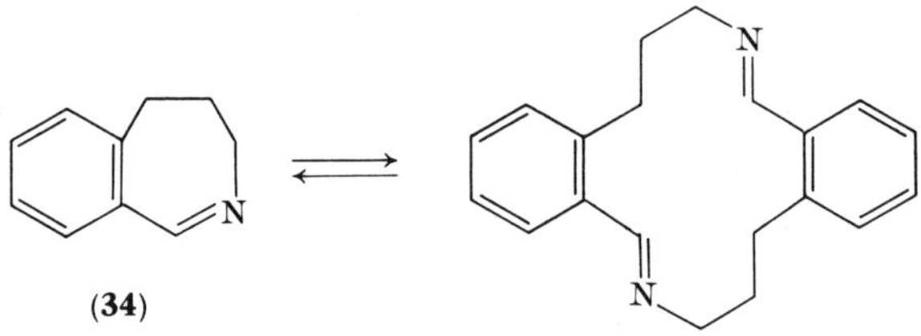

(**34**) ⇌ (**35**)

d. *Photooligomerization.* On irradiation with ultraviolet light, cyclohexanone thioacetal produces the radical (**36**), which gives intermediary cyclohexanethione by the elimination of a $HSCH_2CH_2$ radical from the side chain; this thione readily dimerizes to afford the spiro-1,3-dithi-

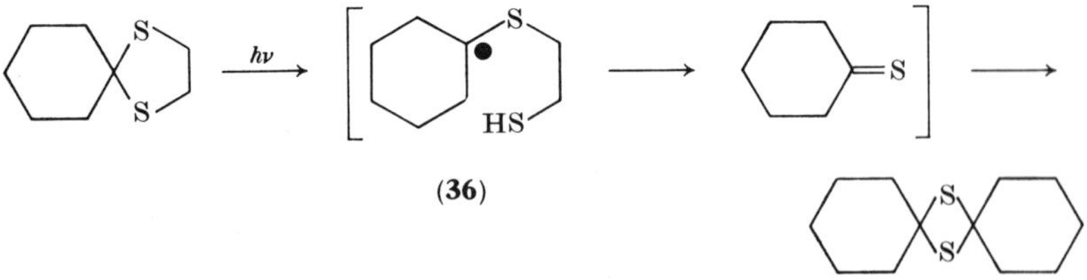

(**36**) → (2)

(**37**)

[51] R. M. Acheson, "An Introduction to the Chemistry of Heterocyclic Compounds," 2nd ed., p. 30. Wiley (Interscience), New York, 1967.
[52] Acheson,[51] p. 9.
[53] G. Stefanovic, L. Lorenc, R. I. Mamuzić, and M. L. Mihailović, *Tetrahedron* **6**, 304 (1959).
[54] I. M. Goldman, J. K. Larson, J. R. Tretter, and E. G. Andrews, *J. Amer. Chem. Soc.* **91**, 4941 (1969).

etane (**37**)[55] [Eq. (2)]. Compound **38** gives the dibenzodithiocindione (**40**) photochemically at room temperature in an inert solvent; **39** was proved to be the intermediate[56] [Eq. (3)].

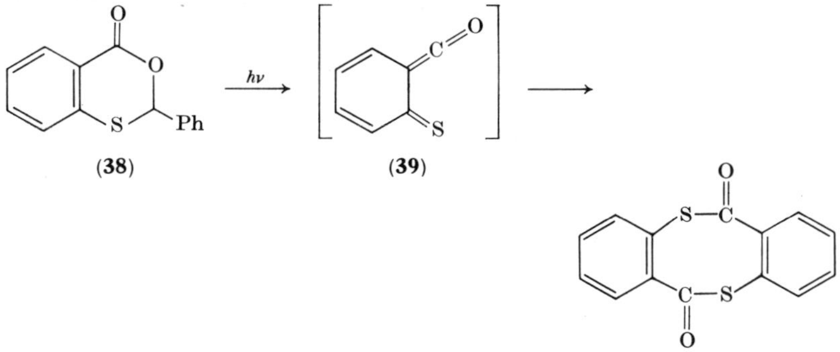

(**38**) → (**39**) → (**40**)   (3)

e. *Miscellaneous Oligomerization.* N-Methylsulfonylaziridine is reported to give a quantitative yield of N,N'-di(methylsulfonyl)piperazine in the presence of sodium iodide in acetone.[57]

Subclass 3. Heterocyclic Co-oligomers from More Than One Kind of Monomer

Co-oligomerization of two or more monomeric units to give mononuclear heterocycles is commonly used in synthesis. A simple example is piperazine,[58] as first prepared by Cloëz in 1853 by the action of alcoholic ammonia on ethylene chloride (2 moles) (see below).

a. *Acid-Catalyzed Co-oligomerization.* Many applications are known of the Schiff condensation for synthesis of heterocyclic co-oligomers. Tröger's base[59] was thought by Tröger, who prepared it in 1887 from p-toluidine (2 moles) and formaldehyde (3 moles) in the presence of acid with loss of water (3 moles), to be an anil (**41**). The correct structure (**42**) was deduced by Spielman in 1935; **42** is reminiscent of part of the structure of the anhydro polymers of 2-aminobenzaldehyde (**5**). In 60% sulfuric acid, Hecht and Henecka[60] prepared tetramethylenedisulfotetramine (**43**), an analog of **50a**, from sulfamide ($H_2N \cdot SO_2 \cdot NH_2$) and

[55] J. D. Willett, J. R. Grunwell, and G. A. Berchtold, *J. Org. Chem.* **33**, 2297 (1968).
[56] O. L. Chapman and C. L. McIntosh, *J. Amer. Chem. Soc.* **92**, 7001 (1970).
[57] H. W. Heine, W. G. Kenyon, and E. M. Johnson, *J. Amer. Chem. Soc.* **83**, 2570 (1961).
[58] See Stecher,[29] p. 836.
[59] J. Tröger, *J. Prakt. Chem.* [2] **36**, 227 (1887); M. A. Spielman, *J. Amer. Chem. Soc.* **57**, 583 (1935).
[60] G. Hecht and H. Henecka, *Angew. Chem.* **61**, 365 (1949).

formaldehyde; **43** has proved to be a violent convulsive poison.[61] In a similar manner to the elegant one-step synthesis of adenine (**12a**), two molecules of formamide react (on heating in a sealed tube in the presence of phosphoryl chloride) with alkyl-substituted acetamides (RCH_2CONH_2) and caprolactams to give pyrimidines **44** ($R = HC_{16}H_{33}$; 20–30% yield) and **45** ($n = 2$–4; 7–9%), respectively.[22] 7-Methyladenine (**12b**) was obtained (but in only 5% yield) by heating *N*-methylformamide with four equivalents of formamide.[22]

(**41**)

(**42**)

(**43**)

(**44**)

(**45**)

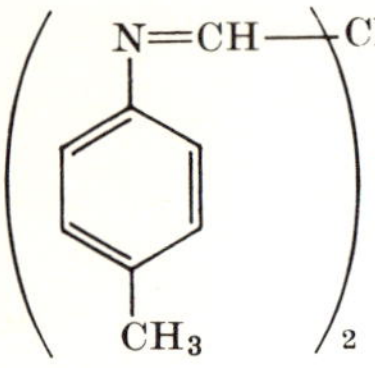

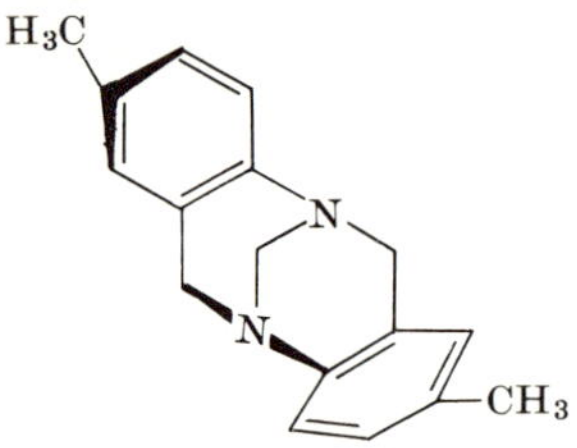

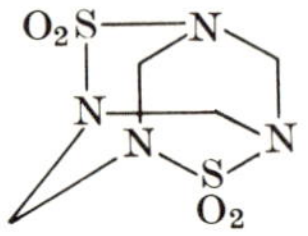

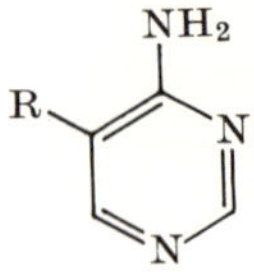

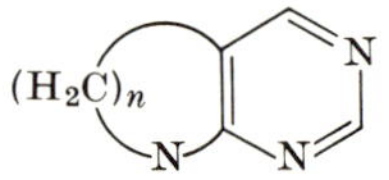

Nitriles can condense cyclically with aldehydes, in the presence of acid, to yield hexahydro-*s*-triazines; e.g., propionitrile and formaldehyde give the hexahydro-1,3,5-tripropionyl-*s*-triazine (**25b**).[62] When benzonitrile (1 mole) and benzoyl chloride (2 moles) were heated to 150° with zinc (or stannic) chloride, a high yield of 2,4,6-triphenyl-3,5-diazapyrylium salt (**46**) was formed,[62a] which was converted into 2,4,6-triphenyl-*s*-triazine by ammonia.

(**46**)

(**47a**) X = O, Y = CH
(**47b**) X = S, Y = N

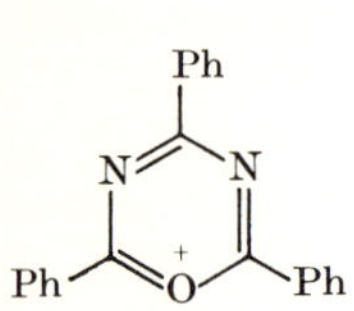

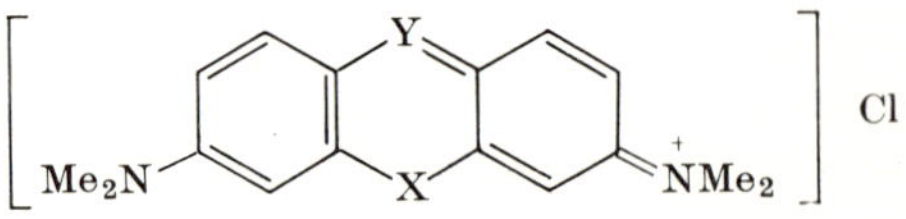

[61] See Stecher,[29] p. 1028.
[62] W. O. Teeters and M. A. Gradsten, *Org. Synth., Coll. Vol.* **4**, 518 (1963).
[62a] R. R. Schmidt, *Chem. Ber.* **98**, 334 (1965).

Many important anthracene analogs containing hetero ring atoms have been synthesized by means of co-oligomerization. The diaminoxanthylium chlorides (e.g., **47a**,[63, 64] an important microscopic stain) are made by condensing m-dialkylaminophenols with formaldehyde and dehydrating the resulting p,p'-tetraalkylamino-o,o'-dihydroxydiphenylmethane to a reduced form of **47a**, which is finally oxidized with ferric chloride. Methylene blue (**47b**) is usually prepared by heating a mixture of p-aminodimethylaniline, thiosulfuric acid, and potassium dichromate;[65, 66] it is a most valuable biological stain. 3,6-Diaminoacridine and its derivatives, similarly synthesized from m-phenylenediamine (2 moles) and formic acid (1 mole) have been extensively reviewed.[67]

(48)

(49)

Thianthrene (**48**) is one of the principal products of the action of sulfur chloride (or of sulfur) on benzene in the presence of aluminum chloride.[68] Ariyan et al.[69] similarly obtained the linear disulfide (**49**) by treating p-diethoxybenzene with sulfur chloride in an inert solvent; the structure was confirmed by X-ray diffraction.[70]

b. *Base-Catalyzed Co-oligomerization.* Many co-oligomers can be made by Schiff condensations under basic conditions; Graymore[71] has prepared hexahydrotriazines (**25**) by the base-catalyzed condensation of three molecules of amines (RNH_2; R = alkyl and aryl) and formaldehyde. Hexamine (**50a**)[72] is formed by evaporating a mixture of aqueous formaldehyde and ammonia. The well-known use of hexamine as a

---

[63] See Stecher,[29] p. 895.
[64] A. Albert, "Heterocyclic Chemistry," 2nd ed., p. 349. Oxford Univ. Press, New York; Athlone, London, 1968.
[65] See Stecher,[29] p. 684.
[66] See Albert,[64] p. 364.
[67] A. Albert, "The Acridines," 2nd ed., p. 103. Arnold, London, 1966.
[68] G. Dougherty and P. D. Hammond, *J. Amer. Chem. Soc.* **57**, 117 (1935).
[69] Z. S. Ariyan and L. A. Wiles, *J. Chem. Soc.*, 4709 (1962); Z. S. Ariyan and R. L. Martin, *Chem. Commun.*, 847 (1969).
[70] J. S. Ricci and I. Bernal, *J. Chem. Soc. B*, 1928 (1971).
[71] J. Graymore, *J. Chem. Soc.*, 39 (1941).
[72] See Smolin and Rapoport,[27] pp. 545–596.

urinary antiseptic depends on its depolymerization to formaldehyde by the natural acidity of urine. It is unchanged by hot aqueous sodium hydroxide. When substituted aromatic aldehydes are suspended in 10% aqueous ammonium carbonate solution (pH 9.2) for 15–20 days, substituted hexamines (**50b**) and their corresponding hydrobenzamides (RCH=N)$_2$CHR are obtained;[72a] the former are easily converted to the latter at pH 12.

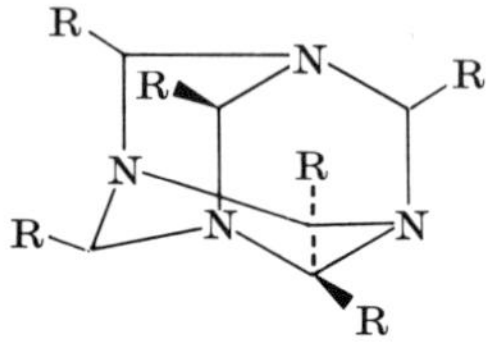

(**50a**) R = H
(**50b**) R = —C$_6$H$_4$X
(**50c**) R = — furfuryl

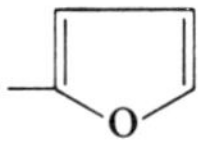

Condensation of α,β-unsaturated carbonyl compounds with ethylenediamine usually gives tetrahydrodiazepines,[73] but Hideg and Lloyd[74] have reported a high yield (80%) of co-oligomers by the condensation of equimolar amounts of α,β-unsaturated ketones (e.g., benzylideneacetone) with ethylenediamine in the presence of sodium carbonate, and without producing tetrahydrodiazepines; the structure of the products was assigned as either **51a** or **51b**. Glyoxal and urea give[75] acetyleneurea (**52**) in the presence of sodium carbonate at 0°. The 3,6-dialkyl-substituted s-tetrazines (**53**) have been conveniently prepared[76] by condensation of hydrazine (2 moles) and aldehydes, followed

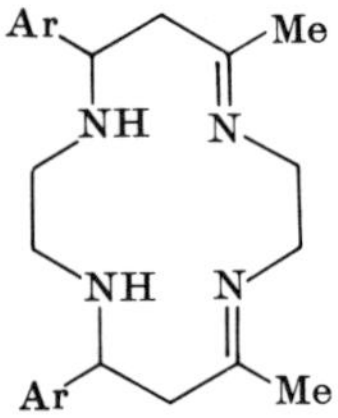

(**51a**) 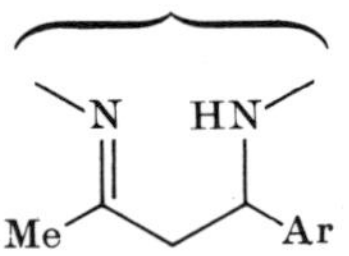   (**51b**) 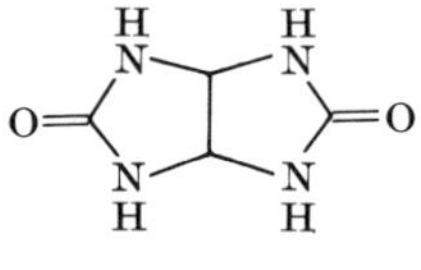   (**52**) 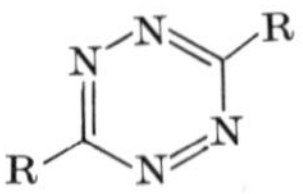   (**53**)

[72a] A. Kamal, A. Ahmad, and A. A. Qureshi, *Tetrahedron* **19**, 869 (1963).
[73] N. F. Curtis and R. W. Hay, *Chem. Commun.*, 524 (1966).
[74] K. Hideg and D. Lloyd, *J. Chem. Soc. C*, 3441 (1971).
[75] B. v. Reibnitz, U.S. Patent 2,731,472; *Chem. Abstr.* **50**, 13999 (1956).
[76] W. Skorianetz and E. sz. Kováts, *Helv. Chim. Acta* **53**, 251 (1970); **54**, 1922 (1971).

by the stepwise oxidation of the resulting dialkyl-substituted hexahydro-s-tetrazine (first with oxygen over platinium oxide, to the 2,3-dihydro-s-tetrazines, then with nitric acid).

Condensation by means of dehydrohalogenation (under basic conditions) is often utilized; e.g., piperazine can conveniently be made in the laboratory by heating aniline with 1,2-dibromoethane followed by dearylation of the resulting N,N'-diphenylpiperazine with nitrous acid.[58] A sulfur analog, 1,4-dithiazine, was made from ethylene chloride and sodium sulfide.[77] Gattermann and Skita[78] synthesized 2,6-dihydroxypyridine derivatives by condensing two molecules of ethyl sodiomalonate with one of dichloromethylamine, with the loss of sodium chloride and alcohol; for another example of pyridine synthesis by co-oligomerization, see Section III,F,2,a. Allen et al.[79] have prepared the large macrocyclic tetrathio compound (**54**) as the only product (10% yield) by co-oligomerization of α,α'-dimercapto-o-xylene (2 moles) with α,α'-dibromo-p-xylene (2 moles) in the presence of sodium.

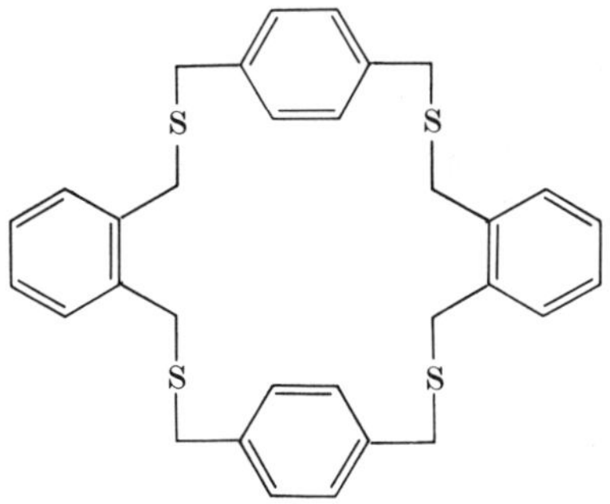

(**54**)

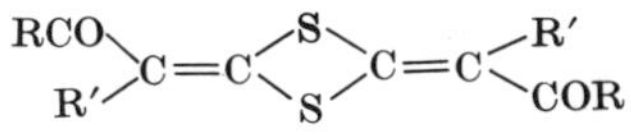

(**55**)

Upon heating carbonyl compounds such as $RCOCH_2R'$ ($R = Ph$, Me; $R' = Ph$, H) with carbon bisulfide in the presence of potassium hydroxide, the dimeric compounds (**55**) (desaurins) are produced in moderate yields.[80]

c. *Thermal Co-oligomerization.* Thieno[2,3-b]thiophene (**56**), called "liquid thiophthene," is obtained[81] in poor yield by distillation of

---

[77] R. C. Elderfield (ed.), "Heterocyclic Compounds", Vol. 6, p. 92. Wiley, New York, 1956.
[78] L. Gattermann and A. Skita, *Ber.* **49**, 494 (1916).
[79] D. W. Allen, P. N. Braunton, I. T. Millar, and J. C. Tebby, *J. Chem. Soc. C*, 3454 (1971).
[80] P. Yates, D. R. Moore, and T. R. Lynch, *Can. J. Chem.* **49**, 1456, 1467 (1971).
[81] A. Biedermann and P. Jacobson, *Ber.* **19**, 2444 (1886).

citric acid with phosphorus trisulfide. The isomeric thieno[3,2-b]thiophene (**57**) ("solid thiophthene") is made (together with **56**) by passing acetylene into boiling sulfur.[82]

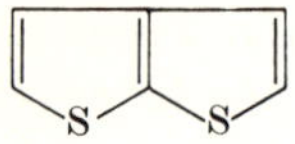

(**56**)

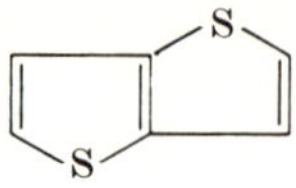

(**57**)

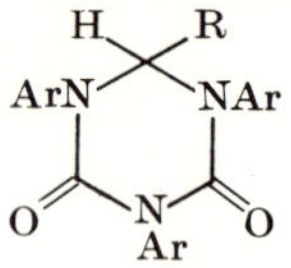

(**58a**) R = NMe$_2$
(**58b**) R = N(Ar)CONMe$_2$

Arylisocyanates and dimethylformamide condense on heating at 80° to give a mixture of 3:1-adduct (**58a**), 4:1-adduct (**58b**), and the spiro compound (**203**);[83] the mechanism for this co-oligomerization has been studied.[83]

d. *Photoco-oligomerization.* Many types of photoaddition reactions have been reviewed,[1] among which an example of co-oligomerization has been reported; the photoaddition of vinyl chloride to phenanthraquinone (**59**) is accompanied by the elimination of HCl to give the intermediate (**60**), which condenses with another molecule of **59** to yield the 1:2-adduct (**61**)[84] [Eq. (4)].

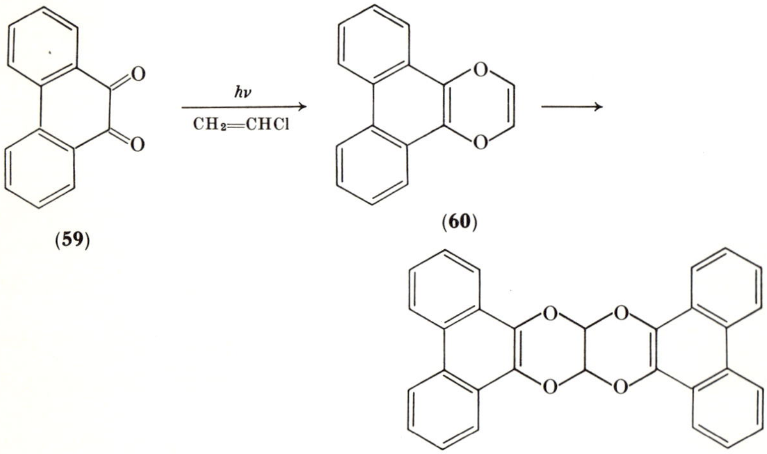

(**59**) $\xrightarrow[\text{CH}_2\!=\!\text{CHCl}]{h\nu}$ (**60**) $\longrightarrow$

(**61**)   (4)

[82] F. Challenger, *Sci. Progr.* **41**, 593 (1953).
[83] R. Richter and H. Ulrich, *J. Org. Chem.* **36**, 2005 (1971).
[84] G. Pfundt and G. O. Schenck, "1,4-Cycloaddition Reactions" (J. Hamer, ed.), p. 359. Academic Press, New York, 1967.

Sec. III.A.] HETEROCYCLIC OLIGOMERS 21

Subclass 4. Naturally Occurring Heterocyclic Oligomers (Including Their Synthesis)

Ellagic acid (62), isolated from the kino of *Eucalyptus maculata* Hook,[85] has also been prepared by sodium persulfate oxidation of gallic acid and by acid hydrolysis of crude tannin from walnuts.[86, 87]

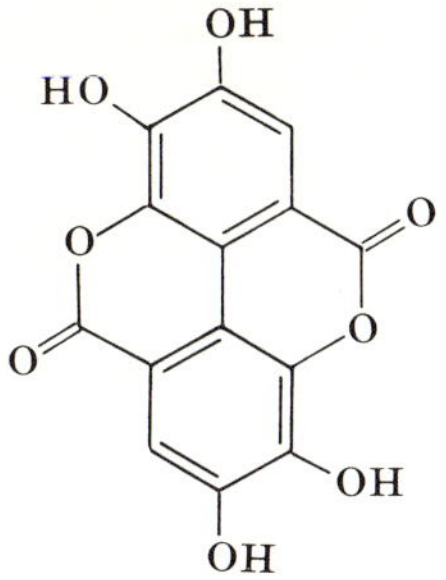

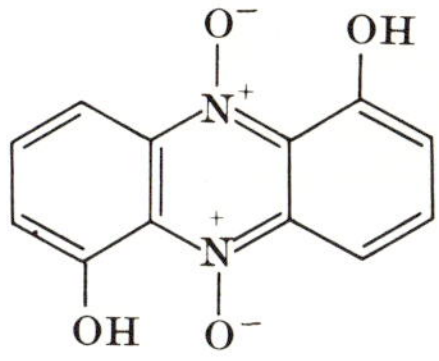

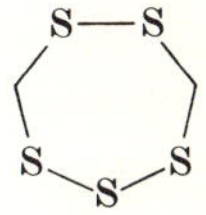

Iodinin, an antibiotic pigment isolated from *Chromobacterium iodinum*, has been shown[88] to be 1,5-dihydroxyphenazine di-*N*-oxide (63), which probably arises biosynthetically by oxidative coupling of two molecules of anthranilic acid.[88a]

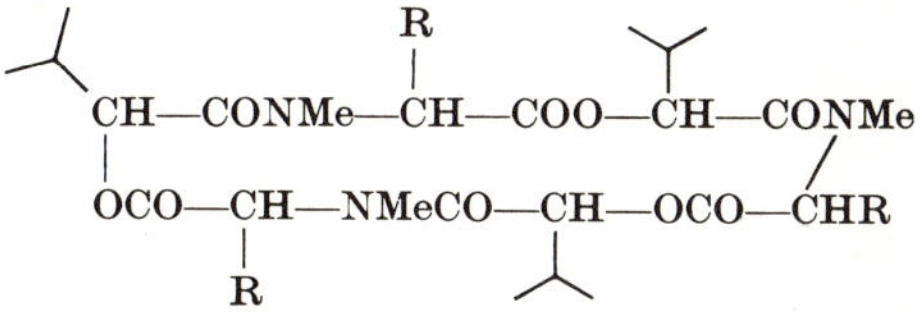

The odorous principle, lenthionine, of the edible mushroom *Shiitake lentinus edodes* (Berk.) Sing., has structure 64; it has been synthesized by condensation of sodium polysulfide with formaldehyde or methylene chloride.[89]

Many antibiotics have turned out to be macrocyclic polypeptides, among which enniatin (65) is a trimer produced by the fungus *Fusarium*

---

[85] R. J. Gell, J. T. Pinhey, and E. Ritchie, *Aust. J. Chem.* **11**, 372 (1958).
[86] A. G. Perkin and M. Nierenstein, *J. Chem. Soc.* **87**, 1412 (1905).
[87] L. Jurd, *J. Amer. Chem. Soc.* **78**, 3445 (1956); **79**, 6043 (1957).
[88] G. R. Clemo and A. F. Daglish, *J. Chem. Soc.*, 1481 (1950), and references therein.
[88a] W. B. Whalley, "Biogenesis of Natural Compounds" (P. Bernfeld, ed.), p. 820. Pergamon, Oxford, 1963.
[89] K. Morita and S. Kobayashi, *Tetrahedron Lett.*, 573 (1966).

orthoceras var. *enniatum* and other *Fusaria*.[90] Another example, gramicidin S, is produced by a strain of *Bacillus brevis*; it has a dimeric structure with a much larger macrocyclic peptide ring.[91]

B. Formation of Oligomers That Have Two (or More) Heterocyclic Nuclei Connected by a Side Chain

Subclass 1. Heterocyclic Oligomers from Nonheterocyclic Monomers

Relatively few oligomers of this type have been reported.

a. *Acid-Catalyzed Oligomerization.* The tetramer (**15**) of anthranilic acid, formed by refluxing the trimer (**14**) with the monomer in xylene, provides the best known example of this process.

b. *Base-Catalyzed Oligomerization.* The bispyrimidinylbutane (**66**) is obtained [together with the pyrimidine (**17**)] by trimerization of adipinonitrile in the presence of sodium.[31]

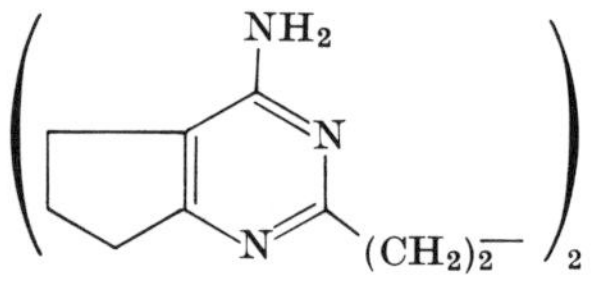

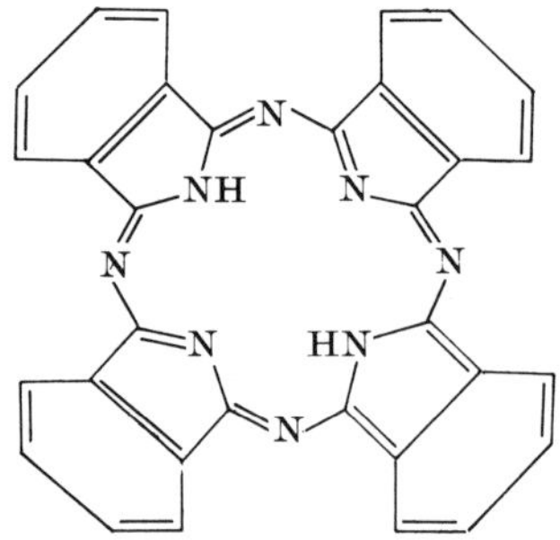

c. *Thermal Oligomerization.* Phthalocyanine (**67**), of which the copper complex is a highly valued greenish blue pigment used in household paints and printing inks, was first made[92] by heating phthalonitrile (*o*-dicyanobenzene) with magnesium oxide at 200°, followed by treating the resulting magnesium complex with sulfuric acid at $-3°$. Phthalocyanine compounds have been reviewed.[93]

Subclass 2. Heterocyclic Oligomers from Heterocyclic Monomers

An oligomer of this type can be formed by condensation of a side chain of one molecule with either the side chain or the nucleus of another

---

[90] See Stecher,[29] p. 409.
[91] See Stecher,[29] p. 506.
[92] R. P. Linstead and A. R. Lowe, *J. Chem. Soc.*, 1022 (1934).
[93] F. H. Moser and A. L. Thomas, "Phthalocyanine Compounds." Amer. Chem. Soc., Washington, D.C., 1963.

molecule. Condensation reactions of the side chains of heterocycles are analogous to those of corresponding aromatic (or alicyclic) compounds except for the strongly polar influence of the heteroatom on the reactivity.

a. *Acid-Catalyzed Oligomerization.* An oligomer formed by Schiff condensation of side chains of the monomers will be exemplified first. Treatment of 2-amino-3-formylpyrazine with acetic formic anhydride gave a mixture of the 2-formamido derivative (26% yield) and the monoformyl trimer (**68**; 46%).[94] As with the analogous 2-aminobenzaldehyde, N-formylation competes with self-condensation (cf. the formation of **4c**). Because of its low solubility in most organic solvents, **68** separates as a precipitate during the reaction, thus causing the termination of oligomerization.

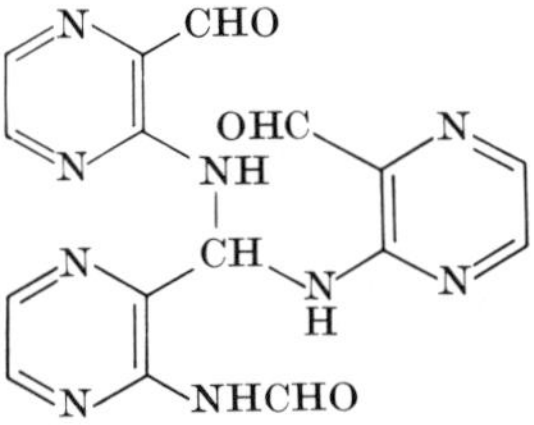

(**68**) 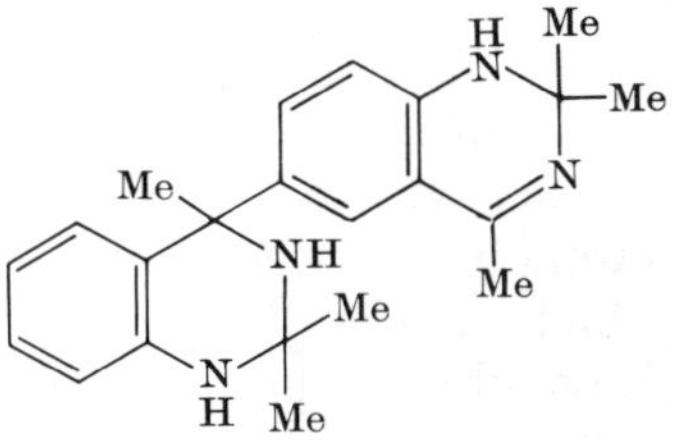   (**69**)

By a different type of side-chain condensation, the dimer **69**[95] was formed from 1,2-dihydro-2,2,4-trimethylquinazoline in 0.1 N hydrochloric acid at 100°. Bimolecular reduction products have been produced when hydrogenating 3-acetamido-6-methoxy-5-nitropyridazine 2-oxide over palladium/charcoal; two different products (**70** and **71**) have been isolated depending on the solvent used and the amount of hydrogen absorbed.[96]

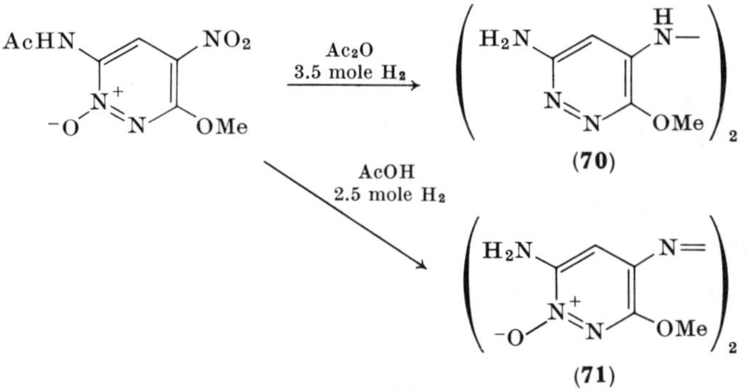

(**70**)

(**71**)

[94] A. Albert and H. Yamamoto, *J. Chem. Soc. C*, 2289 (1968).
[95] J. P. Brown and B. K. Tidd, *J. Chem. Soc. C*, 1075 (1968).
[96] T. Horie, *Chem. Pharm. Bull.* **11**, 1157 (1963).

A simple Michael-type addition of the methyl group of a monomer to the C=N bond of another molecule gives a dimer of type B2; e.g., dimers **72** and **73** are formed by the acid-catalyzed self-condensation of 2-hydroxy- (or 2-mercapto-)4-methyl-6-phenyl-1,6-dihydropyrimidine[97]

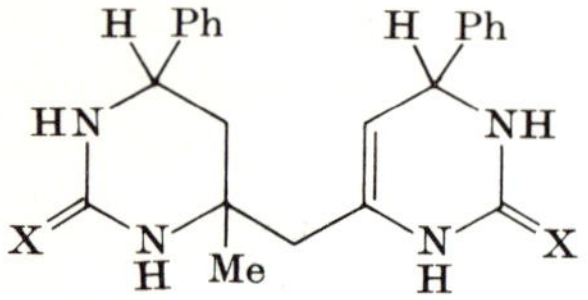

(72)

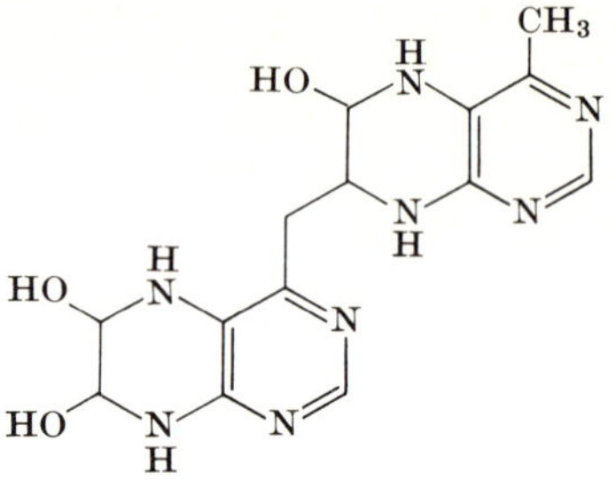

(73)

and 4-methylpteridine,[98] respectively. Formula (**74**) for an orange pigment, "2-methylquinoxaline orange", produced by aeration of 2-methylquinoxaline in 2 N hydrochloric acid,[99] was later revised to (**75**).[99a]

Intermolecular twofold amide formation (i.e., dehydration) to give a dimer is possible; on heating with acetic anhydride, pyrrole-2-carboxylic

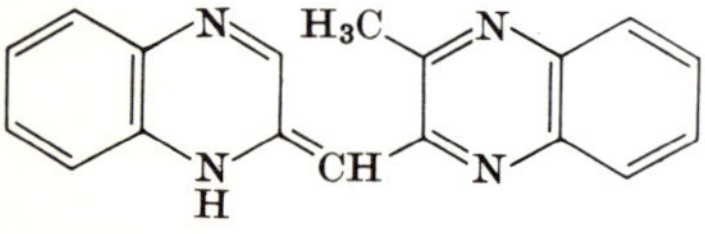

(74)

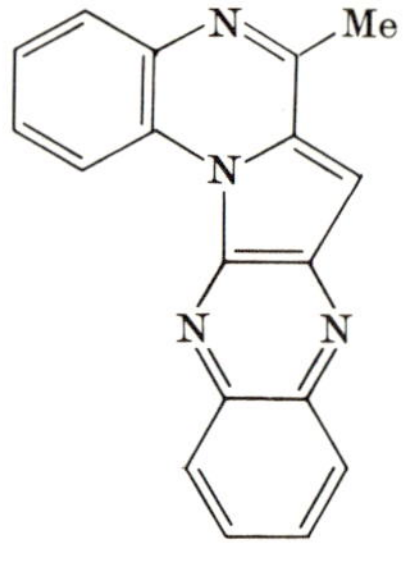

(75)

acid gives a cyclic compound (**76**)[100] in a similar manner to the formation of diketopiperazine from glycine. The indolocarbazole (**77a**) was obtained by intermolecular dehydrohalogenation from 1-methylindole-2-carboxyl chloride on heating in dichloromethane in the presence of aluminum chloride.[101] The action of thionyl chloride on 2-hydroxymethyl-5,6-

[97] G. Zigeuner, H. Brunetti, H. Ziegler, and M. Bayer, *Monatsh. Chem.* **101**, 1767 (1970).
[98] A. Albert and H. Yamamoto, *J. Chem. Soc. C*, 1181 (1968).
[99] C. L. Leese and H. L. Rydon, *J. Chem. Soc.*, 303 (1955).
[99a] G. W. H. Cheeseman and B. Tuck, *Tetrahedron Lett.*, 4851 (1968).
[100] See Acheson,[51] p. 82.
[101] J. Szmuszkovicz, *J. Org. Chem.* **28**, 2930 (1963).

methylenedioxybenzo[b]thiophene followed by attempted crystallization of the product from ethanol yielded a mixture of the dimeric compound (**78**) and 2-ethoxymethyl-5,6-methylenedioxybenzo[b]thiophene.[102]

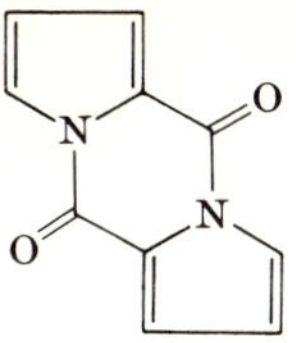

(**76**)

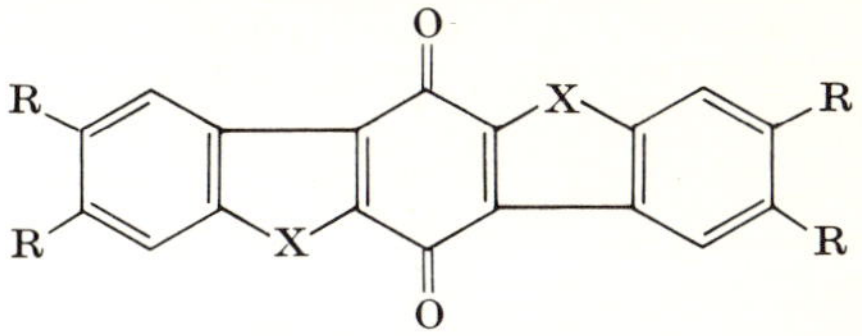

(**77a**) R = H, X = NCH$_3$
(**77b**) R = OH, X = O

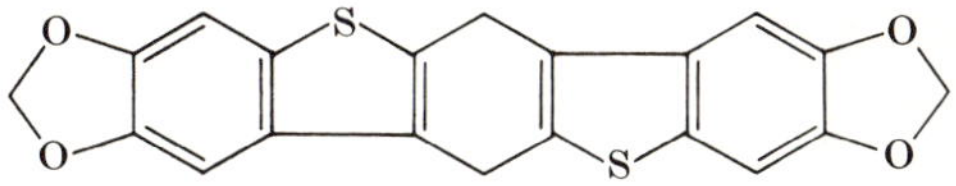

(**78**)

b. *Base-Catalyzed Oligomerization.* The "benzoin condensation" of heterocyclic compounds with potassium cyanide affords dimers of the B2 type; thus, the 2,2'-thenoins (**79a**; from thiophene-2-aldehydes), 3,3'-thenoins (from thiophene-3-aldehydes), and 3,3'-benzo[b]thenoins (from benzothiophene-3-aldehydes) have been obtained.[103] Similarly furfural and 2-cyanoquinoline give furoin (**79b**)[104] and 2-quinaldoin,[105] respectively; oxidation of these products affords the diketones (**80**) and (**81**).

(**79a**) X = S
(**79b**) X = O

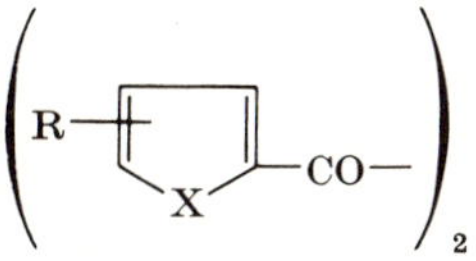

(**80**) 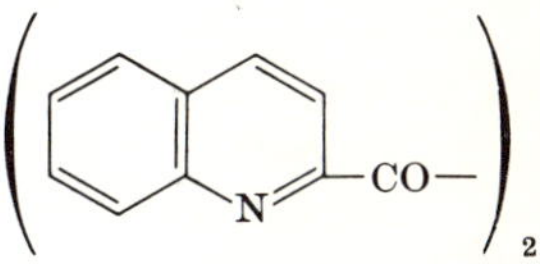   (**81**)

---

[102] E. Campaigne and E. S. Neiss, *J. Heterocycl. Chem.* **2**, 100 (1965).
[103] R. D. Schuetz and G. P. Nilles, *J. Org. Chem.* **36**, 2486 (1971), and references therein.
[104] Acheson,[51] p. 114.
[105] For a review, see B. R. Brown, *Quart. Rev.* **5**, 131 (1951).

During attempts to make 5,7-dihydroxy-1,3,8-triazanaphthalene from 4-amino-5-ethoxycarbonylpyrimidine and ethyl acetate by heating in the presence of sodium, Bredereck et al.[106] noticed that the dimer (**82**) was formed by amide-like links of the monomer side chains (the desired triazanaphthalene was finally obtained using sodium ethoxide as the catalyst).

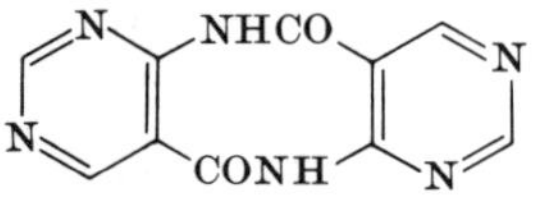

(**82**)

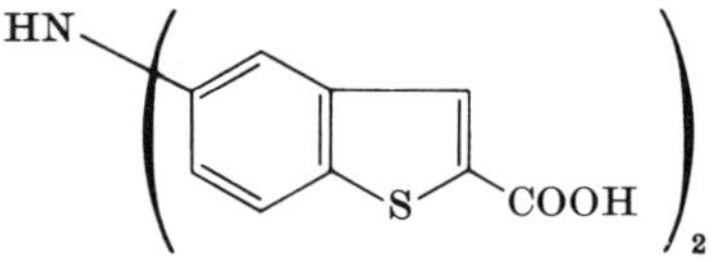

(**83**)

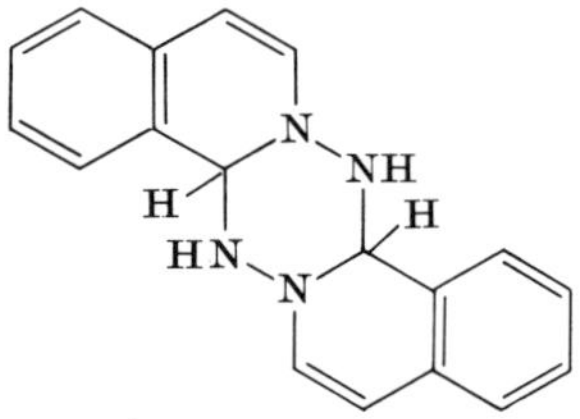

(**84**)

The dicarboxylic acid (**83**) was formed as a by-product during the preparation of 5-hydroxybenzo[b]thiophene-2-carboxylic acid from the corresponding amino compound by means of the Bucherer reaction.[107]

The Michael-type condensation of the side chain of a monomer with the nucleus of another molecule under basic conditions is exemplified by the dimer (**84**) formed when 2-aminoisoquinolinium picrate is passed through an anion-exchange column.[108]

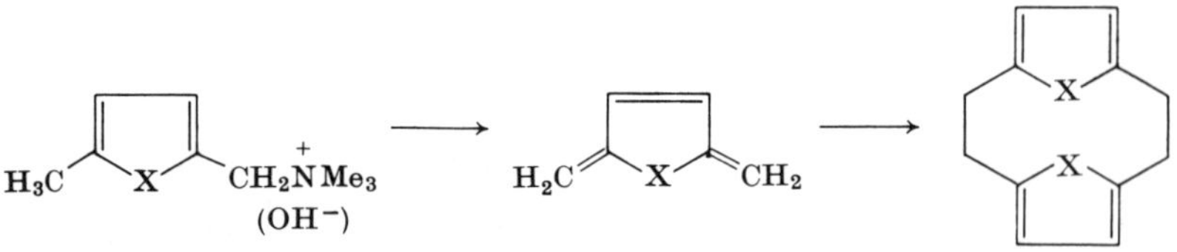

(**85**) (**86a**) X = O   (**87**)   (5)
(**86b**) X = S

---

[106] H. Bredereck, F. Effenberger, E. Henseleit, and E. H. Schweizer, *Chem. Ber.* **96**, 1868 (1963).
[107] M. Martin-Smith and S. T. Reid, *J. Chem. Soc.*, 938 (1960).
[108] Y. Tamura, N. Tsujimoto, and M. Uchimura, *Chem. Pharm. Bull. (Tokyo)* **19**, 143 (1971).

c. *Thermal Oligomerization.* Thermal condensation of side chains of the monomer can give oligomers of this type; for example, a Hofmann elimination from the 5-methyl-2-furfuryl salt (**85a**) gave the dimethylene-dihydrofuran (**86a**), which dimerized to a heterocyclophane (**87a**)[109] [see Eq. (5)]. The sulfur analog (**87b**) was obtained in a similar manner from **85b** (but in lower yield).[109]

Many aliphatic and alicyclic nitrones have been reported to give dimers; thus 2,4,4-trimethyl-1-pyrroline-1-oxide dimerized, on standing, to **88** by the nucleophilic addition of a 2-methyl group of the monomer across the C=N bond of another molecule (the chemistry of "nitrones" has been reviewed by Delpierre and Lamchen[110]). Oxidation of 1-hydroxypiperidine does not give the expected cyclic nitrone but a product to which the dimeric structure **89** has been assigned.[111]

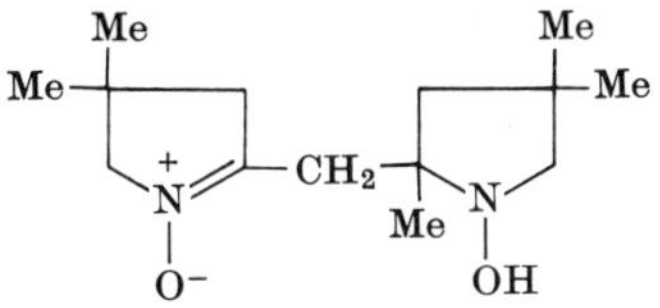

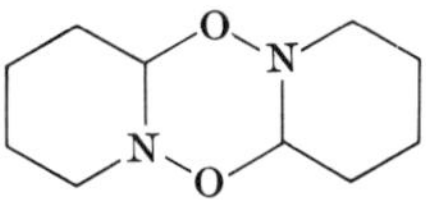

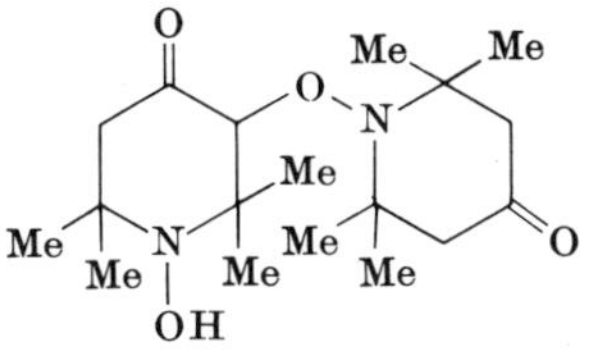

Another example of oligomers formed by condensation of nucleus with side chain is the dimer (**90**) formed when 2,2,6,6-tetramethyl-4-oxopiperidin-1-oxyl (**91**) is set aside at room temperature for six months;[112] the mechanism for the formation of **90** has been assumed[112] to be hydrogen abstraction, followed by the coupling reaction of the N-oxyl radical with the C-radical derived from **91**.

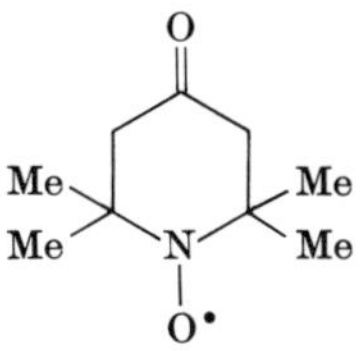

d. *Photooligomerization.* Photodimerization ($2\pi \rightarrow 2\sigma$) of a side-chain double bond in the monomer can give type B2 products, such as the

[109] H. E. Winberg, F. S. Fawcett, W. E. Mochel, and C. W. Theobald, *J. Amer. Chem. Soc.* **82**, 1428 (1960).
[110] G. R. Delpierre and M. Lamchen, *Quart. Rev.* **19**, 329 (1965).
[111] J. Thesing and H. Mayer, *Chem. Ber.* **89**, 2159 (1956).
[112] T. Yoshioka, S. Higashida, S. Morimura, and K. Murayama, *Bull. Chem. Soc. (Jap.)* **44**, 2207 (1971).

cyclobutane-type dimer (**92**)[113] formed by solid-state irradiation of
β-(3-pyridyl)-*trans*-acrylic acid; the trans head-to-head configuration
of **92** has been interpreted in terms of crystal packing.[113] Already in
1937, a similar cyclobutane-type photodimer had been obtained[114]
from 2-(β-styryl)quinoline (solid state), although it is still uncertain
whether the dimer is head-to-head or head-to-tail. George and Roth[115]
have recently reported a photodimer of an α,β-unsaturated cyclanone
(**93**), to which the structure with an anti-trans form (**94**) has been reliably
assigned.

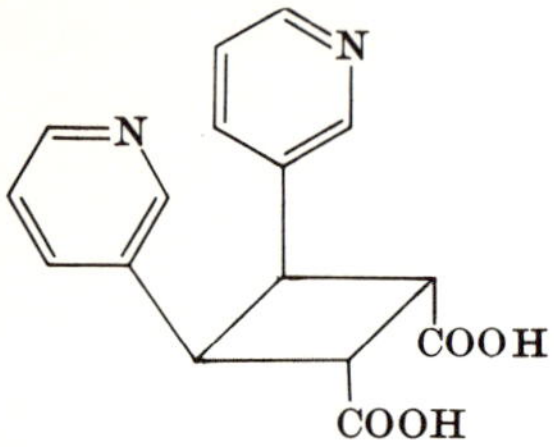

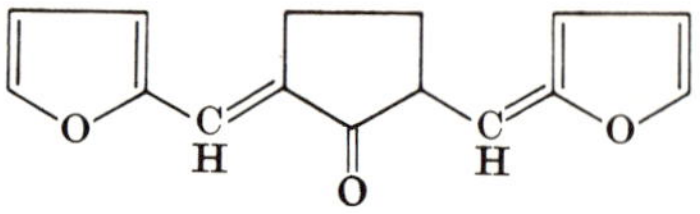

Étienne[116,117] reported a few photodimers of 1-aza- and 2-aza-
anthracene; the structure of the dimer from the former was assumed
to be **95**.

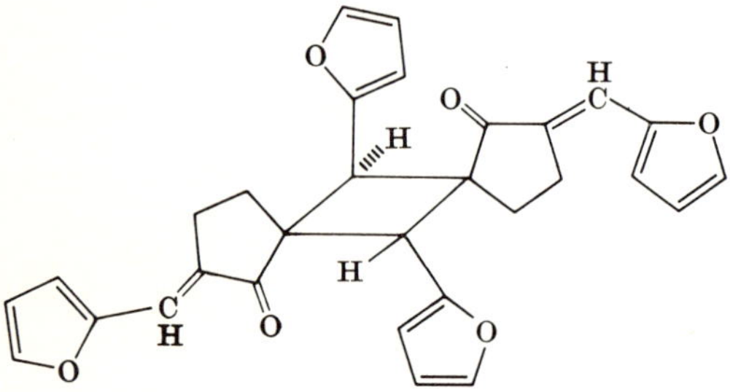

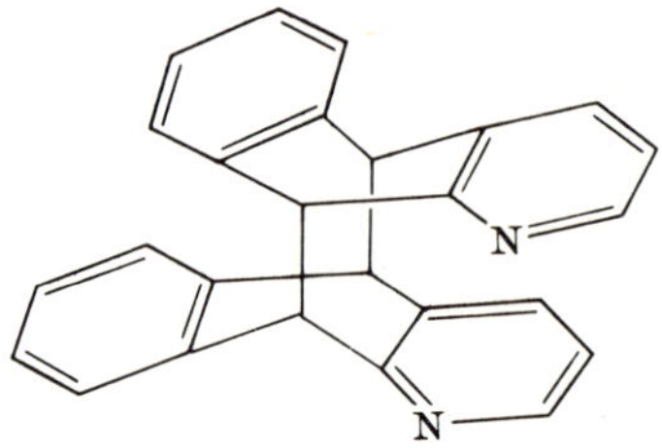

The 4-acyl-1,2,3-triazole (**96**) gave an interesting pinacol-type
dimer (**97**) by photoreduction in isopropanol[118] [Eq. (6)]. When a
heterocycle has an exocyclic C=C bond, a photodimer of type B2 can be
formed; for example, the dimer (**99**) was obtained[118a] from the pseudo-
oxazolone (**98**) [Eq. (7)].

[113] M. Lahav and G. M. J. Schmidt, *J. Chem. Soc. B*, 239 (1967).
[114] M. Henze, *Ber.* **70**, 1273 (1937).
[115] H. George and H. J. Roth, *Tetrahedron Lett.*, 4057 (1971).
[116] A. Étienne, *Ann. Chim. (Paris)* [12] **1**, 5 (1946).
[117] A. Étienne, *C. R. Acad. Sci.* **219**, 622 (1944).
[118] J. van Thielen, T. V. Thien, and F. C. de Schryver, *Tetrahedron Lett.*, 3031 (1971).
[118a] R. Filler and E. J. Piasek, *J. Org. Chem.* **28**, 221 (1963).

(96) (97) (6)

(98) (99) (7)

(100) (101) (102)

(103) (8)

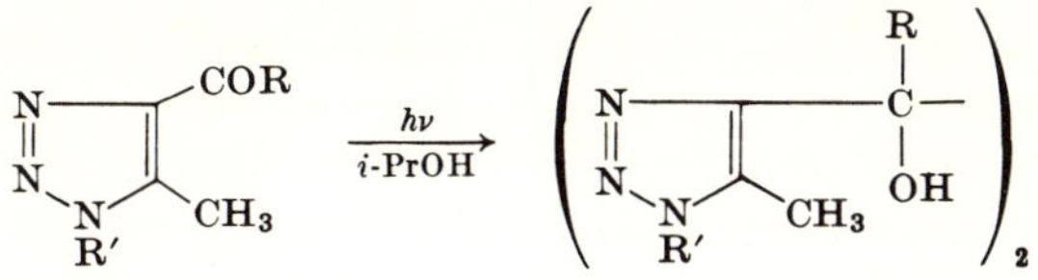

Besides the cyclobutane-type dimers (see type E2 oligomers), addition dimers such as (**103**) have been produced by photolysis (254 nm) of a frozen solution of thymine (**100**, R = H)[119] and thymidine.[120] These biologically significant dimers, which also arise from partial photodestruction of DNA, are thought to be formed by combination of two radicals (**101** and **102**), both derived from two molecules of the starting material by the action of light[120] [Eq. (8)].

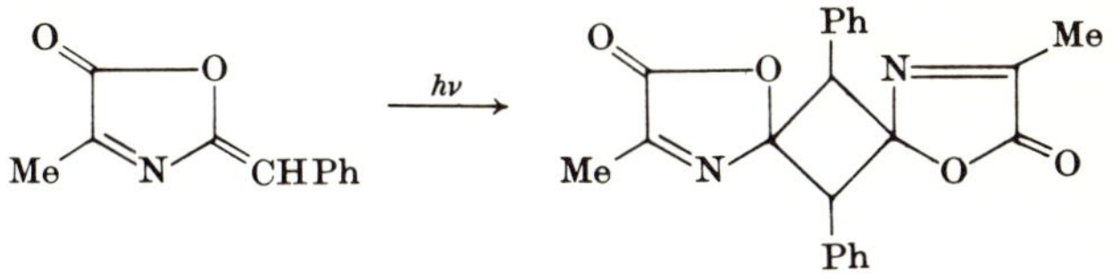

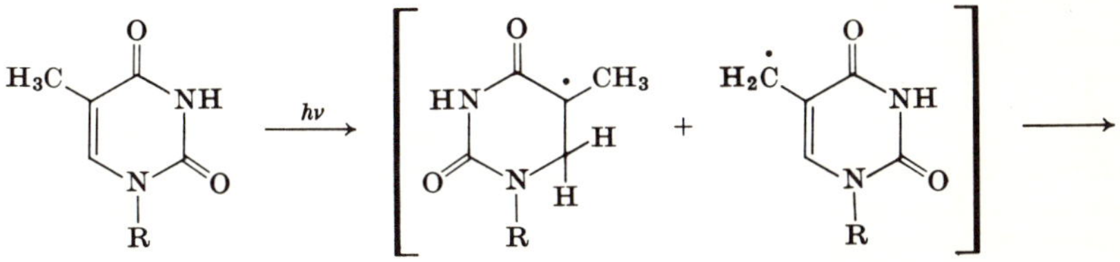

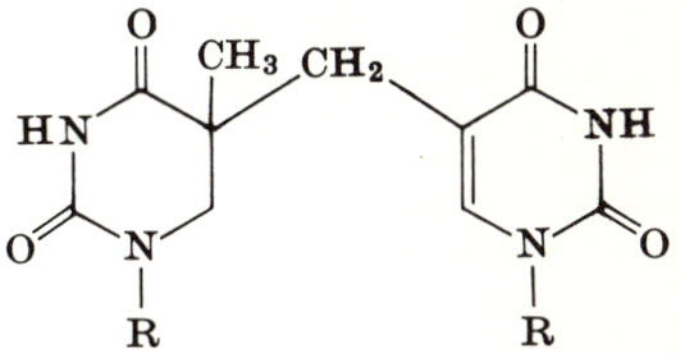

e. *Miscellaneous Oligomerization.* Oxidative condensation of *N*-amino groups of some heterocycles has been reported to give dimeric compounds of type B2; thus, a dimer (**104**) has been obtained by oxidation of *N*-aminophthalamide with $C_6H_5I(OAc)_2$.[121] Similar oxidative dimers were

119 A. J. Varghese, *Biochem. Biophys. Res. Commun.* **38**, 484 (1970).
120 A. J. Varghese, *Biochemistry* **9**, 4781 (1970).
121 D. J. Anderson, T. L. Gilchrist, and C. W. Rees, *Chem. Commun.*, 800 (1971).

produced from 1-amino-3-isopropenyl-2-benzimidazolone (with the same oxidizing reagent),[121] 1-aminoimidazo[1,2-*a*]pyridinium salts (with bromine),[122] and 2-amino-5,5-dimethyl-1-pyrroline 1-oxide (with alkaline ferricyanide or neutral permanganate).[123]

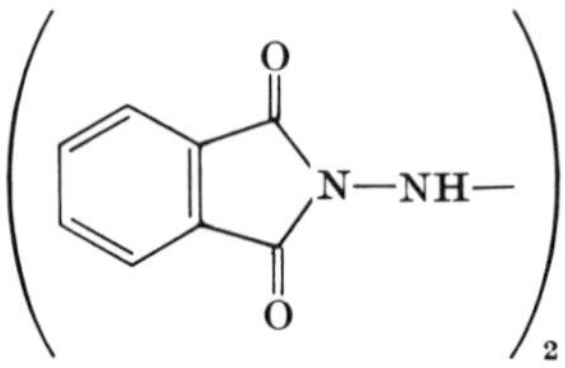

(104) 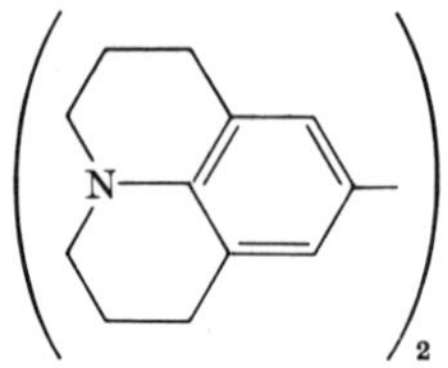   (105)

Oxidation of julolidine with peroxyacetic acid (at pH 4.8) has been reported[124] to give bis-9,9'-julolidyl (105) by oxidative coupling of the benzene ring. A similar oxidative dimer was formed[125] when the isolable 4-bromomethylene compound, produced by bromination of 2,2,4-trimethyl-1,2-dihydroquinoline in chloroform, was allowed to react with bromine in the same solvent.

3. *Heterocyclic Co-oligomers from More Than One Kind of Monomer*

Many substances of this type, produced by co-oligomerization, are known and have a wide variety of important commercial uses (e.g., medicines and photosensitizing agents).

a. *Acid-Catalyzed Co-oligomerization.* When heated with thionyl chloride at 45°, terephthalodihydroxamic acid (*p*-HONHCOC$_6$H$_4$CONHOH) gave the dioxathiazole derivative (106) in 96% yield;[126] this is an example of co-oligomerization of nonheterocyclic monomers. The same compound was formed[126] in 80% yield from terephthalodinitrile dioxide by treatment with sulfur dioxide at about $-15°$. On heating at a higher temperature, 106 was converted into terephthalodiisocyanate almost quantitatively.[126]

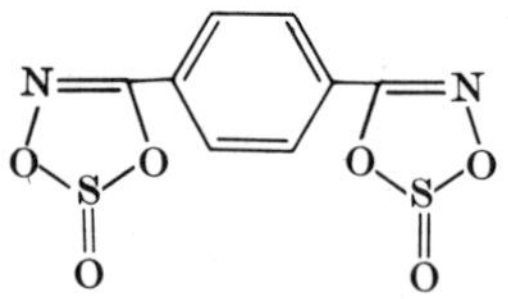

(106)

[122] E. E. Glover and M. Yorke, *J. Chem. Soc. C*, 3280 (1971).
[123] A. R. Forrester and R. H. Thomson, *J. Chem. Soc.*, 1224 (1965).
[124] V. R. Holland and B. C. Saunders, *Tetrahedron* 27, 2851 (1971).
[125] J. P. Brown and O. Meth-Cohn, *J. Chem. Soc. C*, 3631 (1971).
[126] E. H. Burk and D. D. Carlos, *J. Heterocycl. Chem.* 7, 177 (1970).

1,3-Di-6-quinolylurea, an intermediate in manufacturing the veterinary medicine quinuronium sulfate,[127] can be prepared by condensation of two molecules of 6-aminoquinoline and one of phosgene (or more conveniently urea in the presence of HCl).[128] 1,3-Bis(4-amino-2-methyl-6-quinolyl)urea, similarly prepared, has antiseptic properties.[127] These are examples of type B3 co-oligomers made by condensation of a side chain of the heterocyclic monomer with a different molecule.

Several dimeric compounds of type B3 have been formed by condensation of the nucleus of the heterocyclic monomer with a different molecule. Besides being made by the benzoin condensation (cf. type B2 oligomers), some 2,2'-thenils (**79**; X = S, R = 5-MeO, 5-i-PrO) can also be prepared by the condensation of oxalyl chloride (1 mole) with thiophenes (2 moles) using stannic chloride in carbon disulfide.[103] Schmidt[129] described how furfural reacts with two molecules of 2-methylfuran in the presence of a trace of acid to give a trimeric furan, for which the constitution **107** was proposed; Acheson,[130] however, preferred the more reasonable structure **108a** for the trimer, because a similar sulfur analog (**108b**) is formed from benzaldehyde (1 mole) and 2-methylthiophene (2 moles) in the presence of zinc chloride. Moreover, condensation of 3(2H)-pyridazinones and aromatic aldehydes (in acetic anhydride) gave similar products (**109**).[131]

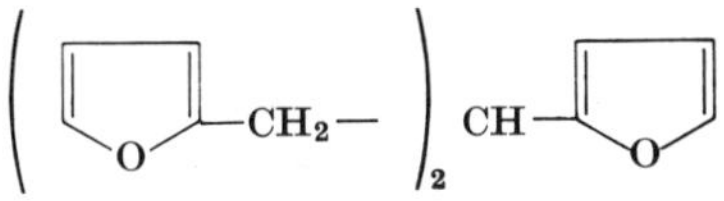

(**107**)

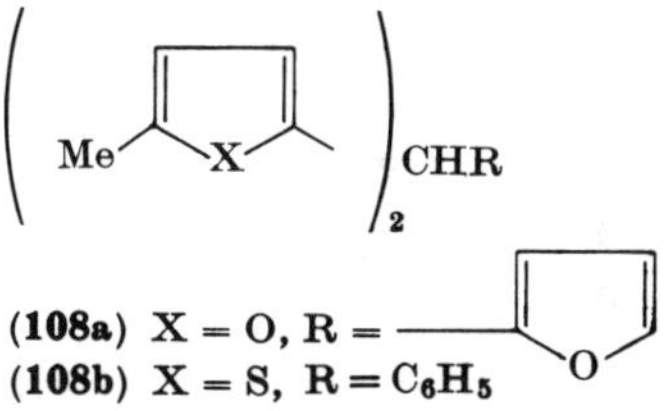

(**108a**) X = O, R =
(**108b**) X = S, R = C$_6$H$_5$

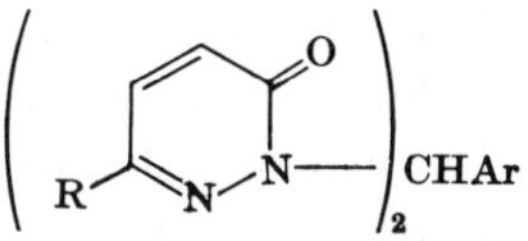

(**109**)

---

[127] See Stecher,[29] pp. 151, 392.
[128] L. Haskelberg, *J. Org. Chem.* **12**, 434 (1947), and references therein.
[129] C.-H. Schmidt, *Angew. Chem.* **67**, 317 (1955).
[130] See Acheson,[51] p. 106.
[131] H. Gregory, J. Hills, and L. F. Wiggins, *J. Chem. Soc.*, 1248 (1949).

With hydrazine hydrochloride, thioindoxyl-1,1-dioxide has been reported to afford the azine (110) by prior formation of a bishydrazone.[132] Murexide (111), a purple indicator used in compleximetric titrations, is formed from ammonium acetate, glacial acetic acid, and alloxantin (152; R = H). The "murexide test" is well known for detecting uric acid and other purines.[133]

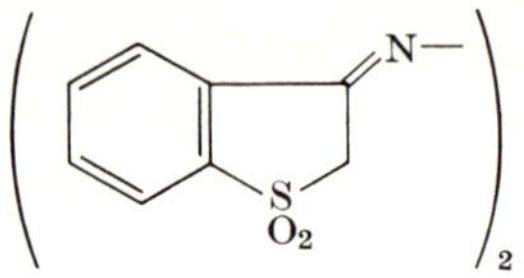

(110)

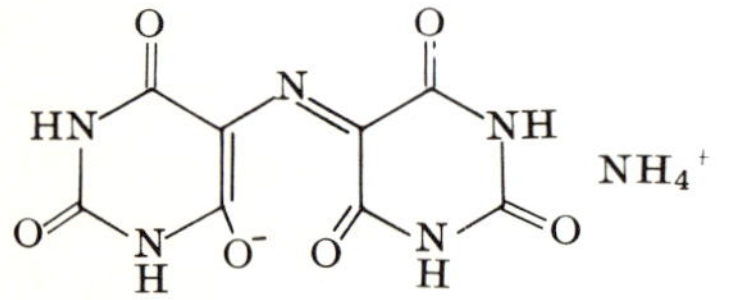

(111) 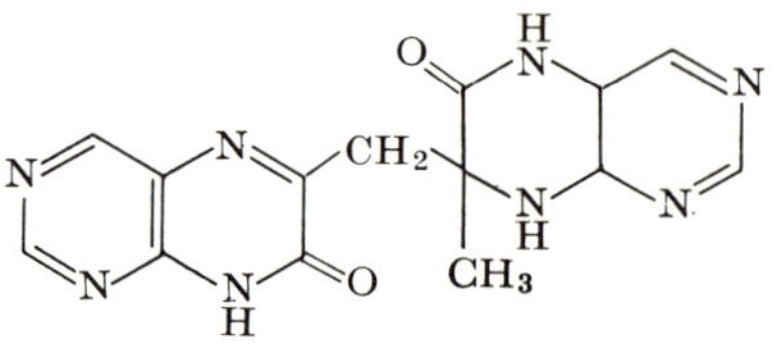   (112)

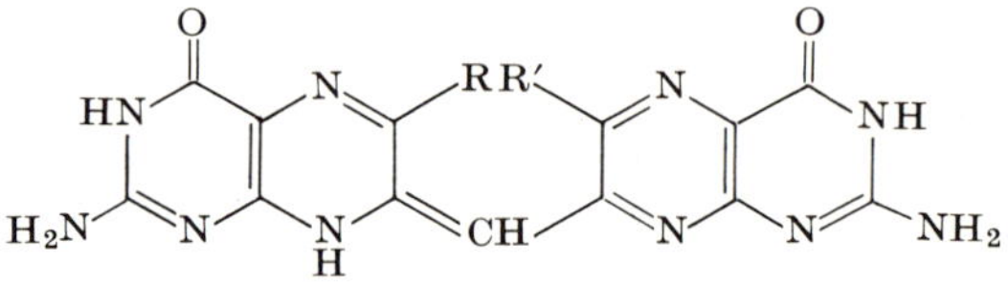

(113a) R = CH₃, R' = H
(113b) R = R' = OH

The dipteridylmethane (112) was obtained[134] by the reaction of 4,5-diaminopyrimidine with ethyl pyruvate at pH 1–3; this is thought to be formed by the addition of 6-methyl-7-hydroxypteridine to 7-methyl-6-hydroxypteridine, because these two pteridines readily give 112 when heated in 0.5 N sodium hydroxide.[134] The red dimer (113a) was likewise formed by condensation (followed by aerial oxidation) between 7-methyl-

---

[132] M. A. Matskanova and G. Ya. Vanags, *Dokl. Akad. Nauk SSSR* **132**, 615 (1960); *Chem. Abstr.* **54**, 24636 (1960).

[133] For a review, see D. J. Brown, "The Pyrimidines," p. 263. Wiley (Interscience), New York, 1962.

[134] A. Albert and E. P. Serjeant, *J. Chem. Soc.*, 3357 (1964).

and 6-methylpterin in 20% sulfuric acid.[135, 136] Some naturally occurring dimeric pteridines have been synthesized similarly (see type B4 oligomers).

Many aromatic N-heterocycles add nucleophiles across one or more of the C=N bonds,[137] thus the adduct (114) is formed[98] by condensation of barbituric acid (2 moles) and 4-methylpteridine (1 mole) at pH 2.

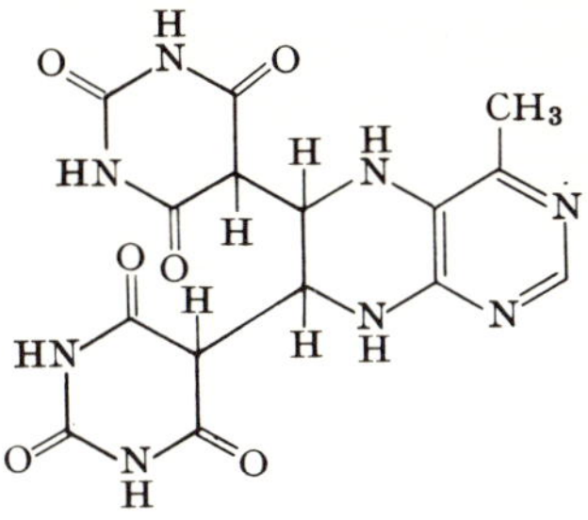

(114)

b. *Base-Catalyzed Co-oligomerization.* The formation of co-oligomers of type B3 from nonheterocyclic monomers is exemplified by the trioxetanylamine (115) obtained from β-tosyloxy-*t*-valeraldehyde (TsOCH$_2$CMe$_2$CHO) and sodium amide in benzene.[138]

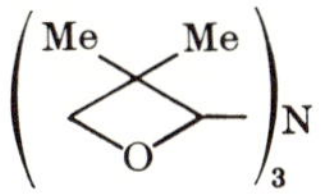

(115) 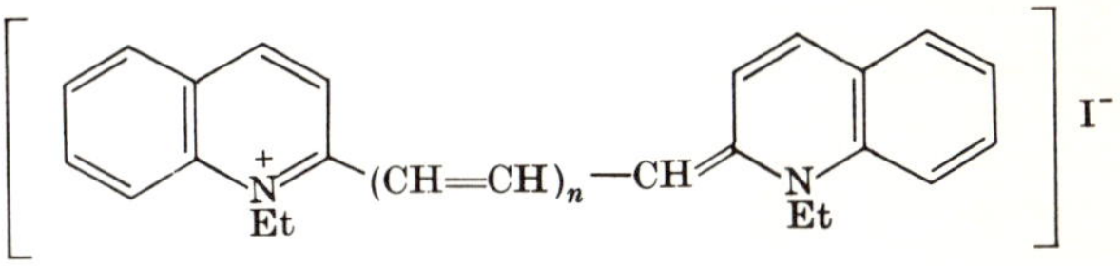   (116)

A mixture of the hexamine (50c) and the corresponding hydrofuramide [(RCH=N)$_2$CHR; R = 2-furyl] is obtained by the Schiff condensation of furfural and ammonia when set aside for 15–20 days at pH 9.2[72a] (the hydrofuramide had been isolated previously[139]); this is an example of side-chain co-oligomerization of heterocyclic monomers. Some cyanine dyes (e.g., 116), which are of great importance in photography, are made

[135] P. Karrer, R. Schwyzer, and B. J. R. Nicolaus, *Helv. Chim. Acta* **33**, 557, 1233 (1950).
[136] P. B. Russell, R. Purrmann, W. Schmitt, and G. H. Hitchings, *J. Amer. Chem. Soc.* **71**, 3412 (1949).
[137] A. Albert, *Angew. Chem. Int. Ed. Engl.* **6**, 919 (1967).
[138] K. Lucas, P. Weyerstahl, H. Marschall, and F. Nerdel, *Chem. Ber.* **104**, 3607 (1971).
[139] W. N. Hartley and J. J. Dobbie, *J. Chem. Soc.* **73**, 598 (1898).

by condensation of 1-ethyl-2-methylquinolinium iodide (2 moles) with an appropriate reagent (e.g., ethyl orthoformate for **116**, $n=1$). The cyanine dyes, including the similarly prepared benzothiazoline series, have been reviewed by Hamer.[140]

N-Heterocycles possessing a C=N bond (in the ring) highly reactive toward nucleophilic attack show a tendency on basification to give a dimeric ether, e.g., the 2,2'-ether from N-methylquinolinium chloride,[141] also the 2,2'-bispyrazinyl ether[142] of similar origin. The bipteridyl amine and sulfide (**117**) were formed from pteridin-6-one with aqueous ammonia and sodium sulfide, respectively.[143]

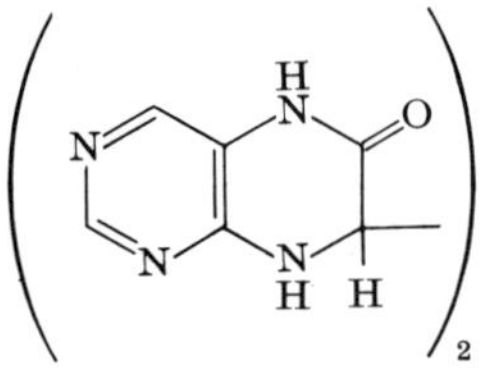

(**117**) X = NH or S

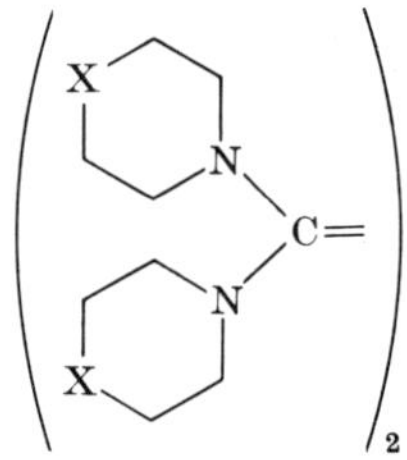

(**118**)

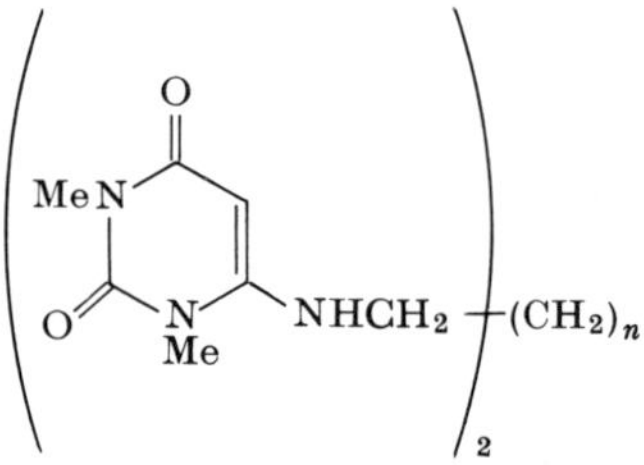

(**119**)

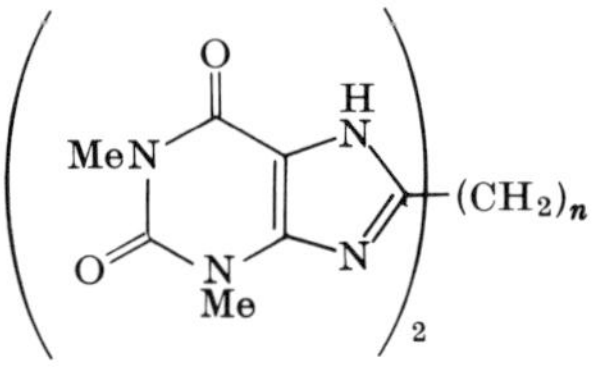

(**120**)

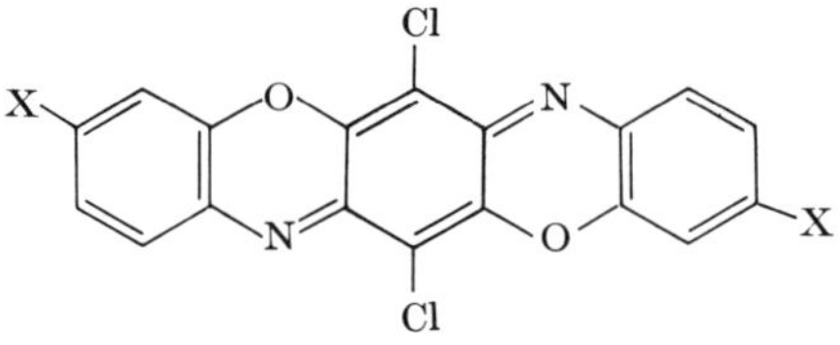

(**121**)

Many examples have been reported of the replacement of a halogen by an amine to produce type B3 co-oligomers, e.g., the tetrasubstituted ethylenes (**118**; X = CH$_2$, O)[144] obtained in high yields by heating chlorotrifluoroethylene with piperidine and morpholine (respectively) at 80°. The reaction of 1,3-dimethyl-4-chlorouracil with 0.5 mole of an alkylenediamine (in the presence of excess triethylamine) yielded the bisuracilyl-

[140] F. M. Hamer, *Quart. Rev.* **4**, 327 (1950); "The Cyanine Dyes and Related Compounds." Wiley (Interscience), New York, 1964.
[141] A. Hantzsch and M. Kalb, *Ber.* **32**, 3109 (1899).
[142] J. G. Aston, *J. Amer. Chem. Soc.* **53**, 1448 (1931).
[143] A. Albert, *J. Chem. Soc.*, 2690 (1955); A. Albert and J. Clark, *ibid.* 27 (1965).
[144] O. Tsuge, K. Yanagi, and M. Horie, *Bull. Chem. Soc. (Jap.)* **44**, 2171 (1971).

aminoalkane (**119**);[145] nitrosation and then intramolecular ring closure (by dehydration) yielded 8,8'-alkylenebistheophyllines (**120**; $n = 2$–10).[145] Similar condensation of N-methyl (saturated or unsaturated) heterocyclic tertiary salts (BH$^+$; B = e.g., N-methylpyrrolidine) with less than 0.5 mole of polymethylene dihalides [X(CH$_2$)$_n$X] has afforded many bis compounds, of the type {X$^-$[B$^+$(CH$_2$)$_n$B$^+$]X$^-$}, on heating in alcohol.[146] The Sirius Brilliant Blues (**121**; X = SO$_3$Na, etc.), direct dyes for cotton of excellent fastness, are obtained by condensing one molecule of chloranil and two of an aromatic amine (the product is then sulfonated).[147]

(**122**)   (**123**)

Two molar proportions of the cation (**122**; X = NMe, O, S) react with one of hydrazine in the presence of triethylamine, to give the corresponding bis compound (**123**).[148] Bisnicotinoylhydrazine was prepared[149] from nicotinic acid chloride (2 moles) and hydrazine hydrate (1 mole); 2-pyridoyl, 4-pyridoyl, 2-quinoloyl, and 4-quinoloyl analogs were made similarly.[149]

1-Methylindol-2(3H)-one(1-methyloxindole), having active hydrogen atoms, reacts with various substituted benzaldehydes in methanol containing piperidine, yielding a mixture of the expected cis/trans aldol condensation products[150] and the diindolylmethane (**124a**).[151] The latter, sometimes the major product (depending on the substituents of benzaldehyde), is thought to be produced by Michael addition of the monomer anion to 3-benzylidene-1-methylindol-2(3H)-one.[151] 4-Hydroxycoumalone (2 moles) and various aldehydes (1 mole) have afforded the same type of di(4-hydroxycoumalon-3-yl)methanes.[152] Pyrrole, a

---

[145] H. Fuchs, Dissertation, Univ. Stuttgart (1967); H. Fuchs and W. Pfleiderer, personal communication (1970).
[146] D. D. Libman, D. L. Pain, and R. Slack, *J. Chem. Soc.*, 2305 (1952).
[147] H. E. Fierz-David, J. Brassel, and F. Probst, *Helv. Chim. Acta* **22**, 1348 (1939).
[148] S. Hünig, G. Kiesslich, F. Linhart, and H. Schlaf, *Ann. Chem.* **752**, 182 (1971).
[149] R. Graf, *J. Prakt. Chem.* **138**, 289 (1933).
[150] See March,[32] p. 692.
[151] R. W. Daisley and J. Walker, *J. Chem. Soc. C*, 3357 (1971).
[152] M. A. Stahmann, I. Wolff, and K. P. Link, *J. Amer. Chem. Soc.* **65**, 2285 (1943), and references therein; W. R. Sullivan, C. F. Huebner, M. A. Stahmann, and K. P. Link, *ibid.* p. 2288.

π-excessive heteroaromatic,[153] readily undergoes electrophilic substitution. Thus, four molecules of pyrrole condense with four of an aldehyde to give *meso*-tetrasubstituted porphyrins (**125**). This synthesis, known as the Rothemund reaction,[154] has found wide applications with aldehydes in which R is aromatic; the reaction mechanism has been studied by Badger *et al.*[155] For naturally occurring porphyrins, see B4 oligomers.

(**124a**) R = OH, R' = H, R" = C_6H_4X
(**124b**) R = R' = R" = H
(**124c**) R = H

(**125**)

(**126**)

c. *Thermal Co-oligomerization.* Diethynyl ketone [CO(C≡CH)$_2$] reacted with diazomethane or diazopropane in ether under reflux to give the bispyrazolyl ketone (**126**; R = H, Me, X = CO);[156] 1,4-dipropioloylbenzene afforded similar compounds (**126**; M = H, Me, X = *p*-COC$_6$H$_4$CO).[156]

3,3'-Diindolylmethane (**124b**), isolated[157] when indole was heated in 0.5 equivalent of aqueous formaldehyde at 75°–80°, was obviously formed by a similar mechanism (although not base-catalyzed) to that responsible for the formation of **124a**; various aldehydes (R'CHO) and some ketones (R'R"CO) afford similar compounds (**124c**).[158]

d. *Miscellaneous Co-oligomerization.* The Grignard reactions of indole[159] with certain reagents also yield type B3 oligomers; for example, indole magnesium bromide, the structure of which is thought to be **127**, reacted with oxalyl chloride (or its diesters) to give **128**. 2-Substituted indole Grignards form dimers (**129**). Treatment of **127** with aliphatic, aromatic, and heterocyclic aldehydes (or aliphatic ketones) produces **124c**; using oxygen and sulfur, di(3-indolyl) ether and thio ether, respectively, can also

---

[153] See Albert,[64] p. 211.
[154] P. Rothemund, *J. Amer. Chem. Soc.* **57**, 2010 (1935); **58**, 625 (1936); **61**, 2912 (1939).
[155] G. M. Badger, R. A. Jones, and R. L. Laslett, *Aust. J. Chem.* **17**, 1028 (1964).
[156] G. Maneck and H.-U. Schenk, *Ber.* **104**, 3395 (1971).
[157] J. Thesing, *Ber.* **87**, 692 (1954).
[158] G. O. Burr and R. A. Gortner, *J. Amer. Chem. Soc.* **46**, 1224 (1924).
[159] R. A. Heacock and S. Kašpárek, *Advan. Heterocycl. Chem.* **10**, 43 (1969).

be made. Okamura et al.[160] have prepared compound **130** by the Grignard reaction of 2-thienylmagnesium bromide with methyl 3-(1-methylpiperidyl)carboxylate, followed by dehydration of the resulting carbinol with 10 N hydrochloric acid; **130**, known as Asverin, is used as an antitussive.

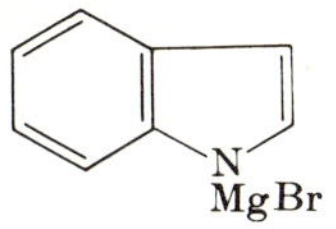

(**127**) 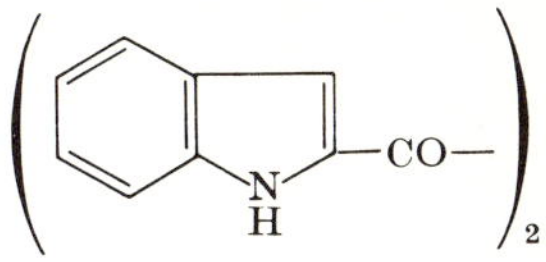 (**128**)

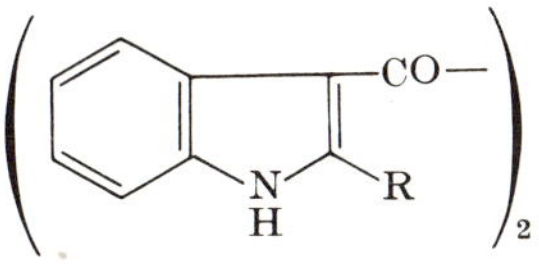

(**129**) 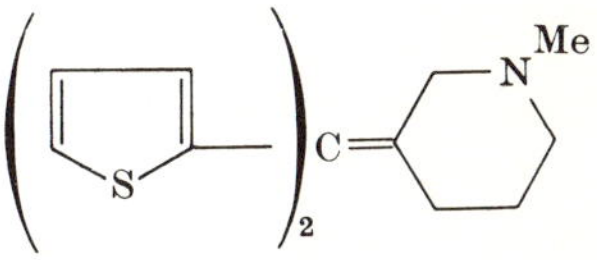 (**130**)

The reaction of n-butyllithium with 1-methylindazole gave the 1-lithiomethyl derivative, which formed the bisindazolyl compound by the action of carbon dioxide, followed by hydrolysis with acid.[161] The 2,2'-thenils (**79**; X = S, R = 5-F, or 5-adamantyl) were made by the reaction of two molecules of the corresponding 2-lithiothiophene with dimethyl oxalate.[103]

Subclass 4. Naturally Occurring Heterocyclic Oligomers (Including Their Synthesis)

Many naturally occurring oligomers of this type have been reported, among which well-known examples are presented.

Absinthin, the active principle of wormwood, *Artemisia absinthium* L. (*Compositae*), is a dimeric guaianolide containing two lactone rings and two hydroxy groups per molecule.[162] Thelephoric acid (**76b**), isolated from fungi of several *Thelephora* species, has been synthesized by oxidative condensation of two molecules of sodium 3,4-dimethoxyphenoxide with one of chloranil in the presence of hydrobromic acid.[162a]

[160] K. Okamura, T. Tanaka, S. Saito, H. Kugita, and N. Sugimoto, *Tanabe Seiyaku Nempo* **3**, 30 (1958); *Chem. Abstr.* **53**, 10214 (1959).
[161] B. A. Tertov and P. P. Onishchenko, *Zh. Obshch. Khim.* **41**, 103 (1971).
[162] L. Novotoný, V. Herout, and F. Šorm, *Chem. Ind.* (*London*), 465 (1958); *Collect. Czech. Chem. Commun.* **25**, 1492 (1960), and references therein.
[162a] J. Gripenberg, *Tetrahedron* **10**, 135 (1960).

Cuscohygrine, 1,3-bis(1-methyl-2-pyrrolidinyl)propan-2-one, one of the best known pyrrolidine alkaloids found in *Erythroxylon coca*, was synthesized by condensation of acetonedicarboxylic acid with two molecules of N-methyl-2-hydroxypyrrolidine.[163] Arcamore *et al.*[164] have obtained an antibiotic substance called distamycin A (**131**) from

$$\left[\text{OHCHN} - \underset{\underset{CH_3}{N}}{\diagdown} - \text{CONH} \right]_3 - \text{CH}_2\text{CH}_2\text{C} \diagup_{NH}^{NH_2}$$

(**131**)

Streptomyces distallicus; the structure elucidation and synthesis have also been made by them.[165] A wide variety of porphyrins which are derivatives of porphin (**125**; R = H) occur in nature and exert three biological functions: (*a*) oxygen storage and transport, as in hemoglobin, (*b*) cellular respiration as in the cytochromes, and (*c*) photosynthesis, as in the chlorophylls; for detailed discussion of structures and synthesis of naturally occurring porphyrins, see references.[166] Recent developments in the chemistry of polypyrrolic compounds have been reviewed.[167] Several naturally occurring indole alkaloids have dimeric structures of type B,[168] an example of which is *c*-toxiferine-I, a calabash curare alkaloid found in South American species of *Strychnos*.[169]

Numerous oligosaccharides occur naturally,[170] an example being amygdalin (D-mandelonitrile-β-glucosido-6-β-D-glucoside) obtained from

---

[163] For a review, see T. Robinson, "The Biochemistry of Alkaloids," p. 24. Springer-Verlag, Berlin and New York, 1968.

[164] F. Arcamore, F. Bizioli, G. Canevazzi, and A. Grein, German Patent, 1,039,198 (1958); *Chem. Abstr.* **55**, 2012 (1961).

[165] F. Arcamore, S. Penco, P. Orezzi, V. Nicolella, and A. Pirelli, *Nature (London)* **203**, 1064 (1964).

[166] See, e.g. (a) Albert,[64] p. 236; (b) J. E. Falk, "Porphyrins and Metalloporphyrins." Elsevier, Amsterdam, 1964.

[167] (a) R. L. N. Harris, A. W. Johnson, and I. T. Kay, *Quart. Rev.* **20**, 211 (1966); (b) K. M. Smith, *ibid.* **25**, 31 (1971).

[168] For a review, see Robinson,[163] p. 77.

[169] C. G. Casinovi, G. B. Marini-Bettolo, and N. G. Bisset, *Nature (London)* **193**, 1178 (1962).

[170] For monographs, see (a) Fieser and Fieser,[18] p. 955; (b) W. Z. Hassid and M. Doudoroff, *Progr. Chem. Org. Natur. Prod.* **5**, 101 (1948); (c) R. L. Whistler (ed.), "Methods in Carbohydrate Chemistry," Vol. 5, pp. 65–181. Academic Press, New York, 1965.

the seeds of *Rosaceae*, principally from almonds.[171] Dicoumarol, 3,3'-methylenebis(4-hydroxycoumarin), a blood anticoagulant substance, is found in spoiled hay or in silage from the common sweet clover.[172] It is now synthesized by the action of formaldehyde on 4-hydroxycoumarin.[173] Ergoflavin, the principal yellow pigment in ergot, has the 3,3'-bound dimeric structure (132); the constitution and the absolute stereochemistry has been defined by X-ray diffraction analysis of tetra-*O*-methylergoflavin di-*p*-iodobenzoate.[174]

(132)

Many dimeric piperidine alkaloids also occur in nature, e.g., carpaine which has been isolated from papaya leaves (*Carica papaya*);[175] the absolute configuration has been determined.[176] Naturally occurring isoquinoline alkaloids[177] exist in dimeric forms in some cases, e.g., the bisbenzylisoquinoline alkaloids which have two benzylisoquinoline nuclei joined together through one, two, or even three ether bridges; an example is epistephanine isolated from *Stephania japonica*.[178]

Nucleic acids in which various purine and pyrimidine bases are united by bridges of D-2-deoxyribose or D-2-ribose phosphoric acid play vital roles in the metabolism of living cells.[179] Many oligonucleotides (which are heterocyclic co-oligomers) have been synthesized in studying the biochemistry of nucleic acids.[180]

---

[171] See Stecher,[29] p. 76.
[172] H. A. Campbell and K. P. Link, *J. Biol. Chem.* **138**, 21 (1941).
[173] K. P. Link, *Fed. Proc.* **4**, 176 (1945).
[174] A. T. McPhail, G. A. Sim, J. D. M. Asher, J. M. Robertson, and J. V. Silverton, *J. Chem. Soc. B*, 18 (1966).
[175] See Stecher,[29] p. 213.
[176] J. L. Coke and W. Y. Rice, Jr., *J. Org. Chem.* **30**, 3420 (1965).
[177] For monographs, see (a) Robinson,[163] p. 54; (b) T. Kametani, "The Chemistry of the Isoquinoline Alkaloids," p. 45. Elsevier, Amsterdam, 1969.
[178] D. H. R. Barton, G. W. Kirby, and A. Wiechers, *J. Chem. Soc. C*, 2313 (1966), and references therein.
[179] For a monograph, see E. Chargaff and J. N. Davidson (eds.), "The Nucleic Acids," 3 vols. Academic Press, New York, 1955, 1960.
[180] For reviews, see (a) F. Cramer, *Angew Chem. Int. Ed. Engl.* **5**, 173 (1966); (b) M. Ikehara, *Kagaku no Ryoiki* **25**, 300 (1971).

Pterorhodine (**113b**)[181] and drosopterin (**133**)[182] are naturally occurring dimeric pteridines; the former was synthesized from xanthopterin and acetone or acetaldehyde in the presence of ammonium sulfate or hydrogen peroxide,[136] and the latter from 7,8-dihydropterin and β-keto-α-hydroxybutyric acid.[182b]

(**133**)

Peptidelike macrocyclic antibiotics of dimeric and trimeric form are often found, an example being echinomycin A (quinomycin A) produced by *Streptomyces echinatus* from the soil of Cuanza (Angola). The complete structure of a large peptide macrocycle bearing two 2-quinoxalinyl substituents has been elucidated by Prelog and co-workers.[183]

C. Formation of Oligomers That Have Two (or More) Heterocyclic Nuclei Joined to One Another by a Single Bond

Subclass 1. Heterocyclic Oligomers from Nonheterocyclic Monomers

Only a few dimers of this type seem to have been reported so far, although many co-oligomers from nonheterocyclic monomers are known (cf. Section III,C,3).

An example is a trimer (**134**) with an *o*-terphenyl skeleton, obtained in small amount from the sulfuric acid polymerization products of *p*-benzoquinone.[184] During Baeyer's first synthesis[185] of indigo (**178a**; see below), the isatogen **135a** was made by treating di(*o*-nitrophenyl)-diacetylene (obtained by oxidation of *o*-nitrophenylacetylene with a ferricyanide in the presence of copper) with concentrated sulfuric acid; ammonium sulfide reduction of **135a** gave indigo.

---

[181] (a) A. Kühn and A. Egelhaaf, *Z. Naturforsch.* B **14**, 654 (1959); (b) W. Pfleiderer, *ibid.* B **18**, 420 (1963).

[182] (a) M. Viscontini, in "Pteridine Chemistry" (W. Pfleiderer and E. C. Taylor, eds.), p. 267 and references therein. Pergamon, Oxford (1964); (b) H. Schlobach and W. Pfleiderer, *Angew. Chem. Int. Ed. Engl.* **10**, 414 (1971).

[183] W. Keller-Schierlein, M. L. Mihailović, and V. Prelog, *Helv. Chim. Acta* **42**, 305 (1959).

[184] H. Erdtman and N. E. Stjernström, *Acta Chem. Scand.* **13**, 653 (1959).

[185] A. Baeyer, *Ber.* **15**, 50 (1882).

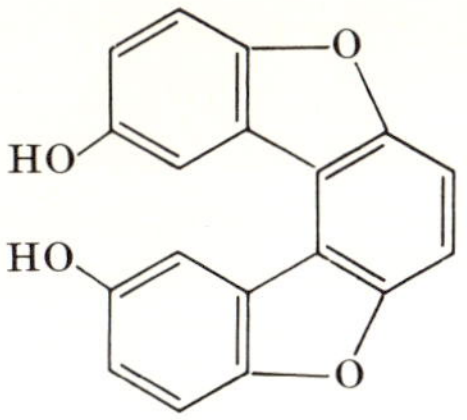

Sec. III.C.]   HETEROCYCLIC OLIGOMERS   41

(134)

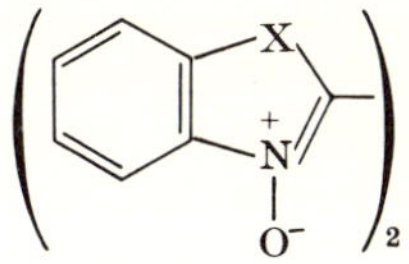

(135a) X = CO
(135b) X = NH

Subclass 2. Heterocyclic Oligomers from Heterocyclic Monomers

a. *Acid-Catalyzed Oligomerization.* Michael-type addition condensations frequently give oligomers of this type. Well-known examples are the polymerization products of pyrroles and indoles;[186] the trimer of pyrrole and the dimer of indole have structures **136** and **137**, respectively. With mild polymerizing reagents such as 100% orthophosphoric acid, thiophene gives a trimer (**138**) and a pentamer of unknown constitution.[187] Variously substituted 2- and 3-methylbenzo[b]furans are also dimerized with sulfuric acid in ethanol to give 2,3'- or 3,2'-bound dimers (depending on the substituents at the 2- and 3-position);[188] the ease of dimerization and dissociation into monomer have been discussed.[188] The mercuric acetate oxidation of variously substituted

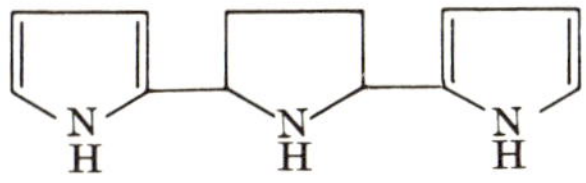

(136)

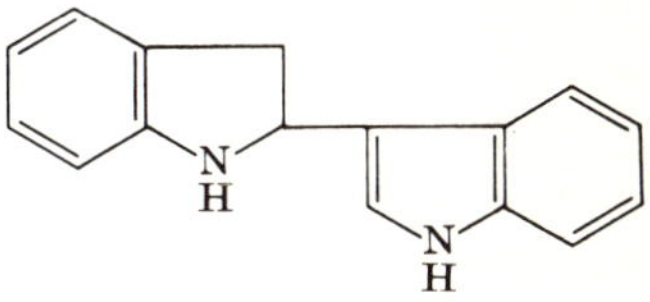

(137)

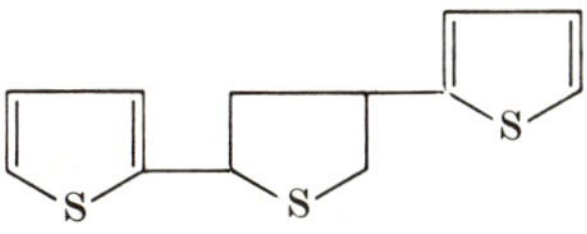

(138)

---

[186] For a review, see G. F. Smith, *Advan. Heterocycl. Chem.* **2**, 287 (1963).
[187] For a review, see H. D. Hartough, "Thiophene and Its Derivatives," p. 165. Wiley (Interscience), New York, 1952.
[188] T. Abe and F. Shimizu, *Nippon Kagaku Kaishi* **91**, 753 (1970); *Chem. Abstr.* **73**, 120436 (1970).

N-methylpiperidines in 5% acetic acid produces much 1,1'-dimethyl-Δ^2-tetrahydroanabasine (3,2'-bound dimeric piperidine) in addition to 1,2,3,4-tetrahydropiperidines.[189]

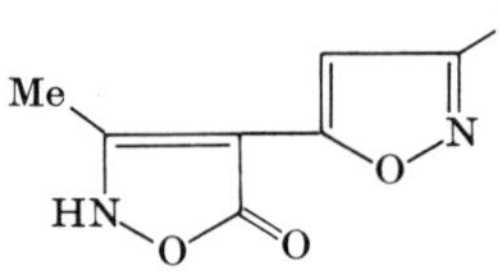

(139)

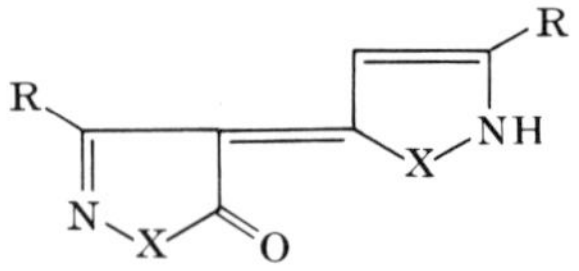

(140a) X = O
(140b) X = NMe

Hantzsch[190] obtained an anhydro dimer (139) by the action of acetic anhydride or acetyl chloride on β-oximidobutyric acid anhydride. A spectroscopic study[191] showed that the two tautomeric forms (139 and 140a) were predominant; the formula 140a is classified as a type D oligomer (see below).

4-Chloro- and 4-bromopyridine have a great tendency to self-quaternization, giving 1,4'-bound dimers; such condensation is caused by acid-catalyzed substitution.[192]

Reduction of pyridine with zinc dust in acetic anhydride gave a mixture of dihydro compounds 141a and 177 (see below), which were converted into 4,4'-bipyridyl by oxidative hydrolysis.[193] 4-Substituted pyridines gave the bipyridyl (142; R = i-Pr, t-Bu) by similar reduction in methyl chloroformate.[194]

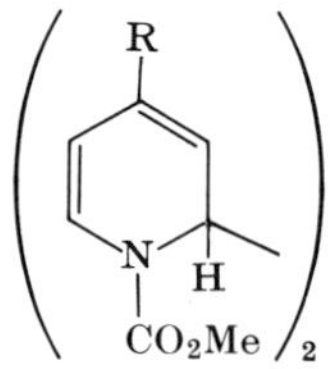

(142)

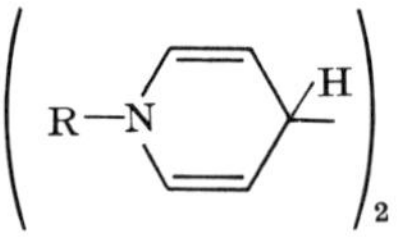

(141a) R = Ac
(141b) R = Me

[189] N. J. Leonard and F. P. Hauck, *J. Amer. Chem. Soc.* **79**, 5279 (1957).
[190] A. Hantzsch, *Ber.* **24**, 495 (1891).
[191] T. Nishiwaki, *J. Chem. Soc. C*, 245 (1969).
[192] For a review, see K. Thomas and D. Jerchel, *Angew. Chem.* **70**, 719 (1958).
[193] R. L. Frank and P. V. Smith, *Org. Synth., Coll. Vol.* **3**, 410 (1955).
[194] P. M. Atlani and J. F. Biellmann, *C. R. Acad. Sci., Ser. C* **271**, 688 (1970); *Chem. Abstr.* **74**, 22667 (1971).

b. *Base-Catalyzed Oligomerization.* The reduction of 1-methylpyridinium chloride with sodium amalgam afforded the relatively stable dimer (**141b**) through coupling of an intermediate neutral radical. 1-Ethylquinolinium iodide gave the 3,4′-bound dimer (apocyanine)[195] by the action of potassium methoxide. There are many other examples of such dimerization of π-deficient N-heteroaromatics in the presence of a metal. This reaction is usually the dimerization of a radical anion formed by the metal, but a metal can sometimes form a radical anion from the dimer itself.[196] The 3,3′-pyridazines were obtained by treating a 3-halogenopyridazine with hydrazine hydrate in the presence of Pd–CaCO$_3$ and alkali;[197] this reaction is probably caused also by reductive radical coupling.

An example of C2 oligomers produced by intermolecular dehydrohalogenation (under basic conditions) is the cyclic trimer (**143**) formed when 2-chlorobenzimidazole was heated with urethane at 180°–200° in a sealed tube (although a minor amount of the normal substitution product, ethyl 2-benzimidazolylcarbamate, was also produced).[198]

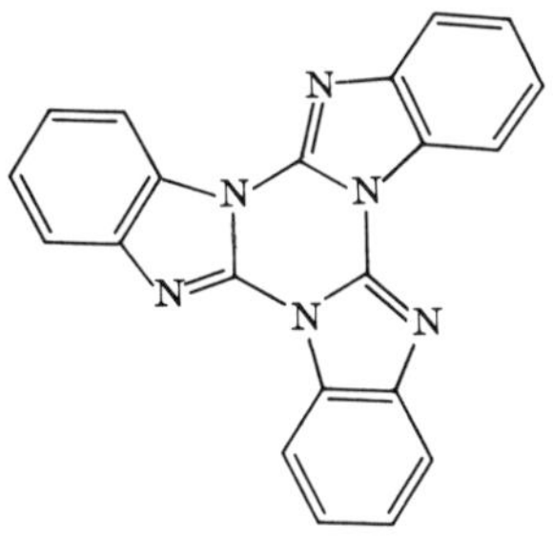

(**143**) 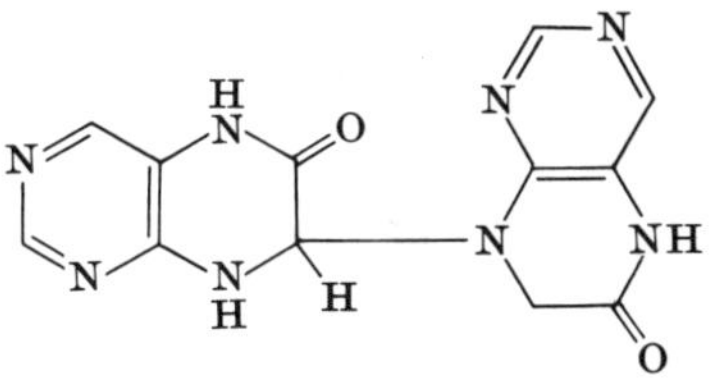   (**144**)

The dimer (**144**) is formed by an unusual disproportionation of pteridin-6-one on briefly boiling in 1.0 N sodium hydroxide;[143] this dimer is also readily produced by the addition of 6-hydroxy-7,8-dihydropteridine to this pteridinone in 0.1 N sodium hydroxide at 20° (2 days).[143]

Alkaline dimerization of o-aminonitriles is exemplified by the compound (**82**) formed by heating 2-amino-3-cyanopyridine in aqueous ammonia.[33]

[195] W. H. Mills and H. G. Ordish, *J. Chem. Soc.*, 81 (1928).
[196] For a monograph, see Albert,[64] p. 149.
[197] H. Igeta, T. Tsuchiya, M. Nakajima, C. Okuda, and H. Yokogawa, *Chem. Pharm. Bull.* **18**, 1228 (1970).
[198] G. I. Gofen, C. S. Kadyrov, and M. N. Kosyakovskaya, *Khim. Geterotsikl. Soedin.* **2**, 282 (1971).

c. *Thermal Oligomerization.* 1-Pyrroline is thought to exist in the trimeric form (**145**) in equilibrium with a small amount of monomer or dimer.[199] 3,3-Dimethyl-3H-indole forms a similar cyclic trimer on storage;[200] NMR evidence shows that the monomer is regenerated above 120°, whereas at lower temperatures the compound is trimeric.[201]

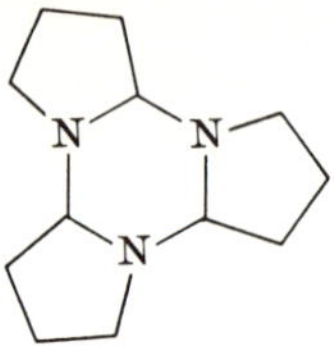

(**145**)

On heating a suspension of 4-hydroxylaminoquinoline 1-oxide in water at 200°–240°, two dimers (**147** and **148**) are formed possibly through the intermediate diradical (**146**)[202] [Eq. (9)].

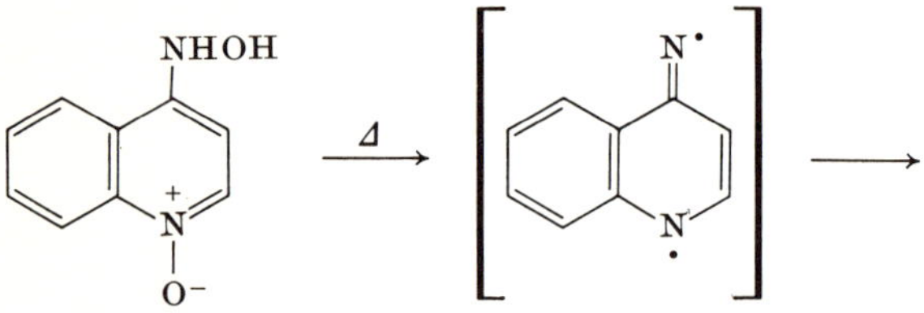

(**146**)

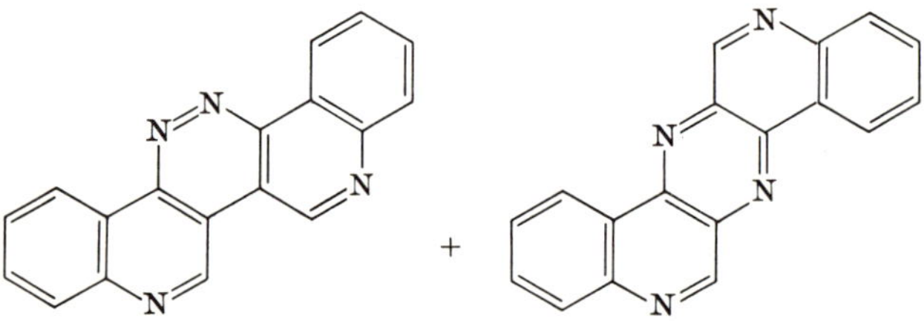

(**147**)   (**148**)   (9)

d. *Photooligomerization.* Photooxidation of some π-excessive heteroaromatics produces dimers of type C; e.g., on exposure to daylight or a mercury vapor lamp, benzo[b]thiophene seems to undergo self-condensa-

---

[199] D. W. Fuhlhage and C. A. Vander Werf, *J. Amer. Chem. Soc.* **80**, 6249 (1958).
[200] A. H. Jackson and A. E. Smith, *Tetrahedron* **21**, 989 (1965).
[201] H. Fritz and P. Pfaender, *Chem. Ber.* **98**, 989 (1965).
[202] T. Kosuge, H. Zenda, and H. Sawanishi, *Chem. Pharm. Bull.* **19**, 1291 (1971).

tion to give a significant quantity of 2,2'-bibenzo[b]thiophenyl.[203, 204] 2-Hydroxy-3-phenylbenzo[b]furan and 3-hydroxy-2-phenylbenzo[b]thiophene have been reported to yield the 3,3'- and 2,2'- dimers, respectively (with the dioxo-dihydro structure in the moiety), by the action of sunlight in the presence of air.[205] Recent results show[206] that indole undergoes a different type of photooxidative oligomerization (in aqueous solution) and produces mainly the trimer (150), together with a minor amount of indigo (178a), through the most likely common intermediate indoxyl (149) [Eq. (10)].

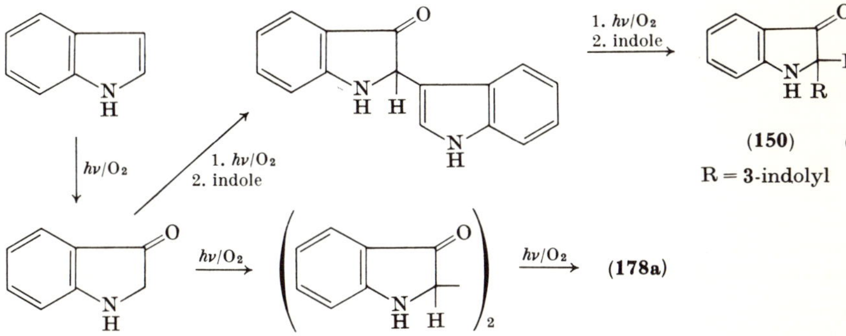

Photoaddition of nucleophiles to heterocycles is often observed.[1] Irradiation of acridine and its quaternary salt in ethanol produced 9,9'-bisacridan (151) as the major product besides a little acridan and 9α-hydroxyethylacridan.[207] The reaction mechanism is most likely a hydrogen abstraction from the alcohol by the excited molecule, followed by competitive radical combination to yield 151. Irradiation of an aqueous solution of alloxan monohydrate and its derivatives produces an alloxantin-type dimer (152; R = H, Me, Et) by combination (at the 5-position) of a radical intermediate.[208]

---

[203] W. E. Haines, R. V. Helm, G. L. Cook, and J. S. Ball, *J. Phys. Chem.* **60**, 549 (1956).
[204] W. E. Haines, G. L. Cook, and J. S. Ball, *J. Amer. Chem. Soc.* **78**, 5213 (1956).
[205] A. Schönberg and A. Mustafa, *J. Chem. Soc.*, 657 (1945).
[206] B. Iddon, G. O. Phillips, K. E. Robbins, and J. V. Davies, *J. Chem. Soc. B*, 1887 (1971).
[207] (a) H. Göth, P. Cerutti, and H. Schmid, *Helv. Chim. Acta* **48**, 1395 (1965); (b) F. Mader and V. Zanker, *Chem. Ber.* **97**, 2418 (1964); (c) V. Zanker and P. Schmid, *Z. Phys. Chem. (Frankfurt)* **17**, 11 (1958).
[208] Y. Otsuji, S. Wake, and E. Imoto, *Tetrahedron* **26**, 4139, 4293 (1970).

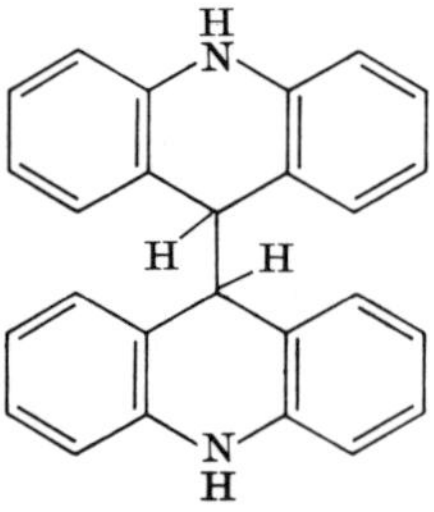

(151) 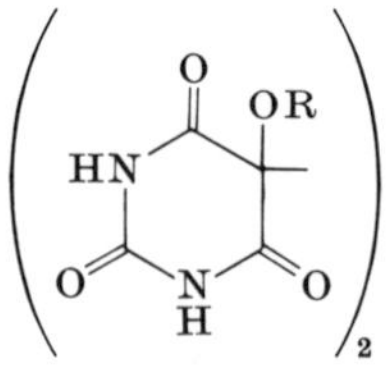 (152) 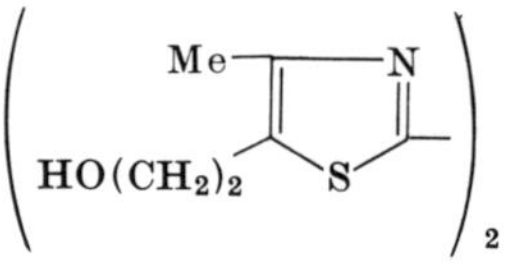 (153)

2-Chloro-5β-hydroxyethyl-4-methylthiazole formed an intensively blue fluorescing substance (153) under UV light,[209] the structure being proved by unambiguous synthesis (see C3 oligomers).

Irradiation of cytosine, cytidine, and 2'-deoxycytidine has recently been reported to produce 4,5'-linked dimers (154; R = H, β-D-ribose, 2-deoxy-β-D-ribose, respectively);[210] this reaction is accompanied by deamination. Photodimerization (in aqueous solution) of uracil[211] and thymine[212] affords the 4,6'-linked dimer (155; R = R' = H and Me, respectively); the oxetane (156) (probably formed by 1,2-cycloaddition of a triplet carbonyl to an olefinic bond) is proposed as the precursor.[212]

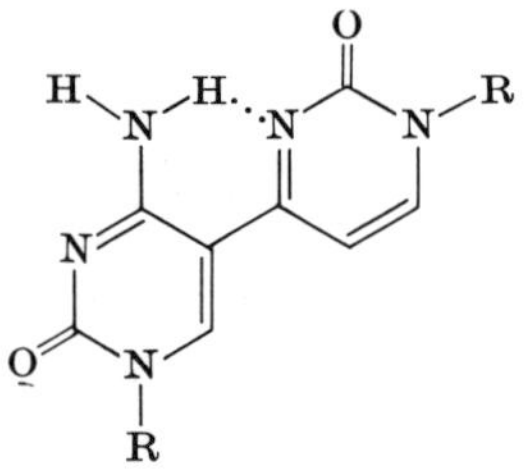

(154) 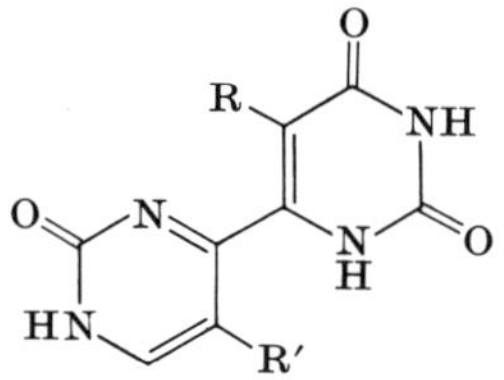 (155)

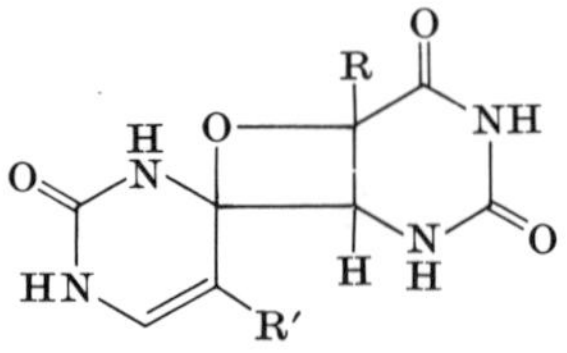

(156)

[209] P. Karrer and M. C. Sanz, *Helv. Chim. Acta* **27**, 619 (1944).
[210] D. F. Rhoades and S. Y. Wang, *J. Amer. Chem. Soc.* **93**, 3779 (1971).
[211] D. F. Rhoades and S. Y. Wang, *Biochemistry* **9**, 4416 (1970); M. N. Khattak and S. Y. Wang, *Science* **163**, 1341 (1969).
[212] A. J. Varghese and S. Y. Wang, *Science*, **160**, 186 (1968).

The similarly linked dimer (**155**; R = Me, R' = H) has been isolated[213] after irradiation of DNA *in vitro* and *in vivo* possibly by combination of cytosine and thymine (accompanied by deamination). The macrocyclic tetramer (**157**) has also been isolated on irradiation of DNA;[214] it is thought to be the dimer of the above product (**155**; R = Me, R' = H). The trans-syn structure of **157** has been confirmed by X-ray diffraction analysis of the hexa-*N*-methyl derivative.[215] These are all biologically significant photoreactions as described in Section III, B, 2. Further irradiation of the trans-syn and cis-syn cyclobutane-type dimers (see Section III, E, 2) of 1,3-dimethyluracil with a sensitizer has been found to produce the 5,5'-linked dimer (**158**).[216]

e. *Miscellaneous Oligomerization.* With mild oxidizing reagents (e.g., potassium ferricyanide or bromine), some π-excessive heteroaromatic monomers form free radicals, which combine at low temperatures to give a singly bonded dimer. On heating, such dimers may dissociate into the monomer radicals; reduction of the dimer (e.g., with hydroquinone) reproduces the monomer. For example, the oxidation of 4,5-diphenyl-2-arylimidazole produces a mixture of the 4,4'-dimer (**159**) and the 2,1'-dimer, which exist in equilibrium through the dissociated monomer radical.[217] Similarly 2,2'-linked dimers are obtained from 3-acetyl-4,5-diaryl-2-methylpyrrole.[218] That ring protons of these heterocycles are

---

[213] S. Y. Wang and A. J. Varghese, *Biochem. Biophys. Res. Commun.* **29**, 543 (1967); *Science* **156**, 955 (1967).
[214] S. Y. Wang and D. F. Rhoades, *J. Amer. Chem. Soc.* **93**, 2554 (1971).
[215] J. L. Flippen, R. D. Gilardi, I. L. Karle, D. F. Rhoades, and S. Y. Wang, *J. Amer. Chem. Soc.* **93**, 2556 (1971).
[216] D. Elad, I. Rosenthal, and S. Sasson, *J. Chem. Soc. C*, 2053 (1971).
[217] L. A. Cescon, G. R. Coraor, R. Dessauer, E. F. Silversmith, and E. J. Urban, *J. Org. Chem.* **36**, 2262 (1971).
[218] K. Tomita and N. Yoshida, *Tetrahedron Lett.*, 1169 (1971).

replaced by (aromatic) substituents obviously contributes to the stabilization of their monomer radicals, producing a high yield of the singly bound dimers.

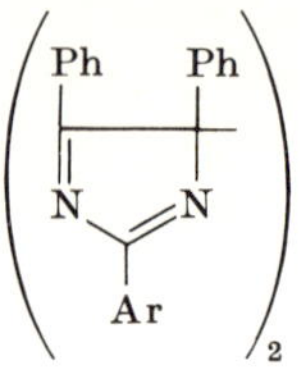

(159)

Oxidation of certain heteroaromatic *N*-oxides has been reported to give dimers of type C, e.g., the dimer (135b) obtained on heating an aqueous solution of benzimidazole 3-oxide in a sealed tube in the presence of oxygen.[219] A similar 2,2′-dimer was formed when 2-(2-naphthyl)thioindoxyl-1,1-dioxide was oxidized with bromine or potassium ferricyanide.[220]

Reductive dimerization has also been reported; isatide (160a) and sulfisatide (160b) were prepared by treatment of isatin with ammonium hydrosulfide under different conditions, the structures were assigned by Bergmann.[221] Yoshida *et al.*[222] recently reported formation of the dimer

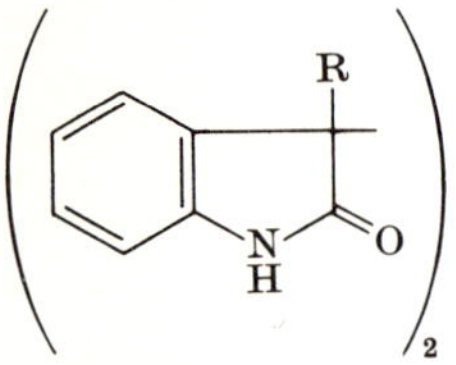

(160a) R = OH
(160b) R = SH

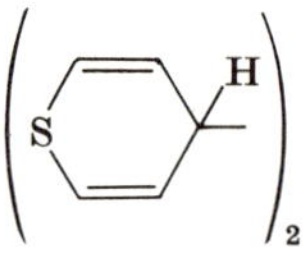

(161)

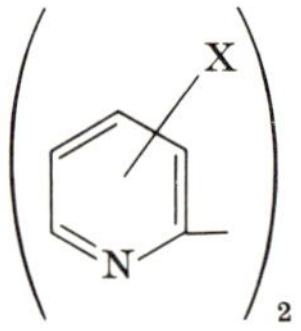

(162)

(161) by reducing thiopyrylium iodide with zinc in acetonitrile, conditions which allowed coupling of the resulting monomer radical. Elimination of hydride ion from 161 with $Ph_3C^+X^-$ yielded a 4,4′-bisthiopyrylium salt.[222] A convenient preparation[223] of bipyridyls and bibenzopyridyls utilizes a special, degassed Raney nickel catalyst, but requires the starting

[219] R. Kuhn and W. Blau, *Ann. Chem.* **615**, 99 (1958).
[220] A. H. Lamberton and J. E. Thorpe, *J. Chem. Soc. C*, 2571 (1967).
[221] E. D. Bergmann, *J. Amer. Chem. Soc.* **77**, 1549 (1955).
[222] Z. Yoshida, S. Yoneda, T. Sugimoto, and O. Kikukawa, *Tetrahedron Lett.*, 3999 (1971).
[223] For a review, see G. M. Badger and W. H. F. Sasse, *Advan. Heterocycl. Chem.* **2**, 179 (1963).

material not to be strongly hindered. Thus, when pyridine is refluxed with this nickel, a free radical is formed by acceptance of an electron from the nickel, and two such radicals unite to form dihydro-2,2'-bipyridyl, which the nickel then dehydrogenates to give a good yield of 2,2'-bipyridyls (162). A similar reaction (i.e., reduction → radical combination → oxidation) was observed when a 4-alkylpyridine was refluxed with 5% palladium–charcoal for 3 days, producing a mixture of the 2,2'-bipyridyl (mainly) and the terpyridyl (163).[224] 9,9'-Biacridyl (164) is best prepared by mixing cold aqueous solutions of 9-chloroacridine hydrochloride and chromous chloride in an atmosphere of hydrogen; the chemical synthesis and properties of binuclear acridines have been fully reviewed.[225] Alloxantin (152; R = H) is usually made either by air oxidation of dialuric acid, or by reduction of alloxan; in both cases the alloxantin is formed by combination of two molecules of an intermediate radical.[226]

(163)

(164)

An application of the Ullmann reaction[227] to prepare bipyridyls and bipyrimidyls has been reported; variously substituted 2-bromopyridines, with copper in dimethylformamide, gave[228] a good yield of 2,2'-bipyridyls (162). A similar reaction afforded good yields of 2,2'-, 4,4'-, and some other bipyrimidyls.[229] For synthesis of 2,2'-bithienyl by this method, see Section III, C, 4.

Kauffmann et al.[230] have prepared the tetraarene (165) by the reaction of 2-lithio-5-(pyrid-2-yl)thiophene with $CuCl_2/O_2$. Attempted addition

[224] P. E. Rosevear and W. H. F. Sasse, J. Heterocycl. Chem. 8, 483 (1971).
[225] See Albert,[67] pp. 391–399.
[226] See Brown,[133] p. 262.
[227] For a review, see P. E. Fanta, Chem. Rev. 38, 139 (1946); 64, 613 (1964).
[228] Z. Kulicki and W. Karminski, Zeszyty Nauk Politech. Slask., Chem. (Poland) 16, 11 (1963); Chem. Abstr. 62, 4001 (1965).
[229] M. P. L. Caton, D. T. Hurst, J. F. W. McOmie, and R. R. Hunt, J. Chem. Soc. C, 1204 (1967).
[230] T. Kauffmann, E. Wienhöfer, and A. Woltermann, Angew. Chem. 83, 799 (1971).

of a Grignard reagent to 5,5-dimethyl-1-pyrroline formed the dimer (**166**), probably as a result of the initial abstraction of a proton at the 3-position, followed by addition to a second pyrroline molecule.[231]

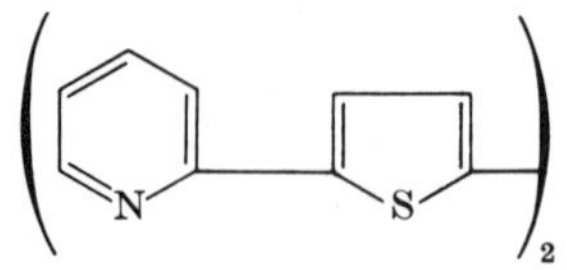

(**165**) 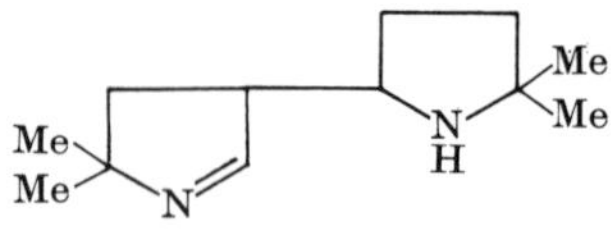   (**166**)

Subclass 3. Heterocyclic Co-oligomers from More Than One Kind of Monomer

a. *Acid-Catalyzed Co-oligomerization.* A new type of heterocyclic dimer (**167**; R = Me, Cl) of type C3 was found when dimethyl- or dichloromaleic anhydride was allowed to react with hydrazine in cooled acetic acid.[232]

2,2′-Bibenzimidazolyl can be made by refluxing o-phenylenediamine with trichloroacetic acid in 4 N hydrochloric acid.[233]

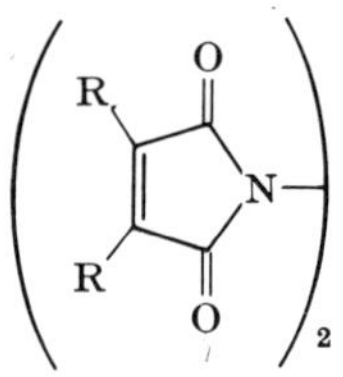

(**167**) 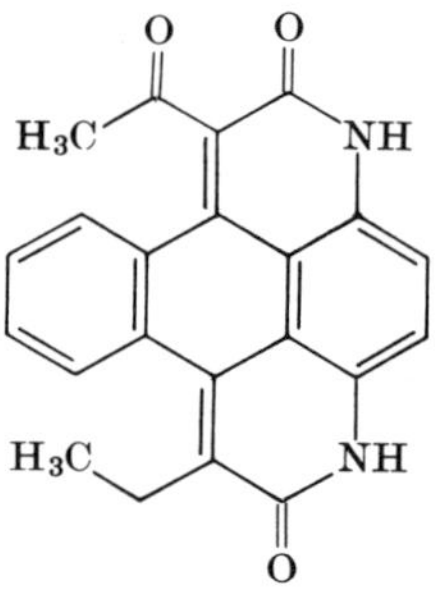   (**168**)

b. *Base-Catalyzed Co-oligomerization.* The action of an excess of hydrazine hydrate on bis(2-oxocyclohexyl)glyoxal (obtained by condensing cyclohexanone and diethyl oxalate) gives a 3,3′-bipyrazole which can be aromatized to 3,3′-biindazolyl by prolonged heating with palladium catalyst.[234a] For compound **168**, which is relevant here,[234] see Section III, D, 3, b.

[231] R. Bonnett, V. W. Clark, A. Giddey, and A. Todd, *J. Chem. Soc.*, 2087 (1959), and references therein.
[232] E. Hedaya, R. L. Hinman, and S. Theodoropulos, *J. Org. Chem.* **31**, 1317 (1966).
[233] K. H. Buechel, *Z. Naturforsch.* B **25**, 945 (1970).
[234] N. A. Kirzner, F. B. Yuĭ, E. A. Poraĭ-Koshits, *Zh. Obshch. Khim.* **30**, 890 (1960).
[234a] J. H. M. Hill, D. M. Berkowitz, and K. J. Freese, *J. Org. Chem.* **36**, 1563 (1971).

c. *Thermal Co-oligomerization.* Compound **153** has been synthesized by condensing two molecules of 3-chloro-5-hydroxypentan-2-one and one of dithiooxamide at 120°, without solvent.[209]

d. *Photoco-oligomerization.* Irradiation of 1-methyl-3,4-dihydroisoquinoline in methanol gives a mixture of the *cis*-imidazolidine (**169**) and the trans isomer.[235]

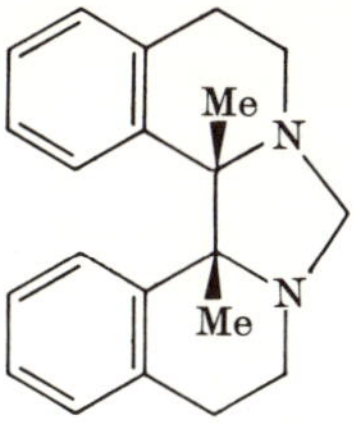

(**169**)

Subclass 4. Naturally Occurring Heterocyclic Oligomers (Including Their Synthesis)

Anemonin (**170**) has been found in *Anemone pulsatilla* L. and other *Ranunculaceae*;[236] its precursor in plants is protoanemonin (5-methylene-2-oxodihydrofuran).

2,2'-Bithienyl was isolated[237] from flowers of the Indian marigold, *Tagetes erecta* L., and has been shown to be the nematicidal principle in the roots of *Tagetes* plants.[238] Synthesis of the bithienyl was made either by the reaction of 2-thiophenylmagnesium bromide with cupric chloride in ether[239] or by the Ullmann reaction of 2-iodothiophene with copper bronze in dimethylformamide as solvent.[240] α,α'-Terthienyl (**171**) has also been isolated from the flowers of another Indian marigold *Echinops sphaerocephalus*.[82, 241]

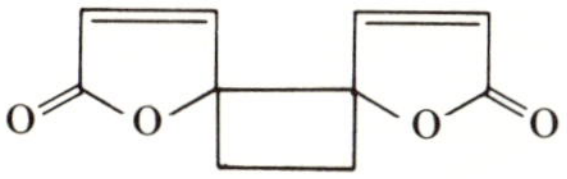

(**170**) 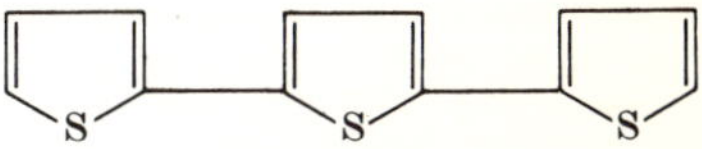   (**171**)

[235] P. Cerutti and H. Schmid, *Helv. Chim. Acta* **47**, 203 (1964).
[236] R. M. Moriarty, C. R. Romain, I. L. Karle, and J. Karle, *J. Amer. Chem. Soc.* **87**, 3251 (1965).
[237] J. W. Sease and L. Zechmeister, *J. Amer. Chem. Soc.* **69**, 270, 273 (1947).
[238] J. H. Uhlenbroek and J. D. Bijloo, *Rec. Trav. Chim. Pays-Bas* **77**, 1004 (1958).
[239] W. Steinkopf and J. Roch, *Ann. Chem.* **482**, 251 (1930).
[240] H. Wynberg and A. Logothetis, *J. Amer. Chem. Soc.* **78**, 1958 (1956).
[241] F. Challenger and J. L. Holmes, *J. Chem. Soc.*, 1837 (1953), and references therein.

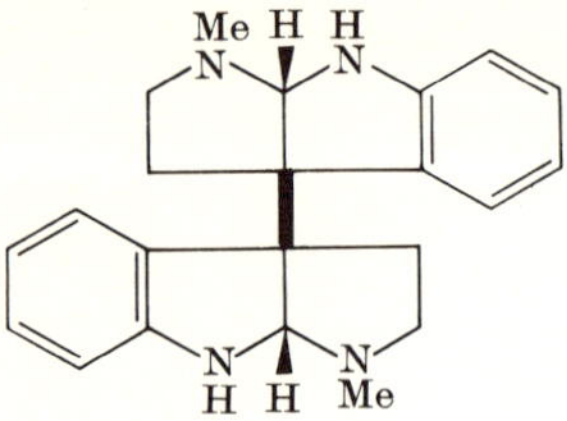

(172)

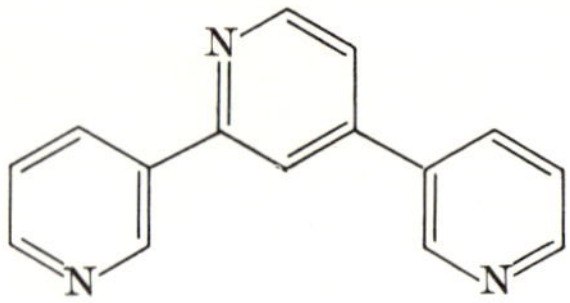

(173)

Chimonanthine (172),[242] isolated from leaves of *Chimonanthus fragrans* Lindle (*Calicanthaceae*), is an example of naturally occurring dimeric indole alkaloids of type C.

The well-known piperidine alkaloid anabasine, found in *Anabasis aphylla* L. (*Chenopodiaceae*) and *Nicotiana glauca* Graeb. (*Solanaceae*), is thought to be synthesized in plants through dimerization of $\varDelta^1$-piperideine (followed by oxidation).[243] A terpyridine nicotelline (173) was found in tobacco leaf,[244] and its structure was proved by synthesis.[245]

D. Formation of Oligomers That Have Two (or More) Heterocyclic Nuclei Joined by a Double Bond

Because these substances have two heterocyclic nuclei joined by a double bond, geometrical isomers should exist owing to prohibited

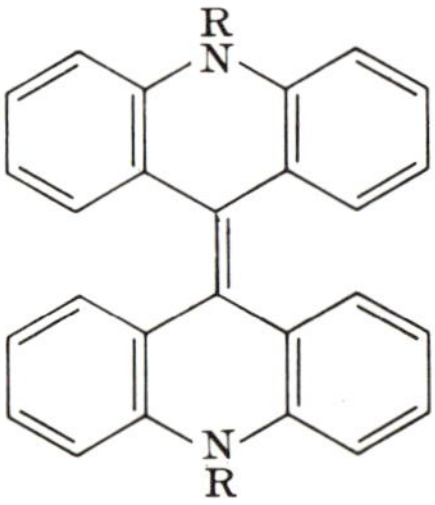

(174a) R = H
(174b) R = Me

[242] H. F. Hodson, B. Robinson, and G. F. Smith, *Proc. Chem. Soc.*, 465 (1961).
[243] For a monograph, see Robinson,[163] p. 35.
[244] A. Pictet and A. Rotschy, *Ber.* **34**, 696 (1901).
[245] J. Thesing and A. Müller, *Angew. Chem.* **68**, 577 (1956); *Chem. Ber.* **90**, 711 (1957).

rotation about the connecting bond. However, several heterocyclic compounds of type D possess ring atoms and substituents (e.g., amino, hydroxy, or mercapto) at such positions that tautomerism (or resonance) lends a single-bond character to the connecting bond; thus a transition between type C and D oligomers can occur, making, in some cases, the geometrical isomers of type D oligomers interconvertible; see e.g., dimer (140) and indole-type compounds (178).

The interconversion of oligomers of types C and D can also be effected by chemical reactions, the best example being the transformations of binuclear acridines[225] by oxidation and reduction:

$$164 \xleftarrow{\text{oxidize}} 174a \xrightarrow{\text{reduce}} 151$$

Although oligomers of type D can be made by such chemical conversions from the corresponding type C compounds (as in the indole series), more commonly used methods are oxidative and reductive condensation of heterocyclic monomers (cf. D2 oligomers).

Subclass 1. Heterocyclic Oligomers from Nonheterocyclic Monomers

Oligomers of this type have not often been made from nonheterocyclic monomers by direct condensation and ring closure, but the following is an example. Catalytic reduction of *o*-cyanobenzophenone over Raney nickel yields 1-phenylisoindole accompanied by the oxidative coupling product (175); this structure has been confirmed by X-ray analysis.[246]

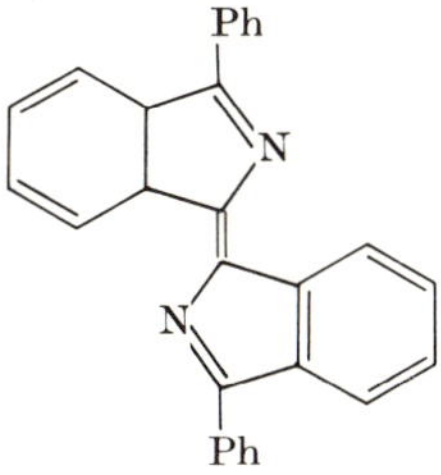

(175)

[246] E. Carstensen-Oeser, *Chem. Ber.* **104**, 3108 (1971).

Subclass 2. Heterocyclic Oligomers from Heterocyclic Monomers

a. *Acid-Catalyzed Oligomerization.* 9,9'-Biacridylidene (**174a**) was obtained by reducing the biacridyl (**164**) with zinc in 5 N hydrochloric acid,[225] but a more convenient preparative method under basic conditions is mentioned below. N,N'-Dimethyl-9,9'-biacridylidene (**174b**) is best made by the reductive action of zinc on N-methylacridone in ethanol containing hydrochloric acid.[225] Treatment of 4,5-diphenyl-1,3-dithiolium perchlorate with an alcohol gave the 2-alkoxy-1,3-dithiol, which yielded the dimer (**176**) upon heating with acetic acid at 80°.[247] Reduction of pyridine with zinc in acetic anhydride produced **141a** and a small amount of a further oxidized dimer (**177**),[193] whose structure has been confirmed.[248]

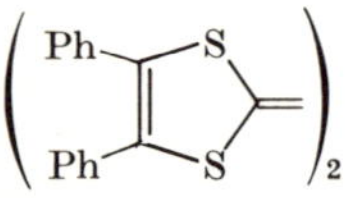

(**176**) 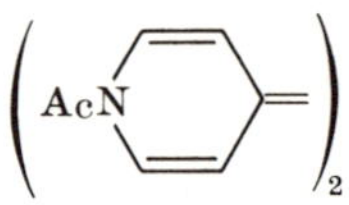   (**177**)

b. *Base-Catalyzed Oligomerization.* It is well known that alkaline solutions of indoxyl (**149**) are readily oxidized in air to indigo (**178a**). Thioindigo (**178b**) is similarly obtained from 3-hydroxythianaphthene (thioindoxyl) by using alkaline ferricyanides. X-Ray diffraction analysis has confirmed[249] the presence of the planar, trans form in **178a**. Other indigos (**178**; X = O, S, Se, NMe) have been isolated in both cis and trans forms[250] which are interconvertible owing to the single-bond

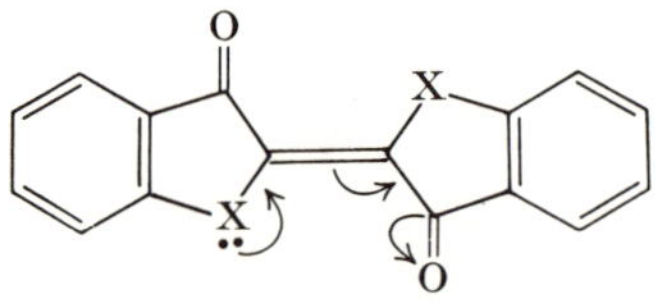

(**178a**) X = NH
(**178b**) X = S

[247] K. M. Pazdro and W. Polaczkowa, *Rocz. Chem.* **45**, 1249 (1971).
[248] A. T. Nielsen, D. W. Moore, G. M. Muha, and K. H. Berry, *J. Org. Chem.* **29**, 2175 (1964).
[249] H. v. Eller, *Bull. Soc. Chim. Fr.*, 1426 (1955).
[250] R. Pummerer and G. Marondel, *Chem. Ber.* **93**, 2834 (1960); H. Gusten, *Chem. Commun.* 133 (1969).

(179)

(180a) R = H, X = N
(180b) R = Me, X = CH

character as shown by arrows in formula **178** (but *cis*-indigo itself has never been isolated). Thorough reviews on indigos[251] and thioindigos[252] have appeared. Reduction (+ 2H) of indigos in alkaline solution gives the dienol ("indigo white"), which undergoes autoxidation to indigo in the air, as in the dyeing process. Similarly the air oxidation of 1,4-dihydroxyisoquinoline in alkali gives the doubly bound 3,3'-dimer "carbindigo," which can be reduced to leucocarbindigo (a type C dimer) by ammonium sulfide solution.[253] A simpler, indigo-like dimer (**179**) has been prepared by heating 4,4-dimethylpyrrolidin-3-one in dimethylformamide at 150° in the presence of potassium cyanide.[254] Biacridylidene (**174a**) is conveniently obtained by refluxing acridine with sodium carbonate in ethylene glycol.[255] Further examples of this "ylidene" type are the orange-colored 7,7'-dimer (**180a**) obtained by heating pteridin-6-one with sodium carbonate in dimethylformamide (95% yield),[256] and the purple 4,4'-dimer formed from pteridin-2-one in hot 1.0 N sodium hydroxide (under nitrogen).[257] Quinoxaline and 6-hydroxy-7,8-dihydropterin have been reported also to give dimers of type E2; the former dimer has the 2,2'-bound trans configuration,[256] and the 7,7'-bound trans structure has been proposed for the latter dimer.[258]

[251] W. C. Sumpter and F. M. Miller, "Compounds with Indole and Carbazole Systems," p. 170. Wiley (Interscience), New York, 1954.
[252] H. D. Hartough and S. L. Meisel, "Compounds with Condensed Thiophene Rings," p. 175. Wiley (Interscience), New York, 1954.
[253] S. Gabriel and J. Colman, *Ber.* **35**, 2421 (1902).
[254] E. Wille and W. Lüttke, *Angew. Chem.* **83**, 853 (1971).
[255] A. Albert and G. Catterall, *J. Chem. Soc.*, 4657 (1965).
[256] A. Albert and H. Rokos, *in* "Chemistry and Biology of Pteridines" (K. Iwai *et al.*, eds.), p. 95. Int. Acad. Print Co., Tokyo, 1970.
[257] A. Albert and F. Reich, *J. Chem. Soc.*, 1370 (1960).
[258] W. Pfleiderer, *in* "Chemistry and Biology of Pteridines (K. Iwai *et al.*, eds.), p. 7. Int. Acad. Print. Co., Tokyo, 1970.

Isoxazolin-5-ones and pyrazolin-5-ones have been shown to give the anhydro dimers (**140a** and **b**, respectively), when treated with a base such as piperidine.[191, 259]

A different type of trimerization is exemplified by 1,4-dithiintetracarboxylic acid diimide which, on heating at 120° in pyridine, gave a mellitic acid triimide (**181**; R = H, Me, CH$_2$Ph) probably through the highly strained intermediate[260] [Eq. (11)].

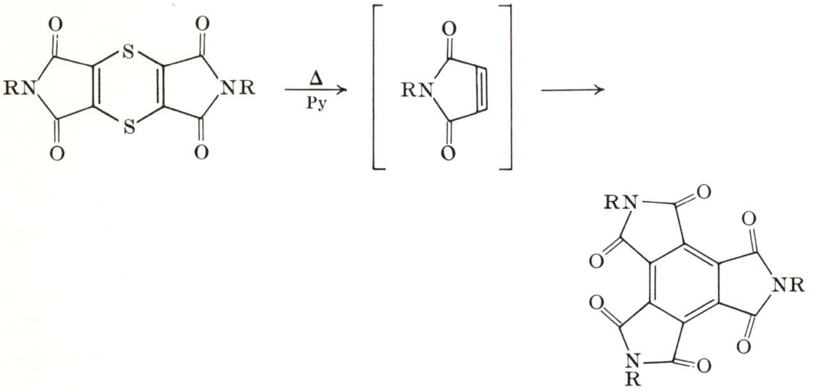

c. *Thermal Oligomerization.* On heating the 4:1 adduct (**58b**) of arylisocyanate and dimethylformamide at 300° for 7 minutes, the dimer (**182**) was formed; a carbene intermediate has been postulated.[83]

Thermal decarboxylation of 1-methyl-2-oxo-1,2-dihydro-1,4,5-triazanaphthalene-3-carboxylic acid is accompanied by the formation of a red compound as a by-product, to which the dimeric structure **180b** has been assigned.[261] Acid or alkaline hydrolysis of the ethyl ester of the monomer yields the same dimer.

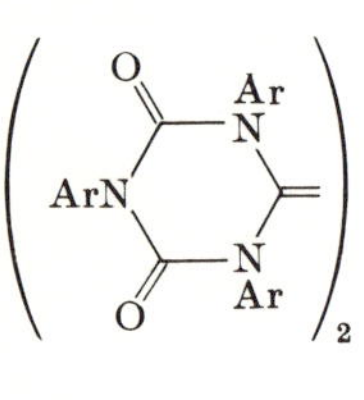

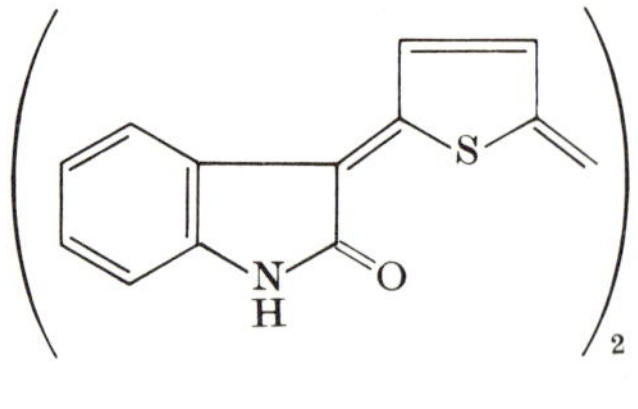

(**182**)   (**183**)

[529] R. H. Wiley and P. Wiley, "Pyrazolones, Pyrazolidones, and Derivatives." Wiley (Interscience), New York, 1964.
[260] W. Draber, *Angew. Chem. Int. Ed. Engl.* **6**, 75 (1967).
[261] J. W. Clark-Lewis, and M. J. Thompson, *J. Chem. Soc.*, 430 (1957).

Subclass 3. Heterocyclic Co-oligomers from More Than One Kind of Monomer

a. *Acid-Catalyzed Co-oligomerization.* Isatin and thiophene condense in sulfuric acid to a deep blue compound and this color was used for the detection of thiophene in coal tar even in the last century. Structure **183** for the product ("indophenine") was established later by Steinkopf and Hanske.[262] Several geometrical isomers are possible.

b. *Base-Catalyzed Co-oligomerization.* The condensation of 1,4-diaminoanthraquinone with two molecules of diketene (in pyridine), followed by ring closure with dilute alkali, gave[234] substance **168**.

Subclass 4. Naturally Occurring Heterocyclic Oligomers

Indigo (**178a**) was originally prepared from an indoxyl glucoside occurring in various species of *Indigofera* (*Leguminosae*), until the industrial synthesis was developed.

The pigment called violacein (**184**), a co-oligomer, has been isolated from *Chromobacterium violaceum*.[263]

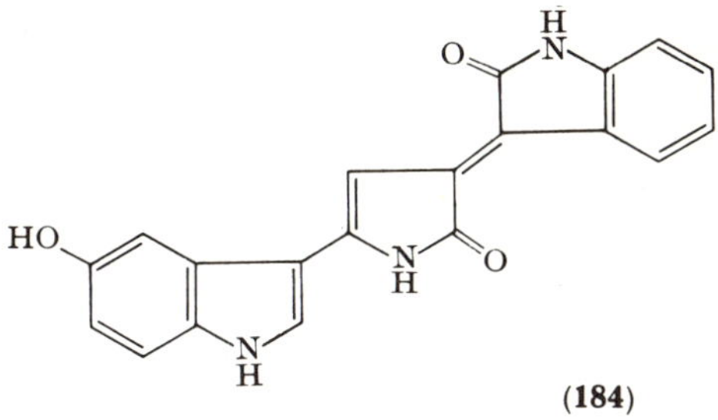

(**184**)

E. Formation of Oligomers That Have Two (or More) Heterocyclic Nuclei Joined by Two (or More) (Separate) Bonds

This type of condensation is of great interest in connection with the Woodward–Hoffmann selection rules for symmetry-allowed concerted *suprafacial* and *antarafacial* cycloaddition reactions.[264] The generalized rules for cycloaddition of an m- to an n-electron system predict that the concerted *supra-supra* or *antara-antara* dimerization is allowed in the excited state (i.e., photochemically) when $m + n = 4q$, and in the ground state (i.e., thermally) when $m + n = 4q + 2$, where m and n are the numbers

---

[262] W. Steinkopf and W. Hanske, *Ann. Chem.* **541**, 238 (1939).
[263] J. A. Ballantine, R. J. S. Beer, D. J. Crutchley, G. M. Dodd, and D. R. Palmer, *J. Chem. Soc.*, 2292 (1960).
[264] See Woodward and Hoffmann,[2a] pp. 65–113.

of π electrons in the molecule of the starting material and q is an integer, 1, 2, 3, ... For the *supra-antara* or *antara-supra* dimerization, the requirement for the total number of the electrons is reversed, i.e., photochemical addition is favored when $m+n=4q+2$, and thermal addition when $m+n=4q$. A more detailed discussion will accompany the examples below.

Subclass 2. Heterocyclic Oligomers from Heterocyclic Monomers

a. *Acid-Catalyzed Oligomerization.* Quinaldine (2-methylquinoline), when reduced with zinc in hydrochloric acid, gave a dimeric compound through dimerization of the dihydro monomer;[265] X-ray diffraction analysis of the 6,6'-dibromo derivatives of the dimer has shown its correct structure to be the trans form (**185**).[266] Electrolysis of quinaldine at a mercury cathode in alcoholic potassium hydroxide, followed by catalytic reduction afforded the analogous cis dimer.[266] This type of dimerization probably proceeds through a two-step mechanism (i.e., nonconcerted).

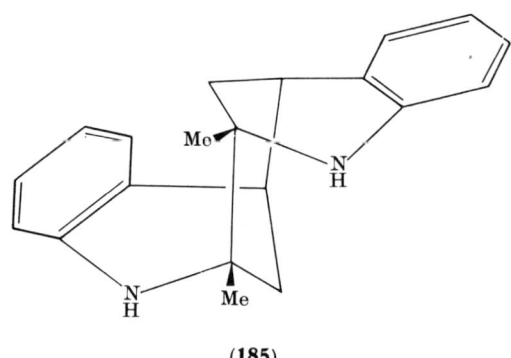

(**185**)

b. *Base-Catalyzed Oligomerization.* Reduction of *N*-methyl-4-cyanopyridinium iodide with sodium borohydride in methanolic sodium hydroxide gave the trans-syn dimer (**186**) at $-45°$ (60%), but at a higher temperature (0°) another dimer (**187**) was isolated.[267] These products are assumed to be derived by two-step dimerization of the 1,2-dihydropyridine intermediate.[267] On treatment with aqueous

[265] H. Dunathan, I. W. Elliott, and P. Yates, *Tetrahedron Lett.*, 781 (1961); I. W. Elliott, E. S. McCaskill, M. S. Robertson, and C. H. Kirksey, *ibid.* 291 (1962).

[266] I. W. Elliott, W. T. Bowie, and D. Wong, presented at the 3rd International Conference on Heterocyclic Chemistry, Sendai, Japan, 1971.

[267] F. Liberatore, A. Casini, V. Carelli, A. Arnone, and R. Mondelli, *Tetrahedron Lett.*, 2381, 3892 (1971).

potassium cyanide at 0°–5°, 1-methylpyridazinium salts afforded a mixture of the trans dimer (**188**) and its cis isomer;[268] the reaction is probably initiated by covalent addition of a cyanide anion [to the C=N (6,1) and C=C (3,4) bond] to give 1,6- and 1,4-dihydropyridazine, which dimerizes similarly to the reaction that produced **186**.

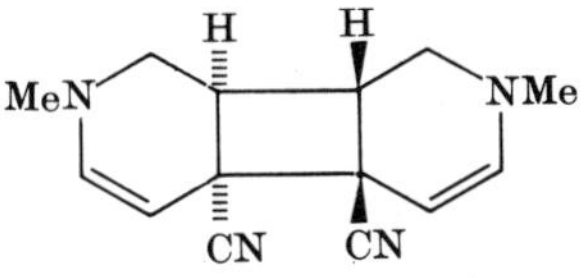

(**186**)

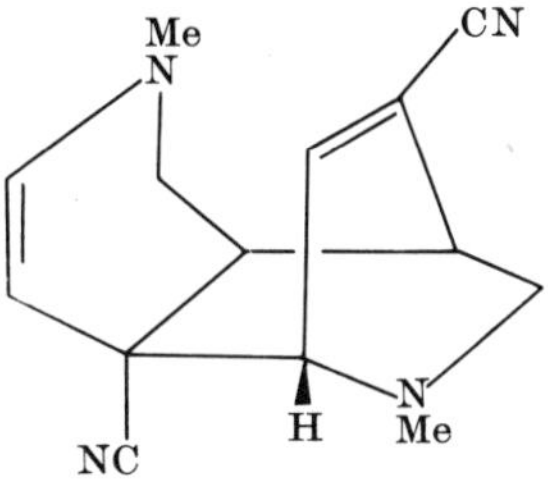

(**187**) 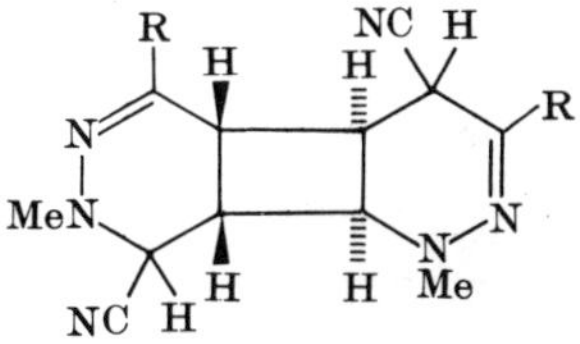   (**188**)

c. *Thermal Oligomerization*. An example of a symmetry-allowed $(6_s + 4_s)\pi(suprafacial)$-concerted cycloaddition is provided by the thermal exo-dimerization products of 1*H*-azepines (**189**; R = COOR',[269] SO$_2$Me[270]) at moderate temperatures. At higher temperatures, many 1-substituted 1*H*-azepines produced symmetrical $(6 + 6)\pi$-cycloaddition products (**191**; R = COOR',[269] CN,[271] Me[272]). These were thought to be formed by thermal rearrangement, perhaps by way of a stabilized diradical (**190**)[270] [Eq. (12)]. Procházka has reported spontaneous dimerization $(4_s + 2_s)$ of thiophene 1-oxide to the exo dimer (**192**).[273] On heating at 180°–200° in an inert solvent (e.g., tetralin), benzo[*b*]-thiophene 1,1-dioxide afforded dihydrobenzonaphthothiophene dioxide,

[268] H. Igeta, T. Tsuchiya, and C. Kaneko, *Tetrahedron Lett.*, 2883 (1971).
[269] L. A. Paquette and J. H. Barrett, *J. Amer. Chem. Soc.* **88**, 2590 (1966).
[270] L. A. Paquette, in "Nonbenzenoid Aromatics" (J. P. Snyder, ed.), Vol. 1, p. 249. Academic Press, New York, 1969.
[271] A. L. Johnson and J. E. Simmons, *J. Amer. Chem. Soc.* **88**, 2591 (1966); **89**, 3191 (1967).
[272] K. Hafner and J. Mondt, *Angew. Chem. Int. Ed. Engl.* **5**, 839 (1966); G. Habermehl and S. Göttlicher, *ibid.* **6**, 805 (1967); S. Göttlicher and G. Habermehl, *Ber.* **104**, 524 (1971).
[273] M. Procházka, *Collect. Czech. Chem. Commun.* **30**, 1158 (1965).

probably through formation of the intermediate $(4_s+2_s,$ or $8_s+2_s)$ dimer (193), which readily lost sulfur dioxide.[274]

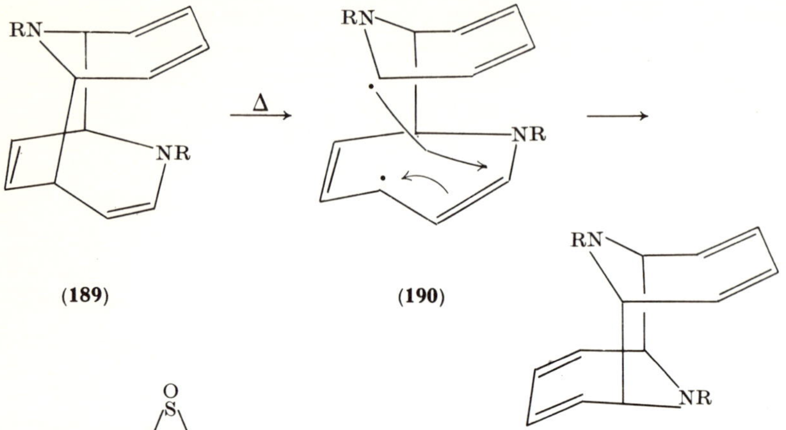

(189)   (190)   (12)

(191)

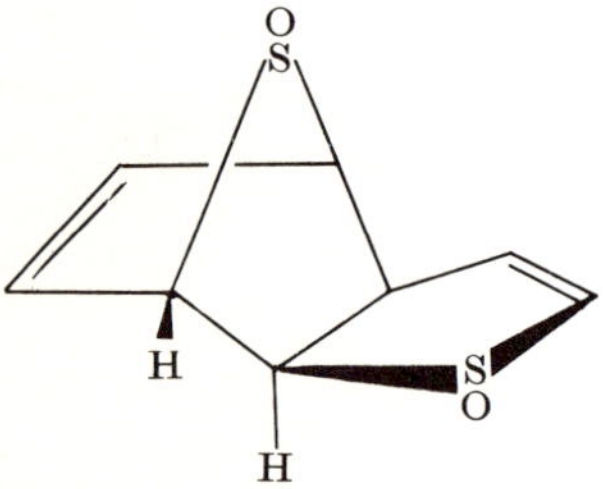

(192)

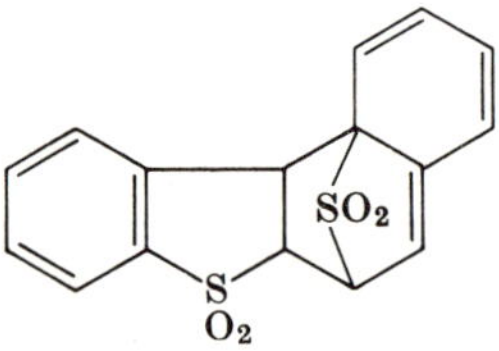

(193)

d. *Photooligomerization*. Many 1,2- and 1,4-photodimerization products, mostly due to (2_s+2_s) and (4_s+4_s), respectively, have been reported and thoroughly reviewed;[1] thus, only a few commonly known examples are presented here.

[274] W. Davies, N. W. Gamble, F. C. James, and W. E. Savige, *Chem. Ind. (London)*, 804 (1952); W. Davies, N. W. Gamble, and W. E. Savige, *J. Chem. Soc.*, 4678 (1952).

Sec. III.E.] HETEROCYCLIC OLIGOMERS 61

The 1,2-photodimer (**194**), of some biological significance, was obtained as the main product on irradiation of a frozen aqueous solution of thymine or DNA (followed by hydrolysis in the latter case).[275] This dimerization between two adjacent thymine residues is considered to be the major change responsible for the photochemical inactivation of DNA; dimerization appeared to occur within one strand (intrastrand).[276] Upon irradiation of 1,3-dimethylthymine,[277] 1,3-dimethyluracil,[216,278] uracil,[279] and thymidine,[280] in frozen aqueous solutions, a mixture of all four possible 1,2-dimerization products (i.e., cis and trans head-to-head, cis and trans head-to-tail) have been produced. A triplet state has recently been proposed as their precursor.[216] It has also been shown[281] that the trans head-to-head photodimer (**195**) of 2-quinolone, obtained in ethanol,[282] is formed by condensation of the triplet state monomer with another unexcited molecule.

3,5-Diethoxycarbonyl-1,4-dihydropyridine initially gives a mixture of a cyclobutane-type anti-cis and anti-trans photodimer, the former being, on further irradiation, transformed into the "cubic" dimer (**196**) by another 1,2-cyclization.[283] On irradiation of 2,6-dimethyl-4-pyrone, a trans cubic dimer with the same skeleton as **196** has been obtained.[283a] Photodimerization of thianaphthene 1,1-dioxide gave a mixture of the

[275] R. Beukers and W. Berends, *Biochim. Biophys. Acta* **41**, 550 (1960); H. Ishihara, *Photochem. Photobiol.* **2**, 455 (1963); G. M. Blackburn and R. J. H. Davies, *J. Chem. Soc. C*, 2239 (1966); R. Anet, *Tetrahedron Lett.*, 3713 (1965).
[276] G. M. Blackburn and R. J. Davies, *J. Amer. Chem. Soc.* **89**, 5941 (1967).
[277] H. Morrison, A. Feeley, and R. Kleopfer, *Chem. Commun.*, 358 (1968).
[278] G. Fürst, E. Fahr, and H. Wieser, *Z. Naturforsch.* **22b**, 354 (1967).
[279] C. H. Krauch, D. M. Krämer, P. Chandra, P. Mildner, H. Feller, and A. Wacker, *Angew. Chem. Int. Ed. Engl.* **6**, 956 (1967).
[280] D. Weinblum and H. E. Johns, *Biochim. Biophys. Acta* **114**, 450 (1966).
[281] T. Yamamuro, I. Tanaka, and N. Hata, *Bull. Chem. Soc. (Jap.)* **44**, 667 (1971).
[282] O. Buchardt, *Acta Chem. Scand.* **18**, 1389 (1964).
[283] U. Eisner, J. R. Williams, B. W. Matthews, and H. Ziffer, *Tetrahedron* **26**, 899 (1970).
[283a] P. Yates and M. J. Jorgenson, *J. Amer. Chem. Soc.* **85**, 2956 (1963).

trans-syn adduct (197) and its trans-anti isomer,[284] in contrast to the thermal dimer intermediate (193).

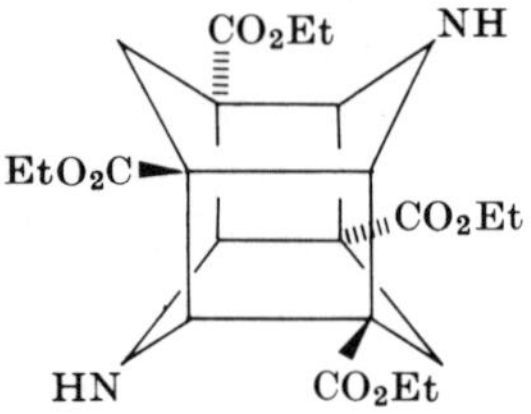

The compound (198), typical of 1,4-photodimerization anti products, was obtained from 2-aminopyridine in aqueous hydrochloric acid.[285] 4,6-Dimethyl-2-pyrone gives a mixture of similar 1,4-adducts (syn and anti) and the unsymmetrical 1,2-dimerization product (199).[286]

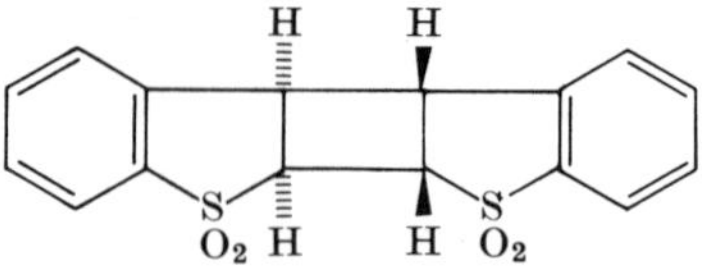

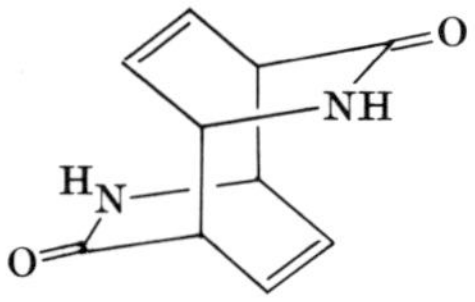

F. Formation of Miscellaneous Oligomers, Including Spiro Compounds

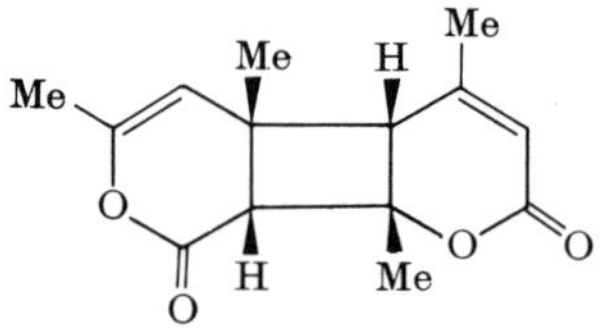

There are substances which do not belong to classes A–E, such as heterocyclic spiro oligomers. A few representative examples are mentioned here.

Subclass 2. Heterocyclic Oligomers from Heterocyclic Monomers

a. *Acid-Catalyzed Oligomerization.* Condensation of diethyl acetonedicarboxylate (sodium salt; 1 mole) with methyl isothiocyanate (2 moles) gave the tetrahydropyridine (200), which was dimerized by the action of bromine in glacial acetic acid to give the spiro dimer (201).[287]

---

[284] D. N. Harpp and C. Heitner, *J. Org. Chem.* **35**, 3256 (1970).
[285] E. C. Taylor and R. O. Kan, *J. Amer. Chem. Soc.* **85**, 776 (1963).
[286] P. de Mayo and R. W. Yip, *Proc. Chem. Soc.*, 84 (1964).
[287] D. E. Worrall, *J. Amer. Chem. Soc.* **62**, 675 (1940).

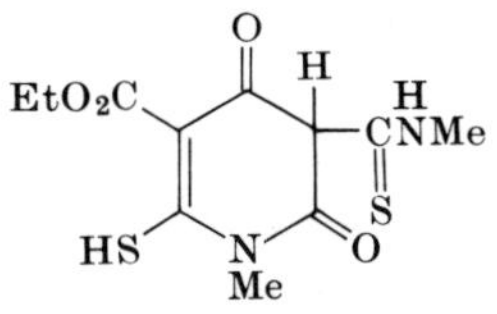

b. *Base-Catalyzed Oligomerization.* The spiro compound (**202**) was obtained on heating 2-(hexahydro-1*H*-azepin-1-yl)ethanol and *p*-toluenesulfonyl chloride in benzene in the presence of sodium carbonate.[288]

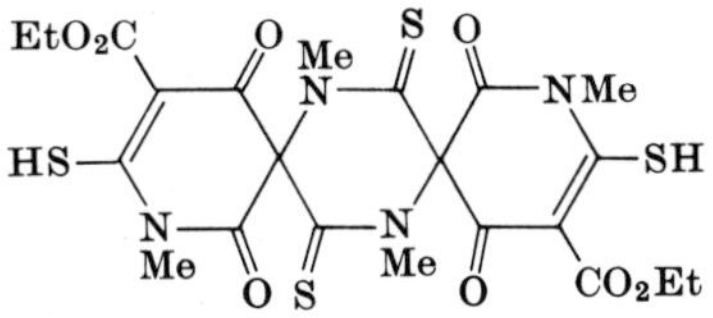

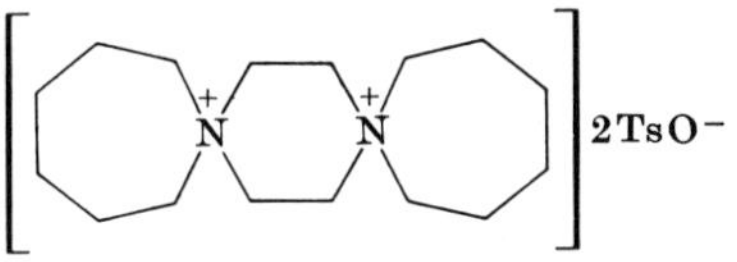

c. *Thermal Oligomerization.* When an aryl isocyanate is heated at 80°, in the presence of dimethylformamide, the unsymmetrical spiro pentamer (**203**) is formed together with the *s*-triazines (**58a** and **b**).[83]

d. *Photooligomerization.* On exposure to sunlight or UV light, compound **204** forms the spiro dimer (**205**) by 1,2-photocycloaddition of the exocyclic C=S bond.[289]

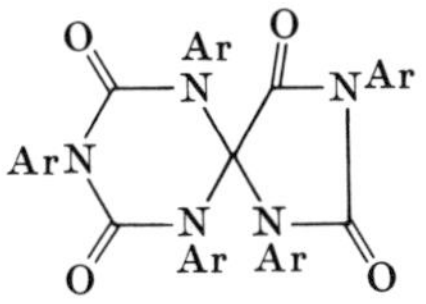

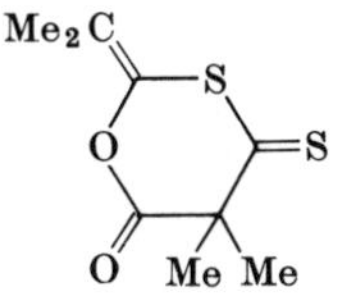

[288] S. Fila-Hromadko and K. Kovačević, *Croat. Chem. Acta* **43**, 93 (1971).
[289] J. C. Martin, R. D. Burpitt, P. G. Gott, M. Harris, and R. H. Meen, *J. Org. Chem.* **36**, 2205 (1971).

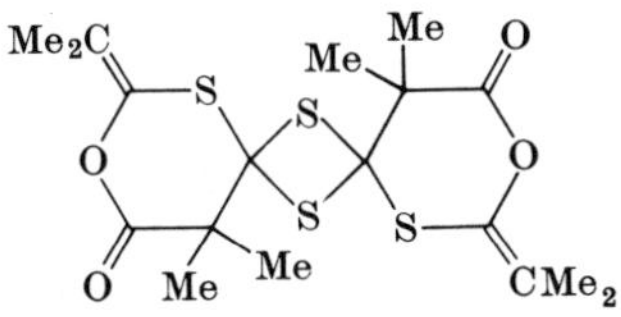

Subclass 3. Heterocyclic Co-oligomers from More Than One Kind of Monomer

a. *Acid-Catalyzed Co-oligomerization.* Condensation of two molecules of o-phenylenediamine with one of diethyl mesoxalate produced the spiroquinoxalone dimer (206) on refluxing in a mixture of 50% acetic acid and ethanol.[290]

The diaminospirothiazoline (207) was obtained by heating thiourea (2 moles) and 1,3-dichloroacetone (1 mole) without solvent;[291] because hydrochloric acid is liberated during the reaction, this is regarded as an acid-catalyzed co-oligomerization.

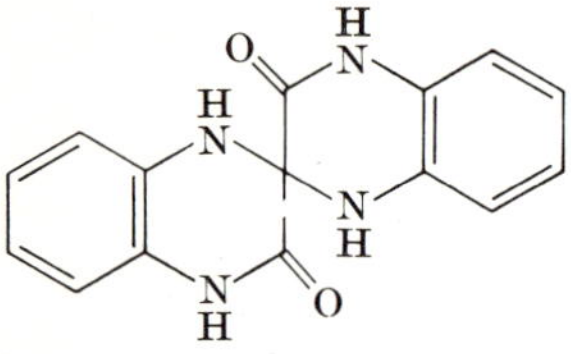

(206)

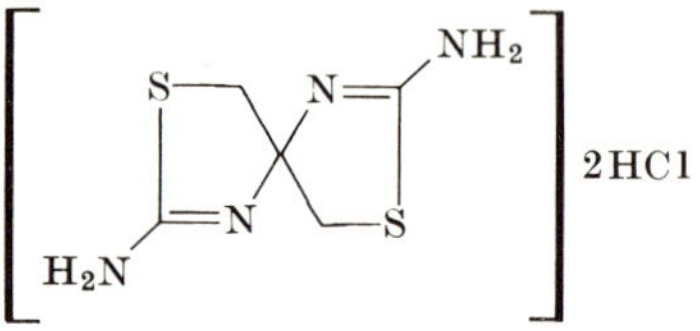

(207)

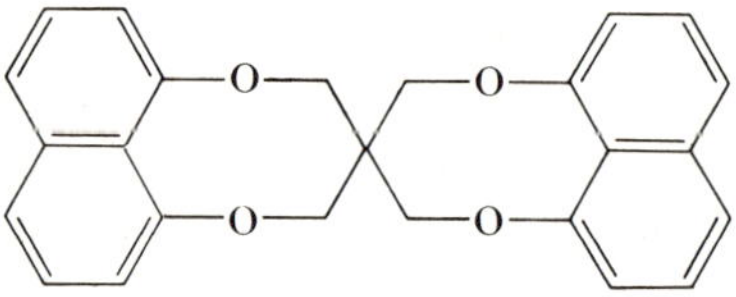

(208)

b. *Base-Catalyzed Co-oligomerization.* Two molecules of peridihydroxynaphthalene condensed with one of pentaerythritol tetrabromohydrin by heating in Cellosolve in the presence of sodium, giving the spiro dimer (208).[292]

IV. Conclusion

This review has illustrated that oligomerization reactions are of frequent occurrence. It is hoped that some useful guidance has been given to a very scattered literature. We believe that the classification of oligomers according to the reaction types, as used here, enables prediction

[290] J. W. Clark-Lewis, *J. Chem. Soc.*, 422 (1957).
[291] J. Harley-Mason, *J. Chem. Soc.*, 323 (1947).
[292] S. Smoliński and J. Jamrozik, *Tetrahedron* **27**, 4977 (1971).

of the products of oligomerization to be expected under various experimental conditions. Thus, it should help in promoting (or else avoiding) desired (or undesired) oligomerization reactions. Moreover, several oligomers, because of their physical and physiological properties, have found a useful place among medicinal chemicals, artificial fibers, chelate reagents, or base materials for paints and varnishes. Thus, there are good reasons for paying more attention to these interesting compounds, once obtainable only as ill-defined by-products, but whose formation and nature are beginning to be well understood.

The Oxidation of Monocyclic Pyrroles

G. P. GARDINI

Istituto Policattedra di Chimica Organica, Università di Parma, Parma, Italy

I. Introduction	. . . .	67
II. Autoxidation	. . . .	68
III. Photo-oxidation	. . . .	72
IV. Oxidation by Chemical Reagents	. . . . . .	79
A. Hydrogen Peroxide	. . . .	79
B. Other Peroxides and Peracids	. . . . .	86
C. Lead Dioxide and Tetraacetate	. . . . .	87
D. Cr^{VI} and Mn^{VII} Salts	. . . . . .	89
E. Nitric and Nitrous Acids	. . . .	91
F. Miscellaneous	. . . .	93
V. Pyrrole Blacks	. . . .	95
VI. Conclusions	. . . .	96
Note Added in Proof	. . . .	97

I. Introduction

It is perhaps unnecessary to emphasize the importance of the pyrrole nucleus in organic chemistry, especially in natural products such as chlorophyll, hemoglobin, bile pigments, and mold metabolites.

Quite surprisingly, the behavior of this nucleus toward oxidizing agents has not yet been studied extensively, and the published literature is sometimes incorrect. Moreover, heterocyclic texts pay little attention to this problem. Since tars, gums, and black compounds were often formed during the oxidation of pyrroles, this reaction was commonly believed almost unapproachable; this opinion does not appear justified according to recent results.

This chapter is a review of reported work and we hope that it will be of assistance to scientists interested in pyrrole natural products.

Literature has been covered until the end of 1970 using the *Chemical Abstracts Index* and until the end of 1971 using the main international journals of organic chemistry and biochemistry.

II. Autoxidation

Pyrroles are unstable in the presence of oxygen. In particular alkylpyrroles undergo rapid autoxidation on exposure to air, even in absence of light. Because of its affinity toward oxygen, pyrrole was proposed as an antioxidant by Ziegler and Ganicke.[1]

It is difficult to establish a clear distinction between autoxidation and photo-oxidation, since it is not always reported in the literature whether the uptake of oxygen by pyrroles required the presence of natural and/or artificial light. Therefore, studies carried out without specific use of light sources are reported in this section, despite the fact that, in some cases, the structures of the products seemed to suggest an attack by singlet oxygen.

The hypothesis that the first step in autoxidation involves a molecular association between pyrrole and oxygen has been proposed.[2] In fact, pyrrole develops a new absorption band at 296.5 nm, when saturated with oxygen at atmospheric pressure; this band disappears when oxygen is removed by a stream of nitrogen or by evacuation.[2] This reversible association has been verified by other authors[3,4] by measuring the variation of the absorbance at 300 nm of freshly distilled pyrrole; the absorbance was found to increase remarkably, as a function of the aging in the presence of air. A decrease of the absorbance by degassing was also confirmed along with the fact that no volatile, highly absorbing oxidation species were present. Infrared spectra did not show any difference between pure pyrrole and aged samples.[3]

Succinimide (1) was found to be a minor product in the air oxidation of unsubstituted pyrrole in water,[5] along with an unidentified product, $C_{12}H_{14}N_2O_5$, and a conspicuous amount of black polymer.

Succinimide (1) was also obtained by allowing pyrrole and air to react in the presence of aluminum powder.[6]

[1] K. Ziegler and K. Gänicke, *Ann.* **551**, 213 (1942).
[2] D. F. Evans, *J. Chem. Soc.*, 345 (1953).
[3] R. H. Linnell and S. Umar, *Arch. Biochem. Biophys.* **57**, 264 (1955).
[4] H. C. Wu, C. C. Chu, and T. C. Ho, *Hua Hsueh Hsueh Pao* **30**, 241 (1964); *Chem. Abstr.* **61**, 11489f (1964).
[5] G. Ciamician and P. Silber, *Chem. Ber.* **45**, 1842 (1912).
[6] A. C. Ray and S. B. Dutt, *J. Indian. Chem. Soc.* **5**, 103 (1928).

The first systematic study of the autoxidation of substituted pyrroles by Metzger and Fischer[7] exposed pyrrole derivatives (having at least one free α position) to air for long periods. Among several organic solvents tested, the best yields of products were obtained by dissolving the pyrroles in mixtures of ethyl ether and ethyl alcohol. Although natural light seemed to speed up reaction, nevertheless, it did not appear to be vital for product formation. Peroxides present in the ether were not involved in the oxidation process;[7] on the other hand, acetic anhydride

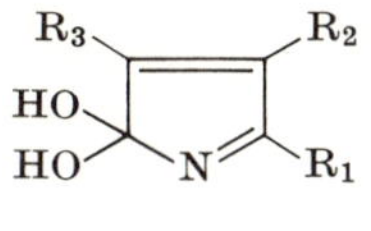

(3) 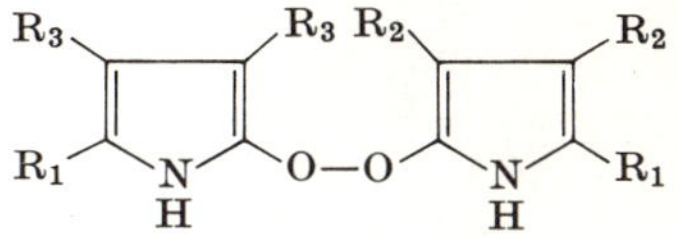　　　(4)

was shown to have a catalytic effect. Simple oxidation products were obtained in yields of up to 70%. The products were formulated by the author as geminal dihydroxy (3) or peroxide compounds (4) (but see later).

Under the same working conditions, pyrroles with electron-withdrawing substituents such as CHO, CN, COR, COOR, and COOH withstand autoxidation; 2-methyl-, 2,3-dimethyl-, and 2-methyl-4-ethylpyrroles gave no isolated oxidation products.[7]

Structure 3 was assigned to the autoxidation product from 2,4-dimethylpyrrole, whereas a peroxidic structure 4 was ascribed to the product from, e.g., 3-methyl-4-ethylpyrrole, although the presence of active oxygen was not found.

The validity of structure 4, with a peroxidic group directly bound to a π system, was questioned by Seebach,[8] who, without experimental proof, proposed the alternative structure 5, which, however, does not agree with the analytical evidence.[9]

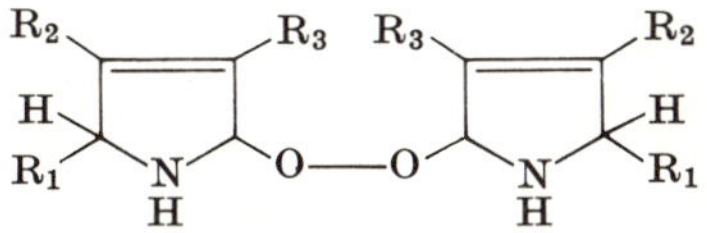

(5)

Later, Atkinson et al.[9] isolated, during the oxidation of 2,3,4-trimethylpyrrole (2; R_1R_2, $R_3 = CH_3$) with hydrogen peroxide, a by-product which was proved to be identical to that obtained by autoxida-

[7] W. Metzger and H. Fischer, Ann. 527, 1 (1937).
[8] D. Seebach, Chem. Ber. 96, 2723 (1963).
[9] J. H. Atkinson, R. S. Atkinson, and A. W. Johnson, J. Chem. Soc., 5999 (1964).

tion,[7] and to which the same structure was assigned. Finally, on the basis of NMR studies, the structure was reformulated as **6a**.[10] Such a product was formed along with a minor amount of a second compound, to which structure **7** has been assigned. Instead, from pyrrole (**2**; R_1, R_3 = CH_3, $R_2 = C_2H_3$) under the same conditions only conpound **6b** has been isolated.[10]

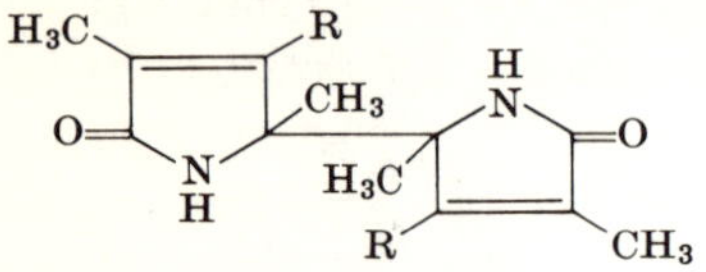

(6a) R = CH_3
(6b) R = C_2H_5

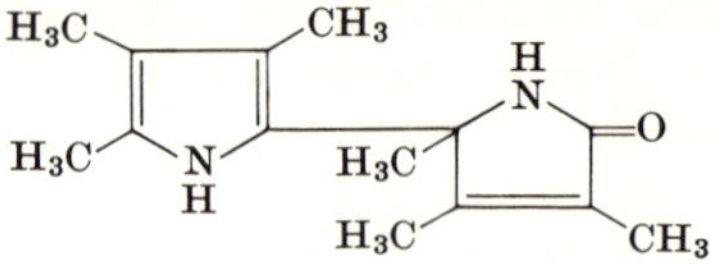

(7)

Autoxidation of **2** (R_1, $R_3 = CH_3$) led to a product $C_6H_9NO_2$ to which Metzger and Fischer[7] assigned structure **3**, later corrected to **8** on the basis of NMR spectrum[10] and the absorption maximum in UV light. The alternative structure **9** was first discarded because of the low value of

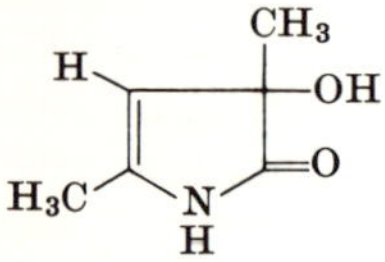

(8) 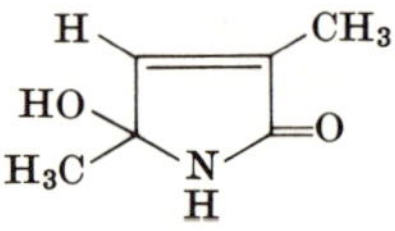 (9)

λ_{max}, but later advanced as the correct one.[11]

The behavior of 1-alkylpyrroles to autoxidation was studied by Smith and Jensen[12] with 1-methyl-, 1-isopropyl-, and 1-n-butylpyrrole. It was found that N-alkylpyrroles reacted much more slowly with oxygen than C-alkylpyrroles. The reactions were characterized by an induction period, during which the colorless liquid turned yellow and no oxygen uptake was detected; successively an autocatalytic reaction took place. The simple oxidation products formed in the case of 1-methylpyrrole were isolated and the structures **10–13** assigned.

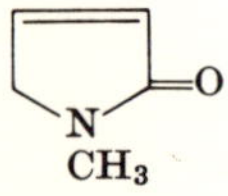

(10) 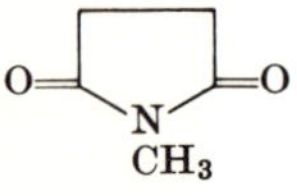 (11) 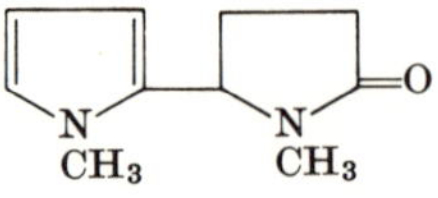 (12) 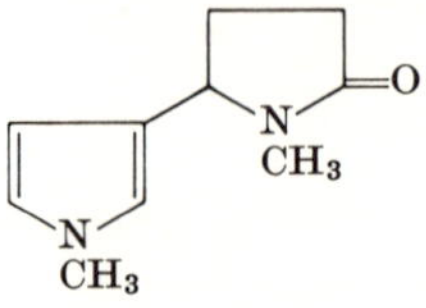 (13)

[10] E. Höft, A. R. Katritzky, and M. R. Nesbit, *Tetrahedron Lett.*, 3041 (1967).
[11] A. R. Katritzky, M. R. Nesbit, and E. Höft, *Tetrahedron Lett.*, 2028 (1968).
[12] E. B. Smith and H. B. Jensen, *J. Org. Chem.* **32**, 3330 (1967).

Peroxides were observed to be present even in very lightly oxidized samples,[12] and this seems to suggest that they are among the earliest reaction products. Since there is evidence that pyrroles exhibit a diene character in the presence of radicals,[13,14] it seems reasonable to assume that the oxidation of 1-alkylpyrroles may occur mainly by radical addition. A possible mechanism was, in fact, postulated.[12] The observed differences in the oxidation rates on varying the alkyl group must be due to steric hindrance of bulky substituents at the electronically preferred α positions.

The oxygen uptake at room temperature by unsubstituted pyrrole is quite slow. However, in the presence of strong bases, e.g. potassium t-butoxide in t-butanol, the oxidation occurred rapidly and two moles of oxygen per mole of pyrrole were absorbed in 10 hours. Unfortunately the products were not identified.[15]

Similar results were obtained with 2,3,4,5-tetraphenylpyrrole which was not affected by bubbling air through its solutions, whereas autoxidation proceeded slowly in benzene in the presence of powdered potassium hydroxide,[16] leading to the ketoamide (14) in good yields and traces of the lactam (15) (see also Section III).

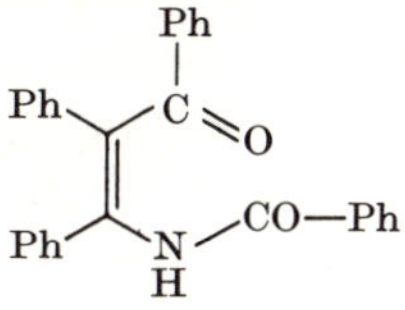

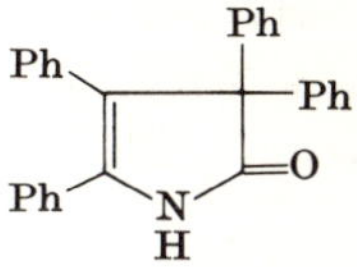

These are the only two examples of anionic autoxidation in the pyrrole series. Many other organic compounds behave similarly with molecular oxygen in solution of strong bases.[15]

Compound 17 was obtained by slow air oxidation from a benzene solution of 16;[17] similarly 2-methoxyindole yields indirubine. So far, no mechanism has been suggested for this interesting reaction.

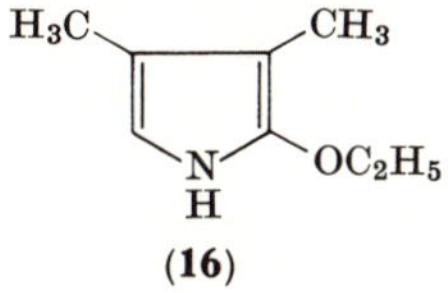

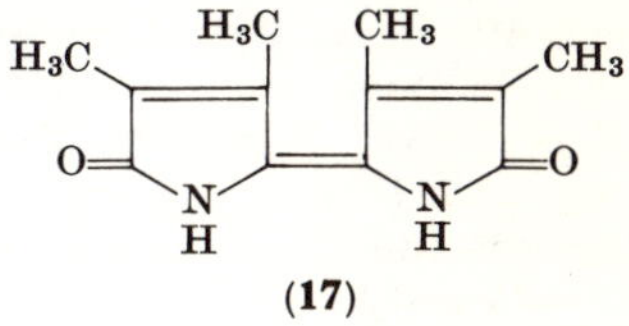

[13] J. B. Conant and B. F. Chow, *J. Amer. Chem. Soc.* **55**, 3475 (1933).
[14] R. J. Gritter and R. J. Chriss, *J. Org. Chem.* **29**, 1163 (1964).
[15] W. Bartok, D. D. Rosenfeld, and A. Schriesheim, *J. Org. Chem.* **28**, 410 (1963).
[16] G. Rio, A. Ranjon, and O. Pouchot, *C.R. Acad. Sci., Ser. C* **263**, 634 (1966).
[17] H. Bauer, *Chem. Ber.* **100**, 1701 (1967).

Pyrrole Grignards also easily undergo autoxidation. The first experiment carried out by Angeli and Pieroni[18, 19] by introducing dry air into pyrrole magnesium bromide, led exclusively to pyrrole blacks. This experiment has been studied by Fischer and co-workers,[20] who confirmed Angeli's results and observed the formation of **18** from cryptopyrrylmagnesium bromide along with a small amount of an unidentified product, $C_{16}H_{24}N_2O$.

$$\begin{array}{c} H_5C_2 \quad\quad CH_3 \quad C_2H_5 \quad\quad CH_3 \\ \diagdown \diagup \quad\quad \diagdown \diagup \\ H_3C \quad N \quad C \quad N \quad H \\ \quad\quad\quad H \quad\quad H \end{array}$$

(**18**)

Several papers relate the autoxidation of pyrroles to important practical problems, such as the storage of distilled petroleum fractions. Substituted pyrrole derivatives (present in small quantities in oil and probably generated by degradation from porphyrins) play an important role in the formation of precipitates with a high content of nitrogen and sulfur.[21–25] The removal of pyrroles by acid washing of fuel oil is certainly one of the reasons why acid treatment improves the quality of unstable oils, but not necessarily the only reason.[24] A number of gum forming reaction paths involving pyrroles have also been suggested.[23]

III. Photo-oxidation

Bernheim and Morgan[26] found that pyrrole in water, ethanol, or acetone, in the presence of methylene blue or eosine absorbed rapidly one mole of oxygen per mole of pyrrole by irradiating with light at 520–580

---

[18] A. Angeli and A. Pieroni, *Gazz. Chim. Ital.* **49** (I), 154 (1919).
[19] A. Angeli, *Atti Accad. Naz. Lincei* [6] **11**, 439 (1930).
[20] H. Fischer, H. Baumgartner, and E. Plötz, *Ann.* **493**, 1 (1932).
[21] Y. G. Hendrickson, 134th *Nat. Meeting Amer. Chem. Soc., Div. Petrol. Chem. Prepr.* **4** (1), 55 (1959).
[22] G. U. Dinneen and W. D. Bickel, *Ind. Eng. Chem.* **43**, 1604 (1951).
[23] A. A. Oswald and F. Noel, *J. Chem. Eng. Data* **6**, 294 (1961); *Chem. Abstr.* **55**, 22793 (1961).
[24] R. B. Thompson, J. A. Chenicek, L. W. Druge, and T. Symon, *Ind. Eng. Chem.* **43**, 935 (1951).
[25] R. D. Offenhauer, J. A. Brennan, and R. C. Miller, *Ind. Eng. Chem.* **49**, 1265 (1957).
[26] F. Bernheim and J. F. Morgan, *Nature (London)* **144**, 290 (1939).

nm. From the reaction mixture a compound was isolated (58%), the structure of which was elucidated as the hydroxylactam (**19a**) by De Mayo and Reid.[27] 1-Methylpyrrole behaves similarly, yielding **19b**.[27]

(**19a**) $R = R_1 = R_2 = R_3 = R_4 = H$
(**19b**) $R = CH_3; R_1 = R_2 = R_3 = R_4 = H$
(**19c**) $R = C_6H_5; R_1 = R_2 = R_3 = R_4 = H$
(**19d**) $R = R_1 = R_2 = R_4 = H; R_3 = CH_3$
(**19e**) $R = R_4 = H; R_1 = R_2 = C_2H_5; R_3 = CH_3$
(**19f**) $R = H; R_1 = R_2 = C_2H_5; R_3 = R_4 = CH_3$
(**19g**) $R = R_3 = H; R_1 = R_2 = C_2H_5; R_4 = CH_3$

Very recently it was found that when carrying out the photo-oxidation in methanol, pyrrole gives mainly **19d** and a low yield of maleimide.[28a] Similarly, 3,4-diethylpyrrole gives diethylmaleimide as the main product with minor quantities of 3,4-diethyl-5-methoxy-Δ^3-pyrrolin-2-one (**19e**). No maleimide was found when photo-oxidation was performed in aqueous solution.

By irradiating 2-methyl-3,4-diethylpyrrole in methanol[28b] under a stream of oxygen, **19g** was obtained as the main product (40%) with minor quantities of **19f** (11%) and diethylmaleimide (3%).

The formation of the latter product is an interesting example of photo-oxidative dealkylation in pyrrole series. The small amount of maleimide was ascribed to its instability in the reaction conditions, and this hypothesis was verified.[28b]

It was suggested[27–28b] that all these compounds would be generated from the unstable endoperoxide (**20**) (originated from a Diels–Alder reaction of singlet oxygen with the dienic pyrrole system) through various paths, as indicated in Scheme I.

ESR studies[29] demonstrated the formation of free radicals in solutions of pyrrole and eosine in ethanol or pyridine on exposure to light; the signal appeared immediately when the light was turned on and remained constant until it was extinguished.

---

[27] P. De Mayo and S. T. Reid, *Chem. Ind.* (*London*), 1576 (1962).
[28a] G. B. Quistad and D. A. Lightner, *Chem. Commun.*, 1099 (1971).
[28b] G. B. Quistad and D. A. Lightner, *Tetrahedron Lett.*, 4417 (1971).
[29] C. Lagercrantz and M. Yhland, *Acta Chem. Scand.* **16**, 508 (1962).

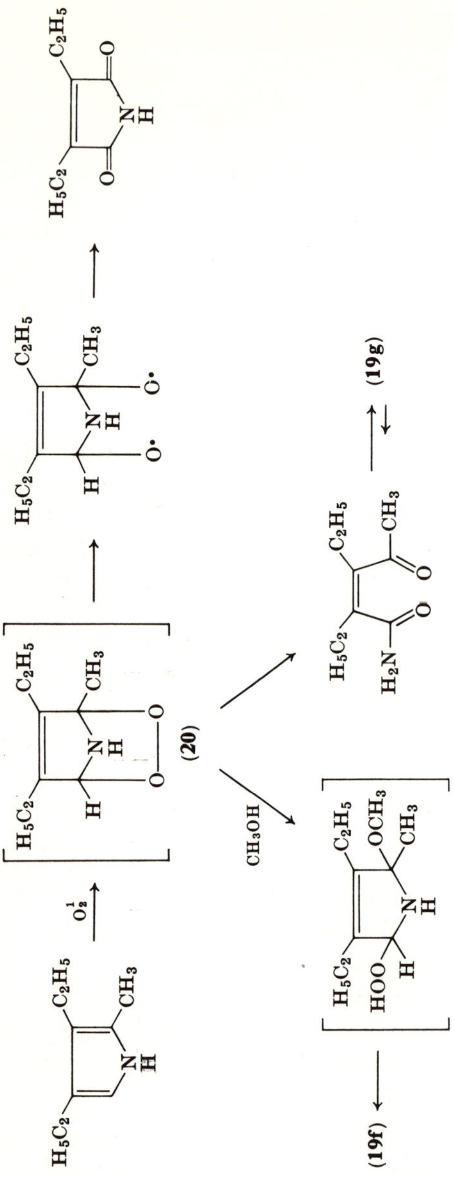

Scheme I

The singlet oxygen oxidation of N-phenylpyrroles has been studied by Frank and Auerbach.[30] N-Phenylpyrrole was oxygenated either photochemically or by using other methods of singlet oxygen generation. In every case low yields of **19** were obtained.

The other pyrroles tested for oxygenation in the same manner were **21** and **23**; **22** and **24**, respectively, were obtained. Oxygenation of **23** took place rapidly, but gave low yields of the lactam (**24**) with tarry residues. The authors supposed that the adduct (**25**) was formed, but that the subsequent opening of the endoperoxide had not occurred efficiently.[30]

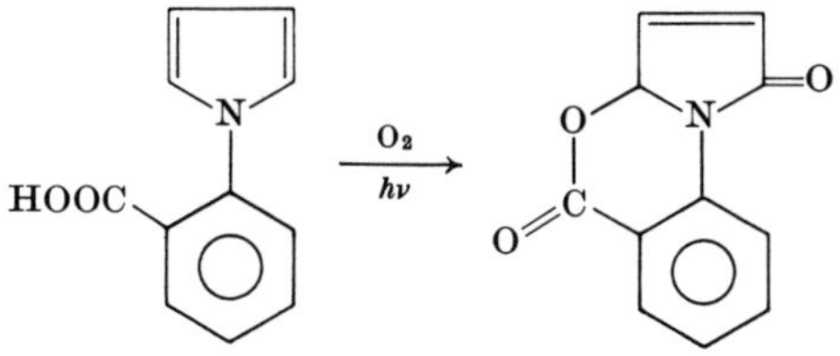

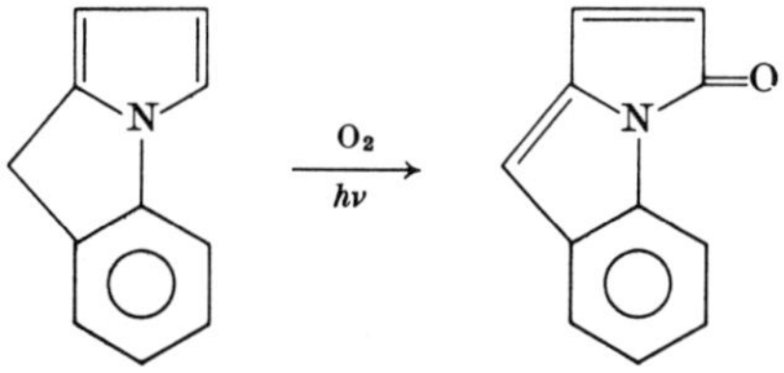

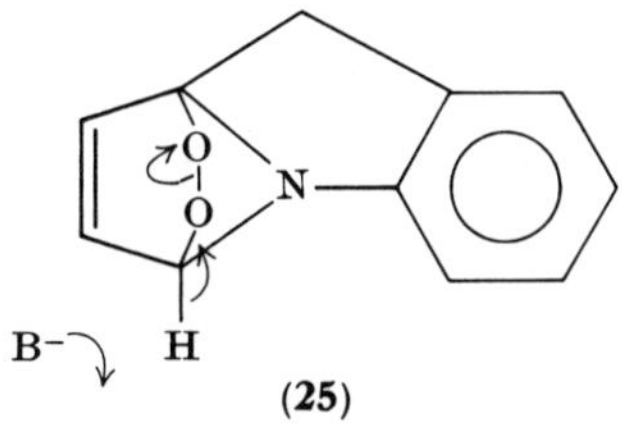

By adding triethylamine to this reaction mixture, no photo-oxidation took place, in accordance with the evidence[31] that tertiary amines quench singlet oxygen. On the other hand, the yield improved to about 70% for **24** on adding pyridine. However, the production of **19c** and **22** was not

---

[30] J. Auerbach and R. W. Frank, *Chem. Commun.*, 991 (1969); R. W. Frank and J. Auerbach, *J. Org. Chem.* **36**, 31 (1971).

[31] C. Ouannes and T. Wilson, *J. Amer. Chem. Soc.* **90**, 6527 (1968).

effected by the addition of pyridine. Deactivated pyrroles, such as pyrrole-2-carboxylic acid or methyl N-phenylpyrrole-2-carboxylate did not react with singlet oxygen.

Several C-polyphenylpyrroles were successfully tested for photo-oxidation. In fact, these compounds can readily be studied since they lead to products which are more stable and easier to isolate and investigate. The first attempt did not produce appreciable results.[32] Later Wasserman and Liberles[33] found that, under very mild conditions, i.e., exposure to air in methanol solution, during protracted irradiation with a 150 watt floodlamp and in the presence of methylene blue, 2,3,4,5-tetraphenylpyrrole underwent oxidation readily, forming products in good yield (55% **26** and 30% **27**).

(26)   (27)   (28)

In the presence of potassium hydroxide the main product was the lactam (**28**) (35%).

Rio and co-workers[16] suggested that **26–28** were not formed directly but via **30**, in the photosensitized oxidation of the tetraphenylpyrrole. The formation of **30** can itself be explained by ring opening of the endo-peroxide (**29**), which should be the first compound to be formed in homocyclic diene systems, owing to the strain in the bicyclic system. Also, the increase in conjugation and strong intramolecular hydrogen bond favor **30**.[34] (See Scheme II.)

By heating **30** (dry or in less polar solvents), high yields of oxygen and starting pyrrole were obtained, along with a small amount of lactam (**28**). Deoxygenation is not normal behavior of the hydroperoxide function, but it had analogy in other five-membered heterocyclic compounds.[35]

[32] J. Martel, Thèse ingénieur-docteur, Paris (1959); *C.R. Acad. Sci., Ser. C* **244**, 626 (1957).
[33] H. H. Wasserman and A. Liberles, *J. Amer. Chem. Soc.* **82**, 2086 (1960); A. Liberles, Ph.D. Thesis, University of Yale, New Haven, Connecticut (1960).
[34] G. Rio, A. Ranjon, O. Pouchot, and M. Scholl, *Bull. Soc. Chim. Fr.*, 1667 (1969); C. Dufraisse, G. Rio, A. Ranjon, and O. Pouchot, *C.R. Acad. Sci., Ser. C* **261**, 3133 (1965).
[35] C. Dufraisse, A. Etienne, and J. Martel, *C.R. Acad. Sci., Ser C* **244**, 970, 3106 (1957); **245**, 457 (1957).

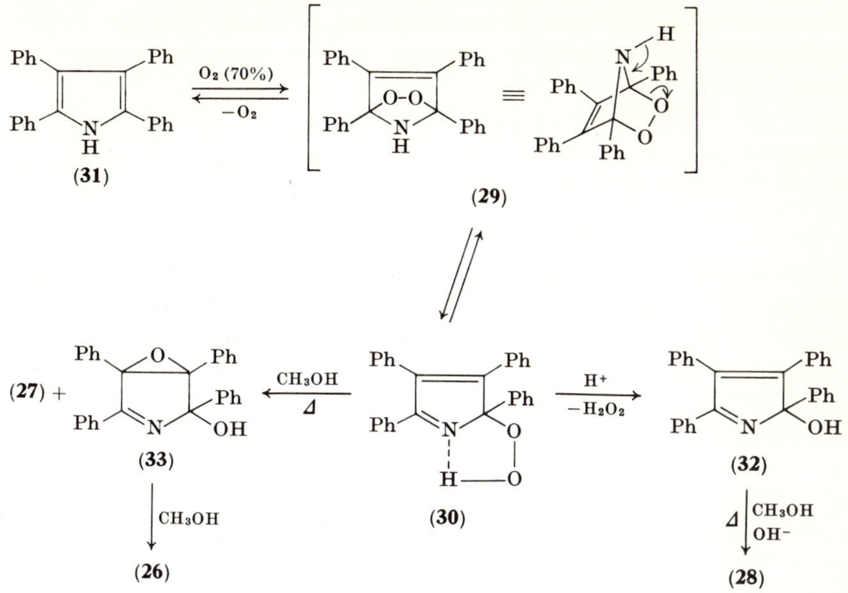

Sec. III.] THE OXIDATION OF MONOCYCLIC PYRROLES 77

SCHEME II

Compound **26** and small quantities of **27** were produced by heating **30** in methanol, whereas if potassium hydroxide was also present, **28** was mainly formed along with a small amount of **32**.

A mechanism has been devised to explain the formation of **27** and **33** from **30**; it suggests a step in which a cyclic peroxide is formed from the hydroperoxide (**30**) (Scheme III). Under basic conditions, and in the

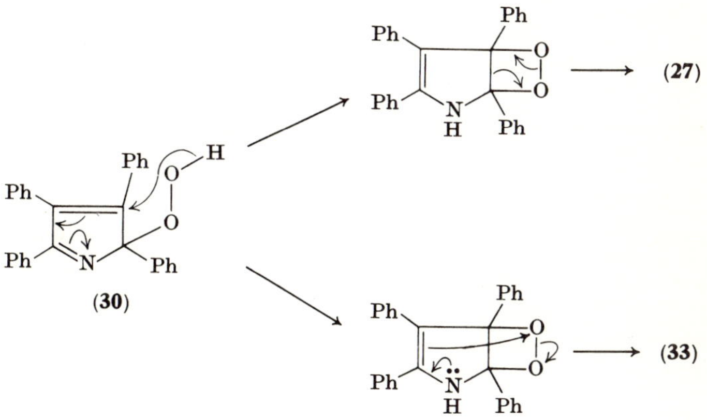

SCHEME III

presence of polar solvents, the hydroperoxide would undergo normal decomposition to the hydroxy derivative (32), whose anion might favor a benzylic type rearrangement to the unsaturated lactam (28).

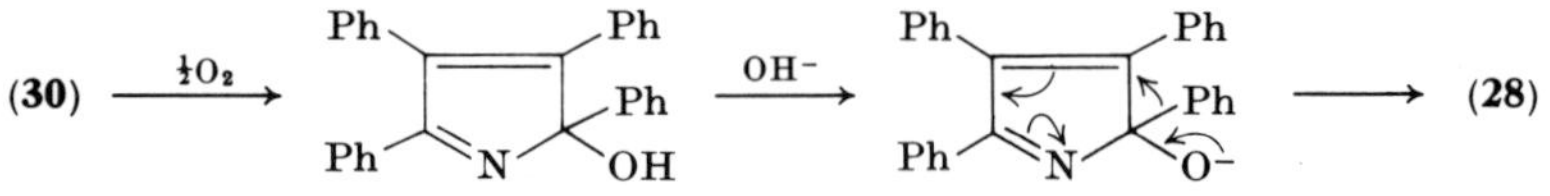

$$(30) \xrightarrow{\frac{1}{2}O_2} (32) \xrightarrow{OH^-} \longrightarrow (28)$$

Further studies on the photo-oxidation of polyphenylpyrroles attempted to verify if the attack by oxygen, forming the hydroperoxide, was an addition at the 2- and 5-positions of the pyrrole ring, rather than the alternative attack at both C-2 and the hydrogen bound to the heterocyclic nitrogen. For this reason the study was extended to pentaphenylpyrrole.[36] Although in this case the nitrogen does not bear a hydrogen atom, this pyrrole does not behave very differently from the tetraphenyl derivative, allowing the rejection of the second hypothesis.

In Scheme IV the course of the oxidation of pentaphenylpyrrole and the products formed are reported. Reversible formation of a 2-hydro-

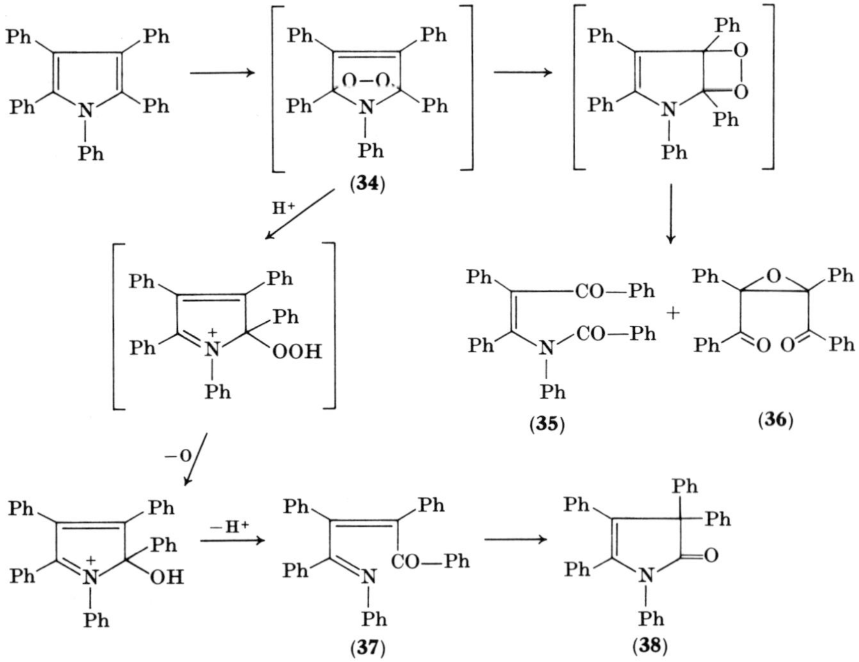

SCHEME IV

[36] C. Dufraisse, G. Rio, and A. Ranjon, *C.R. Acad. Sci., Ser. C* **265**, 310 (1967).

peroxypyrrolenine has also been observed in the photo-oxidation of 2,5-diphenylpyrrole, but this study is complicated by the presence of unidentified polymeric compounds.

Finally, the reactivity of N-methyl-(or N-phenyl)2,3,5-triphenylpyrrole toward oxygen (in methylene chloride solution and in the presence of photosensitizers) has been studied.[37] Peroxide compounds were not noticed, but benzoic acid (12%) and *cis*-dibenzoylstilbene epoxide (**36**) (65%) were obtained from **39**, while a mixture of the Schiff bases (**41**) and (**42**) (70%) was obtained from **40**. By acid hydrolysis **36** was easily produced from **41** and **42**.

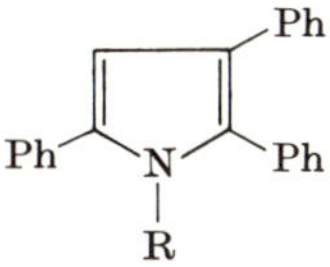

(**39**) R = CH_3
(**40**) R = C_6H_5

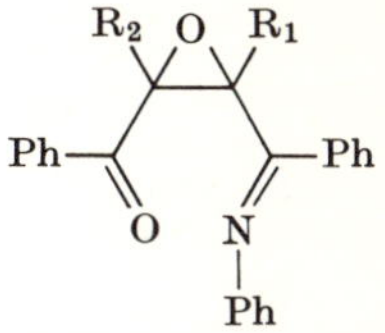

(**41**) R_1 = H, R_2 = Ph
(**42**) R_1 = Ph, R_2 = H

The mechanism that Wasserman proposed to explain these results agrees with that previously suggested by French authors, only as far as the primary formation of a endoperoxide; the successive fate of this intermediate to afford the Schiff bases is then suggested as following this path[37]:

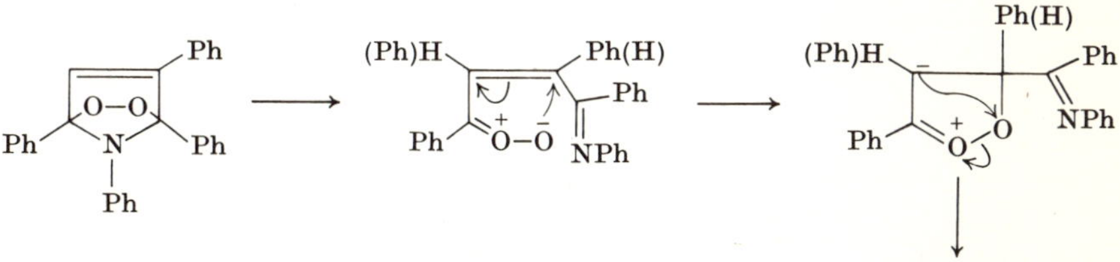

(**41**) or (**42**)

IV. Oxidation by Chemical Reagents

A. Hydrogen Peroxide

Hydrogen peroxide has been widely used in the oxidation of pyrroles. Along with ill-defined polymeric compounds, simple oxidation products were often isolated, with the best yields in neutral or weakly basic media.

[37] H. H. Wasserman and A. H. Miller, *Chem. Commun.*, 199 (1969).

1. Unsubstituted Pyrrole

The first experiments with hydrogen peroxide have been carried out in acetic solution, varying the concentration of pyrrole and the amount of hydrogen peroxide. Pyrrole blacks were obtained in these conditions (Section V) and other oxidation products were isolated as trace components.[38-41] Their structures were recently elucidated as **43**[42] and **44**.[43]

(43)   (44)

Compounds **43** and **44** are not intermediates in the formation of pyrrole blacks. During oxidation with hydrogen peroxide in pyridine, small quantities of succinimide and maleimide were identified by paper chromatography.[44]

Recently, the reaction has been reinvestigated, in buffers at various pH (1.1–9.0), showing that pyrrole blacks are formed only in strongly acidic medium. In neutral media, compound **45** is formed always together with its isomer (**46**) in the ratio 10:1.[45-47] Tautomerism between the isomers was studied by NMR.[48] It was proved that **45** and **46** are intermediates, giving **43** and **44** when reacted with pyrrole in acidic medium.[46]

The chemistry of these γ-lactams has been studied in acidic and basic medium, with interesting results.[49]

(45)   (46)

---

[38] A. Angeli and L. Alessandri, *Gazz. Chim. Ital.* **46**, II, 283 (1916).
[39] A. Angeli and C. Lutri, *Gazz. Chim. Ital.* **50**, I, 128 (1920).
[40] A. Pieroni, *Atti Accad. Naz. Lincei* [5] **30**, II, 316 (1921).
[41] A. Pieroni and A. Moggi, *Gazz. Chim. Ital.* **53**, I, 120 (1923).
[42] L. Chierici and G. P. Gardini, *Tetrahedron* **22**, 53 (1966).
[43] V. Bocchi, L. Chierici, and G. P. Gardini, *Tetrahedron* **23**, 737 (1967).
[44] R. Scarpati and C. Santacroce, *Rend. Accad. Sci. Fis. Mat. Naples* **28**, 27 (1961).
[45] V. Bocchi, L. Chierici, and G. P. Gardini, *Chim. Ind. (Milan)* **49**, 1346 (1967).
[46] V. Bocchi, L. Chierici, G. P. Gardini, and R. Mondelli, *Tetrahedron* **26**, 4073 (1970).
[47] G. P. Gardini, *L'Ateneo Parmense* **39**, Suppl. 5, 7 (1968).
[48] R. Mondelli, V. Bocchi, G. P. Gardini, and L. Chierici, *Org. Magn. Resonance* **3**, 7 (1971).
[49] V. Bocchi and G. P. Gardini, *Org. Prep. Proc.* **1**, 271 (1969); **2**, 65 (1970); *Tetrahedron Lett.*, 211 (1971); V. Bocchi, G. Casnati, and G. P. Gardini, *Tetrahedron Lett.*, 683 (1971).

2. Alkylpyrroles

Several alkylpyrroles have been reacted with hydrogen peroxide. The α,α'-unsubstituted compounds gave pyrrole blacks in acidic medium, while in neutral media (water or organic solvents) appreciable amounts of liquid or crystalline derivatives could be isolated, along with unidentified syrups.

N-Methylpyrrole reacted in diluted aqueous solution to give N-methyl-3-pyrrolin-2-one[50] (**10**), already isolated in small yields by autoxidation.[12]

2-Methylpyrrole in acetic acid gave a dark unidentified product,[51] while in organic solvents, depending on the relative concentrations of oxidant and pyrrole, solid products of peroxide nature were produced,[7] to which structures **47** and **48** have recently been assigned.[50]

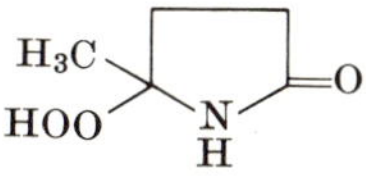

(**47**) 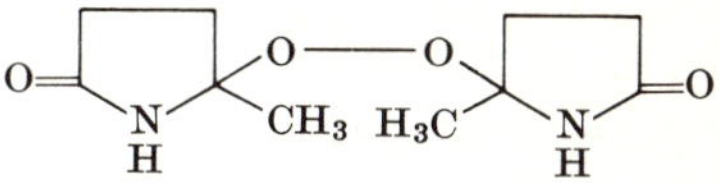   (**48**)

3-Methylpyrrole reacts with hydrogen peroxide in a neutral medium[50] to give compound **49** (>50%). Surprisingly, the isomer (**51**) was not present, suggesting a high specificity of the oxidizing species. Analogously, but less surprising, because of the electrophilic nature of oxidizing agents, 3-methyl-4-pyrrole carboxylic acid is oxidized to **50** (85%) without by-products.[52]

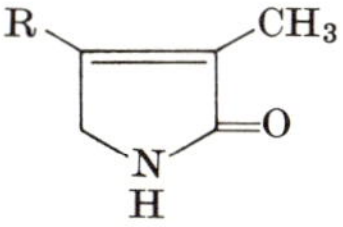

(**49**) R = H   (**51**)
(**50**) R = COOH

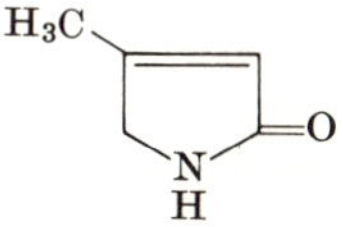

Attempts to oxidize 2,5-dimethylpyrrole were carried out by Angeli[38] and Pieroni,[51] but without significant results. It seems that at least one α position must be free, while β positions are only slightly reactive toward oxidation.

2,3-Dimethylpyrrole was oxidized in organic solvents at room temperature;[50] a Δ^4-pyrrolin-2-one (**52**) was isolated, along with quantities of polymeric oily products.

---

[50] G. P. Gardini and V. Bocchi, *Gazz. Chim. Ital.* **102**, 91 (1972).
[51] A. Pieroni and P. Veremeenco, *Gazz. Chim. Ital.* **56**, 455 (1926).
[52] R. Scarpati and V. Dovinola, *Rend. Accad. Sci. Fis. Mat., Naples* **27**, 503 (1960).

2,4-Dimethylpyrrole behaved like the 2-methyl derivative, giving peroxidic products (**53**) and (**54**) and an addition compound with unreacted pyrrole (**55**).[50]

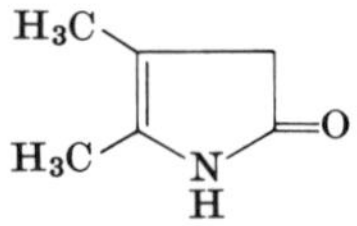

(**52**)

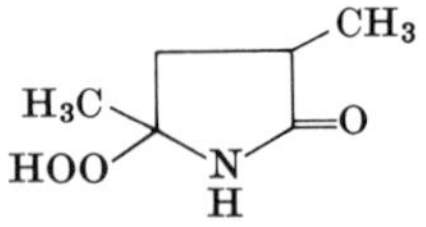

(**53**)

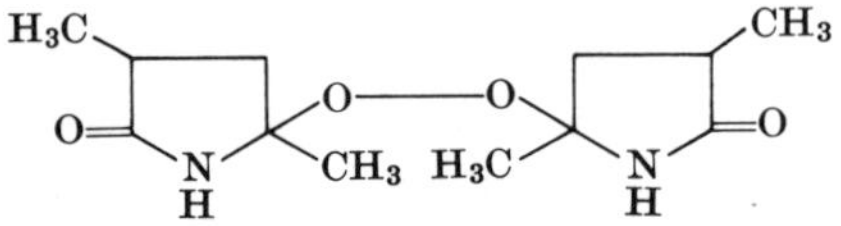

(**54**)

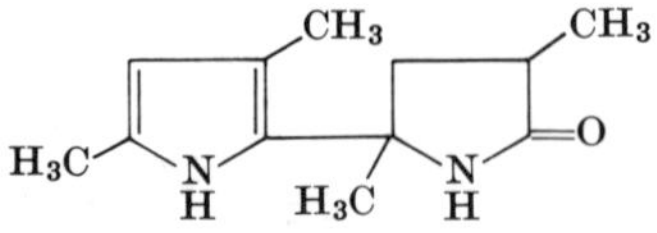

(**55**)

Compounds **53** and **54** were also isolated by Metzger and Fischer,[7] who gave incorrect structures. During the oxidation leading to **53–55**, small quantities of the autoxidation product (**9**) were also recovered.[50]

By oxidation in acetic acid of 2,4-dimethylpyrrole, a compound $C_{12}H_{20}N_2O_5$ (still unelucidated structure), which by hydrolysis gives $CH_3COCH_2CH(CH_3)COOH$, was obtained by Pieroni and Veremeenco.[51]

3,4-Dimethyl- and 2,3,4-trimethylpyrrole were oxidized in pyridine to give the corresponding Δ^3-pyrrolin-2-ones[9] in good yields, according to the method used initially by Fischer and co-workers[53] for cryptopyrrole (Table I).

From 3,4-dimethylpyrrole small quantities of 3,4-dimethylmaleimide were also obtained, while from 2,3,4-trimethyl derivatives, **6a** was isolated as a by-product.[9]

Tetramethyl- and pentamethylpyrroles were treated with hydrogen peroxide (70%) by Seebach.[8] With the latter, a peroxidic compound, to which structure **56** was assigned, was obtained in yields up to 50%.

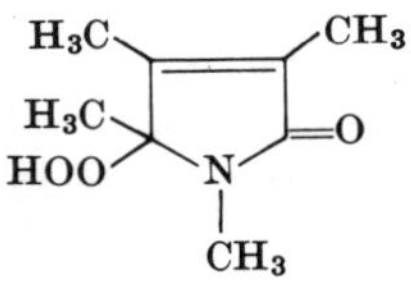

(**56**)

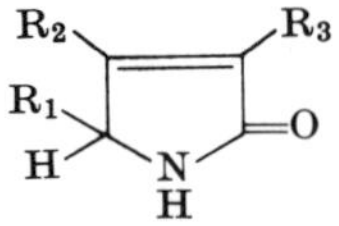

(**57a**) $R_1 = H$, $R_2 = (CH_2)_2COOH$, $R_3 = CH_3$
(**57b**) $R_1 = H$, $R_2 = CH_3$, $R_3 = (CH_2)_2COOH$
(**57c**) $R_1 = H$, $R_2 = CH_3$, $R_3 = C_2H_5$
(**57d**) $R_1 = H$, $R_2 = C_2H_5$, $R_3 = CH_3$
(**57e**) $R_1 = R_3 = CH_3$, $R_2 = C_2H_5$
(**57f**) $R_1 = R_2 = CH_3$, $R_3 = C_2H_5$

[53] H. Fischer, T. Yoshioka, and P. Hartmann, *Z. Physiol. Chem.* **212**, 146 (1932); *Chem. Abstr.* **27**, 99 (1933).

Sec. IV.A.] THE OXIDATION OF MONOCYCLIC PYRROLES 83

In Table I other examples of the oxidation of alkylpyrroles reported in literature are listed.

TABLE I

OXIDATION OF SOME ALKYLPYRROLES WITH HYDROGEN PEROXIDE

R_1	R_2	R_3	R_4	R_5	Products	Ref.
Ph	CH_3	H	H	CH_3	Brown precipitate	51
Isoamyl	CH_3	H	H	CH_3	Red precipitate	51
H	H	CH_3	$(CH_2)_2COOH$	H	(57a)+(57b)	54, 55
H	H	CH_3	C_2H_5	H	(57c)+(57d)	56
H	CH_3	C_2H_5	CH_3	H	(57e)+(6b)	9, 53, 57
H	CH_3	CH_3	C_2H_5	H	(57f)	57

In our opinion, the oxidations with hydrogen peroxide can be summarized as follows.

The products first formed are usually unsaturated γ-lactams ($\varDelta^3$- and/or $\varDelta^4$-pyrrolin-2-ones). If alkyl substituents are present at the 5-position, the structure $\varDelta^4$ is preferred.[50, 58] These can react further in neutral or weakly acidic media[46] to give addition products with unreacted pyrrole (see compounds **43**, **44**, and **55**) or with hydrogen peroxide (see compounds **47**, **48**, **53**, and **54**). On the other hand, substituents at the 3- and/or 4-positions, as well as the lack of alkyl substituents, tends to stabilize the $\varDelta^3$ structure.[50]

All the cases reported seem to agree with this hypothesis, except of the formation of compound **9**, which can be ascribed to oxidation by oxygen derived from decomposition of hydrogen peroxide. Also pentamethylpyrrole is atypical, since it is α,α'-disubstituted. The following mechanism has been proposed by Seebach[8] to explain the formation of the hydroperoxy derivative (**56**).

[54] H. Fischer and W. Lautsch, *Ann.* **528**, 265 (1937).
[55] H. Fischer and H. Plieninger, *Z. Physiol. Chem.* **274**, 231 (1942).
[56] H. Fischer and H. Reinecke, *Z. Physiol. Chem.* **259**, 83 (1939).
[57] H. Fischer and P. Hartmann, *Z. Physiol. Chem.* **226**, 116 (1934).
[58] A. R. Katritzky and J. M. Lagowski, *Advan. Heterocycl. Chem.* **2**, 12 (1963).

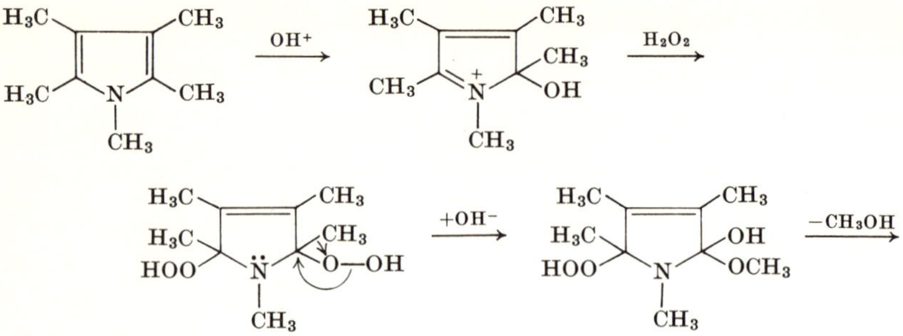

[reaction scheme (56)]

The high selectivity in the α positions which is illustrated in the formation of compounds 49 and 50 has been tentatively explained[50] by a homolytic hydroxylation, a process well known for phenols[59] and also for other heterocyclic compounds.[60] This hypothesis is supported by studies on oxidation of unsubstituted pyrrole with hydrogen peroxide in conditions suitable for the generation of hydroxy radicals: heating, light, or ferrous ions.[50]

3. *Arylpyrroles*

Arylpyrroles, substituted or unsubstituted at the nitrogen, undergo a much more drastic oxidation than alkyl derivatives by the action of hydrogen peroxide. In most cases opening of the heterocyclic ring occurs, both at the bond between the heteroatom and the α-carbon and between the α- and β-carbon atoms. From 2,5-diphenylpyrrole in acetic medium benzoic acid and acetophenone are formed.[51] From 2,3,5-triphenylpyrrole (58, R = H) the product is 59,[61] while from 58 (R = benzyl) the compounds 60 (R = benzyl) and 61 are produced.[62] Other pyrroles that were found to behave similarly are listed below.

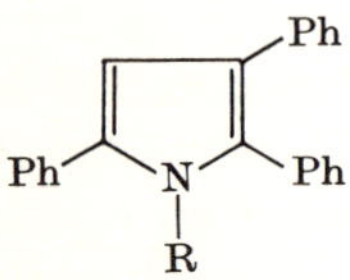

[structures (58), (59), (60), (61)]

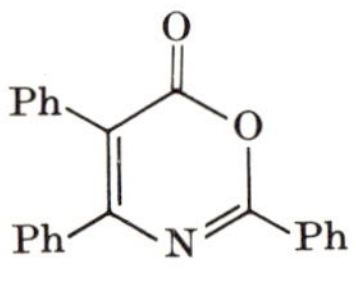

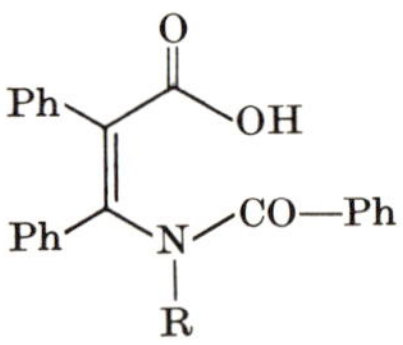

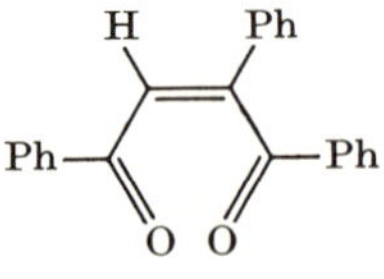

R = H, CH_3, CH_2–CH=CH_2, CH_2–CH_2–OH, CH_2–Ph, Ph

[59] C. R. E. Jefcoate, J. R. Lindsay Smith, and R. O. C. Norman, *J. Chem. Soc. B*, 1013 (1969) and references cited therein.
[60] R. O. C. Norman and G. K. Radda, *Advan. Heterocycl. Chem.* **2**, 163 (1963).
[61] V. Sprio, *Gazz. Chim. Ital.* **85**, 569 (1955).
[62] V. Sprio and P. Madonia, *Gazz. Chim. Ital.* **86**, 101 (1956).

In the oxidation of **62a** and **62b** by hydrogen peroxide, Lutz and Boykin[63] isolated the ketoamide (**63**) along with the epoxide of *cis*-dibenzoylstilbene (**64**).

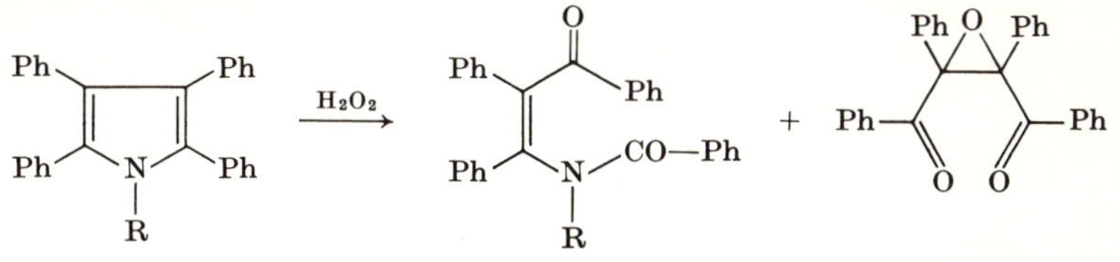

(**62a**) R = H
(**62b**) R = CH$_3$

To explain the formation of the compounds, which involves breaking of the C-α–C-β bond of the pyrrole nucleus, the authors proposed the following mechanism.

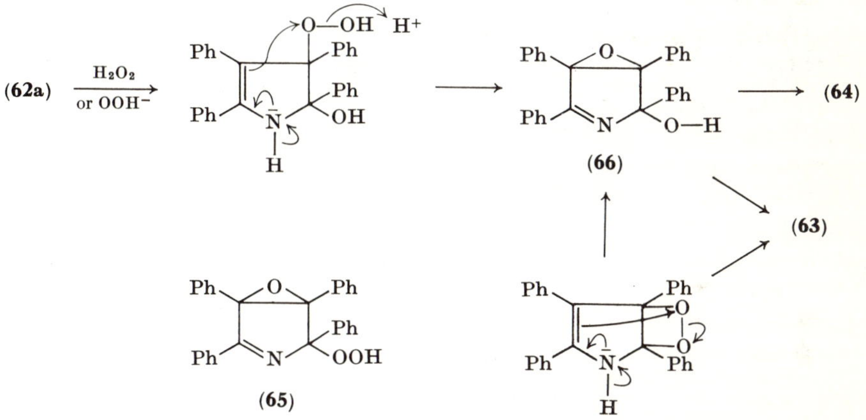

Very recently, by performing the oxidation of **62** in milder conditions and for a shorter time[64] it has been possible to detect the hydroperoxide (**65**), arising from the hydroxypyrrolenine epoxide (**66**), also identifiable in the reaction medium.

4. *Pyrrolealdehydes and Pyrroleketones*

Pyrrolealdehydes show unusual behavior with hydrogen peroxide. In most cases tested to date, the elimination of the formyl group was observed. Although one explanation could be that oxidation of the

---

[63] R. E. Lutz and D. W. Boykin Jr., *J. Org. Chem.* **32**, 1179 (1967); D. W. Boykin, M.S. Thesis, 1963; Ph.D. Dissertation, University of Virginia, 1965.
[64] G. Rio and M. Scholl, *Bull. Soc. Chim. Fr.*, 826 (1972).

formyl group to carboxyl occurs, it is not apparent why decarboxylation should occur, considering the stability of pyrrolecarboxylic acids toward oxidizing agents.[65]

Huni and Frank[66] found that, by oxidation of 2-formyl-3-ethyl-4-methylpyrrole a mixture of two pyrrolinones are obtained, although in small yields. By carrying out the oxidation of 2-formylpyrrole in water, succinimide was obtained.[44]

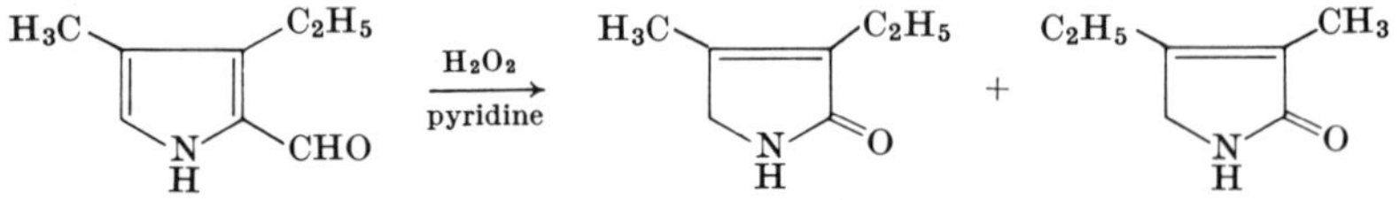

The pyrrole nucleus carrying a keto group is quite stable to oxidation; 2-acetyl- and 2-benzoylpyrroles do not react at all.[51]

5. Pyrrolecarboxylic Acids and Esters

Particular attention has been paid to the oxidation of pyrrolecarboxylic acids, since it is related to the studies about the oxidative degradation of several natural pigments,[67] which give mainly pyrrole acids among the identifiable products. Thus, the stability of the acids toward the oxidizing agents, and also hydrogen peroxide, has been tested.[65]

In the absence of alkyl groups in the ring, pyrrolecarboxylic acids are not attacked, thus confirming the stabilizing effect of electron-withdrawing substituents on the pyrrole ring. From 2,4-dimethyl-3-pyrrolecarboxylic acid and its ethyl ester the corresponding Δ^3-pyrrolinones were produced in high yields,[7] whereas with the tetrasubstituted pyrrole (67) the oxidation occurs in the side chain.[66]

(67)

B. Other Peroxides and Peracids

Pyrrole reacts with di-t-butyl peroxide at 150° to give the pyrrylpyrroline (68); N-methylpyrrole under the same conditions undergoes a hydrogen abstraction from the methyl group.[14] The carbon radical

[65] R. A. Nicolaus, R. Scarpati, and C. Forino, *Rend. Accad. Sci. Fis. Mat., Naples* **26**, 51 (1959); R. Scarpati and R. A. Nicolaus, *ibid*. **26**, 9 (1959).
[66] A. Hüni and F. Frank, *Z. Physiol. Chem.* **282**, 96 (1947).
[67] R. A. Nicolaus, "Melanins," p. 213. Hermann, Paris, 1968.

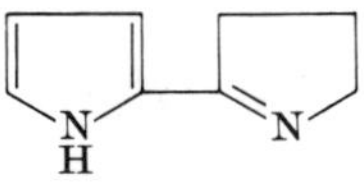

Sec. IV.C.]   THE OXIDATION OF MONOCYCLIC PYRROLES

(68) 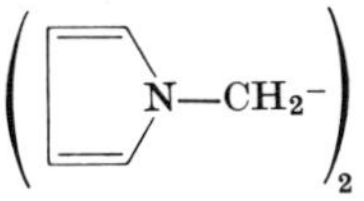   (69) 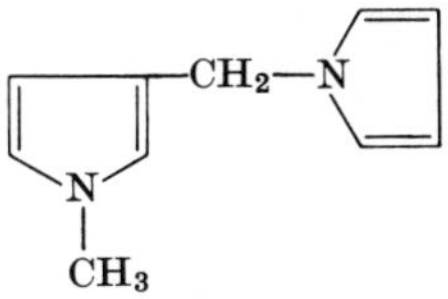   (70)

formed can either dimerize to **69** or β-substitute into another molecule of substrate, leading to **70**.[14] Since compounds **69** and **70** are produced only in low yields (less than 1%), it is not certain that this is the most important route in the oxidation process.

Benzoyl peroxide reacts with N-methylpyrrole giving N-methylsuccinimide;[68] N-methylcarbazole does not react at all.

A curious result was obtained by Kuhn and Kainer[69] on oxidation of tetraphenylpyrrole with perbenzoic acid in chloroform. The only product isolated in appreciable quantity was tetraphenylpyrazine (**71**), from which the starting pyrrole may be regenerated by reaction with metal and acids.

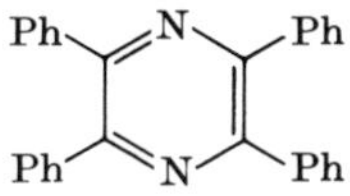

(71)

C. Lead Dioxide and Tetraacetate

Nearly all the studies on oxidation with lead dioxide and lead tetraacetate have been carried out with tetrasubstituted pyrroles. The reactions of tetraacetate in acetic acid with various 2-methyl-substituted pyrroles (**72**) have been investigated. When the reaction is carried out at low temperature with equimolecular quantities of oxidizing agent, 2-hydroxymethylpyrroles are obtained in high yields.[70]

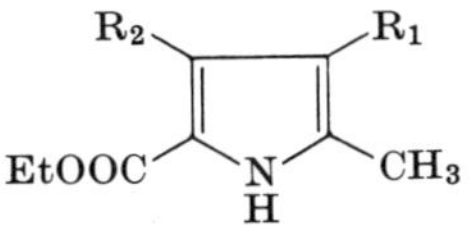

R_1	R_2
H	H
H	CH_3
CH_3	CH_3
CH_3	C_2H_5
C_2H_5	CH_3
C_3H_7	C_3H_7
$(CH_2)_2COOH$	CH_3

(72)

[68] I. Nabih and E. Helmy, *J. Pharm. Sci.* **56**, 649 (1967).
[69] R. Kuhn and H. Kainer, *Ann.* **578**, 226 (1952).
[70] W. Siedel and F. Winkler, *Ann.* **554**, 162 (1943).

At higher temperatures with two moles of oxidizing agent, further oxidation of the hydroxymethyl to the formyl group occurs.[70]

Tetraphenylpyrrole in chloroform affords tetraphenylpyrazine (**71**),[69] with lead tetraacetate; in acetic acid only 2-hydroxytetraphenylpyrroline (**32**) is formed, in high yield.[63] Tetraphenylpyrrole, and some of its phenyl-substituted derivatives, shaken with lead dioxide, generate a red color[71] which disappears with time or on exposure to air. A tetraphenylpyrryl radical (**73**) is formed, as demonstrated by ESR.[72] The radical is stable except in the presence of oxygen, when the color disappears, probably owing to the formation of peroxide (**74**).[16]

(**73**) → O_2 → (**74**)

Several substituted polyphenylpyrroles have been oxidized under the same conditions used by Kuhn and Kainer[71] generating compounds which showed colored solutions.[73] The decolorization rate was studied as a function of the substituents present on phenyl groups.

Recent papers on the oxidation of N-hydroxypyrroles with lead dioxide in benzene have produced a new series of stable pyrryl radicals (**75**).[74-76] In some cases it was possible to isolate them as deep blue crystalline substances, which could be reduced to the starting hydroxypyrrole with hydrogen and metals.[75]

(**75**)   (**76**)

---

[71] R. Kuhn and H. Kainer, *Biochim. Biophys. Acta* **12**, 325 (1953); *Chem. Abstr.* **48**, 11391 (1954).

[72] S. M. Blinder, M. L. Peller, N. W. Lord, L. C. Aamodt, and N. S. Ivanchukov, *J. Chem. Phys.* **36**, 540 (1962).

[73] B. S. Thanaseichuk, S. L. Vlasova, A. N. Sunin, and V. E. Gavrilov, *Zh. Org. Khim.* **5**, 144 (1969); *Chem. Abstr.* **70**, 87433 (1969).

[74] R. Ramasseul and A. Rassat, German Patent 1,917,048; *Chem. Abstr.* **72**, 3367 (1970).

[75] R. Ramasseul and A. Rassat, *Bull. Soc. Chim. Fr.*, 4330 (1970).

[76] E. G. Rozantsev, A. A. Medzhidov, and M. B. Neiman, *Bull. Acad. Sci. USSR, Div. Chem. Sci.*, 1729 (1963); *Chem. Abstr.* **60**, 2878 (1964).

The highest stabilities are shown by the radicals (75) and (76). These compounds have been extensively studied by ESR.[77] The structures of the leuco derivatives formed by further oxidation were clarified only later (see Section IV, F, 1).

If methyls or hydrogens are present in 76 instead of the two t-butyl groups, the radicals are quite stable in solution, but are not isolable.[75]

The oxidation of the pyrroleamine (77) gives 78 with lead tetraacetate,[78, 79] while the ether (79) in benzene with lead dioxide is transformed into the dimeric product (80).[17]

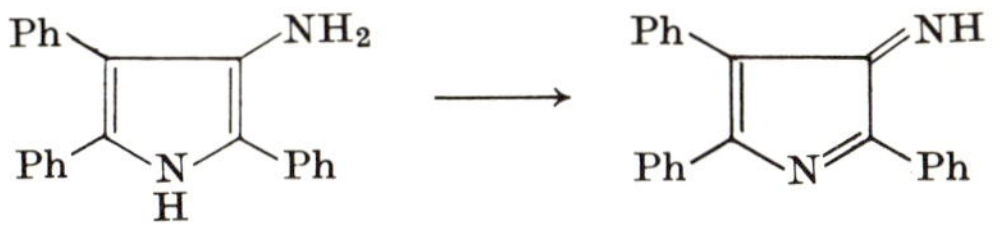

Further study of other pyrroles with free α positions would be interesting to verify analogies between Pb^{IV} and other oxidizing agents.

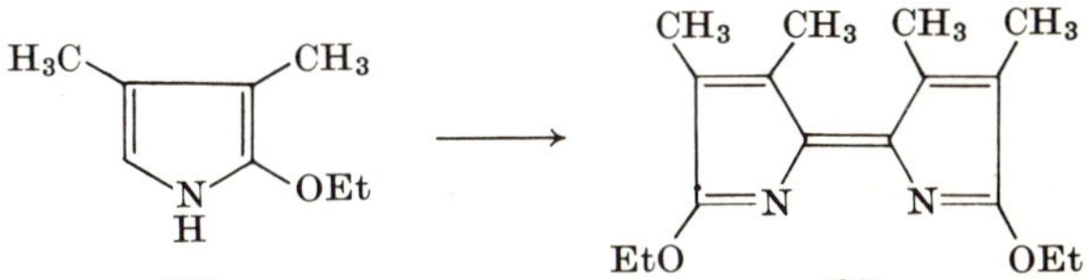

D. Cr^{VI} AND Mn^{VII} SALTS

1. *Potassium Permanganate*

Because of its high oxidizing power in neutral or basic medium, potassium permanganate has been employed for carbonyl- or carboxyl-substituted pyrroles. In a series of α-pyrrylketones of general formula (81) oxidation occurs at the side chain to give the corresponding pyrrylglyoxylic acids[80-82] which do not affect any 1-methyl group. The oxidation of the side chain reflects the strong stabilization of the pyrrole nucleus by electron-withdrawing substituents. Studies of a series of

---

[77] R. Ramasseul, A. Rassat, G. Rio, and M. Scholl, *Bull. Soc. Chim. Fr.*, 215 (1971).
[78] T. Ajello, *Gazz. Chim. Ital.* **68**, 602 (1938).
[79] T. Ajello and G. Sigillò, *Gazz. Chim. Ital.* **68**, 681 (1938).
[80] G. Ciamician and M. Dennstedt, *Chem. Ber.* **16**, 2348 (1883).
[81] G. De Varda, *Gazz. Chim. Ital.* **18**, 451 (1888).
[82] B. Oddo, *Chem. Ber.* **43**, 1012 (1910).

(81)   (82)

R = H, CH$_3$
R$_1$ = CH$_3$, CH$_2$Ph, C$_3$H$_7$

pyrrole carboxylic acids by Nicolaus and co-workers[65] verify their stability toward permanganate. The stability of the pyrrole nucleus increases according to the number of carboxylic groups present. For instance, pyrrole-2,3,4-tricarboxylic acid does not react with permanganate in basic solution, even with a free α position.

However, phenyl groups on the nucleus favors its degradation; from 1-acetyl-2,3,4-triphenylpyrrole, benzoic acid was obtained exclusively.[83]

The pyrrole (82) was oxidized to pyrrole-2,3,4,5-tetracarboxylic acid with dilute potassium permanganate.[84]

2. *Chromic Anhydride and Potassium Dichromate*

By the action of chromic anhydride in sulfuric acid, pyrrole[85] and alkylpyrroles[86] produce maleimide derivatives. For instance, from 2-methyl-4-ethylpyrrole, 3-ethylmaleimide (83) was obtained;[87, 88] 3-methyl-4-ethylpyrrole affords 3-methyl-4-ethylmaleimide in satisfactory yield.[89]

By oxidation with chromic anhydride in acetic acid, 3-amino-2,4,5-triphenylpyrrole leads to compound 84,[78, 90] while 1-acetyl-2,3,5-triphenylpyrrole undergoes bond opening between the pyrrole α- and β-carbons.[83] However, if the nitrogen in 2,3,5-triphenylpyrrole bears alkyl or aryl substituents, different behavior is observed in acetic acid —the pyrrole nucleus is still opened, but between the α-carbon and the heteroatom.[91]

Lutz and Boykin[63] heated tetraphenylpyrrole in acetic acid with chromic anhydride and obtained the hydroxytetraphenylpyrrolenine

[83] T. Ajello and S. Giambrone, *Boll. Sci. Fac. Chim. Ind., Bologna* **11**, 93 (1954).
[84] R. A. Nicolaus and G. Oriente, *Gazz. Chim. Ital.* **84**, 230 (1954).
[85] G. Plancher, *Atti Accad. Naz. Lincei* [5] **12**, I, 490 (1904).
[86] G. Plancher and F. Cattadori, *Atti Accad. Naz. Lincei* [5] **12**, I, 10 (1903).
[87] H. Fischer, G. Hummel, and A. Treibs, *Z. Physiol. Chem.* **185**, 42 (1929).
[88] W. Rüdiger and W. Klose, *Tetrahedron Lett.*, 5893 (1966).
[89] O. Piloty and J. Stock, *Ann.* **392**, 215 (1912).
[90] T. Ajello, *Gazz. Chim. Ital.* **69**, 453 (1939).
[91] V. Sprio and P. Madonia, *Gazz. Chim. Ital.* **85**, 965 (1955).

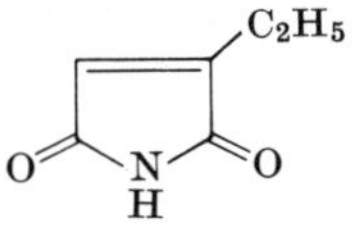

(32) (58%). This compound rearranged on refluxing in acetic acid to the pyrrolinone (15), which is obtained directly by oxidation in boiling acetic acid (85%). With the 1-methyltetraphenylpyrrole, ring opening occurred with formation of *cis*-dibenzoylstilbene epoxide (36), which is not obtainable by direct oxidation of *cis*-dibenzoylstilbene.[63]

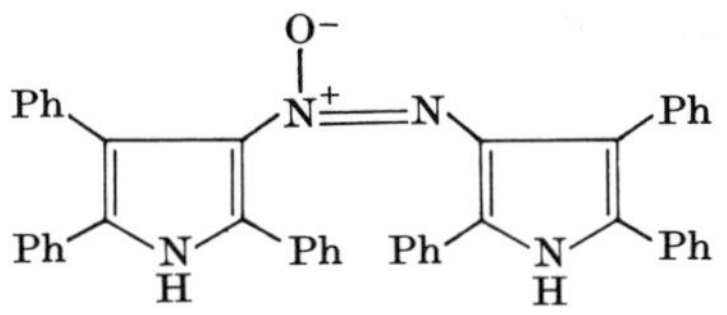

Hence, with Cr^{VI} as oxidizing agent ring opening frequently occurs, if the pyrrole nucleus bears phenyl substituents. It is not yet possible to draw firm generalizations because few cases have been studied and because pyrroles of similar structure sometimes behave differently.

E. Nitric and Nitrous Acids

As with other strong oxidizing agents, nitric acid converts alkylpyrroles into maleimides, removing any α-alkyl groups. In this way 2,3-dimethylpyrrole[92] and 2-methyl-3,4-diethylpyrrole[93] form the corresponding maleimides. Analogous behavior is observed with tetrabromo and tetrachloropyrroles, which lead to dihalomaleimides.[94]

A dimeric structure (85) has been proposed for the compound obtained from 3-bromo-2,4,5-triphenylpyrrole.[95]

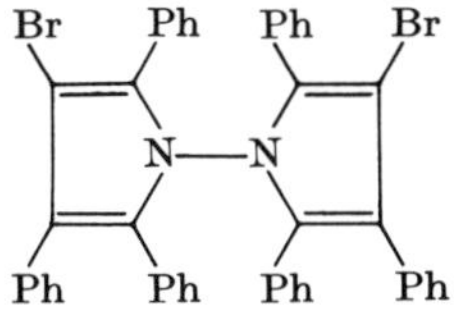

Several authors have recently observed that, whereas pyrrole[96] and acylpyrroles[97] react with nitric acid in acetic anhydride to give 2- or 3-nitro products with the same acid in dilute aqueous solution,

---

[92] O. Piloty, *Ann.* **377**, 314 (1910).
[93] H. Fischer and R. Baumler, *Ann.* **468**, 58 (1929).
[94] G. Ciamician, *Chem. Ber.* **37**, 4200 (1904).
[95] T. Ajello and S. Giambrone, *Gazz. Chim. Ital.* **81**, 276 (1951).
[96] A. R. Cooksey, K. J. Morgan, and D. P. Morrey, *Tetrahedron* **26**, 5101 (1970).
[97] K. J. Morgan and D. P. Morrey, *Tetrahedron* **27**, 245 (1971).

nitroacylpyrroles undergo mainly oxidation in the aliphatic part of the ketone, affording pyrrylglyoxylic acids (**86**)[98] or substituted furoxans (**87**).[97]

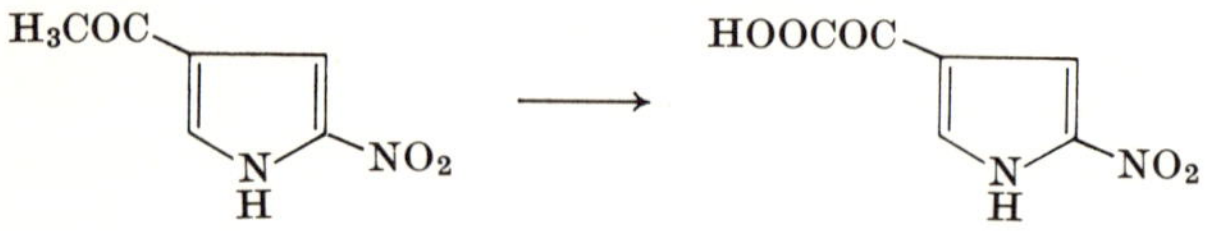

Nitrous acid reacts with unsubstituted pyrrole to give pyrrole blacks.[99] With tetraphenylpyrrole hydroxytetraphenylpyrrolenine (**32**) is formed at low temperature[100] while at higher temperatures ring opening occurs, leading to *cis*-dibenzoylstilbene.[63] The latter product is also formed with nitric acid. Some evidence was proposed for the following mechanism.[63]

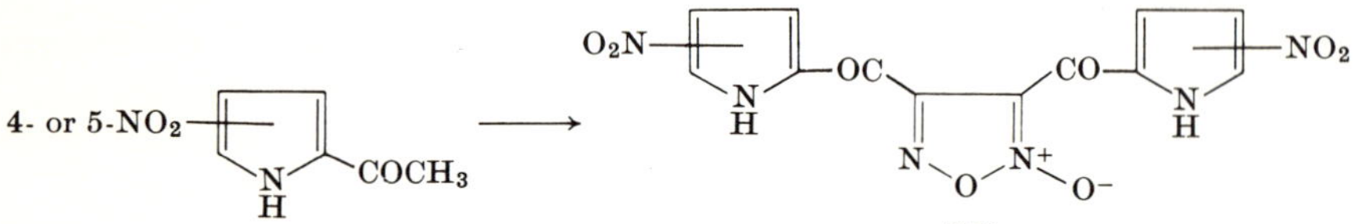

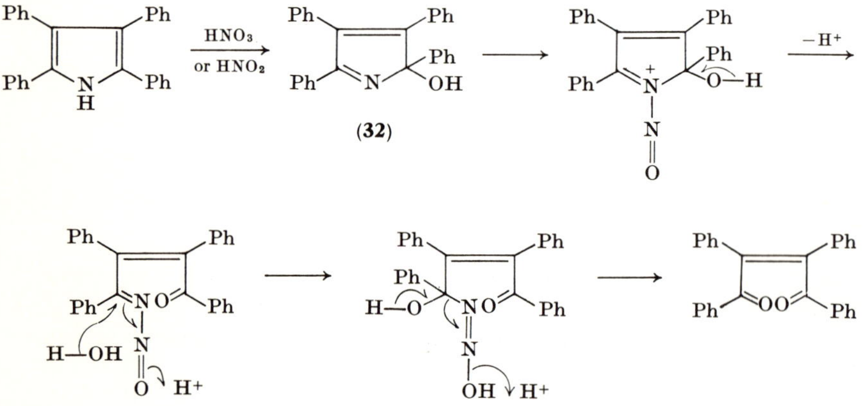

[98] I. J. Rinkes, *Rec. Trav. Chim., Pays Bas* **52**, 538 (1933).
[99] A. Angeli and G. Cusmano, *Gazz. Chim. Ital.* **47**, I, 207 (1917).
[100] R. Kuhn and H. Kainer, *Ann.* **578**, 227 (1952).

F. Miscellaneous

1. Ferric Salts

Ferric chloride oxidizes α,α'-unsubstituted pyrroles to pyrrole blacks;[101, 102] if substituted with electron-withdrawing groups, pyrroles withstand oxidation quite well.[102]

Recently, aqueous potassium ferricyanide has been used to oxidize a series of tetrasubstituted pyrroles of the type (88) to their dimers (89), which may be dissociated on heating into two pyrryl radicals (90).[103]

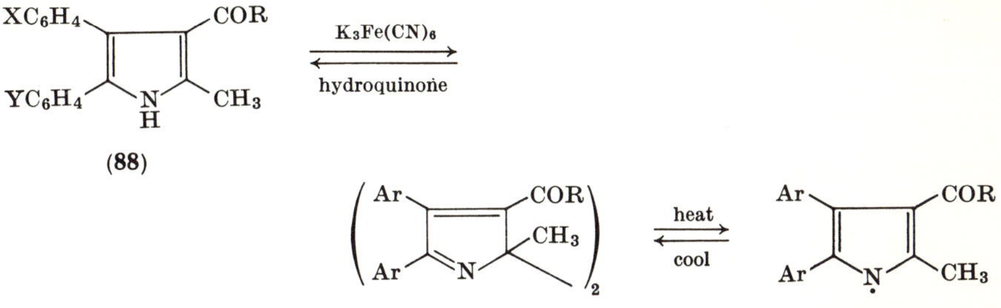

This work has clarified the structure of some earlier dimeric leuco-derivatives from pyrryl radicals.[71, 72]

2. Phosphorus Pentachloride

The title reagent, studied as an oxidizing agent for furans,[104] converts tetraphenylpyrrole into a phosphorus complex, which is hydrolyzed by water to the hydrochloride of 2-hydroxypyrrolenine (32).[63] The methyl ether (91) is obtained when the crude complex is treated with sodium methoxide.

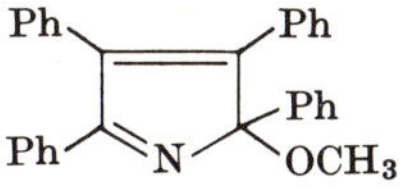

3. Sodium Periodate

Ethyl 3,4,5-trimethylpyrrole-2-carboxylate (92) treated with sodium periodate in dilute methanol gave (93) as the only identifiable product.[105]

---

[101] P. Pratesi, *Gazz. Chim. Ital.* **67**, 183 (1937).
[102] P. Pratesi, *Gazz. Chim. Ital.* **67**, 199 (1937).
[103] K. Tomita and N. Yoshida, *Tetrahedron Lett.*, 1169 (1971).
[104] R. E. Lutz and W. J. Welstead, *J. Amer. Chem. Soc.* **85**, 755 (1963).
[105] L. J. Dolby and R. M. Rodia, 159th *Amer. Chem. Soc. Nat. Meeting, Abstr. Papers*, 1970, ORGN 16; R. M. Rodia, Ph.D. Thesis, Oregon State University.

Similarly, 2,3,4,5-tetramethylpyrrole gave 2-formyl-3,4,5-trimethylpyrrole, but 2,3,4-trimethylpyrrole afforded the pyrrolinone (**94**).

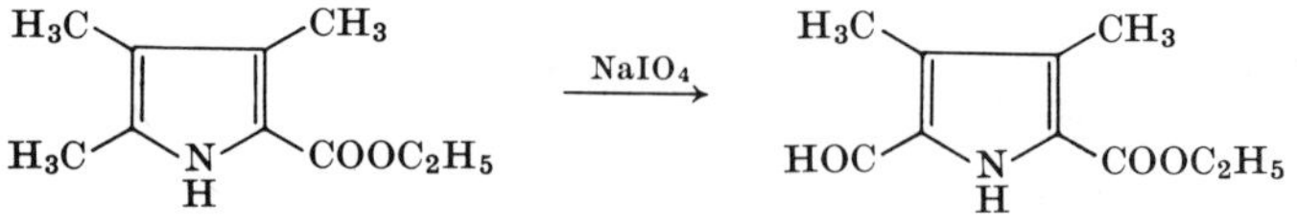

Tetraphenylpyrrole also reacts with periodate under the same conditions, to give the pyrrolinone (**28**).[105]

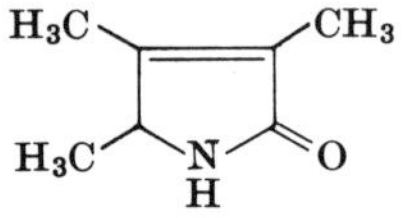

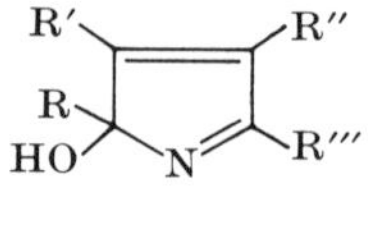

On the basis of these data and previous studies of an analogous oxidation of indoles,[106] a mechanism was proposed,[105] via the hydroxypyrrolenine (**95**)

4. Ozone

First attempts to ozonize pyrrole and alkylpyrroles gave unidentifiable tarry products.[107] Wibaut and co-workers,[108, 109] obtained ozonization products from various methylpyrroles which cannot be accounted for by cleavage of the double bonds in the normal structures, and were explained by Wibaut on the basis of ionic structures of type (**96**). This hypothesis was questioned by Bailey and Colomb[110] on the

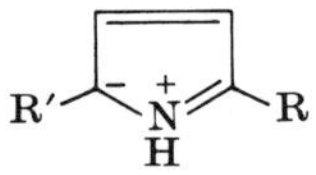

basis of their experiments on 2,5-diphenylfurans. Later work by Wibaut[111] reasserted the importance of **96**.

---

[106] L. J. Dolby and R. M. Rodia, *J. Org. Chem.* **35**, 1493 (1970).
[107] M. Freri, *Gazz. Chim. Ital.* **62**, 600 (1932); **63**, 281 (1933).
[108] J. P. Wibaut and A. R. Guljé, *Kon. Ned. Akad. Wetensh., Proc. Ser. B* **54**, 330 (1951); *Chem. Abstr.* **47**, 6934 (1953).
[109] J. P. Wibaut, *Adv. Chem. Ser.* **21**, 153 (1959); *Chem. Abstr.* **54**, 1511 (1960).
[110] P. S. Bailey and H. O. Colomb, *J. Amer. Chem. Soc.* **79**, 4238 (1957).
[111] J. P. Wibaut, *Kon. Ned. Akad. Wetensh., Proc. Ser. B* **74**, 38 (1965); *Chem. Abstr.* **63**, 11306 (1965).

5. Anodic Oxidation, Quinones, and Diazonium Salts

The only products obtained by carrying out oxidation of pyrroles with the above reagents were pyrrole blacks (see Section V).

V. Pyrrole Blacks

Angeli[38, 112] first observed that pyrrole with hydrogen peroxide in acetic acid developed a dark color and later a black precipitate. This product, which contains oxygen, was named "pyrrole black". Its physical characteristics were very similar to those of natural melanins and this fact stimulated research on the structure of this product, considered an interesting model for natural pigments.

Pyrrole blacks are obtained by other oxidizing reagents, which were tested mainly on unsubstituted pyrrole. Among these are nitrous acid,[99] lead dioxide,[112] ferric chloride,[101, 102] quinones,[39, 51, 113–115] diazonium salts,[116] and ozone.[107]

The black powder obtained from pyrrole and hydrogen peroxide in acetic acid is insoluble in all organic solvents and only slightly soluble in aqueous alkali. Because of this, spectroscopic techniques found little or no application to the structural study. The only useful experiments which have been carried out so far to elucidate the structure of pyrrole blacks are (a) the analysis, which suggests a composition approximately $C_{4.0-4.5}H_{3.0-4.5}N_1O_{1.0-1.5}$; (b) a value of 800–1,000 was proposed for the molecular weight; (c) the oxidative degradation with potassium permanganate or dichromate, or hydrogen peroxide, yielding small quantities of various pyrrolecarboxylic acids. Among these pyrrole-2,5-dicarboxylic acid was the most abundant, thus suggesting that polymerization must take place in the α positions. In fact, pyrroles substituted in the 2- and/or 5-positions do not afford black substances on oxidation.[38, 51] Indeed, pyrrole-2,3-dicarboxylic and pyrrole-2,3,5-tricarboxylic acids are prevalent in the degradation of natural pigments.[67]

Mention might be made of some recent work, in which the reaction between pyrrole and hydrogen peroxide in acetic acid was studied by

---

[112] A. Angeli, *Gazz. Chim. Ital.* **46**, II, 279 (1916); **48**, II, 21 (1918).
[113] A. Angeli and C. Lutri, *Gazz. Chim. Ital.* **51**, I, 31 (1921).
[114] R. Möhlau and A. Redlich, *Chem. Ber.* **44**, 3605 (1911).
[115] P. Pratesi, *Gazz. Chim. Ital.* **66**, 215 (1936).
[116] A. Quilico and M. Freri, *Atti Accad. Naz. Lincei* [6] **11**, 296, 409 (1930); [6] **13**, 282, 377 (1931).

the electron spin resonance (ESR) technique.[117] An ESR signal appeared some time after the reagents were mixed together; until then a signal could be registered only by freezing the reaction mixture. This fact was connected with the short lifetime of the first-formed radicals. Later, a signal appeared from the liquid, and after reaching a maximum in intensity, decreased with formation of a black precipitate. No hyperfine structure was found for the signal, suggesting lack of interaction between the nitrogen atoms and free electrons. A similar signal was found in the precipitate, due to trapped electrons.[117]

Another ESR study[118] was concerned with the pyrrole blacks obtained by anodic oxidation of pyrrole in dilute sulfuric acid.[119] This substance showed a high conductivity, and an ESR signal, again due to trapped electrons and free from hyperfine structure, but with an exceptionally low g factor (2.0026 ± 0.0001).

Black products were also obtained from pyrrole in the presence of sulfur dioxide.[120]

VI. Conclusions

The literature of the oxidation of pyrroles does not yet provide a clear general picture of the behavior of the pyrrole nucleus toward oxygen and other oxidizing agents.

The only generalities that we discern are: (a) pyrroles being very activated toward electrophiles in the α positions, the attack by oxidants proceeds exclusively in those positions, or in the side chain if both α positions are occupied; (b) in strongly acidic media pyrrole blacks are the main products from 2,5-unsubstituted substrates, while brown tars and syrups are formed from other pyrroles; (c) in weakly acidic or basic media, in water or organic solvents, simple oxidation products (unsaturated γ-lactams, maleimides, and succinimides) are obtained from alkylpyrroles, while aryl derivatives often undergo ring opening; and (d) electron-withdrawing groups on the pyrrole ring show a strong stabilizing effect toward oxidation.

The oxidation mechanisms need much more research to be well defined; at present only the photo-oxidation mechanism has been studied in any detail.

[117] G. Dascola, C. Giori, V. Varacca, and L. Chierici, *C.R. Acad. Sci., Ser. C* **262** 1617 (1966).
[118] A. Dall'Olio, G. Dascola, V. Varacca, and V. Bocchi, *C.R. Acad. Sci., Ser. C* **267**, 433 (1968).
[119] L. Chierici, G. C. Artusi, and V. Bocchi, *Ann. Chim.* (*Rome*) **58**, 903 (1968).
[120] G. Illari, *Gazz. Chim. Ital.* **64**, 883 (1934); **65**, 453 (1935).

Note Added in Proof

Since the completion of this review, a number of papers on the subject have been published.

Lightner and co-workers studied the behavior of various simple alkylpyrroles by irradiation in methanolic solution in the presence of a photosensitizer and a stream of oxygen.

From 2-methylpyrrole, the hydroxylactam (**19**, $R = R_1 = R_2 = R_3 = H$; $R_4 = Me$) was obtained along with corresponding methyl ether. Similarly, 2,4-dimethylpyrrole gave the hydroxylactam and the methyl ether.[121]

Quite surprisingly, in addition to forseeable products, photo-oxidation of 2,5-dimethyl[122] and 2,5-dimethyl-3,4-diethyl-pyrrole[123] afforded lactams like (**19**) with $R_4 = CH_2OCH_3$, the formation of which is not easily rationalized. It was assumed that all these products can be derived from an endo-peroxide (**20**).

Particularly interesting was the case of 3-methylpyrrole: in addition to the expected hydroxy- and methoxylactams, compounds **97** and **98** were found.[124] This fact was presented as the first evidence for the formation of a cyclic dioxetane intermediate (**99**) arising from 1,2-cycloaddition of singlet oxygen to the enamine-like 2,3 bond of the pyrrole.

(97)   (98)   (99)   (100)

High yields of hydroperoxypyrrolenines like (**30**) (Section III) were obtained by photo-oxidation of hindered pyrroles, such as 2,5-di- and 2,3,5-tri-*t*-butyl-pyrroles.[125] The autoxidation of 2,4-dimethyl-pyrrol-5-ethyluretane yielded 50% of the hydroxyderivative (**100**).[126] A series of tetra-arylpyrroles carrying substituents on the 2 and 5 aryl groups were oxidized with lead dioxide and the stability of corresponding free

---

[121] D. A. Lightner and L. K. Low, *J. Heterocycl. Chem.* **9**, 167 (1972).
[122] L. K. Low and D. A. Lightner, *Chem. Commun.*, 116 (1972).
[123] D. A. Lightner and G. B. Quistad, *Angew. Chem. (Int.)* **11**, 215 (1972).
[124] D. A. Lightner and L. K. Low, *Chem. Commun.*, 625 (1972).
[125] R. Ramasseul and A. Rassat, *Tetrahedron Lett.*, 1337 (1972); R. Ramasseul, Thèse, Université de Grenoble, 1968.
[126] A. Treibs and D. Grimm, *Ann.* **752**, 44 (1971).

radicals related to Hammett's σ constants.[127] Vapor phase catalytic oxidation of pyrroles at 330–350°, using V–Mo–P catalyst was investigated.[128]

Finally, it was found that the biological inactivation of pyrrolnitrin (**101**) *in vivo* by rat liver microsomes is an oxidative process leading to four products (**102–105**);[129]

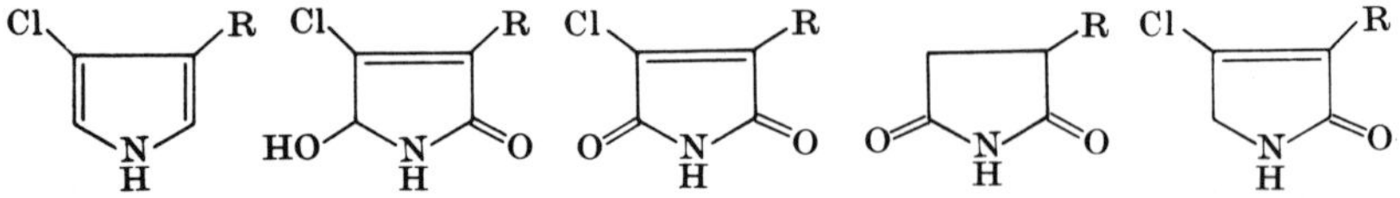

(101) (102) (103) (104) (105)

R = 3-chloro-4-nitrophenyl-

102 and **103** are also formed by oxidation of **101** with chromic anhydride in acetic acid, while **104** and **105** could be obtained by oxidation of **101** with m-chloroperbenzoic acid.

---

[127] B. S. Tanaseichuk, S. L. Vlasova, and E. N. Morozov, *Zh. Org. Khim.* **7**, 1264 (1971); *Chem. Abstr.* **75**, 110127 (1971).

[128] V. A. Slavinskaya, D. Kreile, D. Eglite, O. I. Starkova, and V. Sates, *Latv. PSR Zinat. Akad. Vestis, Khim. Ser.* 533 and 630 (1971); *Chem. Abstr.* **76**, 24405 and 24414 (1972).

[129] P. J. Murphy and T. L. Williams, *J. Med. Chem.* **15**, 137 (1972).

The Chemistry of 4-Oxy- and 4-Keto-1,2,3,4-tetrahydroisoquinolines

J. M. BOBBITT

*Department of Chemistry, University of Connecticut
Storrs, Connecticut*

I. Introduction	99
II. Synthesis of 4-Oxy-1,2,3,4-tetrahydroisoquinolines	103
A. Synthesis via Benzylaminoacetals	104
B. Synthesis via 4-Keto-1,2,3,4-tetrahydroisoquinolines	111
C. Synthesis via β-Hydroxy-β-phenylethylamines	114
D. Miscellaneous Syntheses	115
III. Reactions of 4-Oxy-1,2,3,4-tetrahydroisoquinolines	116
A. Hydrogenolysis to 1,2,3,4-Tetrahydroisoquinolines	116
B. Aromatization to Isoquinolines	117
C. Dehydration to 1,2-Dihydroisoquinolines	118
D. Nucleophilic Displacement at C-4	119
IV. Naturally Occurring 4-Oxy-1,2,3,4-tetrahydroisoquinolines	122
V. The State of the Art of Isoquinoline Synthesis	123
A. Substitution in the Benzene Ring	124
B. Substitution at the 1-Position	124
C. Substitution on Nitrogen	126
D. Substitution at the 3-Position	127
E. Substitution at the 4-Position	127
VI. Tables of Specific Compounds	128

I. Introduction

The Pomeranz–Fritsch synthesis [Eqs. (1) and (2)][1] is the only isoquinoline synthesis involving a simple two-step sequence from common starting materials. Furthermore, it is one of the few methods which can be used to prepare isoquinolines substituted in the 7- and 8-positions. The first step, Schiff base formation [Eq. (1)], takes place readily, but the ring closure [Eq. (2)] is difficult. The yields vary markedly with the concentration of H_2SO_4 and are generally low. Frequently the reaction fails completely. Most of the work described in this chapter was undertaken to circumvent these problems and to realize the potential promise of the synthesis.

[1] W. J. Gensler, *Org. React.* **6**, 191 (1951).

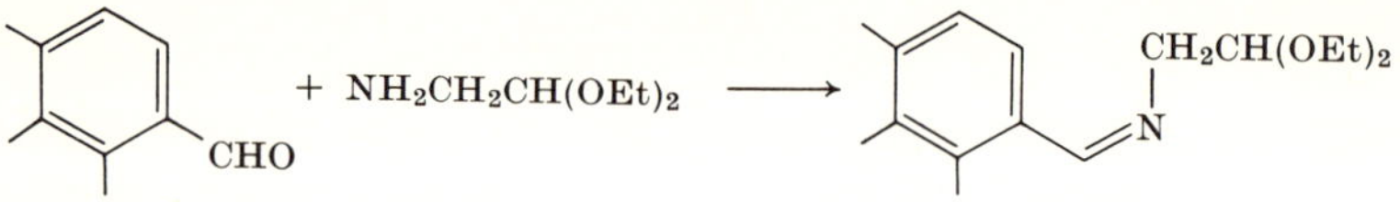

It has long been apparent that when the acid concentration is too low, the Schiff base is hydrolyzed back to its components and, when it is too high, extensive decomposition results. In order to obviate the decomposition problem, numerous acidic reagents have been tried. Some of these are polyphosphoric acid–phosphorus oxychloride,[2] polyphosphoric acid,[3] and boron trifluoride–trifluoroacetic anhydride.[3] Some improvement has been observed, but many yields are still poor.

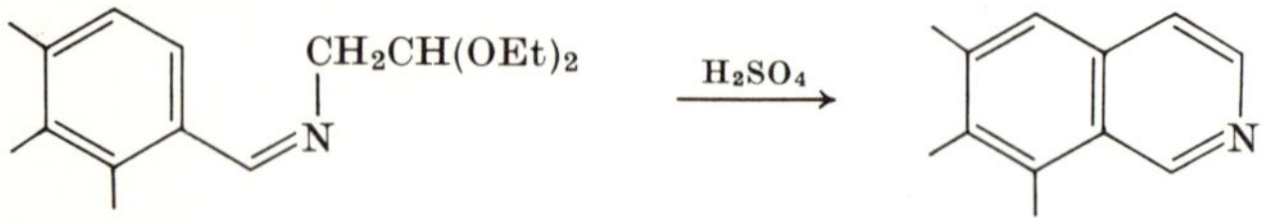

Attempts to circumvent the problem by reduction of the Schiff base to prevent hydrolysis and to ring close the benzylaminoacetals were made long ago by Fischer.[4] He used fuming H_2SO_4 which served as the cyclizing agent as well as an oxidizing agent to yield an aromatic product [Eq. (3)]. Others used As_2O_5 as an oxidizing agent,[5,6] but the results were poor. This work is well summarized by Frank and Purves,[7] who attempted to make (benzylamino)acetaldehydes and cyclize them, again with limited success.

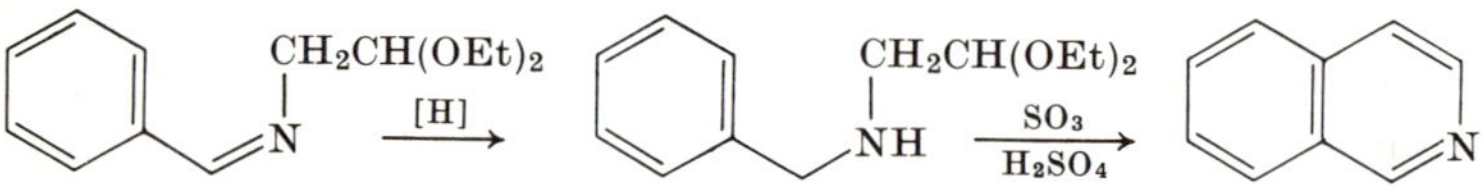

In 1933, Young and Robinson[8] treated piperonylmethylaminoacetal (1) with mild acids (phosphorus oxychloride and 10% aqueous HCl in different experiments) and isolated several products in low yields. Two of these were isolated as picrates and tentatively assigned structures 2 and 3. This is the first example of the synthesis of compounds of the type described in this chapter. Treatment of 1 with hot 10% acid yielded

---

[2] F. D. Popp and W. E. McEwen, *J. Amer. Chem. Soc.* **79**, 3773 (1957).
[3] M. J. Bevis, E. J. Forbes, N. N. Naik, and B. C. Uff, *Tetrahedron* **25**, 1585 (1969); **27**, 1253 (1971).
[4] E. Fischer, *Ber. Deut. Chem. Ges.* **26**, 764 (1893).
[5] L. Rügheimer and P. Schön, *Ber. Deut. Chem. Ges.* **42**, 2374 (1909).
[6] R. Forsyth, C. I. Kelly, and F. L. Pyman, *J. Chem. Soc.* **127**, 1659 (1925).
[7] A. W. Frank and C. B. Purves, *Can. J. Chem.* **33**, 365 (1955).
[8] P. C. Young and R. Robinson, *J. Chem. Soc.*, 275 (1933).

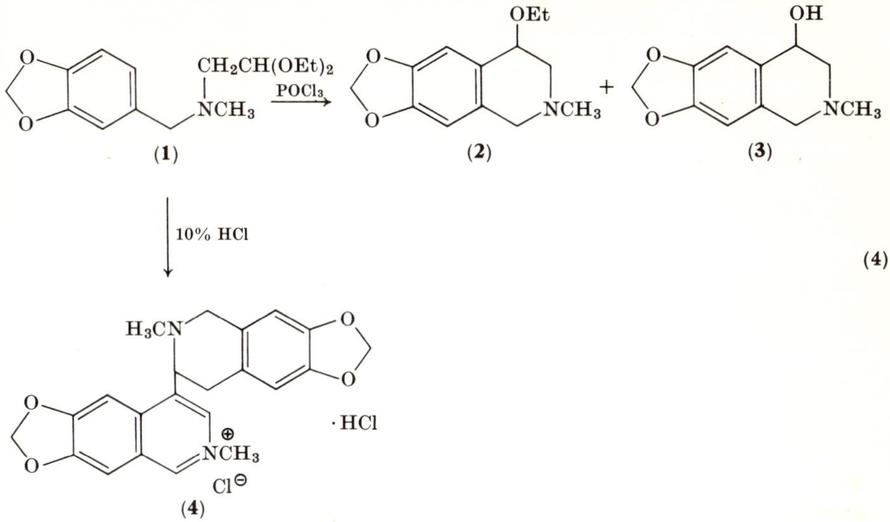

a salt which was clearly not the hydrochloride of **2**. In view of its solubility properties and recent work from this laboratory[9] the substance may be a dimer such as **4** or its 1,2-dihydro derivative.[10]

In 1955, Guthrie, Frank, and Purves[11] reported that the treatment of **5** with concentrated HCl at room temperature probably gave the pyrrolidine (**6**). Compound **6** was not isolated and the experiment is not described. Treatment of **5** with hot acid was thought to yield **7** which was characterized. The structure of **7** was subsequently corrected to an *iso*pavine skeleton (**25**, in Section III,D). In light of the content of this chapter,[49] **6** should probably be corrected to **8**, the 4-hydroxytetrahydroisoquinoline. When the acid cyclization was carried out in the presence of ethyl mercaptan a compound was isolated, characterized, and assigned structure **9**. However, a corrected structure (**10**) could be well explained as resulting from a nucleophilic attack at the 4-hydroxy group of **8** as described below (Section III, D). This work, so interpreted, may then represent one of the earliest syntheses of 4-hydroxytetrahydroisoquinolines and the first example of a nucleophilic attack at C-4.

The first really successful ring closures of the benzylaminoacetals were brought about with BF_3–etherate [Eq. (6)] yielding 4-ethoxy-

---

[9] J. M. Bobbitt, J. M. Kiely, K. L. Khanna, and R. Ebermann, *J. Org. Chem.* **30**, 2247 (1965).

[10] Compound **4** would have 6.2% N and 15.7% Cl. The reported values are 6.1% N and 14.8% Cl.

[11] D. A. Guthrie, A. W. Frank, and C. B. Purves, *Can. J. Chem.* **33**, 729 (1955).

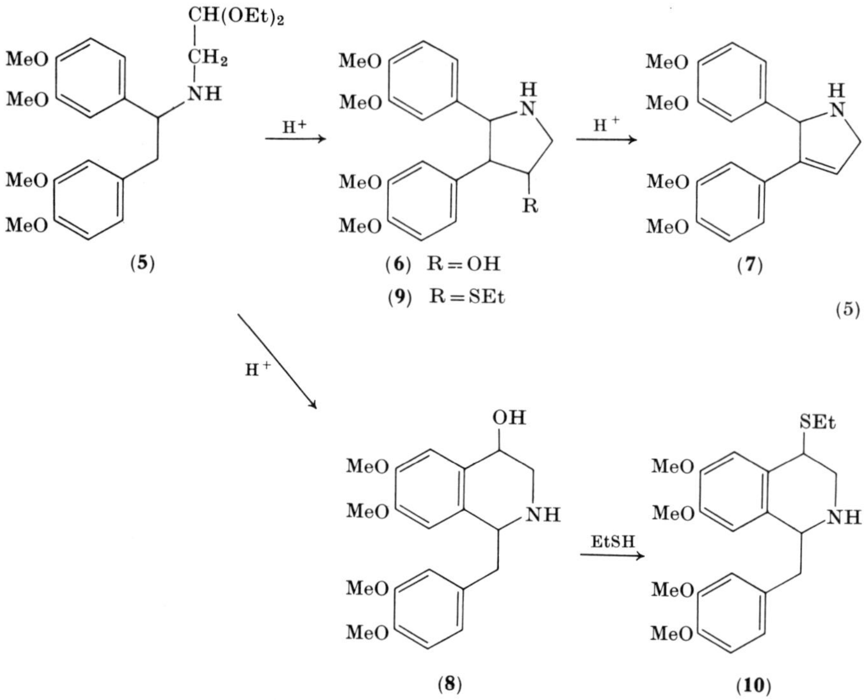

tetrahydroisoquinolines. In an outstanding series of papers,[12-18] Vinot and Quelet developed the reaction into a general method for tetrahydroisoquinolines of various substitution patterns. The method suffers from one major flaw: it does not work well when the benzene ring is oxygenated,[17] probably owing to extensive complex formation with the BF_3. This work will be described more completely below (Section II, A).

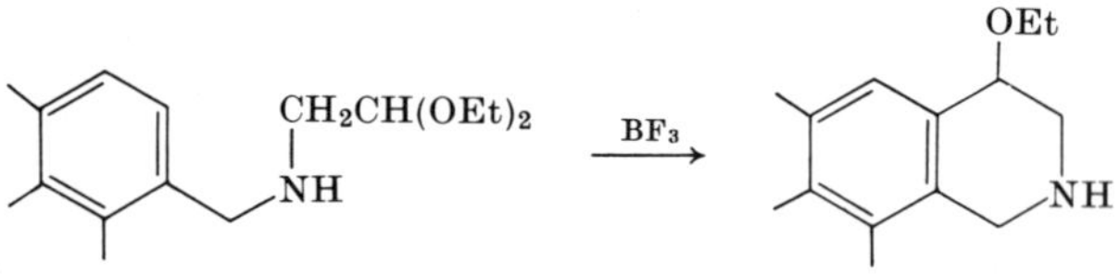

[12] R. Quelet, J. Hoch, and N. Vinot, *C. R. Acad. Sci.* **241**, 1583 (1955).
[13] R. Quelet, J. Hoch, C. Borgel, M. Mansouri, R. Pineau, E. Tchiroukine, and N. Vinot, *Bull. Soc. Chim. Fr.* 26 (1956).
[14] R. Quelet and N. Vinot, *C. R. Acad. Sci.* **244**, 909 (1957).
[15] N. Vinot, *Ann. Chim. (Paris)* **3**, 461 (1958).
[16] N. Vinot and R. Quelet, *C. R. Acad. Sci.* **246**, 1712 (1958).
[17] N. Vinot and R. Quelet, *Bull. Soc. Chim. Fr.*, 1164 (1959).
[18] N. Vinot, *Bull. Soc. Chim. Fr.*, 617 (1960).

Ring closures of benzylaminoacetals with dilute HCl have been extensively investigated in this laboratory.[9,19-26] The products are 4-hydroxytetrahydroisoquinolines [Eq. (7)], and the reaction has been

$$\text{(structure with CH(OEt)}_2\text{, CH}_2\text{, N—, (O) substituents)} \xrightarrow{\text{6 N HCl}} \text{(4-hydroxytetrahydroisoquinoline with OH, NH, (O) substituents)} \quad (7)$$

developed along the general lines pioneered by Vinot and Quelet. In this case, the reaction fails *unless* there is an oxygen ortho or para to the point of ring closure leading to a 5- or a 7-substituent. Thus, the two reactions [Eqs. (6) and (7)] complement one another.

Since the 4-oxytetrahydroisoquinolines are key compounds in both ring closures, they were chosen as the major focal point of this chapter. Although they are stable, isolable compounds with a unique structure, they are probably more interesting as intermediates leading to other isoquinoline derivatives. The 4-keto-1,2,3,4-tetrahydroisoquinolines[27] are also important, both in their own right as synthetic intermediates and in their facile conversion into the 4-oxytetrahydroisoquinolines. Thus, the ketones were adopted as a second focal point and are discussed in detail in Section II,B.

II. Synthesis of 4-Oxy-1,2,3,4-tetrahydroisoquinolines

The 4-oxytetrahydroisoquinolines have been prepared by three major routes and a few minor ones. The major routes involve the ring closures of benzylaminoacetals [Eqs. (6) and (7)], the reduction of

---

[19] J. M. Bobbitt, K. L. Khanna, and J. M. Kiely, *Chem. Ind. (London)*, 1950 (1964).
[20] J. M. Bobbitt, D. P. Winter, and J. M. Kiely, *J. Org. Chem.* **30**, 2459 (1965).
[21] J. M. Bobbitt, D. N. Roy, A. Marchand, and C. W. Allen, *J. Org. Chem.* **32**, 2225 (1967).
[22] J. M. Bobbitt and J. C. Sih, *J. Org. Chem.* **33**, 856 (1968).
[23] J. M. Bobbitt and T. E. Moore, *J. Org. Chem.* **33**, 2958 (1968).
[24] J. M. Bobbitt and C. P. Dutta, *J. Org. Chem.* **34**, 2001 (1969).
[25] J. M. Bobbitt, A. S. Steinfeld, K. H. Weisgraber, and S. Dutta, *J. Org. Chem.* **34**, 2478 (1969).
[26] J. M. Bobbitt and S. Shibuya, *J. Org. Chem.* **35**, 1181 (1970).
[27] The 4-keto-1,2,3,4-tetrahydroisoquinolines are more systematically called 2,3-dihydro-4(1H)-isoquinolones, but the keto nomenclature would seem to be more descriptive of their properties.

4-ketotetrahydroisoquinolines [Eqs. (8) and (9)], and the condensation of an aldehyde or ketone with a β-hydroxy-β-phenylethylamine [Eq. (10)].

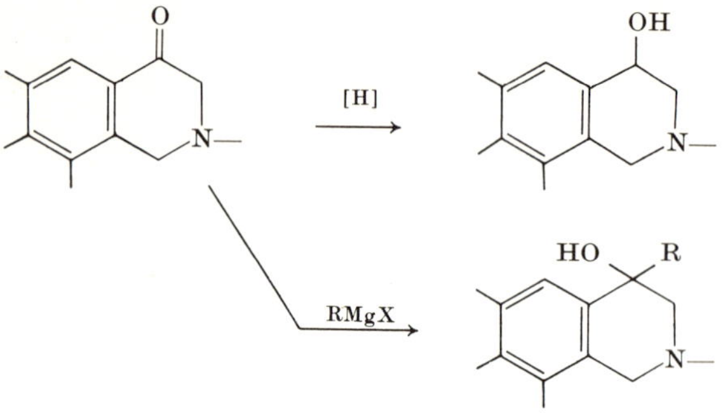

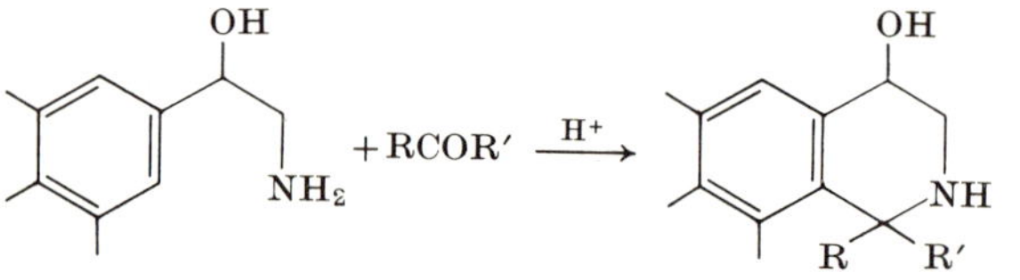

A. Synthesis via Benzylaminoacetals

1. *Synthesis of Benzylaminoacetals*

   a. *Via the Mannich Reaction.* The condensation of phenols and certain heterocyclic systems with formaldehyde and amines to give benzylamines is one form of the well-known Mannich condensation.[28] When the amine is aminoacetal[29] or methylaminoacetal, the resulting benzylaminoacetals can be cyclized to tetrahydroisoquinolines [Eq. (11)].[24, 26] Neither the acetals nor the 4-hydroxy intermediates were isolated. The major disadvantage of this route is that the Mannich condensation can take place ortho or para to a phenol giving rise to isomer mixtures. Thus, guiacol yields both the 8-hydroxy-7-methoxy- and the 6-hydroxy-7-methoxytetrahydroisoquinolines. As it happens, these are easily separable. This reaction should be applicable to heterocyclic systems.

   b. *From Aromatic Aldehydes.* The condensation of substituted benzaldehydes with aminoacetal leads to Schiff bases. When the Schiff base itself is to be isolated, it is prepared by heating the aldehyde and

---

[28] F. F. Blicke, *Org. React.* **1**, 303 (1942).

[29] Both the diethyl acetal and the dimethyl acetal are commercially available. We have used the less expensive one at any given time and have found little difference.

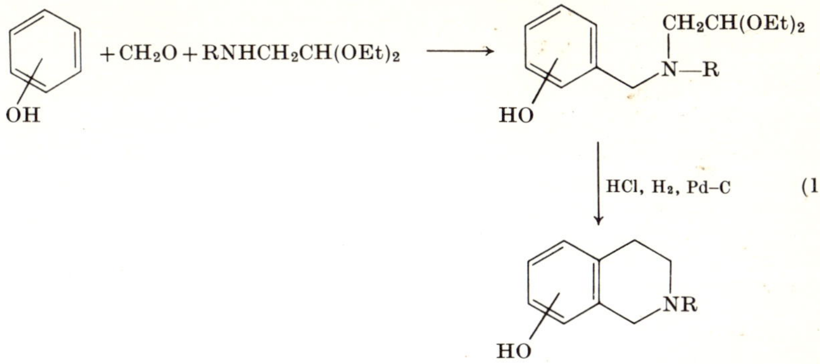

Sec. II.A.] 4-OXY- AND 4-KETO-1,2,3,4-TETRAHYDROISOQUINOLINES

$$\text{PhOH} + CH_2O + RNHCH_2CH(OEt)_2 \longrightarrow \text{Ar(OH)-CH}_2\text{-N(R)-CH}_2CH(OEt)_2$$

$$\downarrow \text{HCl, H}_2\text{, Pd-C} \quad (11)$$

[hydroxy-tetrahydroisoquinoline with NR]

the acetal together and slowly distilling the excess acetal,[17] or by azeotropic removal of water from a refluxing mixture of the two starting materials in benzene.[25] The bases can then be treated in two ways. They can be reduced to benzylaminoacetals [Eq. (12)] which eventually lead to the simple tetrahydroisoquinolines or they may be allowed to react with Grignard reagents [Eq. (13)] leading to the formation of 1-alkyltetrahydroisoquinolines.

The simple reduction has been used frequently and is well documented.[30] It may be carried out electrolytically,[8] catalytically over platinum[9] or palladium-on-carbon,[31] and with $NaBH_4$.[32,33] In most cases, the aldehyde and amine are mixed in alcohol and reduced directly

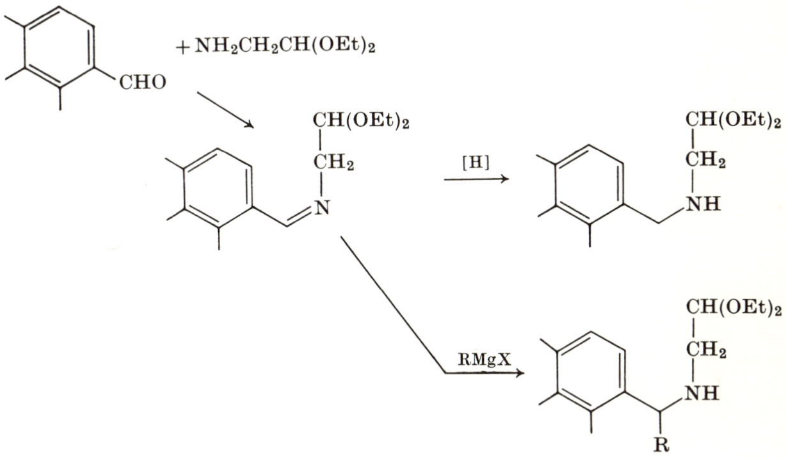

$$\text{ArCHO} + NH_2CH_2CH(OEt)_2 \longrightarrow \text{ArCH=N-CH}_2\text{-CH(OEt)}_2$$

$$\xrightarrow{[H]} \text{ArCH}_2\text{-NH-CH}_2\text{-CH(OEt)}_2 \quad (12)$$

$$\xrightarrow{RMgX} \text{ArCH(R)-NH-CH}_2\text{-CH(OEt)}_2 \quad (13)$$

---

[30] W. S. Emerson, *Org. React.* **4**, 174 (1948).
[31] O. Wacker and H. Fritz, *Helv. Chim. Acta* **50**, 2481 (1967).
[32] M. Takido, K. L. Khanna, and A. G. Paul, *J. Pharm. Sci.* **59**, 271 (1970).
[33] This has become the method of choice in our laboratory.

without isolation of the Schiff bases. The reduction using a Grignard reagent [Eq. (13)] has been carried out by several groups.[13, 15, 25, 34, 35]

In at least one case, a substituted aminoacetal has been used. Wacker and Fritz[31] reported the use of D-glucosamine [Eq. (14)] to form derivatives of tetrahydroisoquinoline. Carbons 1 and 2 of the sugar became carbons 3 and 4 of the isoquinoline, and the acetal oxygen became the 4-oxy substituent. This area of work has been retarded by the lack of readily available 2-aminoaldehyde acetals other than the simple acetaldehyde derivative.

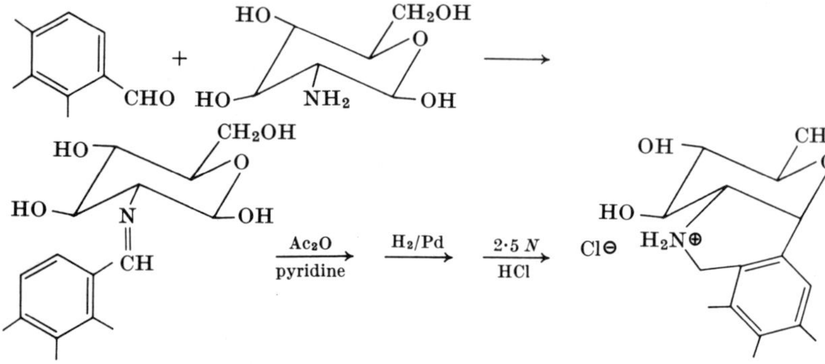

c. *From Ketones*. The synthesis of benzylaminoacetals from aminoacetal and ketones is similar to the preparation from aldehydes, but somewhat more difficult experimentally [Eq. (15)]. In the case of the simple acetophenones[32] the synthesis was carried out in absolute ethanol with $NaBH_4$ at room temperature. In other cases,[36, 37] the Schiff base

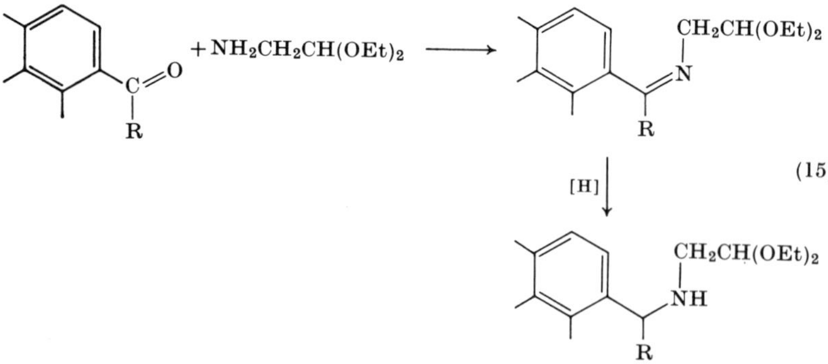

---

[34] G. J. Kapadia, M. B. E. Fayez, and M. L. Sethi, *Chem. Commun.*, 856 (1970).
[35] M. Sainsbury, S. F. Dyke, D. W. Brown, and R. G. Kinsman, *Tetrahedron*, **26**, 5265 (1970).
[36] A. R. Battersby and D. A. Yoewell, *J. Chem. Soc.*, 1988 (1958).
[37] S. M. Kupchan and A. Yoshitake, *J. Org. Chem.* **34**, 1062 (1969).

Sec. II.A.] 4-OXY- AND 4-KETO-1,2,3,4-TETRAHYDROISOQUINOLINES 107

was prepared by heating the reactants, without solvent, to 105°–125° under nitrogen followed by catalytic reduction over platinum. In one case,[38] a ketone (11) failed to give a Schiff base under any conditions. In another example (12),[39] difficulty was experienced until it was realized that the Schiff base was extremely sensitive to moisture.

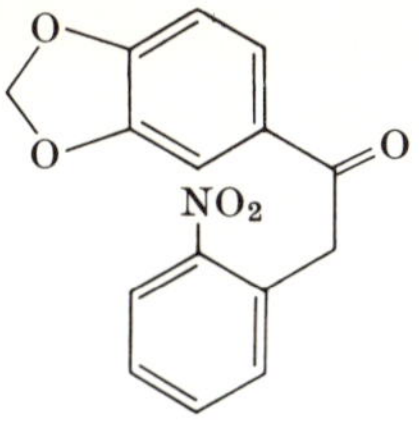

(11) 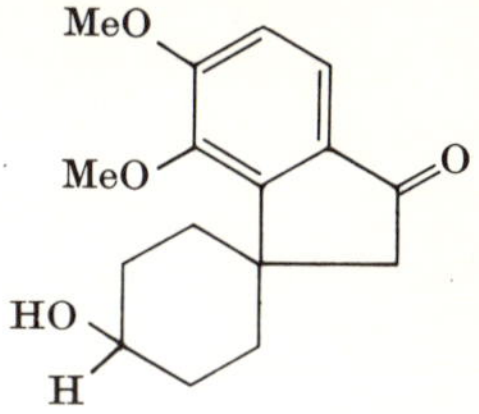   (12)

An additional route [Eq. (16)][13] involves benzonitriles as starting materials. The Schiff base is formed from the imine derived from the nitrile and a Grignard reagent. It may be reduced to the desired benzylaminoacetal, or it may be allowed to react with a second Grignard reagent to lead, eventually, to a 1,1-dialkyltetrahydroisoquinoline.

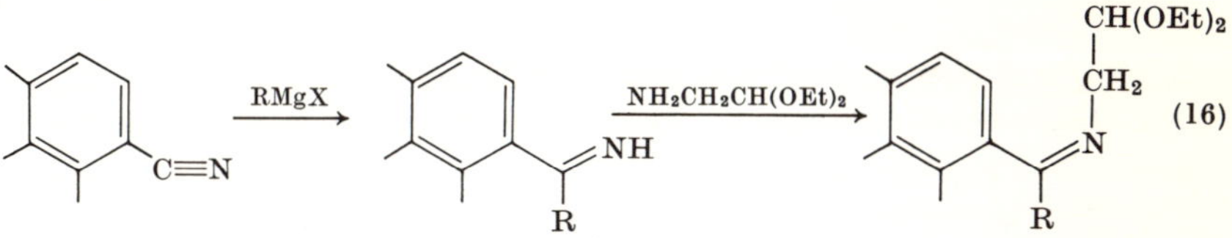

d. *From Benzylamines.* Benzylamines can be converted into benzylaminoacetals by essentially three methods. The oldest of these involves the heating together of the amine and bromoacetaldehyde diethyl acetal [Eq. (17)]. It was used by Young and Robinson[8] and has been applied recently.[35] Substituted bromoacetals have been used[18,40] in the synthesis of benzazepine derivatives [Eq. (21)] and offer one solution

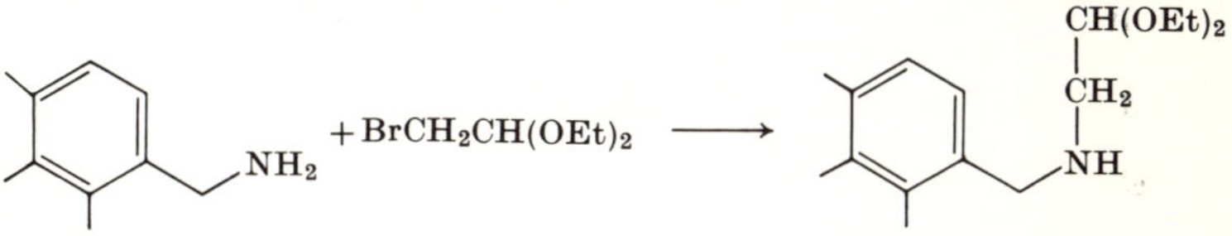

[38] D. R. Dalton, S. I. Miller, C. K. Dalton, and J. K. Crelling, *Tetrahedron Lett.*, 575 (1971).
[39] J. W. Huffman and C. E. Opliger, *Tetrahedron Lett.*, 5243 (1969); *J. Org. Chem.*, **36**, 111 (1971).
[40] J. Likforman and J. Gardent, *C. R. Acad Sci. Ser. C* **268**, 2340 (1969).

to the problem of preparing 3-substituted tetrahydroisoquinolines. A reaction has recently been described[40a] between amines and α-haloaldehydes which may be pertinent to these reactions. Under certain conditions, the α-aminoaldehydes are stable.

The second route [Eqs. (18) and (19)] involves the condensation of a benzylamine with glyoxal semiacetal. This was devised by Schlittler and Müller[41] as an improvement on the Pomeranz–Fritsch synthesis.[1]

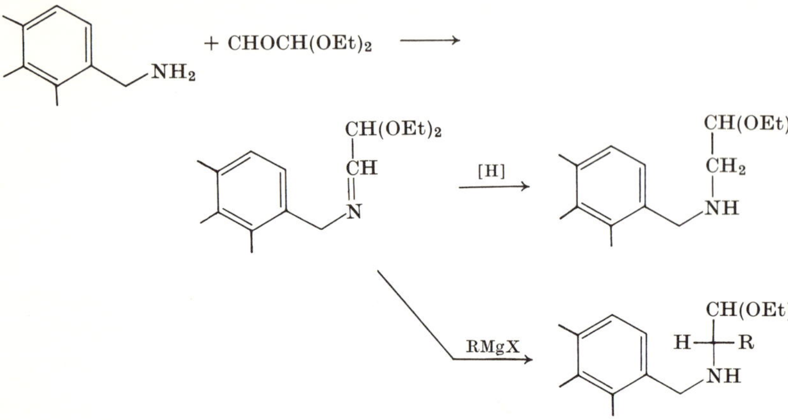

Originally the Schiff base was treated with H_2SO_4 in a typical Pomeranz–Fritsch synthesis. However, more recently, the Schiff bases have been reduced[38] [Eq. (18)] or allowed to react with a Grignard reagent[18] [Eq. (19)] to yield benzylaminoacetals which were subsequently converted into tetrahydroisoquinolines. The latter route [Eq. (19)] is one of the few general methods for placing substituents in the 3-position. The major difficulty of this route is in the preparation of glyoxal semiacetal.[42,43] The preparation from acrolein diethyl acetal or crotonaldehyde diethyl acetal is a two-step sequence with an 18% yield.

The third and newest method was devised by Frank and Purves[7] and in this laboratory[21] and is the most suitable for preparing tertiary amines [Eq. (20)]. The secondary benzylamine is allowed to react with glycidol[44] to give an amino glycol which can be cleaved with periodate to the (benzylamino)acetaldehyde, which is not isolated, but cyclized

[40a] P. Duhamel, L. Duhamel, and J.-Y. Valnot, *C. R. Acad. Sci. Ser. C* **272**, 966 (1971).
[41] E. Schlittler and J. Müller, *Helv. Chim. Acta* **31**, 914 (1948).
[42] H. O. L. Fischer and E. Baer, *Helv. Chim. Acta* **18**, 514 (1935).
[43] L. A. Yanovskaya, R. N. Stepanova, G. A. Kogan, and V. F. Kucherov, *Izv. Akad. Nauk SSSR, Otd. Khim Nauk* 857 (1963); *Chem. Abstr.* **59**, 7368 (1963).
[44] The glycidol should be taken from a previously unopened bottle and distilled under good vacuum just before use.

Sec. II.A.] 4-OXY- AND 4-KETO-1,2,3,4-TETRAHYDROISOQUINOLINES 109

and reduced to give the tetrahydroisoquinoline. It has been used in more complex situations.[35, 45]

$$\text{Ar-CH}_2\text{-NHR} + \text{H}_2\text{C}\underset{\text{O}}{\overset{}{\diagdown}}\text{CHCH}_2\text{OH} \longrightarrow \text{Ar-CH}_2\text{-N(R)-CH}_2\text{-CHOH-CH}_2\text{OH} \xrightarrow{\text{IO}_4^-} \xrightarrow[\text{H}_2/\text{Pd-C}]{\text{HCl}} \text{tetrahydroisoquinoline-NR} \quad (20)$$

2. *Ring Closure of Benzylaminoacetals*

The methods for causing the ring closure of benzylaminoacetals to isoquinoline skeletons are summarized in the Introduction. Except for the boron trifluoride method and the dilute acid method, they would appear to be of historical interest only.

As in all ring closure methods, the cyclization of benzylaminoacetals can theoretically take place at two points. For example, ring closure of a 3-substituted benzylaminoacetal could give the 7-substituted isoquinoline (para closure) or the 5-substituted isoquinoline (ortho closure). As yet, however, all observed ring closures have been para with none of the other isomer isolated. In the dilute HCl closure [Eq. (7)], this may be due to a thermodynamic control of an essentially reversible reaction. The precise mechanistic details are not known.

a. *The Boron Trifluoride Method*.[12-18] By this method [Eq. (6)] benzylaminoacetals are converted into 4-alkoxytetrahydroisoquinolines. In all recorded cases, the ethyl acetals have been used, and the 4-ethoxy derivatives have been obtained. The experimental procedure[15] involves the passage of BF_3 gas (produced by heating an anisole solution) into a solution of the acetal in dichloroethane at 0° until saturation is achieved. The mixture is stirred for an unspecified time, and the complex is destroyed with 15% NaOH. The solution is warmed to 40°–50° to expel BF_3, cooled, and diluted with ether. After 3 days, the ether layer is separated, and the base is extracted with cold 10% H_2SO_4. The acid solution is neutralized with ammonia, and the base is extracted with ether, dried, concentrated, and distilled under vacuum. The 4-ethoxy

[45] D. W. Brown, S. F. Dyke, and M. Sainsbury, *Tetrahedron* **25**, 101 (1969).

derivatives have always been isolated prior to further reaction, in contrast to the 4-hydroxy derivatives which are not.

The method has several disadvantages.[17] First, it does not work well when there are oxygen substituents in the aromatic ring. Second, it does not work well when the substituent in the aromatic ring is not meta to the point of ring closure. The third and chief disadvantage is the general lack of applications of the method. It was not explored initially with aromatic rings containing more than one substituent, and it has not been used for isoquinoline synthesis by anyone other than the original authors. It has been used for the synthesis of benzazepines by Likforman and Gardent [Eq. (21)].[40]

$$\text{[benzyl-CH(OEt)}_2\text{-R, NR']} \xrightarrow{BF_3} \text{[benzazepine with OEt, R, N-R']} \quad (21)$$

On the other hand, it provides good methods for the introduction of groups into the 1- and 3-positions of simple isoquinolines. The compounds prepared by this method are listed in Table I.

b. *The Dilute Hydrochloric Acid Method.*[8, 9, 19–26] By this method [Eq. (7)], the benzylaminoacetals are converted into 4-hydroxytetrahydroisoquinolines which are generally not isolated, but converted into other isoquinoline derivatives (Section III). In the original example of its application,[8] Young and Robinson probably used too vigorous a condition for the closure (steam-bath temperatures). In the first four papers from this laboratory,[9, 19–21] it was not realized that the 4-hydroxy derivatives were isolable intermediates. The products obtained in a series of reactions were rationalized as arising from 1,2-dihydroisoquinolines. Only after 4,6-dihydroxy-7-methoxy-1,2,3,4-tetrahydroisoquinoline hydrochloride crystallized during a reaction[22] was it realized that the 4-hydroxy derivatives were probably intermediates for all the reactions and, furthermore, that they were reasonably stable.[46] Since then, they have been isolated in several cases.[23, 32, 34, 35, 38, 45, 47–50]

[46] The 4-hydroxytetrahydroisoquinolines are stable as solids, either as hydrochlorides or, in many cases, as free bases. They are unstable to warm acid or solvents which easily provide a nucleophile (alcohols, or amines, see Section III).
[47] W. J. Gensler, K. T. Shamasundar, and S. Marburg, *J. Org. Chem.* **33**, 2861 (1968).
[48] D. W. Brown, S. F. Dyke, G. Hardy, and M. Sainsbury, *Tetrahedron Lett.*, 2609 (1968).
[49] M. Sainsbury, D. W. Brown, S. F. Dyke, and G. Hardy, *Tetrahedron* **25**, 1881 (1969).
[50] F. D. Popp and E. B. Moynahan, unpublished data.

The reaction is normally carried out by dissolving the benzylaminoacetal in 6 N HCl, extracting with some organic solvent, generally ether or benzene, to remove nonbasic compounds, and allowing the solution to stand overnight at room temperature. Acid concentrations used have ranged from 2.5[31] to 8 N[32] and may well vary from one case to another. Concentration of the acid solution on a rotary evaporator under water-pump vacuum frequently causes the 4-hydroxytetrahydroisoquinoline to crystallize in good yield. In some cases, it is necessary to cool the acid solution, either before or after concentration, make basic with ammonia or NaOH and, if it does not crystallize, extract the product. Excessive heat should be avoided, and the bases frequently cannot be recrystallized without appreciable loss. When the hydrochlorides crystallize during the concentration, they are generally quite pure.

The method appears to have one serious limitation. There must be an oxygen, either as a phenol or an ether, in a position ortho or para to the point of ring closure. One failure has been reported[51] when furan was used in place of the benzene ring. Ring closure did take place with a ferrocene ring.[50] Ring closure should be explored with respect to other activating ring substituents, especially the acyloxy and the acylamino group and with other heterocyclic systems. A major advantage for alkaloid synthesis lies in the fact that free phenol groups do not interfere, thus saving blocking and unblocking steps.

B. Synthesis via 4-Keto-1,2,3,4-tetrahydroisoquinolines[27]

1. *Synthesis of* 4-*Keto*-1,2,3,4-*tetrahydroisoquinolines*

These ketones have been prepared by four different routes: a Dieckmann condensation [Eq. (22)], ring closure of benzylglycine derivatives [Eq. (23)], ring closure of substituted nitriles [Eq. (24)], and a Grignard reaction on 4-hydroxy-N-alkylisoquinolinium salts [Eq. (25)]. Earlier unsuccessful attempts at their synthesis have been summarized.[52, 53]

The Dieckmann condensation was first explored in this context by Hinton and Mann,[53] where R was CH_3 and the aromatic ring was unsubstituted. The reaction was extended by Grethe and co-workers[54] to the oxygenated 4-ketotetrahydroisoquinoline series, where R was benzyl and therefore removable by hydrogenolysis.

The preparation of 4-ketotetrahydroisoquinolines by ring closure of

[51] M. P. Mertes, R. F. Borne, and L. E. Hare, *J. Org. Chem.* **33**, 133 (1968).
[52] D. N. Harcourt and R. D. Waigh, *J. Chem. Soc. C*, 967 (1971).
[53] I. G. Hinton and F. G. Mann, *J. Chem. Soc.*, 599 (1959).
[54] G. Grethe, H. L. Lee, M. Uskoković, and A. Brossi, *J. Org. Chem.* **33**, 494 (1968).

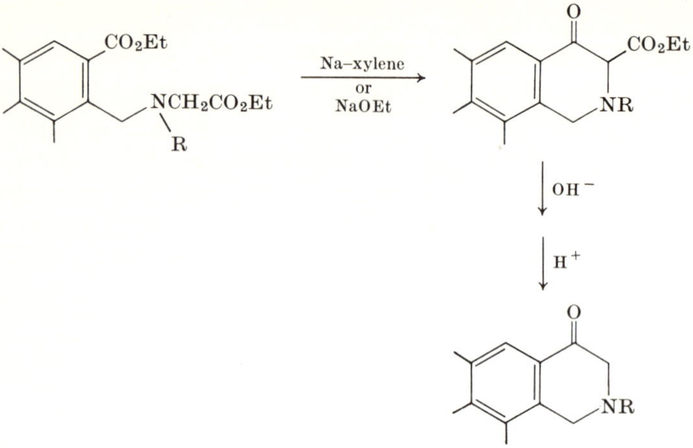

glycine derivatives [Eq. (23)] is a most attractive route because the starting material can be prepared by reductive alkylation of glycine as in Eq. (12). Unfortunately, it does not always work well. The method was first successfully applied by Kametani and Fukumoto[55] to the prepara-

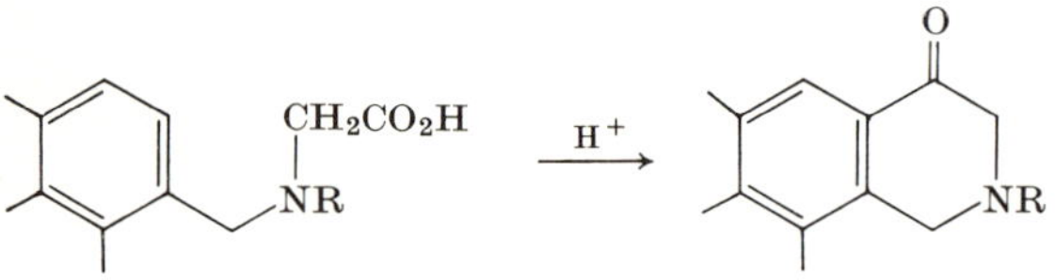

tion of 6,7-dimethoxy-4-keto-1-methyl-1,2,3,4-tetrahydroisoquinoline from the glycine ester and has been used several times since.[56-58] It appears to work best when a single substituent, a methoxy group, is present para to the point of ring closure[56] or when the nitrogen is blocked by an N-formyl[57, 58] or a benzyl group.[56] It has been used successfully when the aromatic ring is furan.[51]

The third method,[52] using the Strecker synthesis to produce a benzyl-aminoacetonitrile which can be closed to a ketone [Eq. (24)], may well be the most versatile method. The variations possible in the carbonyl

[55] T. Kametani and K. Fukumoto, *J. Chem. Soc.*, 4289 (1963).
[56] G. Grethe, H. L. Lee, M. Uskoković, and A. Brossi, *J. Org. Chem.* **33**, 491 (1968).
[57] B. Umezawa, O. Hoshino, and Y. Terayama, *Chem. Pharm. Bull. (Japan)* **16**, 180 (1968).
[58] B. Umezawa, O. Hoshino, and Y. Yamanashi, *Tetrahedron Lett.*, 933 (1969).

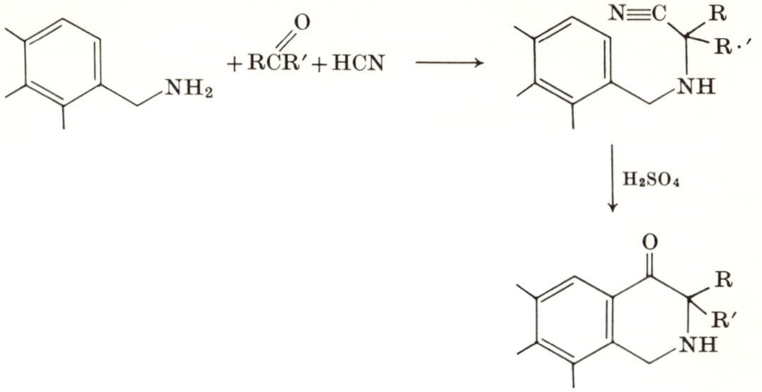

Sec. II.B.] 4-OXY- AND 4-KETO-1,2,3,4-TETRAHYDROISOQUINOLINES

[Reaction scheme (24): benzylamine + RCR'+ HCN → imine intermediate → (H₂SO₄) → 4-keto-1,2,3,4-tetrahydroisoquinoline]

reactant, the reactivity of the 4-keto group itself, and the possibility of using substituted benzylamines support this contention. However, the method is quite new, and has not been extensively used. Ring closure had been attempted previously without success (see citations in ref. 52).

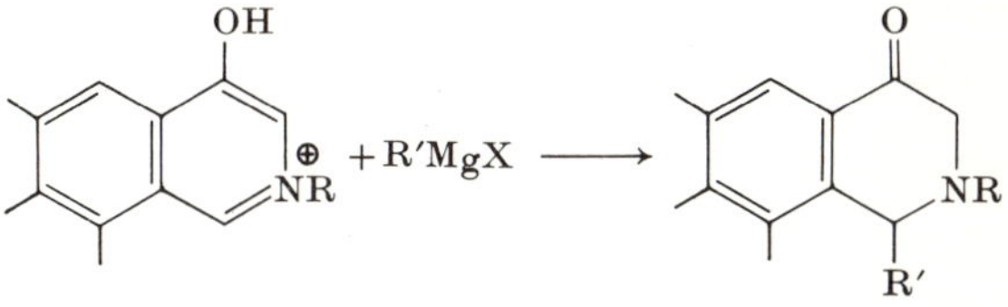

[Reaction scheme (25): N-alkyl-4-hydroxyisoquinolinium + R'MgX → 1-R'-4-keto-2-alkyl-1,2,3,4-tetrahydroisoquinoline]

Finally, the addition of a Grignard reagent to an N-alkyl-4-hydroxyisoquinolinium ion[59] yields a 4-ketotetrahydroisoquinoline with a new alkyl group in the 1-position.

In general, the 4-ketotetrahydroisoquinolines are stable as their hydrochloride salts,[52, 53] but the free bases are easily air-oxidized to the 4-hydroxyisoquinolines. The types of 4-ketotetrahydroisoquinoline which have been prepared are outlined in Table III (see p. 132).

2. Reduction to 4-Hydroxy-1,2,3,4-tetrahydroisoquinolines

4-Ketotetrahydroisoquinolines have been converted to 4-hydroxytetrahydroisoquinolines by reduction with $NaBH_4$,[54, 59] $LiAlH_4$,[53] and hydrogen over 10% palladium-on-carbon [Eq. (26)].[56, 59] In addition, the ketones have been allowed to react with Grignard reagents to yield 4-aryl-4-hydroxytetrahydroisoquinolines [Eq. (27)].[53, 59a] The oxygen function may be removed entirely by the ethylene dithioketal method[52, 54]

---

[59] G. Grethe, M. Uskoković, T. Williams, and A. Brossi, *Helv. Chim. Acta* **50**, 2397 (1967).

[59a] A. Brossi, G. Grethe, S. Teitel, W. C. Wildman, and D. T. Bailey, *J. Org. Chem.* **35**, 1100 (1970).

or by hydrogenation over 10% palladium-on-carbon at high pressures (60–90 atm) to yield tetrahydroisoquinolines [Eq. (28)].[54]

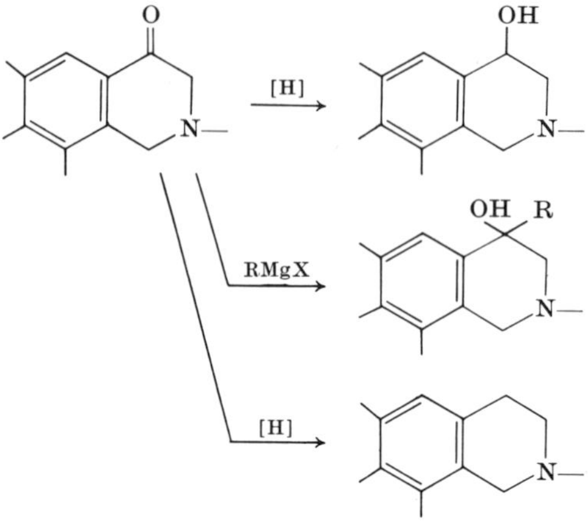

C. Synthesis via β-Hydroxy-β-phenylethylamines

This method [Eq. (10)] was first used in 1930 by Kondo and Tanaka[60] for the preparation of 4-hydroxy-5-methoxytetrahydroisoquinoline from β-hydroxy-β-(o-methoxyphenyl)ethylamine and methylal (mixed in aqueous HCl). Such ethanolamine derivatives are frequently used in the Bischler–Napieralski synthesis of isoquinolines[61] as the Pictet–Gams modification. In this case, the reaction proceeds with dehydration to yield completely aromatic isoquinolines directly. Preparations of the ethanolamines have been reviewed.[62]

More recently, this reaction has been used successfully for the preparation of a number of 4-hydroxytetrahydroisoquinolines (Table II).[63–67a]

[60] T. Kondo and S. Tanaka, *J. Pharm. Soc. Jap.* **50**, 923 (1930); *Chem. Abstr.* **25**, 515 (1931).
[61] W. M. Whaley and T. R. Govindachari, *Org. React.* **6**, 74 (1951).
[62] W. J. Gensler, in "Heterocyclic Compounds" (R. C. Elderfield, ed.), Vol. 4, p. 344. Wiley, New York, 1952.
[63] O. M. Friedman, K. N. Parameswaran, and S. Burstein, *J. Med. Chem.* **6**, 227 (1963).
[64] J. P. Forneau, C. Caignault, R. Jacquier, O. Stoven, and M. Davy, *Chim. Ther.* **4**, 67 (1969); *Chem. Abstr.* **71**, 81108 (1969).
[65] B. Jaques, R. H. L. Deeks, and P. K. J. Shah, *Chem. Commun.*, 1283 (1969).
[66] T. Kametani, K. Fukumoto, H. Agui, H. Yagi, K. Kigasawa, H. Sugahara, M. Hiiragi, T. Hayasaka, and H. Ishimaru, *J. Chem. Soc. C*, 112 (1968).
[67] T. Kametani, K. Kigasawa, M. Hiiragi, and H. Ishimaru, *Chem. Pharm. Bull. (Tokyo)* **17**, 2353 (1969).
[67a] T. Kametani, K. Kigasawa, M. Hiiragi, H. Ishimaru, and S. Saito, *Yakugaku Zasshi* **89**, 1482 (1969); *Chem. Abstr.* **72**, 55212 (1970).

It is thought to be the major reaction involved in a histochemical method for the detection of catecholamines[68] and may play a role in the biochemistry of alcoholism.[69]

D. Miscellaneous Syntheses

Treatment of corypalline (13) with lead tetraacetate leads[57] to the quinol acetate (14) which rearranges in acetic anhydride–sulfuric acid to yield the 4-acetoxytetrahydroisoquinoline (15). The reduction of the lactol (16) to 4-hydroxytetrahydroisoquinoline (17) has been reported.[70]

4-Hydroxytetrahydroisoquinolines have been prepared by the hydroboration of 1,2-dihydroisoquinolines,[71, 71a] thus reversing Eq. (32)

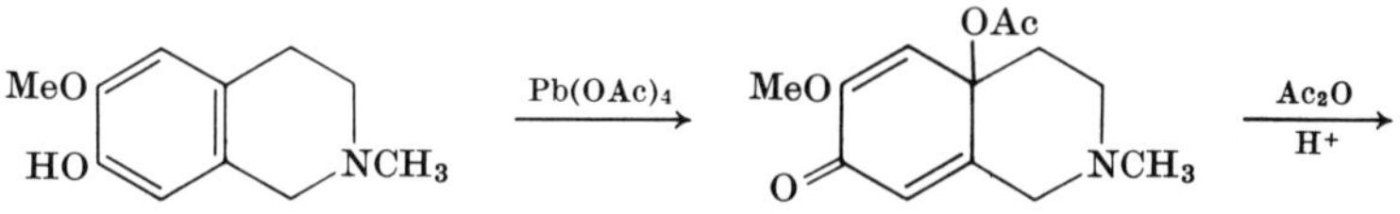

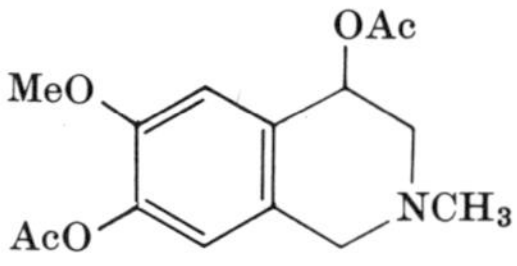

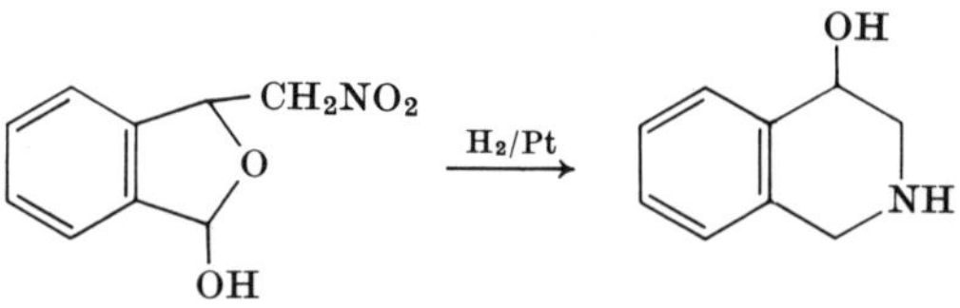

[68] H. Corrodi and G. Jonsson, *J. Histochem. Cytochem.* **15**, 65 (1967).
[69] G. Cohen and M. Collins, *Science* **167**, 1749 (1970).
[70] H. Nakao, Y. Yura, and M. Ito, *Sankyo Kenkyusho Nempo* **18**, 48 (1966); *Chem. Abstr.* **66**, 104886 (1967).
[71] S. F. Dyke, *Advan. Heterocycl. Chem.* **14** (1972).
[71a] S. F. Dyke and A. C. Ellis, *Tetrahedron* **27**, 3803 (1971).

(Section III,C). Finally, the 4-hydroxy-3,4-dihydroisocarbostyrils (**19**) were prepared by a base-catalyzed rearrangement of **18**.[72, 72a]

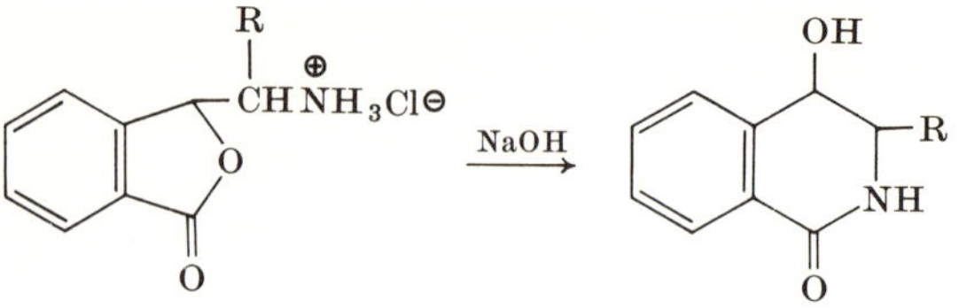

(**18**) →NaOH→ (**19**)

III. Reactions of 4-Oxy-1,2,3,4-tetrahydroisoquinolines

A. Hydrogenolysis to 1,2,3,4-Tetrahydroisoquinolines

The hydrogenation [Eq. (29)] is carried out at room temperature in dilute hydrochloric acid over 5% palladium-on-carbon.[9, 19, 21, 22, 24, 25] The acid solution has been used mainly as a matter of convenience and may not be necessary. Although pressures up to 15 psi can be used, atmospheric pressure is considered preferable, and the amounts of catalyst used are comparable, in weight, to the amount of substrate. On some occasions, hydrogen absorption becomes very slow or may cease. In these cases, the mixture should be filtered and a fresh amount of catalyst added. At other times, the hydrochloride of the product precipitates during the reduction and must be washed out of the catalyst with ethanol or water. In general, this reaction has been carried out directly on the ring-closure reaction mixture [Eq. (7)] without isolation of the 4-hydroxytetrahydroisoquinoline intermediates. Examples are given in Table IV (p. 133).

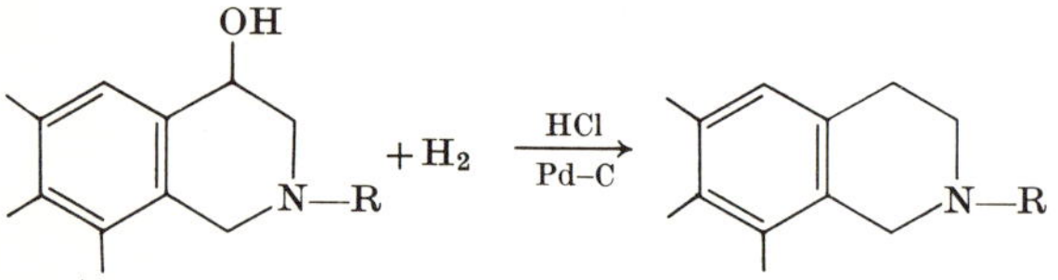

$$+ H_2 \xrightarrow[Pd-C]{HCl}$$ (29)

Although the 4-ethoxytetrahydroisoquinolines [Eq. (6)] should hydrogenolyze equally well to the tetrahydro derivatives, the reaction does not seem to have been attempted.

---

[72] J. W. Wilson, III, E. L. Anderson, and G. E. Ullyot, *J. Org. Chem.* **16**, 800 (1951).
[72a] H. H. Baer and B. Achmatowicz, *J. Org. Chem.* **29**, 3180 (1964).

Sec. III.B.] 4-OXY- AND 4-KETO-1,2,3,4-TETRAHYDROISOQUINOLINES 117

B. AROMATIZATION TO ISOQUINOLINES

This reaction [Eq. (30)] has been carried out classically by heating almost any saturated isoquinoline with 10% palladium-on-carbon (sometimes 5% and sometimes up to 30%)[48] in a high boiling solvent such as xylene or tetralin.[73] It has been applied to 4-hydroxy-[9] and 4-ethoxytetrahydroisoquinolines.[18] The yields are often erratic, however, and the reaction is not very satisfactory.

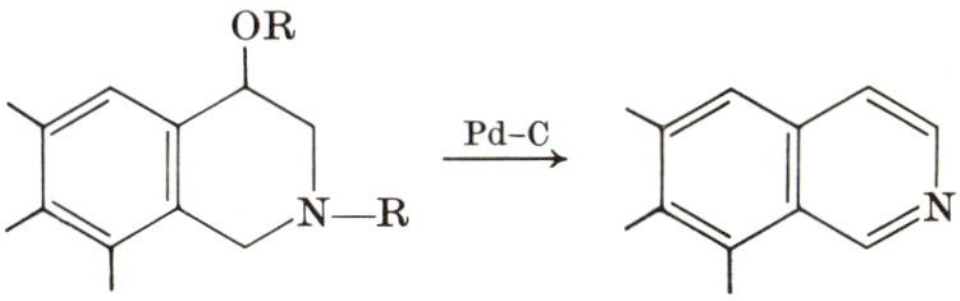

(30)

Two methods have been devised for use with the 4-hydroxytetrahydroisoquinolines. Neither has been well explored, and both probably require that phenols be blocked. In the first [Eq. (31)],[46, 50] 4-hydroxytetrahydroisoquinolines were treated with an equimolar amount of N-bromosuccinimide (NBS) in chloroform to yield a 3,4-dihydro intermediate which was not isolated, but dehydrated in HCl to give the aromatic product.

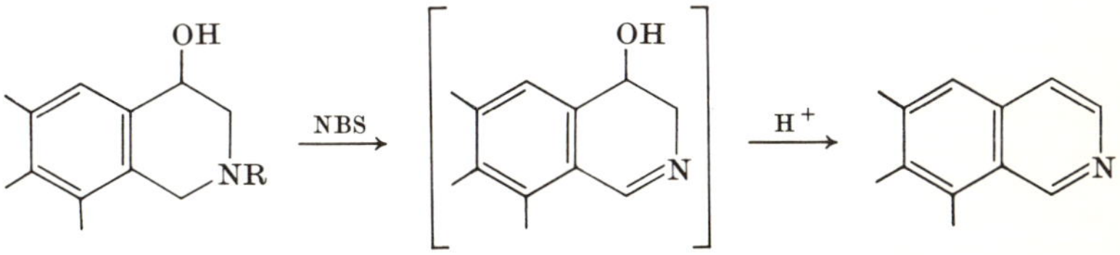

(31)

In an alternative method related to the general theme of this chapter, Jackson and Stewart[74] treated the benzylamino acetal (20) with p-toluenesulfonyl chloride in pyridine to yield 21, which was cyclized in dilute acid to 22 and aromatized with base to the aromatic product (23). Some aromatic isoquinolines are listed in Table V (p. 134).

[73] M. P. Cava and M. V. Lakshmikantham, *J. Org. Chem.* **35**, 1867 (1970).
[74] A. H. Jackson and G. W. Stewart, *Chem. Commun.*, 149 (1971).

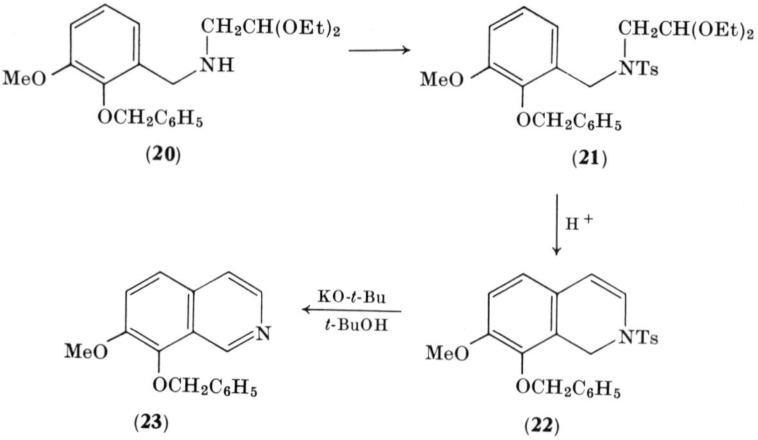

C. Dehydration to 1,2-Dihydroisoquinolines

The dehydration of 4-hydroxytetrahydroisoquinolines [Eq. (32)] is probably the most important single reaction of these compounds. Since 4-hydroxyisoquinolines are not normally isolated prior to further reaction, it therefore becomes the most important application of the dilute acid-catalyzed ring closure [Eq. (7)]. The products, 1,2-dihydroisoquinolines, are highly reactive cyclic enamines[20, 75] and undergo a number of interesting electrophilic and nucleophilic substitutions and rearrangements. Since these compounds have been thoroughly discussed in the preceding volume of this series,[71] the subject will be considered here only briefly.

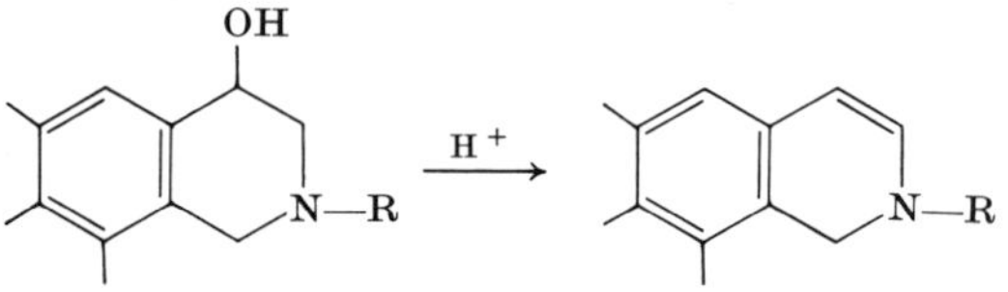

The conversion of benzylaminoacetals into 1,2-dihydroisoquinolines involves three reactions: hydrolysis of the acetal, ring closure [Eq. (7)], and dehydration [Eq. (32)]. The hydrolysis and ring closure are much faster than the dehydration, but, in the absence of kinetic studies, it is difficult to say just how much faster they are. In cold concentrated HCl, nuclear magnetic resonance studies have shown[45] that the hydrolysis

---

[75] K. Bláha and O. Červinka, *Advan. Heterocycl. Chem.* **6**, 147 (1966).

and closure are very fast. However, the ring closures are generally done in dilute HCl. Normally, 12 hours at room temperature (25°–30°) will bring about hydrolysis and closure in 6 N HCl, and dehydration does not become an important side reaction for several days.[9] In one specific case [Eq. (33)],[23] treatment of benzylaminoacetals with cold 6 N HCl yielded the 4-hydroxytetrahydroisoquinolines, whereas hot concentrated (12 N) HCl led to the formation of 1,2-dihydroisoquinolines and the final product, a quinolizidine (**24**).

D. NUCLEOPHILIC DISPLACEMENT AT C-4

An oxygen function in the 4-position of a tetrahydroisoquinoline is benzylic and in a ring. On the basis of its electronic and steric situation, it should be, and is, readily replaced by nucleophiles. The nucleophiles which have been found to be effective thus far are activated (by oxygen) aromatic rings and alcohols. In a related reaction, 4,7-diacetoxytetrahydroisoquinolines have been shown to undergo a number of base-catalyzed nucleophilic reactions.

The reactions with aromatic rings were discovered first and are quite interesting. Battersby and Yoewell[36] correctly interpreted the structure given earlier as **7**[11] to be the bicyclic isopavine (**25**). It was later suggested[50] that **6** should be the 4-hydroxytetrahydroisoquinoline (**8**) and that **25** was formed by a nucleophilic attack on the 4-hydroxyl group. Several alkaloids have now been synthesized by this same route.[37, 71a]

The formation of this general type of bicyclic compound has been further developed by Dyke and co-workers[49, 76] into a synthesis of compounds with the skeletal structure (**26**) [Eq. (35)] and in this laboratory[26] into a synthesis of the skeletal type (**27**) [Eq. (36)]. It is quite significant that these reaction sequences (acetal hydrolysis, ring closure, and displacement) were carried out both in concentrated acid[49, 76] and dilute acid[26] and that no dehydration of the intermediate 4-hydroxy compound

[76] D. W. Brown, S. F. Dyke, G. Hardy, and M. Sainsbury, *Tetrahedron Lett.*, 1515 (1969).

took place. Such a dehydration would have yielded a 1,2-dihydroisoquinoline which should have undergone nucleophilic attack at C-3[71] [cf. Eq. (33)]. In fact, attempts to bring about such a dehydration followed by ring closure failed.[49]

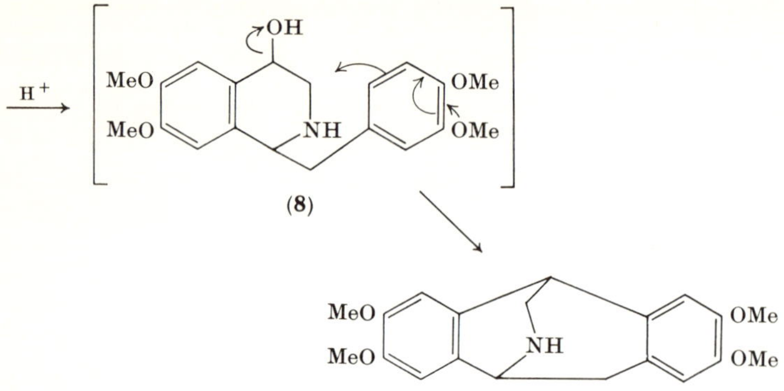

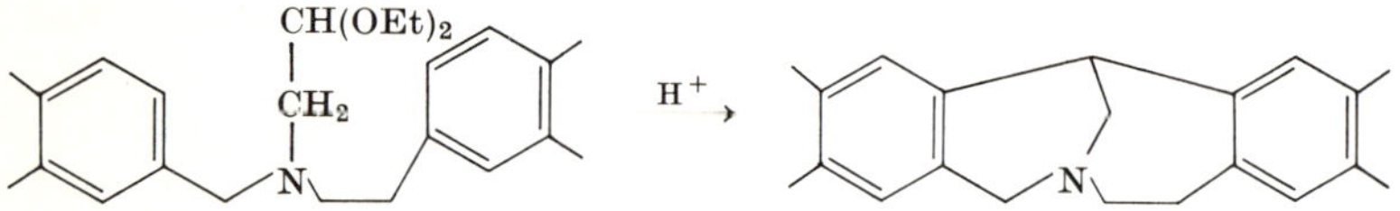

The nucleophilic reaction with aromatic compounds in an intermolecular reaction has been briefly explored in this laboratory[26] with several phenols and alkoxybenzenes [Eq. (37)]. This reaction could well be extended to a number of interesting heterocyclic systems.

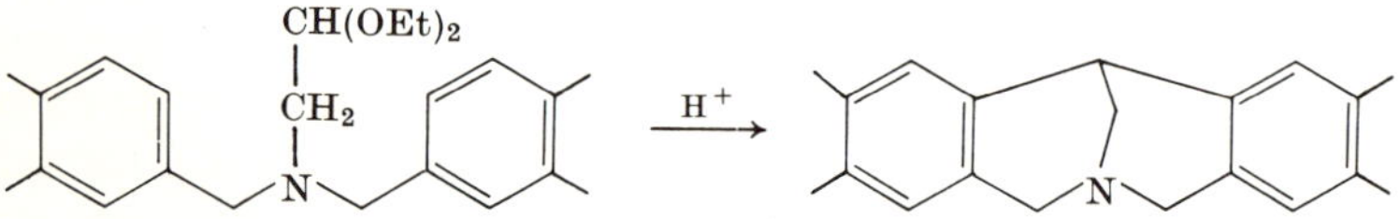

Simpler nucleophiles give similar results. For example, when a 4-methoxytetrahydroisoquinoline (**28**) was crystallized from ethanol, the 4-ethoxy derivative (**29**) was obtained.[65] The facility of these simple displacements accounts in part for the instability of 4-hydroxytetrahydroisoquinolines and certainly explains the losses encountered on

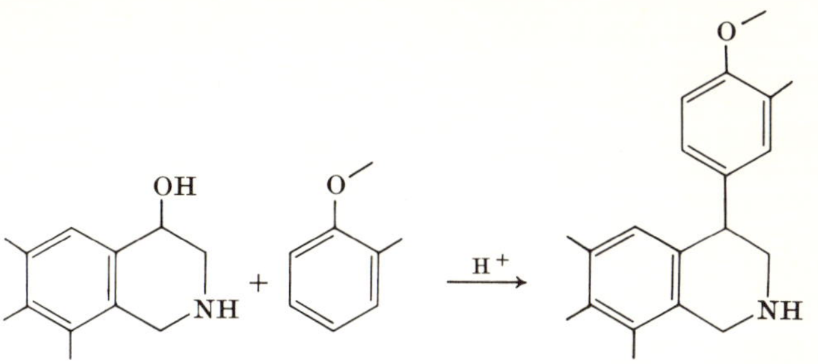

Sec. III.D.] 4-OXY- AND 4-KETO-1,2,3,4-TETRAHYDROISOQUINOLINES 121

(37)

crystallization from protonic solvents.[22] It may prevent the extension of the general dilute acid-catalyzed ring closure to very reactive aromatic systems such as furan[52] or rings with two or more phenols. Such systems would polymerize easily.

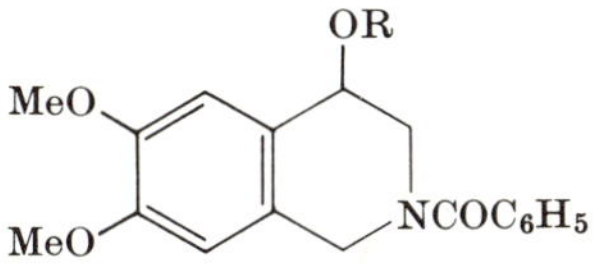

(28) R = Me
(29) R = Et

There is some question about the mechanism involved in an extensive series of nucleophilic reactions discovered by Umezawa and Hoshino and their co-workers.[58, 77, 78] In these reactions, the 4,7-diacetoxy compound (30) was treated, in base, with alcohols (to yield 31), amines (to yield 32), mercaptans (to yield 33), KCN (to yield 34), nitromethane (to yield 35), malononitrile (to yield 36), malonic ester (to yield 37), cyclohexanone (to yield 38), acetophenone (to yield 39), and dimethyl sulfoxide (to yield 40). The reaction with cyclopentanone was anomalous in that a dimer of the ketone reacted with the isoquinoline. Since the acetoxy group on C-7 was almost always lost and the reaction failed when a 7-methoxy group was present, a quinone methide intermediate (41) was proposed. Some of the reactions were done in aqueous or alcoholic base and some were done under anhydrous conditions with NaH.

Considerable work must still be done before the scope and limitations of the nucleophilic substitution reaction are completely defined.

[77] O. Hoshino, Y. Yamanashi, and B. Umezawa, *Tetrahedron Lett.*, 937 (1969).
[78] O. Hoshino, Y. Yamanashi, and B. Umezawa, *Chem. Pharm. Bull.* **19**, 2161 (1971).

(30) (31) (41)

(31) X=OR
(32) X=N—R
(33) X=SR
(34) X=CN
(35) X=CH_2NO_2
(36) X=$CH(CN)_2$
(37) X=$CH(CO_2Et)_2$
(38) X=

(39) X=$CH_2COC_6H_5$
(40) X=CH_2SOCH_3

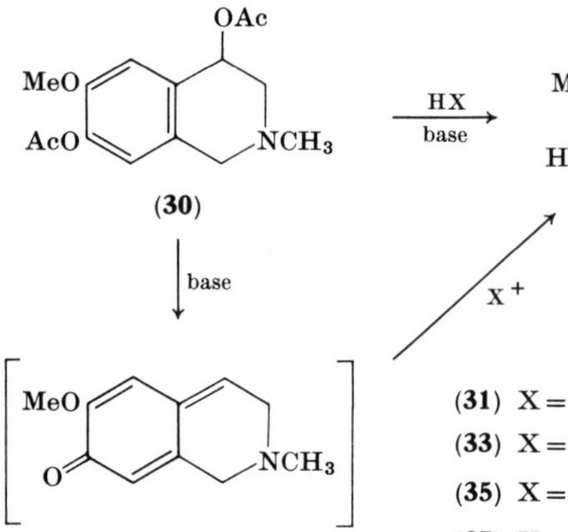

IV. Naturally Occurring 4-Oxy-1,2,3,4-tetrahydroisoquinolines

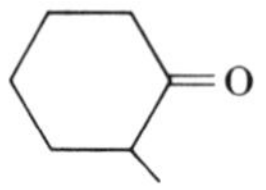

No simple 4-oxytetrahydroisoquinolines have been found thus far in nature. Structure **42** was originally proposed[79] for the alkaloid gigantine. However, the structure was not confirmed by synthesis[59] and a new structure (**43**) has been put forward.[34]

(**42**)

(**43**)

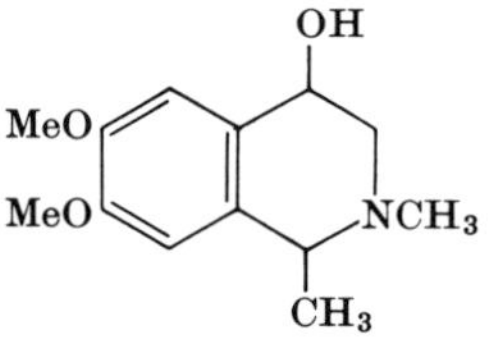

Cohen and Collins[69] have proposed that epinephrine (**44**) and similar compounds common in living tissue (i.e., the catecholamines) react with acetaldehyde formed in the body by oxidation of consumed ethanol to form 4-hydroxytetrahydroisoquinolines (**45**) [see Eq. (10)]. The latter compounds may play a role in alcohol toxicity. Given the presence of

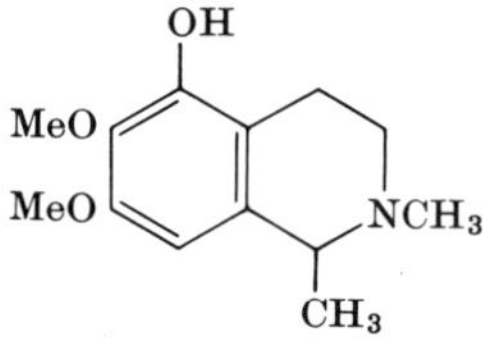

[79] J. E. Hodgkins, S. D. Brown, and J. L. Massingill, *Tetrahedron Lett.*, 1321 (1967).

such amines as **44** in nature, it would be logical to expect to find some of the derived 4-hydroxytetrahydroisoquinolines. The lack of such, thus far, may be due to their great reactivity.

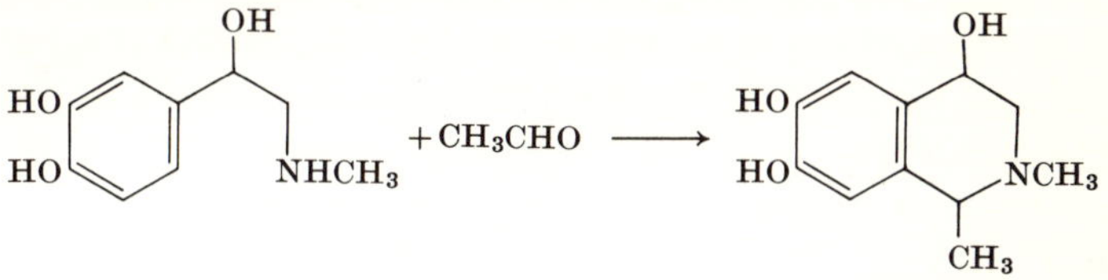

Two erythrine alkaloids, erythristemine (**46**)[80] and erythrinine (**47**),[81] contain 4-methoxy and 4-hydroxy groups, respectively, in a tetrahydroisoquinoline ring, albeit highly substituted.

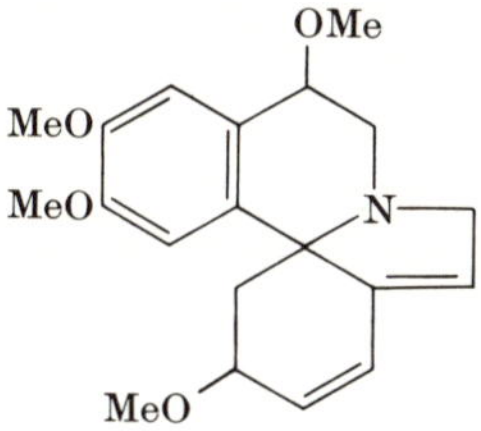

V. The State of the Art of Isoquinoline Synthesis

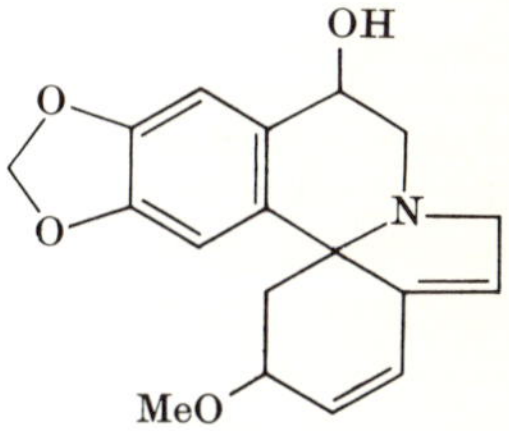

In 1966, Brossi summarized[82] a large amount of the isoquinoline chemistry that he and his group had carried out and, as a part of this summary, he gave something of an overall picture of isoquinoline synthesis. It would seem desirable to summarize again in this chapter the general state of the art of isoquinoline synthesis in as far, at least, as it is related to or could arise from the material given here. In order to present a different perspective from that given previously, this discussion will be organized around substitution in the various positions of the isoquinoline skeleton.

---

[80] D. H. R. Barton, P. N. Jenkins, R. Letcher, D. A. Widdowson, E. Hough, and D. Rogers, *Chem. Commun.*, 391 (1970).
[81] K. Ito, H. Furukawa, and H. Tanaka, *Chem. Commun.*, 1076 (1970).
[82] A. Brossi, *Trans. N.Y. Acad. Sci.* **28**, 685 (1966).

A. Substitution in the Benzene Ring

The methods discussed in this chapter have made possible the facile synthesis of isoquinoline derivatives oxygenated at the 7-position, the 5,8-positions, the 6,7-positions, the 7,8-positions, and the 6,7,8-positions [Eq. (7)] or having carbon substituents at the 6-position [Eq. (6)]. The starting materials are readily available aromatic aldehydes, and the reactions are experimentally simple. These methods are not suitable for the preparation of isoquinolines with oxygen at the 5-position, the 6-position, or the 5,6-positions. These must still be prepared by the Bischler–Napieralski method,[61] the Pictet–Spengler method,[83] or the Dieckmann route shown in Eq. (22).

A new ring closure has recently been described[84] which may be of general value for the preparation of many types of isoquinolines [Eq. (38)]. It involves the conversion of a benzylaminoethanol into a bromide and ring closure with $AlCl_3$. Although it has been tried in only two cases (the tetrahydroisoquinoline and 6-chlorotetrahydroisoquinoline), it should be less sensitive to ring activation requirements than present methods. It may even provide a good method for the synthesis of 8-substituted isoquinolines[3] which is presently lacking.

The NMR spectral properties of a number of the oxygenated tetrahydroisoquinolines have been published.[84a]

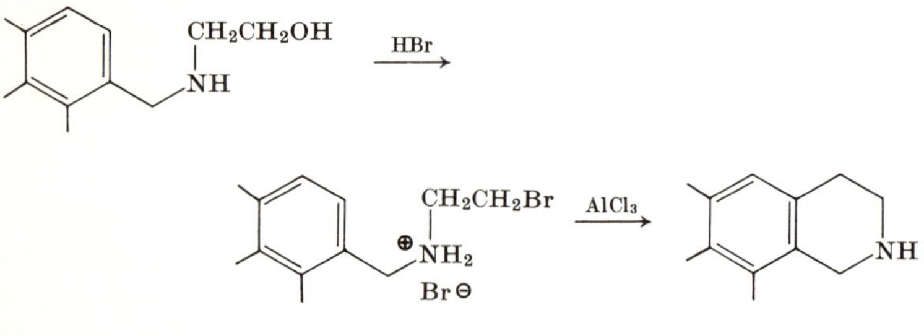

B. Substitution at the 1-Position

Substitution in the 1-position is present in most of the isoquinoline alkaloids and is of considerable importance. These compounds have traditionally been made by the Bischler–Napieralski route,[61] but this method is not suitable for compounds having oxygenation in the 7,8-

---

[83] W. M. Whaley and T. R. Govindachari, *Org. Reactions* **6**, 151 (1951).
[84] L. W. Deady, N. Pirzada, and R. D. Topsom, *Chem. Commun.*, 799 (1971).
[84a] F. Schenker, R. A. Schmidt, T. Williams, and A. Brossi, *J. Heterocyclic Chem.* **8**, 655 (1971).

positions. Substituents can be introduced into the 1-position via Eqs. (10), (13), (15), and (16).

The ready availability of simple isoquinolines, especially 7,8-dioxygenated ones, as described in this chapter has prompted the development or application of three general reactions for placing alkyl groups in the 1-position. 1-Benzyl groups are the most important for alkaloid synthesis, and major effort has been placed on their introduction. The first of these [Eq. (39)], involving a Stevens rearrangement, was developed by Grethe and co-workers.[85, 86] The reaction appears to be quite general.

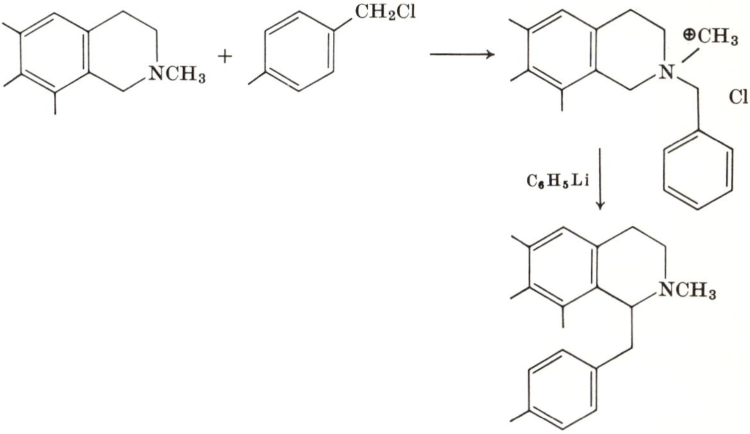

The second route involves the oxidation of a tetrahydroisoquinoline to the 3,4-dihydro derivative and the addition of a Grignard reagent [Eq. (40)].[82, 87] It would appear that a general route through Eqs. (13), (7), and (29) may now provide a simpler sequence.

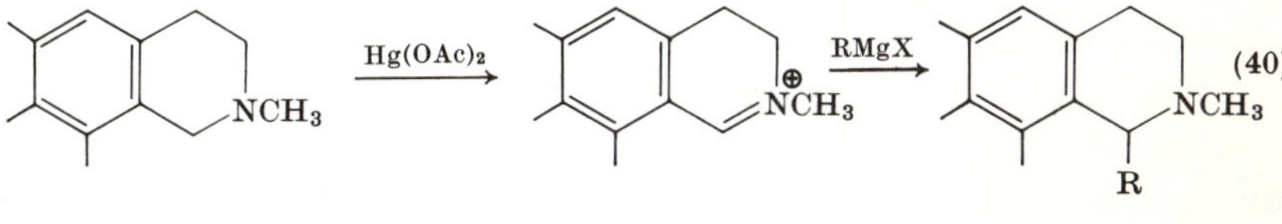

In the third example, the Reissert reaction[88] was used by Jackson and Stewart[74] to prepare still another 1-benzyl-7,8-dioxygenated isoquinoline [Eq. (41)]. The application of a Reissert reaction to a Schiff

[85] G. Grethe, H. L. Lee, and M. R. Uskoković, *Tetrahedron Lett.*, 1937 (1969).
[86] G. Grethe, H. L. Lee, M. R. Uskoković, and A Brossi, *Helv. Chim. Acta* **53**, 874 (1970).
[87] G. Grethe, M. Uskoković, and A. Brossi, *J. Org.Chem.* **33**, 2500 (1968).
[88] F. D. Popp, *Advan. Heterocycl. Chem.* **9**, 1 (1968).

base of the benzylideneaminoacetal type previously described [Eq. (12)] was successful,[71a] but the following ring closure was less successful.

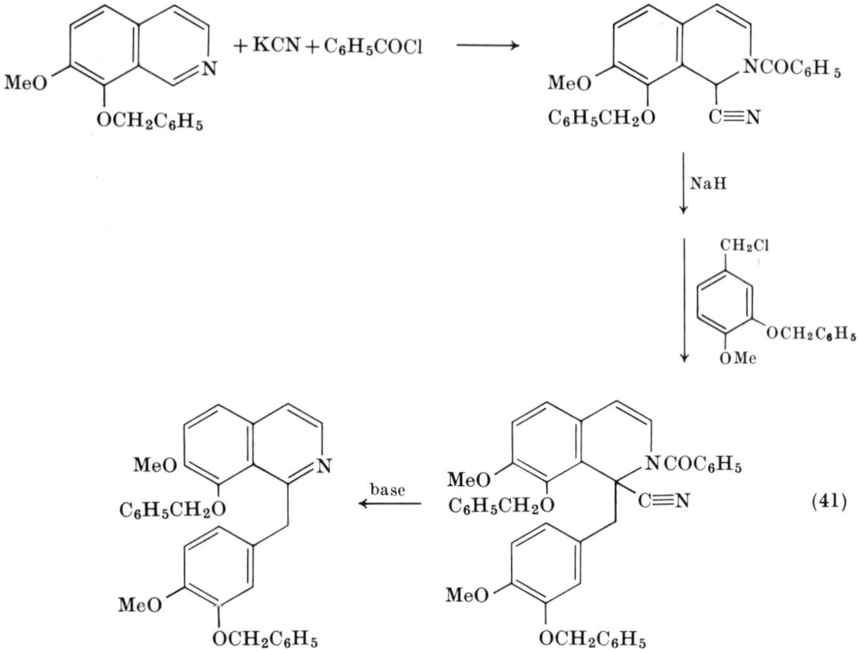

(41)

C. Substitution on Nitrogen

This is a normal N-alkylation reaction, but it may be of interest to summarize those methods which have recently been used in tetrahydroisoquinoline synthesis. N-Methylation is the most common reaction, and it can be brought about by various methods [Eqs. (42) and (43)]. The first is a reductive alkylation[30] using either hydrogenation over platinum[21] or $NaBH_4$[32, 34, 37] as the reducing agent. The second method involves the $LiAlH_4$ reduction of $N-CO_2Et$ groups, and it combines methylation with the removal of a commonly used acyl blocking group.[89] Either method can be used at almost any stage in the various syntheses described in this chapter. Methylation of aromatic isoquinolines by reduction of the methiodides is a classic method.[62]

The glycidol method [Eq. (20)] and the bromoacetal method [Eq. (17)] provide general routes to N-alkyl derivatives containing almost any alkyl group.

[89] M. P. Cava and K. T. Buck, *Tetrahedron* **25**, 2795 (1969).

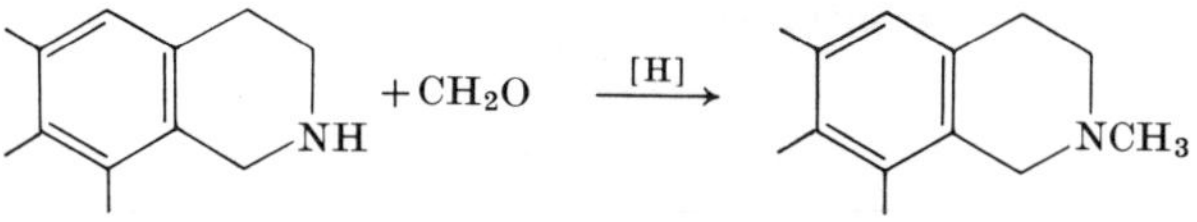

$$\text{[structure]} \xrightarrow{+CH_2O \; [H]} \text{[structure with NCH}_3\text{]} \quad (42)$$

$$\text{[structure with NCO}_2\text{Et]} \xrightarrow{LiAlH_4} \text{[structure with NCH}_3\text{]} \quad (43)$$

D. SUBSTITUTION AT THE 3-POSITION

Four methods are available for placing substituents in the 3-position, at least within the areas related to this paper. The route with greatest promise is through the Strecker synthesis via 4-ketotetrahydroisoquinolines [Eq. (24)].[46] A second route involves the addition of Grignard reagents to the Schiff bases formed from benzylamines and glyoxal hemiacetal [Eq. (19)].[18] Third, this type of substitution can be brought about by using amino acetals[31] [Eq. (14)] or haloacetals [Eq. (21)][40] other than acetaldehyde acetal or by using substituted glycines for the preparation of 4-ketotetrahydroisoquinolines [Eq. (23)].[52, 55-58] Finally, the keto group in 4-ketotetrahydroisoquinolines activates the 3-position sufficiently to undergo condensations with benzaldehyde to give benzylidene derivatives.[53]

Substituents can be placed in the 3-position by nucleophilic attack on 1,2-dihydroisoquinolines,[71] but thus far, only ring closures have been successful.

E. SUBSTITUTION AT THE 4-POSITION

This substitution can be brought about in at least three ways. The most general method may well be the addition of Grignard reagents to 4-ketotetrahydroisoquinolines [Eq. (27)] as originally carried out by Hinton and Mann[53] and more recently by Brossi and co-workers.[59a] The second method, nucleophilic displacement of the 4-oxy group of 4-hydroxytetrahydroisoquinolines [Eqs. (35)–(37) and reactions of **30**] has great promise, but is not completely explored or understood. The third method involving electrophilic attack at the 4-position of 1,2-dihydroisoquinolines has been extensively discussed by Dyke.[71]

VI. Tables of Specific Compounds (pp. 129–136)

In Tables I and II, the known types of 4-acetoxy-, 4-alkoxy-, and 4-hydroxy-1,2,3,4-tetrahydroisoquinolines are indicated. Table III summarizes the known 4-keto-1,2,3,4-tetrahydroisoquinolines. In Tables IV and V, the various classes of 1,2,3,4-tetrahydroisoquinolines and aromatic isoquinolines which have been made *by the methods discussed in this article* are given. In Tables VI and VII, the 4-phenyl-1,2,3,4-tetrahydroisoquinolines and bicyclic compounds made by nucleophilic displacement of the 4-hydroxy group are summarized. Twenty-four compounds e.g., (**31–40**) prepared from **30** with active CH,[78] NH, and SH[77] compounds have been reported; details may be found in the references cited.

In each table, the chief synthetic routes are indicated by the equation numbers in the text.

The various compounds that can be made from 4-hydroxytetrahydroisoquinolines via 1,2-dihydroisoquinolines [Eq. (32)] have been listed by Dyke.[71]

ACKNOWLEDGMENTS

The author would like to thank the National Cancer Institute of the National Institutes of Health (Grants CA-3437 and CA-3905) and the Department of the Army (Contract DA-49-193-MD-2948 of the antimalarials program) for financial support of his isoquinoline work over the years. He would like to thank Professor R. D. Topsom and LaTrobe University for their hospitality while the author wrote a portion of this paper.

TABLE I
4-ALKOXY- AND 4-ACYLOXY-1,2,3,4-TETRAHYDROISOQUINOLINES

			Substituent at position					
1	2	3	4	6	7	8	Routes[a]	Ref.
—	H, Me	H	OEt	H, Me, OMe	H, Me, OMe	H, Me, OMe	(13), (16)[b]	13–17
Ph	Me	H	OEt	H	H	H	[c]	16
H, Ph	H	Me, Et, Ph	OEt	H, Me	H	H	(19), (6)	18
H	Me	H	OEt	–O–CH$_2$–O–		H	$1 \rightarrow 2$	8
H	H, Ac		–glucose–O–[d]	H, OH, OMe	OH, OMe	H	(14)	31
Me	Me	H	OMe	OMe	H	H	(10)[e]	66
H	COPh	H	OMe, OEt	OMe	OMe	H	(10)	65
H	Me	H	O-Alkyl	OMe	OH	H	$30 \rightarrow 31$	58
H	Me	H	OAc	OMe	OH	H	$13 \rightarrow 14 \rightarrow 15$	57
H, Me	Bz	H	OCOEt	H	OMe	H	—	54

[a] See text. Equation number given in parentheses.
[b] 2-Methyl derivatives prepared by methylation of the Schiff base.
[c] Prepared by methylation of the 2H compound.
[d] As in text.
[e] Prepared by methylation of the 4-OH compound.

TABLE II

4-Hydroxy-1,2,3,4-Tetrahydroisoquinolines

			Substituent at position						Route[a]	Ref.
1	2	3	4 (+OH)	5	6	7	8			
H	H	H	H	H	H	H	H		**16→17**	70
H	Me	H	H	H	OMe	OMe	OMe		(10)	63
H	H	H	H	OMe	H	H	H		(10)	60
Me, Et, Ph and spiro-fused systems[b]	H	H	H	H	H or OH	H	H		(10)	66, 67, 67a
CO₂R (R=H, alkyl)	H, Me, or Et	H	H	H	OH	H	H		(12), (7)	64
H	H	H	H	↓	H, OH, OMe, OCH₂O (various)	H	↑		(12),[c] (7)	22
H	MeCO(CH₂)₂	H	H	H	H, OH, OMe, OCH₂O (various) →	OCH₂O (various) →			(12), (7)	23
H	H	H	H	H	H or OMe	OMe	OMe or H		(12), (7)	45, 47, 49
CH₂=CHCH₃	H	H	H	H	OMe	OMe	H		(13), (7)	35
Me	H	H	H	H	OMe	OMe	OH, OMe		(15), (7)	32, 34
o-NO₂–C₆H₄CH₂–	H	H	H	H	–O–CH₂–O–		H		(19), (7)	38

Sec. VI.] 4-OXY- AND 4-KETO-1,2,3,4-TETRAHYDROISOQUINOLINES

H	H or Me	H	H	H	OMe	OMe	H	(23), (26)	57
Me	H	H	H	H	H	OMe	H	(23), (26)	56
H, Me, or Ph	Me, or CH$_2$Ph	H	H	H	H	OMe	H	(22), (26)	54
H	Me	H	H	H	OMe	OH	H	13→14→15; saponification	57
Me	H or Me	H	H	H	OMe	OMe	H	(26)^d	59
CONHNH$_2$ and hydrazones	H, Me, or Et	H	H	H	OH	H	H	(10)	64
H	Me	H	H or aryl	H	H or OMe	H or OMe	H	(22), (26) or (27)	53, 59a
H	Me	H	H	H	—O—CH$_2$—O—		H	1→3	8
=O	H	H	H	H	H	H	H	18→19	72
=O	Ph, various alkyl	H	H	H	Me	Me	H	18→19	72
=O	H	Et	H	H	H	NH$_2$	H	18→19	72
Me	Et	H	H	H	OMe	OMe	H	^{a}	59, 32
H	H	Me	H	H	OMe or H	OMe	H or OH	—	49
H	C$_{10}$H$_{13}$O$_2$^{e}	H	H	H	OH	OH	H or OMe	(12), (7)^f	90

^{a} See text.
^{b} Spiro-C(CH$_2$)$_4$, -C(CH$_2$)$_5$, -C(CH$_2$CH$_2$)$_2$NCH$_2$Ph, and -C(CH$_2$CH$_2$)$_2$S.
^{c} The benzylaminoacetal reacts with methyl vinyl ketone.
^{d} N-Methyl compounds by methylation of N–H.
^{e} β-(3,4-Methylenedioxyphenyl)ethyl.
^{f} Via di-O-benzyl derivative.

90 S. T. Ross, Smith, Kline and French Laboratories, Philadelphia, unpublished results.

TABLE III
4-Keto-1,2,3,4-Tetrahydroisoquinolines

		Substituent at position[a]					Route	Ref.
1	2	3	6	7	8			
H	Me	H	H	H	H		(22)	53
H	PhCH$_2$	H	H	Cl	H		(22)	54
H	PhCH$_2$	H	H or OMe	OMe	H or OMe		(22), (23)	54, 56
Me	PhCH$_2$	H	OMe	OMe	H		(25)	59
Me	H	H	OMe	OMe	H		(23)	55
Me	Me	Me	H	OMe	H		(23)	56
Ph	Me	H	H	OMe	H		(23)	56
H	—	H	OMe	OMe	H		(24)[b]	52, 56
H	Me	H	OMe	OMe	H		(23)	56
H	H	Various aryl, alkyl, spiro-C$_5$H$_{10}$	OMe	OMe	H		(24)	52

[a] Position 5 unsubstituted throughout.
[b] Prepared by debenzylation of the N-benzyl derivative.

TABLE IV

1,2,3,4-TETRAHYDROISOQUINOLINE HYDROCHLORIDES FROM 4-OXY-1,2,3,4-TETRAHYDROISOQUINOLINES

Substituent at position[a]							
1	2	5	6	7	8	Route	Ref.
H	H	H or OMe	H or OMe	OMe or H	H or OH	(12), (7), (29)	9, 19, 87
H	Me	H	OH, OMe or $-O-CH_2-O-$		H	(12), (42), (7), (29)	21
H	Various alkyl	H	$-O-CH_2-O-$		H	(20)	21
H	H or Me	H	H	OMe	H or OMe	(22), (20), (29)	54
H	H	H	H	OMe	OMe	(12), (7), (29)	47
Me or Et	H	H	OH, OMe or $(OMe)_2$		H	(13), (7), (29)	25
H	H or Me	H	H, OH or OMe	OMe or OH	H or OH	(11), (7), (29)	24
H	Me	Me	H	OMe	OH	(11), (7), (29)	24
Me	H or Me	H	OMe	OMe	OH	(15), (7), (29)	32
Me	H or Me	H or OMe	H or OMe	OMe	OMe or H	(13), (7), (29)	34
H	Me	OMe	OMe	OMe	H	(12), (7), (29), (42)	34
Various alkyl	Me	H	OH, OMe or $(OMe)_2$		H	(13), (7), (42), (29)	25
R[b]	H	H	OMe	OMe	R[b]	(15), (7)	39

[a] Positions 3 and 4 unsubstituted throughout.

[b] This is ±-hexahydropronuciferine with complex substitution at positions 1 and 8.

TABLE V
Isoquinolines Prepared From 4-Oxy-1,2,3,4-Tetrahydroisoquinolines

	Substituent at position[a]						Route	Ref.
1	2	3	6	7	8			
Me, Et, Ph, p-tolyl	H	H	H, Me, OMe	H	H	(13), (6), (30)	14, 15
H or Ph	H	Me, Et, Ph	H or Me	H	H	(19), (6), (30)	18
H	H	H	OMe	OH	H	(12), (7), (30)	9
H	H	H	H	OMe	OMe	(12), (7), (30), (31)	47, 49
H	H	H	OMe	OMe	OMe	(30)	73
H	H	H	H	OMe	OCH$_2$Ph	**20→23**	74
H	H	Ph	OMe	OMe	H	(31)	52
H	C$_{10}$H$_{13}$O$_2$[b]	H	H or OMe	OMe	OMe or H	(31)	49

[a] Positions 4 and 5 unsubstituted throughout.
[b] 3,4-Dimethoxyphenethyl quaternary iodide.

TABLE VI

4-PHENYL-1,2,3,4-TETRAHYDROISOQUINOLINE HYDROCHLORIDES

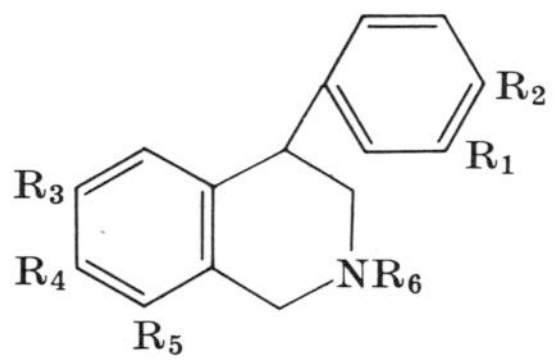

			Substituent				
R_1	R_2	R_3	R_4	R_5	R_6	Route	Ref.
OH	OH	H, OMe	OMe	H, OH	H	(12), (7), (37)	26
Me	OH	H, OMe	OH, OMe	H, OH	H	(12), (7), (37)	26
OMe	OH, OMe	OMe	OH, OMe	H	H	(12), (7), (37)	26
H	OMe	OMe	OMe	H	PhCH$_2$, H, Me	(27), (29)^a	59a

^{a} 4 − OH removed by HCl–NaBH$_4$; R_6 = H and Me prepared by successive debenzylation and methylation of the N-benzyl derivative.

TABLE VII

Bicyclic Isoquinoline Compounds

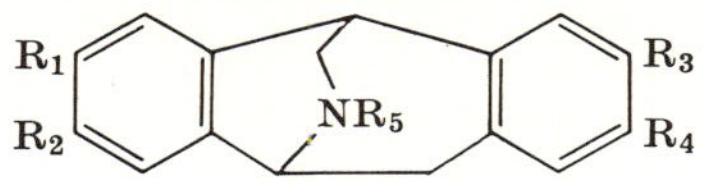

	Substituents					
R₁	R₂	R₃	R₄	R₅	Route	Ref.
OMe	OMe	OMe	OMe	H	(34)	11, 36
OMe	OMe	OMe	OH, OMe	Me	(34) and ^{a}	37, 71
—OCH₂O—		—OCH₂O—		H, Me	(34) and ^{a}	71
—OCH₂O—		OMe	OMe	H, Me	(34) and ^{a}	71
OMe	OMe	OMe	OMe	Me	(34)	76, 50
H, OH or OMe	H or OH	H, OH or OMe	·HCl		(36)	26
OH or OMe	OH or OMe	OH or OH				
H or OMe	H or OH	OMe	·HCl		(35)	49
OH or OMe		OMe				

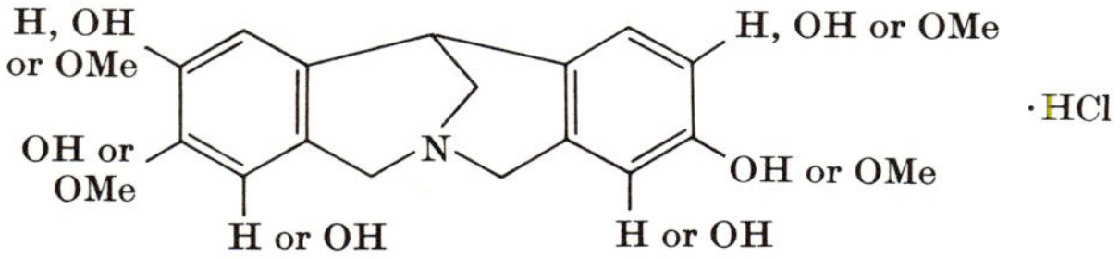

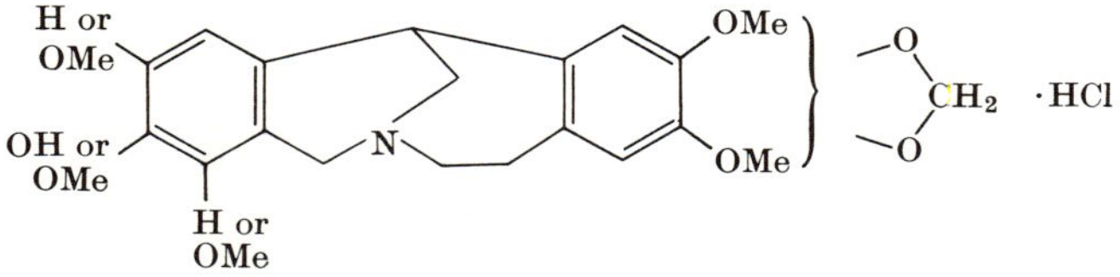

^{a} Prepared by methylation of the N–H derivative.

Isotopic Hydrogen Labeling of Heterocyclic Compounds by One-Step Methods

G. E. CALF and J. L. GARNETT

Chemistry School, The University of New South Wales, Kensington, N.S.W., Australia

I.	Introduction	137
II.	Radiation-Induced Methods	138
	A. Recoil Tritiation	138
	B. Wilzbach Exchange	139
	C. Modified Wilzbach Methods	147
	D. Radiation Synthesis	148
III.	Heterogeneous Metal-Catalyzed Exchange	149
	A. Types of Catalysts	150
	B. Reaction Procedures	152
	C. Mechanisms of Exchange	153
	D. Exchange in Specific Heterocyclic Systems	160
IV.	Current Trends in Labeling	174
	A. Homogeneous Metal Catalysis	175
	B. Acid–Base Catalysis	178
	C. Radiation-Induced Exchange with Tritium Oxide	182
	D. Comparison of Labeling Techniques.	183

I. Introduction

Molecules labeled with deuterium and/or tritium are useful for fundamental investigations in a variety of chemical fields, particularly spin decoupling in NMR,[1] the measurement of coupling constants in ESR,[2] mass spectrometry,[3,4] and general reaction mechanism studies. Such tagged compounds are also valuable as tracers in biochemical and medical research.

It is the purpose of this chapter to review the development of essentially one-step methods of labeling heterocyclic molecules with isotopic

---

[1] J. L. Garnett, L. J. Henderson, W. A. Sollich, and G. V. D. Tiers, *Tetrahedron Lett.*, 516 (1961).
[2] G. H. Aylward, J. L. Garnett, and H. Sharp, *Chem. Commun.*, 137 (1966).
[3] F. W. McLafferty, *Chem. Commun.*, 78 (1966).
[4] H. Budzikiewicz, C. Djerassi, and D. H. Williams, "Interpretation of Mass Spectra of Organic Compounds. Structure Elucidation of Natural Products by Mass Spectrometry," Vols. I and II. Holden-Day, San Francisco, 1964.

hydrogen. Chemical synthetic procedures for such labeling have already been summarized elsewhere.[5] In addition to the obvious advantages of convenience and high yields, one-step methods can produce both specifically and generally labeled molecules. The present labeling methods can be divided into radiation-induced and catalytic procedures. The radiation-induced methods are applicable only to tritium, whereas catalytic techniques can yield deuterated and/or tritiated compounds. Emphasis in the catalytic methods will be placed on both homogeneous and heterogeneous exchange in the presence of metals; however, a brief discussion of relevant acid–base work will also be included.

II. Radiation-Induced Methods

Three types of radiation-induced procedures have been used including (a) recoil tritiation,[6] (b) Wilzbach gas irradiation[7] with its modifications, and (c) radiation synthesis.[8] All these methods utilize high-energy ionizing radiation to achieve tritium incorporation. Wilzbach exchange[7] is the most satisfactory of the three and has been the most widely used. Radiation synthesis is a relatively recent development of limited but valuable specificity.

A. Recoil Tritiation

This method involves neutron irradiation of a mixture containing the organic compound and a lithium salt such as the carbonate or oxalate. Energetic tritons are generated by a $Li(n,\alpha)T$ reaction, the recoiling tritium atom (T) displacing hydrogen in the organic molecule. The activity (A) of the recoil atoms produced in the nuclear reaction [Eq. (1)] may be calculated from Eq. (2)

$$^{6}_{3}Li + ^{1}_{0}n \rightarrow ^{4}_{2}He + ^{3}_{1}H \tag{1}$$

$$A = \frac{wf\sigma\theta N}{M}(1 - e^{-0.693t/T}) \tag{2}$$

---

[5] E. A. Evans, "Tritium and Its Compounds." Butterworths, London, 1966.
[6] R. Wolfgang, F. S. Rowland, and C. N. Turton, *Science* **121**, 715 (1955).
[7] K. E. Wilzbach, *J. Amer. Chem. Soc.* **79**, 1013 (1957).
[8] A. Ekstrom and J. L. Garnett, *J. Label. Compounds* **4**, 43 (1968).

where σ = reaction cross section (945 barns), w = weight of element (lithium) of natural isotopic abundance (grams), f = reactor neutron flux, M = atomic weight of Li, θ = fraction abundance of target isotope, (^{6}Li) in the natural element, N = Avogadro's number, t = irradiation time, and T = half-life of isotope. For crystalline materials, an intimate mixture of the organic compound and inorganic salt can be ground together; however, for liquids, a gel using a substance such as lithium myristate may be used.[9, 10]

For tritiating heterocyclic compounds the recoil method has been used to label nicotinic acid (**1**)[11–13] and also reserpine.[14, 15]

(**1**)

As a preparative labeling tool it possesses very limited potential, since the level of specific activities attainable is relatively low ($\simeq 1$ μCi/mg) and also long irradiation times are necessary in the nuclear reactor leading to significant chemical degradation of the parent compound. Generally, isotopic incorporation is nonuniform and nonspecific within a molecule. The method has now largely been superseded by other procedures including Wilzbach gas irradiation.[7]

B. Wilzbach Exchange

1. Scope of Method

Technically, any organic molecule containing a C–H bond can be labeled in one step by this procedure[7] which involves exposure of an organic compound to tritium gas. The β radiation from the tritium

---

[9] W. G. Brown and J. L. Garnett, *Int. J. Appl. Radiat. Isotopes* **5**, 115 (1959).
[10] H. Krizek, A. J. Verbiscar, and W. G. Brown, *J. Org. Chem.* **29**, 3443 (1964).
[11] F. S. Rowland and R. L. Wolfgang, *Nucleonics* **14** (8), 58 (1956).
[12] R. F. Dawson, D. R. Christman, A. D'Adamo, M. L. Salt, and A. P. Wolf, *J. Amer. Chem. Soc.* **82**, 2628 (1960).
[13] A. N. Nesmeyanov, B. G. Dzantiev, V. V. Pozdeev, and Yu. M. Rumyantsev, *Proc. Conf. Use Radioisotop. Phys. Sci. Ind.*, 1960 **2**, 130 (1962).
[14] F. S. Rowland and P. Numerof, *Int. J. Appl. Radiat. Isotop.* **1**, 246 (1957).
[15] H. Sheppard, W. H. Tsien, A. J. Plummer, E. A. Peets, B. J. Giletti, and A. R. Schubert, *Proc. Soc. Exp. Biol. Med.* **97**, 717 (1958).

induces exchange between the hydrogen of the organic molecule and the tritium gas [Eq. (3)]. The efficiency of labeling can be increased by

$$RH + T_2 \rightarrow RT + HT \qquad (3)$$

agitation;[16] the tritium gas should be free from hydrogen, but the decay product ^{3}He does not affect the reaction in low concentrations.

The main limitation of the technique is the problem of achieving *radiochemical*, as distinct from *chemical*, purification of the final labeled product. Purification of the irradiated sample is necessary to remove radiation-induced degradation products and labile tritium. Many of the degradation products are not only chemically similar to the parent compound, but also possess much higher specific activities. For complete purification it is necessary to use multistage processes, such as gas and column chromatography, countercurrent distribution, and fractional distillation. Distribution of isotope within a molecule is generally random and nonuniform; however, in some circumstances useful specificity can be achieved.[17]

Although organic gases can be labeled by the present technique, for heterocyclic systems the most useful application is in condensed phases. Under these conditions, three principal mechanisms have been proposed.[18-20]

(*a*) Recoil labeling [Eqs. (4) and (5)] or labeling initiated by the decay species (HeT)$^+$.

$$T_2 \rightsquigarrow (HeT)^+ + e \qquad (4)$$

$$(HeT)^+ + RH \rightarrow RT + (HeH)^+ \qquad (5)$$

(*b*) Beta labeling due to initiation by electrons [Eq. (6)].

$$T_2 + RH + e \rightarrow RT + HT \qquad (6)$$

(*c*) Reaction of excited tritium molecules and atoms with the organic compound and/or reaction of excited organic molecules with molecular tritium [Eqs. (7)–(13)].

---

[16] J. Rydberg and A. Hanngren, *Acta Chem. Scand.* **12**, 332 (1958).
[17] S. J. Angyal, C. M. Fernandez, and J. L. Garnett, *Aust. J. Chem.* **18**, 39 (1965).
[18] P. L. Gant and K. Yang, *J. Chem. Phys.* **32**, 1757 (1960).
[19] T. H. Pratt and R. Wolfgang, *J. Amer. Chem. Soc.* **83**, 10 (1960).
[20] J. L. Garnett and S. W. Law, *Aust. J. Chem.* **20**, 1875 (1967).

$$T_2 \longrightarrow T_2^* \longrightarrow 2T \qquad (7)$$

$$2T + RH \longrightarrow RT + HT \qquad (8)$$

$$RH \longrightarrow RH^+ + e \qquad (9)$$

$$RH^+ + T_2 \longrightarrow RT + HT \qquad (10)$$

$$RH \longrightarrow RH^* \qquad (11)$$

$$RH^* + T_2 \longrightarrow RT + HT \qquad (12)$$

$$RH^* \longrightarrow R\cdot + H \xrightarrow{T_2} RT + T \qquad (13)$$

A number of authors[20, 21] have found that both ionic and radical processes contribute to the tritiation mechanism; however, it is obvious why Wilzbach labeling is accompanied by the formation of tritiated by-products of high specific activity, since the energy released in the disintegration of the tritium is absorbed by the system and provides the necessary activation energy required to effect labeling.[22] Generally polymer formation during radiation-induced labeling is higher in heterocyclic systems than with carbocyclic aromatic compounds.[23–25]

The results[5, 20, 22, 26–31] of the Wilzbach labeling of a representative number of heterocyclic compounds are shown in Table I. A significant feature of the data is the relatively high specific activity in the crude product immediately after exposure to tritium gas compared with the ultimate activity in the purified compound. For example, the specific activity of crude benzothiophene (777 μCi/mg) dropped to 155 μCi/mg in the purified compound. Comparable data for N-methylpyrrole and trimethylpyridine were 247, dropping to 110, and 423, dropping to

---

[21] P. Y. Feng and T. W. Greenlee, *Proc. Symp. Tritium Phys. Biol. Sci.*, *1961* **2**, 11 (1962).

[22] M. Wenzel and P. E. Schulze, "'Tritium Markierung' Preparation Measurement and Uses of Wilzbach Labelled Compounds." De Gruyter, Berlin, 1962.

[23] G. Fletcher and J. L. Garnett, unpublished work.

[24] A. Ekstrom and J. L. Garnett, *Proc. 2nd Int. Conf. Methods Preparing Storing Label. Compounds, 1966*, 961 (1968).

[25] S. Apelgot and M. Duquesne, *Advan. Tracer Methodol.* **3**, 1961 (1966).

[26] M. L. Whisman, F. G. Schwartz, and B. H. Eccleston, *U.S. Bur. Mines Rep. Invest* **5717** (1961).

[27] M. Wenzel and P. Kortge, *Naturwissenschaften* **48**, 431 (1961).

[28] C. Rosenblum, *Nucleonics* **17** (12), 80 (1959).

[29] M. Wenzel, H. Wollenberg, and E. Schutte, Lecture, Natur. Substances Symp. Brussels, June, 1962.

[30] M. Wenzel, H. Wollenberg, and P. E. Schulze, *Proc. Symp. Tritium Phys. Biol. Sci.*, *1961* **2**, 37 (1962).

[31] J. Pany, *Naturwissenschaften* **46**, 515 (1959).

TABLE I

WILZBACH TRITIUM LABELING OF HETEROCYCLIC COMPOUNDS

Compound	Weight (gm)	Tritium (Ci)	Time (days)	Specific activity (μCi/mg)	Ref.
Acridine	0.1	2	3	0.11	20
Benzo[b]thiophene	1	2.5	21	155	26
Diethylbarbituric acid	0.09	5	11	1.6	22
6-Dimethylamino-9- (3'-aminoribosylpurine)	0.35	4.5	12	8	27
Imidazole dicarboxylamide	1.95	3.24	17	8	28
N-Methylpyrrole	1	2.18	37	110	26
Nicotinamide	0.06	5	5.5	0.18	22,29
Phenylethylbarbituric acid	0.23	4	11	0.12	30
Thymidine	0.1	4.5	10	174	31
Thymidine	0.3	100	21	322	5
2,4,6-Trimethylpyridine	1	1.83	28	103	26
Tryptamine hydrochloride	0.1	20	8	262	5

103, μCi/mg, respectively. These results emphasize the extreme care required in the radiochemical purification process to remove the highly radioactive impurities.

2. *Orientation of Isotope*

Regarding the labeling of substituted heterocyclics, little work has been done by the Wilzbach process. Trager and Wacker[32] exposed compounds in the pyrimidine series (**2**)–(**4**) to tritium; unfortunately, the number of curies of tritium used was not reported, although it was constant for all compounds. The maximum level of activity incorporated into the molecules is summarized in Table II.

Halogen substitution in the 5-position of the pyrimidines leads to very little tritium incorporation whereas activity at the mCi/mM level

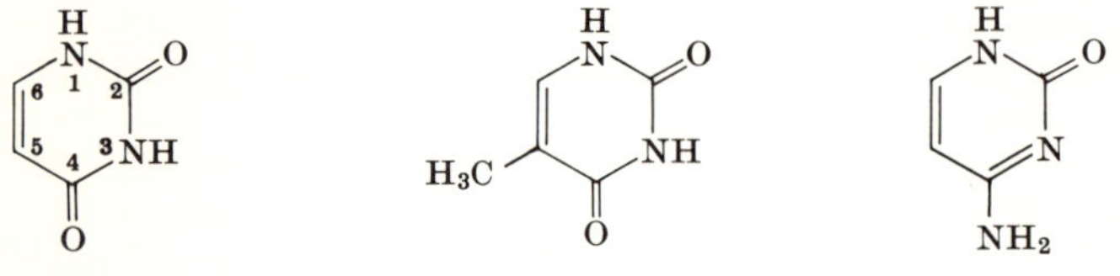

[32] L. Trager and A. Wacker, *Advan. Tracer Methodol.* **3**, 57 (1966).

TABLE II

Wilzbach Labeling of Pyrimidine and Purine Derivatives[32]

Compound	Specific activity (mCi/mmole)
Uracil (2)	5
Thymine (3)	4
Cytosine (4)	0.7
5-Fluorouracil	$< 0.5 \times 10^{-3}$
5-Bromouracil	$< 0.5 \times 10^{-3}$
5-Iodouracil	$< 0.5 \times 10^{-3}$
Xanthine (5)	0.3
Hypoxanthine (6)	1.5
Purine (7)	6
6-Aminopurine (adenine)	8
2-Aminopurine	110

is obtained in the unsubstituted compound. The substituent at C-5 blocks tritium substitution at C-6, rather than inducing reactivity at C-6, since, if tritium-labeled uracil is brominated, tritiated 5-bromouracil is obtained. In the labeling of 5-iodouracil, partial dehalogenation to tritiated uracil is observed, this being consistent with dehalogenation reported during the tritiation of iodobenzene derivatives.[21]

Trager and Wacker[32] also labeled purines (5)–(7) under the same conditions as their pyrimidine series. The large differences in tritium

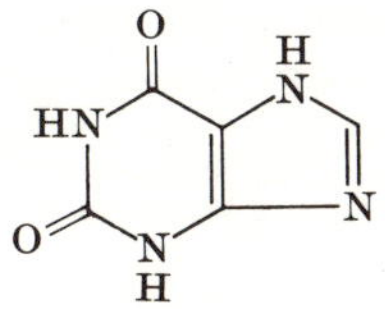

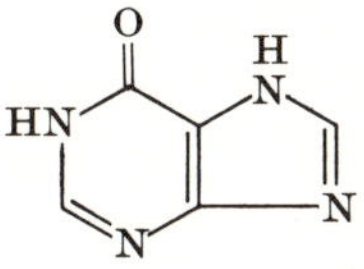

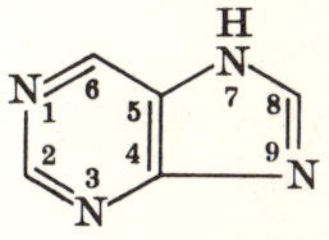

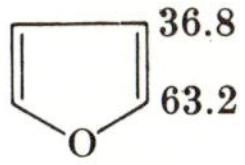

incorporation between 2-aminopurine and 6-aminopurine could not be explained; however, the authors suggested that the contribution from the imino form of 2-aminopurine may be of significance.

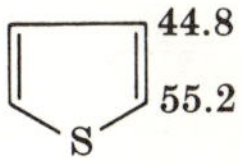

Two other simple heterocyclic compounds which have been labeled by the Wilzbach procedure are furan and thiophene,[33] exchange being faster in the α positions of each molecule (8) and (9).

Despite the paucity of data for orientation studies with substituted heterocyclics labeled by the Wilzbach method, extensive work has been done with analogous aromatic compounds and, thus, by extrapolation, a number of general predictions can be made for heterocyclic systems. For example, in toluene[34] tritium incorporation in the ring predominates over side-chain labeling.

The reactivities differ for ortho, meta, and para positions of toluene. In anthranilic acid (10), positions 3 and 5 in the molecule possess 49.8 and 27.4% of the total activity; thus, tritium incorporation tends to be highest at positions of highest electron density.[35] In isopropanol,[35] the hydroxyl group incorporates approximately three times as much isotope as any other hydrogen position, while tritiation of the secondary C–H is slightly more efficient than the primary C–H. In the aliphatic esters of aromatic carboxylic acids, the aryl hydrogens tritiate ten times faster than alkyl hydrogens.[10, 36]

$$\begin{array}{c} \text{COOH} \\ 11.4 \underset{11.4}{\overset{1}{\underset{6}{\bigcirc}}} \overset{2}{\underset{4}{}} \text{NH}_2 \\ 27.4 \qquad 49.8 \end{array}$$

(10)

3. Effect of Phase and Tritium Pressure on Tritiation

For the tritiation of gases and liquids, exchange occurs efficiently; however, for the labeling of crystalline solids, particle size is important.[20] Studies with phenanthrene show that at room temperature a linear relationship [Eq. (14)] exists between log specific activity (A in μCi/mg) and the inverse square of τ, where τ (in mm) is half the average mesh diameter of the particles.

$$A = 0.232 \exp(2.37 \times 10^{-4}/\tau^2) \tag{14}$$

In other preliminary surface area studies, Rosenblum and Meriwether[37] concluded that surface structure was more important than

---

[33] F. Cacace, *Proc. Symp. Chem. Effects Nucl. Transformations*, **2**, 155 (1961).
[34] H. J. Ache, W. Herr, and A. Thiemann, *Proc. Symp. Tritium Phys. Biol. Sci.*, *1961* **2**, 21 (1962).
[35] J. L. Garnett, S. W. Law, and A. R. Till, *Aust. J. Chem.* **18**, 297 (1965).
[36] B. R. Crawford and J. L. Garnett, *Aust. J. Chem.* **18**, 1951 (1965).
[37] C. Rosenblum and H. T. Meriwether, *Advan. Tracer Methodol.* **1**, 12 (1963).

surface extent. Wenzel and co-workers[30] increased the surface area by adsorbing molecules on charcoal prior to tritiation. The trends observed for three heterocyclic compounds are shown in Table III, where a fivefold

TABLE III

WILZBACH LABELING OF HETEROCYCLES ADSORBED ON CHARCOAL

Compound	Weight (gm)	Tritium (Ci)	Charcoal (gm)	Time (days)	Specific activity (μCi/mg)	Ref.
Diethylbarbituric acid	0.09	5	0	11	1.6	22,29
Diethylbarbituric acid	0.08	5	1.23	11	5.1	22,29
Nicotinamide	0.06	5	0	5.5	0.18	22,29
Nicotinamide	0.06	5	0.436	5.5	0.9	22,29
Phenylethylbarbituric acid	0.15	4	0	11	0.12	30
Phenylethylbarbituric acid	0.228	4	1	11	0.226	30

increase in specific activity for charcoal-adsorbed nicotinamide is obtained, whereas a twofold increase is found with phenylethylbarbituric acid.

Tritium pressure above the organic compound is an important variable in determining the specific activity incorporated into crystalline solids.[20, 38] The specific activity of phenanthrene at constant particle size varies with the square of the tritium pressure. Pascual and Wilzbach[38] have found that in the labeling of benzoic acid, tritiation of the parent acid was pressure-independent, whereas the formation of labeled by-products was pressure-dependent. Subsequent work[36] has shown that in such compounds initial rapid exchange of the carboxylic acid hydrogen occurs to give appreciable HT in the system. Thus, compounds with labile hydrogen (e.g., COOH, NH_2, SH) appear to exhibit a tritium pressure dependency different from other compounds.

4. *Comparison of Wilzbach Labeling of Heterocyclics with That of Aromatic Hydrocarbons*

Although most Wilzbach labeling work published has concerned non-heterocyclic molecules, sufficient heterocyclic compounds have been tritiated by this procedure for satisfactory predictions of labeling patterns

[38] K. E. Wilzbach, *Proc. Symp. Tritium Phys. Biol. Sci.*, 1961 **2**, 7 (1962).

to be made for most heterocyclics. Comparison of the reactivity of acridine with the polycyclic aromatic hydrocarbons, particularly anthracene, is instructive.[20] Under conditions of constant particle size, temperature, and tritium pressure, studies show that acridine is more reactive by an order of magnitude than anthracene. Generally, crystalline heterocyclic compounds are more reactive than the corresponding aromatic hydrocarbons in radiation-induced tritium labeling.

5. Stereochemistry

Stereochemical changes may occur when tritium replaces a hydrogen atom in the condensed phase. Wilzbach[38] has reported 30% racemization during tritiation of mandelic acid. Brown, Gordon, and Intrieri[39] reported the absence of diastereoisomers from the tritiation of 1,2,3,5-tetra-O-acetyl-D-ribofuranose, while other authors[17, 40, 41] obtained a predominance of retention of configuration in crystalline (−)-inositol, but appreciable racemization in liquid inositol hexamethyl ether. The myoinositol (11) obtained from irradiated crystalline (−)-inositol (12) as a by-product was specifically labeled at the carbon on which the configurational inversion had occurred.

In the labeling of solid and liquid alcohols, Crawford and Garnett[36] reported predominant, but not exclusive, retention of configuration with (+)- and (−)-2-octylhydrogen phthalates, whereas octan-2-ol showed 80% racemization.

Of particular importance to heterocyclic chemistry is the labeling of optically active amino acids. Preliminary results[5, 42] with methionine, proline, tryptophan, and glutamic acid show that significant racemization can occur even in the crystalline phase. In a systematic study of the

[39] G. B. Brown, M. P. Gordon, and O. M. Intrieri, *J. Amer. Chem. Soc.* **80**, 5161 (1958).
[40] S. J. Angyal, J. L. Garnett, and R. M. Hoskinson, *Nature (London)* **191**, 485 (1963).
[41] S. J. Angyal, J. L. Garnett, and R. M. Hoskinson, *Aust. J. Chem.* **16**, 252 (1963).
[42] J. H. Parmentier, *J. Label. Compounds* **2**, 367 (1966).

amino acids,[43] including alanine, benzoylalanine, and phthaloylglutamic acid, it has been found that racemization in the group during Wilzbach labeling decreases from alanine (13) to phthaloylglutamic acid (14). This has been related to the presence of the NH bond which is known to be sensitive to rupture by ionizing radiation. In particular, loss of hydrogen to give an imine intermediate [Eq. (15), $R = R_1 = H$] has been shown to be the predominant radiolysis pathway.[43,44]

$$CH_3-CH(R'NR)-COOH \longrightarrow CH_3-C(=NR)-COOH + R'H \quad (15)$$

(13) $CH_3-CH(NH_2)-COOH$

(14) $HOOC-CH_2-CH_2-CH(COOH)-N(\text{phthaloyl})$

C. Modified Wilzbach Methods

A number of methods for increasing the efficiency of Wilzbach labeling have been proposed. A significant reduction in irradiation time from days to minutes can be achieved by passing an electric discharge through the tritium gas during the exposure,[45,46] the simplest source for the electric discharge being a Tesla-coil leak tester.[48] A number of heterocyclic compounds labeled in this manner are shown in Table IV.

Alternative methods for accelerating the Wilzbach procedure include irradiation with ultraviolet light,[47-49] X-irradiation,[45,50] use of microwaves,[51] and addition of iodine.[52] The accelerated labeling is

[43] J. L. Garnett, S. W. Law, J. H. O'Keefe, B. Halpern, and K. Turnbull, *Chem. Commun.*, 323 (1969).
[44] A. J. Swallow, "Radiation Chemistry of Organic Compounds." Pergamon, London, 1960.
[45] R. M. Lemmon, B. M. Talbert, W. Strohmeier, and I. M. Whittemore, *Science* **129**, 1740 (1959).
[46] L. M. Dorfman and K. E. Wilzbach, *J. Phys. Chem.* **63**, 799 (1959).
[47] N. A. Ghanem and T. Westermark, *J. Amer. Chem. Soc.* **82**, 4432 (1960).
[48] F. Cacace, E. Ciranni, G. Cirrani, and G. Montefinale, *Energ. Nucl.* **8**, 561 (1961).
[49] H. J. Ache, W. Herr, and A. Thiemann, *Proc. Conf. Chem. Effects Nucl. Transformations* **2**, 111 (1961).
[50] R. W. Ahrens, M. C. Sauer, and J. E. Willard, *J. Amer. Chem. Soc.* **79**, 3285 (1957).
[51] T. Westermark, H. Lindroth, and B. Enander, *Int. J. Appl. Radiat. Isotop.* **7**, 331 (1960).
[52] H. J. Ache, W. Herr, and A. Thiemann, *Angew. Chem.* **73**, 707 (1961).

TABLE IV

WILZBACH LABELING IN THE PRESENCE OF ELECTRIC DISCHARGE[a]

Compound	Tritium (Ci)	Specific activity (μCi/mg)
2-Isobutylthiophene	2.5	74
2-Methylpyridine	3.28	89
Quinoline	2.88	107

[a] Each heterocycle (1 gm) exposed for 0.04 days.[26]

accompanied by an increase in yield of decomposition products; thus, purification becomes more difficult and in many cases virtually impossible.[53, 54]

D. RADIATION SYNTHESIS

Previous radiation-induced labeling methods discussed normally produce random isotope incorporation, although in isolated compounds useful specificity has been achieved. Recently,[8] direct radiation synthesis has been proposed as a unique method for specific tritium labeling. The procedure will possess only limited usefulness, but may be valuable for specific labeling under conditions where the only alternative would be tedious chemical procedures. Wolf and co-workers[55] have shown how an analogous radiation synthesis procedure can be used for ^{14}C-labeling; thus, the method appears to be of general applicability. Preliminary tritium work has utilized aromatic molecules such as benzene and also derivatives of pyridine.[23]

The principle of the method involves the exposure of a mixture of two compounds (A) and (B) to ionizing radiation when a number of cross-products are formed from the fragments of the parent compounds. If A or B is labeled with tritium prior to irradiation, then the cross-products from the radiolysis will contain tritium in specific parts of the molecule. Ideal systems for the present labeling method are those in which A is an aliphatic molecule and B is aromatic or heterocyclic. Benzene and pyridine are suitable B components since these compounds can be labeled with tritium to very high specific activities by catalytic

[53] K. E. Wilzbach, *Atomlight* **15**, 1 (1960).
[54] P. Klubes and M. O. Schultze, *Int. J. Appl. Radiat. Isotop.* **14**, 241 (1963).
[55] S. Oae, C. S. Redvanly, and A. P. Wolf, *J. Label. Compounds* **4**, 54 (1968).

methods (Section III). Aliphatic molecules are suitable second components since these fragment readily during radiolysis. The fragments thus formed are then scavenged by the more radiation-stable benzene or pyridine.

The yields of scavenging products from direct irradiation are small in terms of a macro effect, typical G^* values ranging from 0.05 to 0.30. However, tritium is radioactive and, thus, possesses great sensitivity in detection. In practice, relatively large volumes of solution can be irradiated (if necessary) and concentrated, or preparative gas chromatography can be used to isolate a pure product in quantities required for a particular chemical reaction. Isotope dilution may also be used to obtain high chemical yields of tritiated compounds.

In the irradiation of pyridine–methanol mixtures, the predominant cross-products are α-picoline, α-pyridylmethanol [Eq. (16)] and β-2-pyridylethanol. In the irradiation of pyridine–cyclohexane mixtures α-pyridylcyclohexane is the cross-product obtained in major yield with smaller quantities of the isomers [Eq. (17)]. If tritiated pyridine is used in the irradiation, then each of these scavenging products will be specifically labeled. Under conditions such as these, the method can be of use for the one-step labeling of individual heterocycles.

$$\text{pyridine} + CH_3OH \longrightarrow \text{2-methylpyridine} + \text{2-(hydroxymethyl)pyridine} \quad (16)$$

$$\text{pyridine} + C_6H_{12} \longrightarrow \text{2-cyclohexylpyridine} + \text{isomers} \quad (17)$$

III. Heterogeneous Metal-Catalyzed Exchange

By contrast with the radiation-induced procedures, isotope exchange reactions catalyzed by Group VIII transition metals are applicable to *both* deuterium and tritium labeling of heterocyclic compounds. Because of recent mechanistic developments in this field, it is possible to predict with some degree of certainty the reactivity of a molecule for deuteration and also for moderate levels of tritiation. If compensation for additional radiation-induced interactions is made, then the theory also satisfactorily explains high specific activity tritiations.

---

* G is the number of molecules formed per 100 eV of energy absorbed.

Two types of metal-catalyzed procedures are now available for these reactions. Extensive studies have been made with *heterogeneous* systems which utilize a wide range of Group VIII transition metals. Recently, *homogeneous* metal-catalyzed systems have also been discovered.[56] In this section, *heterogeneous* metal-catalyzed exchange only will be discussed and current *homogeneous* developments will be treated in the final section of this chapter.

Heterogeneous metal catalysis is the most useful general method for the deuteration and/or tritiation of heterocyclic compounds.[57] It involves exchange between the organic substrate and isotopic hydrogen (as water, usually, or gas) in the presence of a Group VIII transition metal catalyst at temperatures up to 180° [Eq. (18)].

$$C_5H_5N + D_2O \rightleftharpoons C_5H_4DN + HOD \qquad (18)$$

Catalysts used include platinum, palladium, nickel, cobalt, iron, ruthenium, iridium, rhodium, and osmium[58] in supported and unsupported form. For labeling work, particularly with tritium, isotopic water[57] is to be preferred to enriched hydrogen gas[59] as source of isotope, since during exchange with hydrogen gas, competing hydrogenation may occur yielding by-products which are highly active and may render difficult or even preclude radiochemical purification of the parent compound.[60] However, for mechanistic studies of the labeling reaction, exchange with isotopic hydrogen, especially on evaporated metal films,[61] has been valuable. Trends in reactivity on evaporated metal films and prereduced catalysts are similar for large numbers of compounds. The emphasis in the present chapter is on preparative labeling so that only the prereduced procedures will be considered in detail; however, relevant evaporated metal film work will be mentioned to support the discussion.

A. Types of Catalysts

The most efficient catalysts for these exchange reactions, particularly with isotopic water, are the oxides and chlorides of the metals previously mentioned. For simplicity, unsupported catalysts are

---

[56] J. L. Garnett and R. J. Hodges, *J. Amer. Chem. Soc.* **87**, 4546 (1967).
[57] J. L. Garnett, *Proc. 2nd Int. Conf. Methods Preparing Storing Label. Compounds, 1966*, 709 (1968).
[58] J. L. Garnett, *Symp. No. 1, 4th Int. Congr. Catal., 1968*, in press.
[59] C. Kemball, *Proc. 4th Int. Congr. Catal., 1968*, in press.
[60] M. A. Long, A. L. Odell, and J. M. Thorp, *Radiochim. Acta* **1**, 174 (1963).
[61] E. Crawford and C. Kemball, *Trans. Faraday Soc.* **58**, 2452 (1962).

favored; however, certain nickel-on-kieselguhr catalysts possess useful specificity with pyridine,[62] although cobalt in unsupported form is even better.[63] Platinum is the most active catalyst for general exchange in heterocyclic molecules;[64, 65] however, certain of the remaining Group VIII metals are valuable for selective exchange in compounds such as thiophene and furan.[63]

Three methods for catalyst activation are currently in use; these include (*i*) hydrogen or deuterium reduction, (*ii*) borohydride reduction,[66, 67] and (*iii*) self-activation.[68, 69] Self-activation [Eq. (19)]

$$C_5H_5N + PtO_2 \rightarrow Pt(\text{self-activated}) + C_5H_4N - NC_5H_4 + H_2O \quad (19)$$

does not involve prereduction of the catalyst, but *in situ* reduction of the oxide[68] or chloride[70] by the heterocycle during exchange. Some heterocyclic compounds activate catalysts more readily than do others, e.g., pyridine is slower than quinoline or isoquinoline, whereas furan and thiophene are very fast. For efficient self-activation of platinum oxide, temperatures of 90° are required, whereas for the remaining oxides, higher temperatures[71] (up to 180°, ReO_2) are necessary. The process of self-activation has been observed by electron spin resonance spectroscopy for the reaction between pyridine and palladium chloride.[72] The main difficulty with self-activation is that small amounts ($\simeq 1\%$) of dimer with high isotopic incorporation are formed during the activation and, for tritium labeling, this material must be removed during radiochemical purification. However, for most deuteration work, this problem is not so important and the technique is useful. Metal chlorides normally self-activate more readily than the corresponding oxides,[70, 73] however, the resulting hydrochloric acid tends to poison heterogeneous exchange with certain compounds, and so the technique should be used with caution.

[62] C. G. Macdonald and J. S. Shannon, *Aust. J. Chem.* **18**, 1009 (1965).
[63] G. E. Calf and J. L. Garnett, *Chem. Commun.*, 306 (1967).
[64] J. L. Garnett, *Nucleonics* **20**, 86 (1962).
[65] K. Hirota and T. Ueda, *Proc. 3rd Int. Congr. Catal., 1964* 1238 (1965).
[66] C. A. Brown and H. L. Brown, *J. Amer. Chem. Soc.* **84**, 1494, 2827 (1962).
[67] G. E. Calf and J. L. Garnett, *J. Phys. Chem.* **68**, 3887 (1964).
[68] J. L. Garnett and W. A. Sollich, *Aust. J. Chem.* **18**, 993 (1965).
[69] J. L. Garnett and W. A. Sollich, *J. Phys. Chem.* **68**, 436 (1964).
[70] G. E. Calf and J. L. Garnett, unpublished data.
[71] B. D. Fisher and J. L. Garnett, *Aust. J. Chem.* **19**, 2299 (1966).
[72] G. E. Calf, J. L. Garnett, and V. A. Pickles, *Aust. J. Chem.* **21**, 961 (1968).
[73] J. L. Garnett, A. T. T. Oei, and W. A. Sollich-Baumgartner, *J. Catal.* **7**, 305 (1967).

Prior to the development of borohydride reduction methods, most catalyst activation was performed with hydrogen. Platinum and palladium oxides are readily reduced with hydrogen at room temperature; however, the remaining oxides are more difficult and require higher temperatures, iridium requiring 100° and nickel 300°.[74] The problem with reduction at elevated temperatures is that sintering of sites readily occurs with a subsequent lowering of catalyst activity. The relative ease of oxide reduction tends to follow metal–oxygen bond strength,[74] i.e., $Pt \simeq Pd > Rh > Ru > Ir \gg Ni$.

Reduction of aqueous or ethanolic solutions of inorganic salts with sodium or potassium borohydride is now the most useful procedure for catalyst activation. Reduction is fast and efficient at room temperature, and, particularly in the case of nickel, a catalyst more active than Raney nickel can be obtained.[66] Instead of borohydride, silicon hydrides such as tribenzylsilane have been used for the reduction to produce active platinum catalysts.[75]

B. Reaction Procedures

For these exchange reactions, the catalyst (previously prereduced with hydrogen or borohydride, if required) is sealed in a preconstricted ampoule with the organic compound and isotopic water under vacuum (10^{-2} Torr). The tubes are shaken or stood at the required temperature, and after the reaction is completed the products are analyzed by infrared, vapor phase chromatography, NMR or mass spectrometry (if deuteration), or counted (tritiation). With some compounds, the material to be labeled may be refluxed with isotopic water in the presence of catalyst instead of using sealed tube procedures.[76] Solvents such as acetic acid and ethanol have also been used;[77-80] however, unless the solvent is fully labeled prior to exchange, the theoretical percentage of isotope incorporation at equilibrium under these conditions can be lowered by loss of isotope to the solvent.

In these reactions, large variations in catalyst activity from batch to batch may be observed for rates of deuteration. For preparative work, especially with deuterium, this problem is not important since the

[74] J. L. Garnett and W. A. Sollich, *Aust. J. Chem.* **18**, 1003 (1965).
[75] R. W. Bott, C. Eaborn, E. R. A. Peeling, and D. E. Webster, *Proc. Chem. Soc.*, 337 (1962).
[76] J. L. Garnett and W. A. Sollich, *Nature (London)* **201**, 902 (1964).
[77] K. Bloch and D. Rittenberg, *J. Biol. Chem.* **149**, 505 (1943).
[78] D. K. Fukushima and T. F. Gallagher, *J. Biol. Chem.* **198**, 861 (1952).
[79] M. L. Eidinoff and J. E. Knoll, *J. Amer. Chem. Soc.* **75**, 1992 (1953).
[80] J. L. Garnett and W. A. Sollich-Baumgartner, *J. Catal.* **5**, 244 (1966).

system is normally allowed to attain equilibrium. For mechanistic kinetic studies of the labeling process, however, special care must be taken to preserve the activity, reproducibility, and stability of a prereduced catalyst.[81]

C. Mechanisms of Exchange

A discussion of the mechanism of the exchange is useful since then we can make predictions of isotope incorporation and orientation within heterocyclic molecules. Two types of mechanisms have been proposed for isotopic hydrogen exchange.[61, 82] For convenience, these have been termed classical, and, the more recently developed, π-complex mechanisms.

1. *Classical Mechanisms*

Two mechanisms can be distinguished depending on whether the compound is chemisorbed in a dissociative [Eqs. (20) and (21)] or associative [Eqs. (22) and (23)] manner. If heavy water is used as source of isotope, it supplies deuterium to the catalyst surface through dissociative adsorption [Eq. (24)].

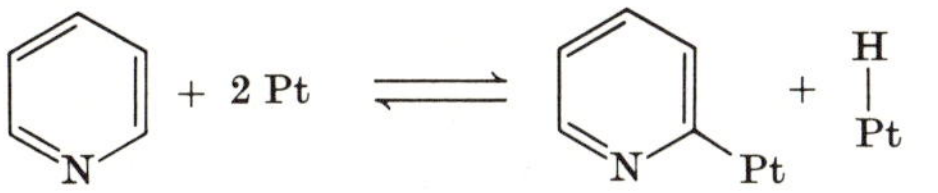

$$\text{pyridine} + 2\,\text{Pt} \rightleftharpoons \text{pyridyl-Pt} + \text{H-Pt} \quad (20)$$

$$\text{pyridyl-Pt} + \text{D-Pt} \rightleftharpoons \text{pyridine-}d + 2\,\text{Pt} \quad (21)$$

$$\text{pyridine} + 2\,\text{Pt} \rightleftharpoons \text{dihydropyridine-diPt} \quad (22)$$

$$\text{dihydropyridine-diPt} + \text{D-Pt} \rightleftharpoons \text{intermediate} \longrightarrow \text{pyridine-}d + 2\,\text{Pt} \quad (23)$$

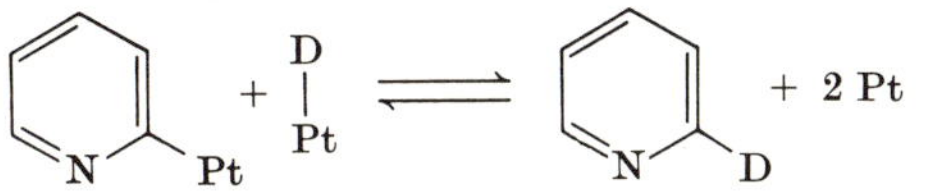

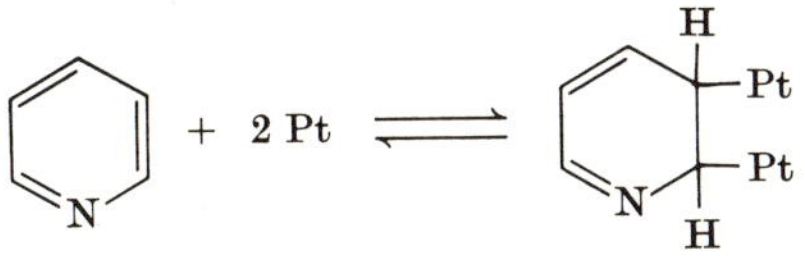

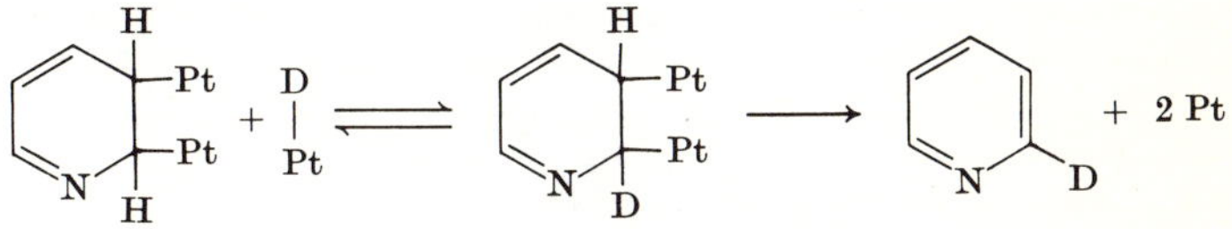

$$\text{D}_2\text{O} + 2\,\text{Pt} \longrightarrow \text{D-Pt} + \text{DO-Pt} \quad (24)$$

---

[81] J. L. Garnett and W. A. Sollich, *J. Catal.* **2**, 339 (1963).
[82] J. L. Garnett and W. A. Sollich-Baumgartner, *Advan. Catal.* **16**, 95 (1966).

In the classical dissociative mechanism[83] pyridine is adsorbed as a radical through carbon–hydrogen bond rupture [Eq. (20)], followed by the exchange step [Eq. (21)]. In the classical associative mechanism, the molecule is adsorbed by the formal opening of the double bond [Eq. (22)], exchange then occurring through the half-hydrogenated state[84] as shown in Eq. (23).

It is generally agreed that saturated compounds exchange by a dissociative process since C–H bond rupture is necessary for initial adsorption prior to exchange; however, no agreement has been reached for explaining the exchange of unsaturated hydrocarbons.[85]

Two main difficulties are associated with an interpretation in terms of classical theories.[86–89] In the classical associative mechanism, there is the problem of explaining the large loss in resonance energy when molecules such as benzene or pyridine are associatively adsorbed [Eq. (22)]. Similarly, classical dissociative adsorption would predict small differences in adsorption strengths and, thus, small differences in exchange rates for different members of an homologous series such as the heterocyclics. However, from Table V, the following order of adsorption strengths is observed:[90] pyridine < acridine ≪ quinoline; thus, in terms of exchange rates pyridine > acridine ≫ quinoline. For this relationship runs 6–8 in Table V are significant since the deuterium content in the benzene reflects the poisoning characteristics and, thus, the relative adsorption strengths of the heterocyclic series.[91, 92] It would also appear that the π-electron properties of these molecules influence the adsorption characteristics of these molecules and, thus, any mechanistic discussion of exchange.

2. π-Complex Mechanisms

Recent extensive evidence[82] supports the proposal that π-complex adsorbed intermediates are responsible for metal-catalyzed exchange

---

[83] A. Farkas and L. Farkas, *Proc. Roy. Soc. Ser.* A **144**, 467, 481 (1934).
[84] J. Horiuti and M. Polanyi, *Nature (London)* **132**, 819, 931 (1933).
[85] T. I. Taylor, *in* "Catalysis" (P. H. Emmett, ed.), Vol. V. Reinhold, New York, 1957.
[86] D. D. Eley, *Advan. Catal.* **1**, 185 (1948).
[87] O. Beeck, *Discuss. Faraday Soc.* **8**, 118 (1950).
[88] G. I. Jenkins and E. K. Rideal, *J. Chem. Soc.*, 2490 (1955).
[89] J. L. Garnett and W. A. Sollich, *Aust. J. Chem.* **15**, 56 (1962).
[90] R. A. Ashby and J. L. Garnett, *Aust. J. Chem.* **16**, 549 (1963).
[91] J. L. Garnett and W. A. Sollich, *Aust. J. Chem.* **14**, 441 (1961).
[92] J. L. Garnett and W. A. Sollich-Baumgartner, *J. Phys. Chem.* **69**, 1850 (1965).

TABLE V

Platinum-Catalyzed Deuterium Exchange Reactions of Heterocyclic Compounds[a]

Run No.	Reaction mixture	Time (hr)	Quantity (moles $\times 10^2$)	Atom % deuterium	% Approach to Statistical equilibrium[b]	% Approach to Instantaneous statistical equilibrium[b]
1	Pyridine	70	2.0	35 ± 2	70	—
	D_2O		5.0	—	—	—
2	Quinoline	70	2.0	28 ± 1	56	—
	D_2O		7.0	—	—	—
3	Isoquinoline	70	2.0	32 ± 2	64	—
	D_2O		7.0	—	—	—
4	Acridine	70	2.0	39 ± 2	78	—
	D_2O		9.0	—	—	—
5	2,2'-Dipyridyl	70	2.0	35 ± 2	70	—
	D_2O		8.0	—	—	—
6	Pyridine	69	2.0	50 ± 2	100	—
	Benzene		2.0	27.8 ± 0.3	—	55.6
	D_2O		11.0	—	—	—
7	Quinoline	69	2.0	38 ± 2	76	—
	Benzene		2.0	1.0 ± 0.1	—	1.8
	D_2O		13.0	—	—	—
8	Acridine	69	2.0	30 ± 2	60	—
	Benzene		2.0	8.7 ± 0.1	—	14.8
	D_2O		15.0	—	—	—
9	2,2'-Dipyridyl	69	1.0	50 ± 3	100	—
	Benzene		1.0	50.0 ± 0.5	—	100
	D_2O		7.0	—	—	—
10	Quinoline	69	1.0	41 ± 2	82	—
	Pyridine		1.0	54.0 ± 0.5	—	100
	D_2O		6.0	—	—	—

[a] All reactions were carried out using 100 mg Adam's catalyst at 150°C.
[b] Terms defined by Garnett and Sollich.[89]

in aromatic and heterocyclic compounds.[58, 59, 93–96] A review article covering π-complex adsorption in heterogeneous catalysis has been published.[82] This summarizes in detail the difficulties of interpreting

[93] J. L. Garnett, *Proc. Roy. Aust. Chem. Inst.* **28**(8), 328 (1961).
[94] G. C. Bond, "Catalysis by Metals." Academic Press, New York, 1962.
[95] J. J. Rooney, *J. Catal.* **2**, 52 (1963).
[96] J. J. Volter, *J. Catal.* **3**, 287 (1964).

the present exchange reactions in terms of classical theories[85] and demonstrates the advantage of the following treatment of π-complex adsorption by Mulliken's charge-transfer theory.[97]

a. *Associative π-Complex Mechanism.* As with the classical mechanisms, the π-complex mechanisms [Eqs. (25)–(28)] involve both associatively

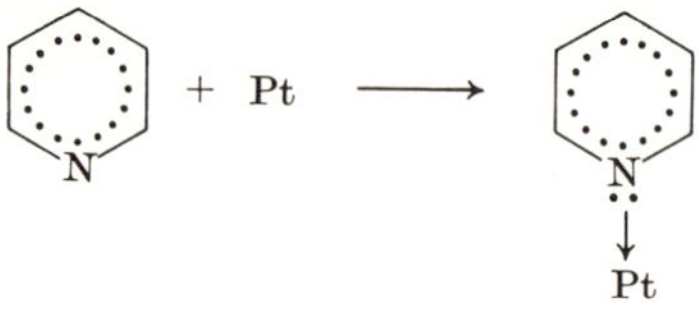

$$\text{pyridine} + \text{Pt} \longrightarrow \text{pyridine-Pt} \tag{25}$$

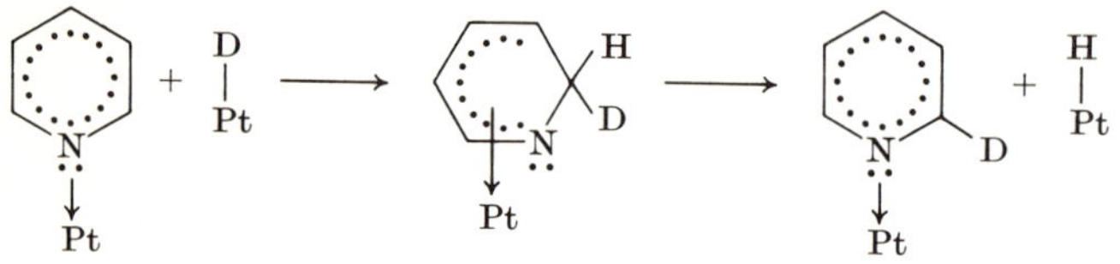

$$\text{pyridine-Pt} + \underset{\text{Pt}}{\text{D}} \longrightarrow \text{intermediate} \longrightarrow \text{product} + \underset{\text{Pt}}{\text{H}} \tag{26}$$

$$\longrightarrow \longrightarrow + \underset{\text{Pt}}{\text{H}} \tag{27}$$

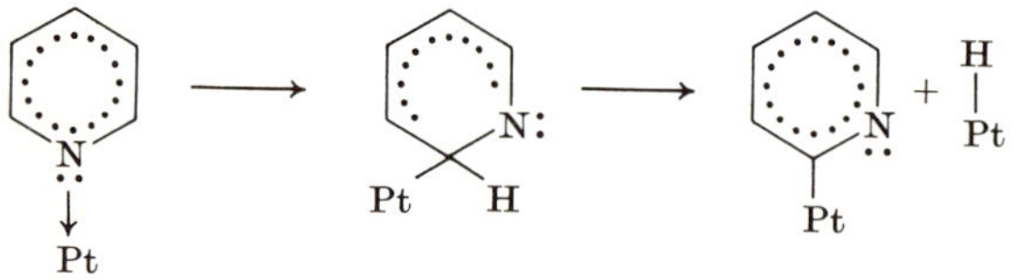

(15)

and dissociatively adsorbed intermediates. In the associative π-complex mechanism [Eqs. (25) and (26)], the π-bonded compound is attacked by a deuterium atom formed by the dissociative adsorption of heavy water or deuterium gas. The transition state differs from conventional substitution reactions by being π-bonded to the catalyst surface. The effect of π-bond adsorption is accentuated when extrapolation is made from benzene to heterocycles such as pyridine and thiophene. A number of transition states analogous to **15** are now possible, depending on isotope orientation because all positions in the heterocyclic molecule are not equivalent. Thus, the additional lone-pair interaction from the nitrogen in pyridine may facilitate an even closer approach between the two exchanging reagents and, hence, decrease the activation energy

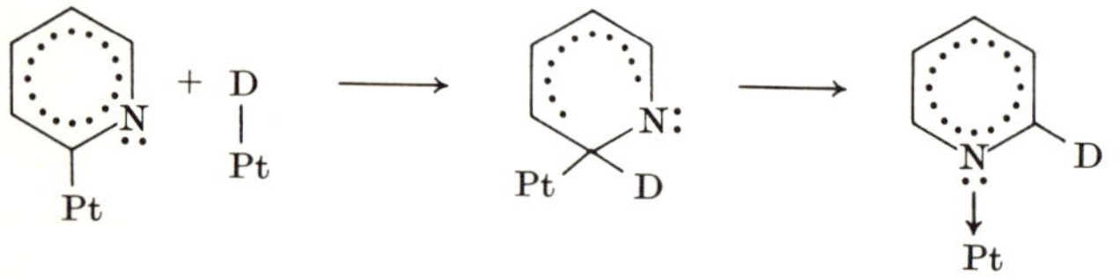

$$+ \underset{\text{Pt}}{\text{D}} \longrightarrow \longrightarrow \tag{28}$$

[97] R. S. Mulliken, *J. Amer. Chem. Soc.* **74**, 811 (1952).

even further than in the case of benzene. To compensate for this, the stronger charge-transfer surface bond with heterocyclic systems may further restrict the ease of formation of the π complex with the attacking radical and, thus, lower the rate of exchange relative to benzene. There are a number of competing factors which will influence both the total isotope incorporation and isotope orientation within a given heterocyclic compound. Because of the presence of the lone pair in heterocycles, it has been proposed that pyridine and thiophene on platinum are not adsorbed in a flatwise manner as is benzene, but are slightly tilted with the lone pair close to the catalyst surface. Current evidence indicates that in crossing the periodic table from platinum to nickel and cobalt, the effect due to the additional participation of the π electrons in the adsorption process is decreased so that ultimately with cobalt the molecule is initially adsorbed in a vertical position on the surface. Furthermore, the adsorption process may be modified if a polar substituent is present. Under some circumstances, substituents such as alkyl groups can reduce the magnitude of the π-complex interaction through a simple steric effect.[82] Because of these steric effects, some authors have found it difficult to interpret their exchange data in terms of the associative π-complex mechanism. Thus, a corresponding dissociative process has been postulated.

b. *Dissociative π-Complex Mechanism.* In the dissociative π-complex mechanism [Eqs. (27) and (28)], the π-complex adsorbed molecule reacts with a metal radical (active site) by a substitution process. During this reaction [Eq. (27)], the molecule rotates through 90° and changes from its horizontal π-complex adsorbed state to a vertically σ-bonded chemisorbed position. The postulated transition state for the π–σ conversion occurs when the plane of the rotating benzene molecule is approximately at a 45° angle to the catalyst surface. While σ-bonded, the aromatic undergoes a second slower substitution reaction at the carbon–metal bond with a chemisorbed isotopic hydrogen atom and returns to the π-bonded state [Eq. (28)]. If this step is repeated several times before desorption occurs, multiply exchanged species can be found. An isotope effect study ($k_D/k_T = 1.7$ at 32° for benzene and 1.4 at 174° for phenanthrene)[82] indicates that reaction (28) is the rate-determining process in the exchange.

Extrapolation of this dissociative mechanism to the prediction of isotope incorporation in heterocyclic molecules shows that the lone pair on the nitrogen atom (pyridine) or sulfur (thiophene) should again assist the formation of the respective transition states (**16**) and (**17**). Owing to this effect, ease of exchange in the positions in pyridine should be $\alpha > \beta > \gamma$.

(16)   (17)

The obvious problem now is to question why a dissociative π-complex substitution mechanism is required in addition to the associative mechanism to interpret these isotope exchange reactions. Inspection of the dissociative mechanism shows that if a substituent such as a methyl group is present in the molecule (i.e., toluene from benzene, or the picolines from pyridine), then it would appear difficult to form a σ bond at the adjacent carbon positions in the picolines (18) owing to steric

(18)

hindrance. Such ortho deactivation effects are experimentally observed with toluene and the picolines. Furthermore, if two methyl groups are meta to each other as in m-xylene or the corresponding lutidines (e.g., 2,4-lutidine), then the position flanked by the two methyl groups is completely deactivated to exchange by metal catalysts.[58, 61, 62, 82, 98] These deactivation data contrast markedly with the results obtained in simple acid-catalyzed exchange of these compounds with D^+, where no such steric effects are observed.[99, 100] These ortho or alpha metal-catalyzed deactivation effects have been interpreted as being due to difficulties in forming a σ bond on the metal surface. Because ortho positions easily exchange with acid, the present author and co-workers[80] suggest that these data indicate difficulty with π bonding and, therefore, the necessity to postulate a dissociative π-complex mechanism. Fraser and Renaud,[101] Hirota,[102] Macdonald and Shannon,[62] and Moyes and co-workers[103] interpret their own metal catalyzed exchange data in terms of a predominance of the dissociative process, whereas Kemball

[98] J. L. Garnett and W. A. Sollich, *J. Catal.* **2**, 350 (1963).
[99] G. E. Calf, B. D. Fisher, and J. L. Garnett, *Aust. J. Chem.* **21**, 947 (1968).
[100] W. M. Lauer and G. Stedman, *J. Amer. Chem. Soc.* **80**, 6433 (1958).
[101] R. R. Fraser and R. N. Renaud, *J. Amer. Chem. Soc.* **88**, 4365 (1966).
[102] K. Hirota, *J. Phys. Chem.*, in press.
[103] C. Horrex, R. B. Moyes, and R. C. Squire, *Proc. 4th Int. Congr. Catal., 1968*, in press.

and Siegel[104] favor the associative mechanism. These last authors do not believe that the steric effect of the methyl group on the surface is large enough to eliminate the formation of the π-complex associative intermediate.

c. *Mechanism of Alkyl Exchange in Alkyl Heterocyclics.* i. *Hydrogen abstraction and carbonium ion processes.* The exchange of the hydrogens in the alkyl group of alkyl heterocyclics is important mechanistically in explaining the pattern of behavior observed with the picolines and lutidines (Section III,D,4). The deuteration of the alkyl hydrogens in the alkyl heterocyclics is similar to the exchange of the corresponding hydrogens in the alkyl aromatics, particularly the alkylbenzenes,[82] where rates of isotope incorporation in the side chain are markedly larger than in the corresponding saturated aliphatics such as hexane and octane.

A dissociative hydrogen abstraction mechanism[57, 105, 106] is proposed to account for the exchange of saturated aliphatics [Eqs. (29) and (30)], whereas increased deuteration of the alkyl hydrogens in the alkylbenzenes is attributed to an analogous process with a lowering of the activation energy by initial π-complex formation from the aromatic

$$M + R\text{-}H \rightarrow M \ldots H \ldots R \xrightarrow{M} M\text{-}H + M\text{-}R \tag{29}$$

$$M\text{-}R + D\text{-}M \rightarrow R\text{-}D + 2M \tag{30}$$

ring. The fact that α-hydrogens react faster than β-hydrogens in these alkyl groups reflects the contribution of hyperconjugation and inductive effects to the exchange process. The transition state for exchange may even be similar to that proposed by Huyser[107] for an electrophilic radical abstraction process (**19**) since this is assisted by electron-donating methyl groups.

$$\begin{array}{c} \text{CH}_3 \\ | \\ \text{Ar} - \text{C} \ldots \text{H} \ldots \text{Pt} \\ | \\ \text{H} \end{array}$$
(**19**)

ii. *Modified π-complex mechanisms—π-allylic intermediates.* In addition to hydrogen abstraction and carbonium ion processes, modified π-complex mechanisms have been suggested to explain alkyl hydrogen

---

[104] R. J. Harper, S. Siegel, and C. Kemball, *J. Catal.* **6**, 72 (1966).
[105] G. E. Calf and J. L. Garnett, *Aust. J. Chem.* **21**, 1221 (1968).
[106] G. E. Calf and J. L. Garnett, *Chem. Commun.*, 373 (1969).
[107] E. S. Huyser, *J. Amer. Chem. Soc.* **82**, 394 (1960).

exchange in the alkylbenzenes.[61, 62, 95] These mechanisms have involved π-allylic intermediates (**20**) of the type shown in Eq. (31), which is analogous to Eq. (27) in the simple π-complex mechanisms. Kemball[59]

$$\text{[4-methylpyridine]} + 2M \longrightarrow \text{[π-allyl complex (20)]} + MH \qquad (31)$$

(**20**)

has reviewed the role of species such as **20** in catalytic reactions, evidence to support these concepts being provided from inorganic complexes.[95]

D. Exchange in Specific Heterocyclic Systems

The application of π-complex theory to the prediction of isotope incorporation in simple heterocyclic compounds will be discussed and the results extrapolated to the more complex heterocycles.

1. *Pyridine*

The exchange of pyridine on hydrogen prereduced Group VIII transition metal oxides and chlorides (Table VI) indicates that platinum oxide is the most efficient catalyst for general labeling. This is consistent with data for aromatic hydrocarbons;[70] however, the reactivity of hydrogen prereduced H_2PtCl_6 is markedly lower for pyridine exchange than with benzene. The oxides of ruthenium and rhodium and the chlorides of rhodium and platinum are reasonably selective in deuteration to the α position, but from the low voltage mass spectra it is obvious that some scrambling to the adjacent β positions has occurred.

Generally, the borohydride-reduced oxides and chlorides produce catalysts which are more active for pyridine exchange than the corresponding hydrogen preparations. Hydrogen reduction of the transition metal oxides produces more active catalysts than when the corresponding chlorides are used. However, with the exception of platinum, the reverse occurs with sodium borohydride preparations which produce more active catalysts from the soluble chlorides. Platinum oxide is again the most active general catalyst; however, the high reactivity exhibited by the borohydride-reduced chlorides of palladium, nickel, and cobalt and the oxide of osmium is significant. The selectivity in isotope orientation with cobalt and osmium is valuable, particularly cobalt, since the results of Table VI (from the low voltage mass spectra) indicate a remarkable specificity for deuteration at the α positions.[63, 72]

TABLE VI

DEUTERATION OF PYRIDINE WITH GROUP VIII METAL CATALYSTS[a]

Catalyst from	Atom % deuterium	Average No. of deuterium atoms at			Deuterium distribution					
		α	β	γ	d_0	d_1	d_2	d_3	d_4	d_5
System A[b]										
$RuO_2 \cdot H_2O$	19.9	0.98	0.03	0.00	27.2	47.7	23.4	1.7	—	—
Rh_2O_3	26.7	1.30	0.04	0.00	12.9	32.2	41.9	1.5	0.5	—
PdO	36.6	1.13	0.37	0.33	8.1	28.1	36.7	20.7	5.7	0.7
$PtO_2 \cdot xH_2O$	47.8	0.98	0.93	0.49	4.4	17.4	31.5	30.6	13.5	2.6
$RhCl_3$	27.0	1.31	0.06	0.00	12.0	43.5	42.4	1.6	0.1	—
K_2PdCl_4	10.9	0.44	0.11	0.00	58.6	36.4	8.4	0.4	—	—
H_2PtCl_6	7.2	0.35	0.00	0.01	73.2	19.2	6.3	0.9	0.4	—
Nickel on kieselguhr[62]	39.0	1.88	0.07	0.00	0.0	8.5	87.2	4.3	—	—
System B[b]										
Rh_2O_3	26.8	1.28	0.05	0.00	13.7	43.1	38.2	4.3	0.7	—
PdO	17.8	0.78	0.13	0.00	33.6	45.9	18.7	1.7	0.1	—
OsO_4	8.2	0.41	0.00	0.00	65.4	28.5	6.0	0.1	—	—
$PtO_2 \cdot xH_2O$	43.1	0.90	0.85	0.42	7.3	21.7	33.0	25.9	10.5	1.6
$CoCl_2$	22.3	1.09	0.04	0.00	20.2	40.5	31.2	0.1	—	—
$NiCl_2$	20.1	0.78	0.25	0.02	32.7	39.9	22.0	4.9	0.5	—
$RuCl_3$	25.5	1.13	0.03	0.12	19.3	42.0	31.5	6.3	0.8	0.1
$RhCl_3$	32.8	0.86	0.46	0.32	15.1	31.7	32.2	16.3	4.3	0.4
K_2PdCl_4	35.5	0.75	0.69	0.33	14.5	32.2	35.4	20.6	6.5	0.8
H_2PtCl_6	27.4	1.06	0.17	0.14	17.8	39.8	31.7	8.9	1.6	0.2

[a] Statistical equilibrium = 50% D; reaction conditions as in refs. 62, 72.
[b] A, hydrogen reduction; B, borohydride reduction; Fe_2O_3, Co_2O_3, NiO, and $IrO_2 \cdot 2H_2O$ are essentially inactive (< 2.0% D).

The cobalt result with pyridine is also significant since the maximum deuteration obtained in benzene and ethylbenzene under equivalent conditions with this catalyst was 0.10 and 0.06% D, respectively.[70]

Orientation of Isotope in Pyridine. From both the NMR and low voltage mass spectrometry data (Table VI) exchange in pyridine on all active catalysts favors α-hydrogen substitution to varying degrees at the expense of the β and γ positions. Platinum and cobalt (as the chlorides) represent the extremes in behavior of those catalysts which are very active (> 20%D). With cobalt, exchange is almost exclusively in the α position. Thus, charge-transfer adsorption of pyridine on this metal presumably involves only the nitrogen lone pair and the molecule is initially adsorbed in a vertical position on the surface [species (21)], whereas in crossing the periodic table from nickel to platinum, there is an increasing participation of the π electrons of the ring in the adsorption process so that the molecule is now tilted [species (22)]. Thus, with catalysts

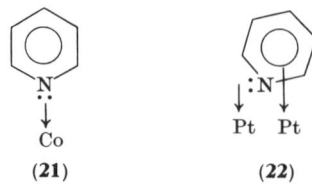

other than cobalt, the β and γ positions are readily capable of forming σ bonds on the catalyst surface. The role of the nitrogen lone pair is then simply to increase the interaction between the organic compound and the catalyst surface such that the α-hydrogens are favorably placed for exchange.

This mode of adsorption is further supported by kinetic studies on platinum which show a marked preference for exchange in the α positions during the first 20 hours of the deuteration at 130°, then randomization or back-exchange of deuteration from these positions is observed in the latter stages of the reaction.

2. *Quinoline and Isoquinoline*

The exchange characteristics of the quinolines with catalysts prepared by borohydride reduction (the most satisfactory reduction method for pyridine) are summarized in Table VII. Platinum oxide is again the most active catalyst for general exchange in both compounds, whereas cobalt catalyzes specific α deuteration. For selective labeling of quinoline, ruthenium and rhodium oxides are faster than cobalt, while with isoquinoline, the chlorides of rhodium and palladium are also very useful. All catalysts are less reactive with the quinolines than with pyridine, this trend being similar to the corresponding aromatic hydrocarbons, naphthalene and benzene, respectively. There the difference is attributed to the lower ionization potential of naphthalene, leading to stronger charge-transfer bonding on the surface and increased catalytic poisoning through displacement of the second reagent (D_2O) in the exchange.[82] Isoquinoline is more reactive than quinoline on platinum oxide, the exchange of isoquinoline being closer to pyridine since isoquinoline possesses two active hydrogen atoms alpha to the nitrogen, whereas quinoline has only one [species (23) and (24)]. The second aromatic ring appears to exert little steric effect on the adsorption of either molecule

TABLE VII

Deuteration of Quinoline and Isoquinoline with Borohydride-Reduced Catalysts[a]

Catalyst from	Atom % deuterium	Average No. of deuterium atoms at				Deuterium distribution				
		C-2	C-3	C-4–C-7	C-8	d_0	d_1	d_2	d_3	d_4
		System A[b]								
$RuO_2 \cdot H_2O$	7.6	0.50	0.00	0.06	0.00	51.7	43.8	4.1	0.4	—
Rh_2O_3	9.2	0.60	0.00	0.08	0.00	42.7	52.7	4.2	0.1	0.3
$RhCl_3$	9.6	0.60	0.01	0.04	0.03	34.1	64.6	1.2	0.1	—
K_2PdCl_4	10.4	0.60	0.02	0.03	0.08	36.7	54.8	7.8	0.7	—
$PtO_2 \cdot 2H_2O$	13.3	0.56	0.12	0.08	0.18	33.9	43.8	18.5	3.1	0.7
H_2PtCl_6	7.9	0.37	0.04	0.12	0.02	56.0	33.9	9.0	1.1	—
Nickel on kieselguhr[62]	36.1[c]	0.89	0.04	0.71	0.89	0.2	6.0	39.0	50.6	4.2
		C-1	C-3	C-4–C-8						
		System B[b]								
$CoCl_2$	3.9	0.27	0.13	0.00		75.6	21.6	2.7	0.1	—
$RuO_2 \cdot H_2O$	5.5	0.17	0.13	0.08		67.4	27.3	5.0	0.2	0.1
Rh_2O_3	10.7	0.34	0.31	0.09		42.9	40.5	16.1	0.4	0.1
$RhCl_3$	15.6	0.57	0.53	0.00		22.2	47.6	29.1	0.7	0.4
K_2PdCl_4	7.9	0.25	0.32	0.00		53.5	38.4	8.0	0.3	—
$PtO_2 \cdot 2H_2O$	21.5	0.65	0.62	0.23		10.2	37.4	44.9	6.4	1.1
H_2PtCl_6	12.4	0.43	0.32	0.11		34.6	45.6	18.7	1.0	0.1
Nickel on kieselguhr[62]	26.9	0.91	0.88	0.09		2.5	17.4	69.8	10.4	—

[a] Reaction conditions as in Table VI; $FeCl_3$, OsO_4, $(NH_4)_2 IrCl_6$, $NiCl_2$, and $RuCl_3$ are inactive (< 3.6% D); $CoCl_2$ inactive with quinoline.

[b] A, quinoline; B, isoquinoline.

[c] Equilibrium = 93.1% D.

with platinum; however, with cobalt chloride the much lower exchange rates of both quinolines in the α position compared with pyridine suggests a significant reduction in the intensity of complexing, presumably due to adsorption in a vertical manner similar to pyridine.

(23)   (24)

a. *Modified Catalysts; Supported versus Unsupported Systems.* Group VIII transition metals, particularly nickel, in supported form such as adsorption on kieselguhr give high α selectivity in the deuteration of pyridine and the quinolines (Tables VI and VII), the authors postulating that such heterocyclic compounds are adsorbed exclusively through the heteroatom.[108, 109] Vapor phase exchange of pyridine with heavy water catalyzed by palladium-on-asbestos also yields pyridine-d_2, although temperatures up to 225° for 36 hours are required.[110] However, unsupported catalysts, particularly cobalt, give even better selectivity and are to be preferred because of better reproducibility and simplicity in preparation.

b. *Self-Activated Catalysts.* A useful type of catalytic reduction already discussed is self-activation (Section III,A) or *in situ* reduction by the organic of the metal salt, since no prereduction of the catalyst prior to exchange is necessary. For the deuteration of pyridine, quinoline, and isoquinoline, self-activated catalysts are less active than borohydride or hydrogen prereduced systems (Table VIII). With potassium chloropalladite, however, exchange is selective to the α position with pyridine so that useful specificity is obtained with this catalyst.[72] Using ESR spectroscopy, the nature of the intermediate charge-transfer complex[111] in this palladium reaction can be identified.

In terms of preparative advantage, particularly when compared with Raney nickel,[112] self-activated catalysts would appear to be satisfactory for specifically labeling pyridine or generally deuterating isoquinoline. With platinum oxide the trend is pyridine < quinoline < isoquinoline in self-activation efficiency. This presumably reflects the repulsive effect of the oxygen in the platinum oxide on the nitrogen lone pair. In isoquinoline, the hydrogens participating in the self-activation appear to originate predominantly from the nonheterocyclic ring, thus explaining the relatively fast exchange of this compound on self-activated platinum oxide.

3. *Substituted Pyridines*

The effect of substituents in pyridine on the exchange rate is important since the results may be extrapolated to the deuteration behavior of

---

[108] C. G. Macdonald and J. S. Shannon, *Tetrahedron Lett.*, 3351 (1964).
[109] W. M. Lauer and L. A. Errede, *J. Amer. Chem. Soc.* **76**, 5162 (1954).
[110] L. Corrsin, B. J. Fox, and R. C. Lord, *J. Chem. Phys.* **21**, 1170 (1953).
[111] I. T. Ernst, J. L. Garnett and W. A. Sollich-Baumgartner, *J. Catal.* **3**, 568 (1964).
[112] P. J. Collin and C. G. Macdonald, *Aust. J. Chem.* **19**, 513 (1966).

TABLE VIII
Deuteration of Pyridine, Quinoline, and Isoquinoline on Self-Activated Catalysts[a]

Compound	Catalyst from	Atom % deuterium	Average No. of deuterium atoms			Deuterium distribution				
			α	β	γ	d_0	d_1	d_2	d_3	d_4
Pyridine[b]	$PtO_2 \cdot xH_2O$	1.5	—	—	—	94.1	4.4	1.3	0.2	—
	K_2PdCl_4	7.3	0.35	0.00	0.02	68.0	27.4	0.5	0.1	—
			C-2	C-3	C-4–C-7					
Quinoline	$PtO_2 \cdot xH_2O$	4.7	0.25	0.01	0.01	71.8	23.8	3.9	0.5	—
			C-2	C-3	C-4–C-8					
Isoquinoline	$PtO_2 \cdot xH_2O$	20.1	0.64	0.55	0.22	13.1	40.7	39.7	5.6	0.9

[a] Reaction conditions as in Table VI.
[b] Fe_2O_3, Co_2O_3, NiO, $RuO \cdot H_2O$, Rh_2O_3, PdO, $IrO_2 \cdot 2H_2O$, $FeCl_3$, $RhCl_3$, H_2PtCl_6, $CoCl_2$, $NiCl_2$, $RuCl_3$, and $(NH_4)_2 IrCl_6$ are inactive (<0.8% D).

other substituted heterocyclic molecules. From the mechanistic discussion in Section III,C, and also from a prior review of relevant catalytic theory,[82] it can be seen that substituents can exert either steric effects leading to difficulty in both π- and σ-bond formation on the surface or polar effects giving enhanced adsorption strength. Alkyl substituents are catalytically inert (i.e., no polar effect); hence, steric hindrance only is observed. Thus, positions adjacent to small alkyl groups such as methyl are *strongly* deactivated, whereas positions adjacent to large bulky groups (t-butyl, isopropyl) or flanked by two methyl groups (2,4-lutidine) are *completely* deactivated. The substituent effect in heterocyclic molecules is unique. In the adsorption of pyridine two extremes are observed: (*a*) tilted (as on platinum), so that both lone pair and π electrons participate in the charge-transfer bond, and (*b*) vertical (as on cobalt), utilizing only the lone pair. An alkyl group causes increased adsorption from the nitrogen lone pair and also steric hindrance.

4. *Picolines*

The results for the deuteration of 2-, 3-, and 4-picolines are summarized in Table IX. With 2-picoline, specific deuteration in the α-position only is observed on cobalt, whereas platinum is catalytically active for all positions,[105] as is nickel chloride which catalyzes deuteration of the methyl group in particular. As with pyridine, borohydride-reduced oxide catalysts are more active than self-activated preparations. The much higher reactivity in 2-picoline compared with pyridine in self-activation indicates that the methyl group compensates for deactivation from the nitrogen lone pair. Hydrogen for self-activation in the former compound originates predominantly from the methyl group.

With the exception of the palladium salt, self-activation of the metal chlorides produces hydrochloric acid and the resulting exchange is exclusively acid-catalyzed (compare the results with the HCl run). With self-activated potassium chloropalladite, deuteration occurs in all positions.

In the deuteration of 3-picoline, borohydride-reduced cobalt is again specific to the α position, whereas platinum exchanges in all positions; however, incorporation in the methyl group is very low. Thus, for the general labeling of 3-picoline, nickel chloride may be a superior catalyst since, although the β and γ positions are less active than with platinum, the methyl group has a relatively high isotopic content. With the self-activated catalysts, potassium chloropalladite deuterated in all positions, whereas platinum oxide is only as active as with pyridine. This suggests that the nitrogen lone pair controls the self-activation process in 3-picoline at the expense of the hydrogens in the methyl group.

TABLE IX
DEUTERATION OF THE PICOLINES WITH GROUP VIII TRANSITION METAL CATALYSTS[a]

Picoline	Catalyst from	Atom % deuterium	Average No. of deuterium atoms at				Deuterium distribution							
			α	β	γ	Me	d_0	d_1	d_2	d_3	d_4	d_5	d_6
			Sodium borohydride reduction										
2–	$CoCl_2$	10.6	0.79	0.01	0.00	0.05	40.1	47.0	11.7	1.1	0.1	—	—
	$NiCl_2$	19.1	0.52	0.17	0.18	0.49	22.7	37.8	26.3	10.1	2.6	0.4	0.1
	$PtO_2 \cdot 2H_2O$	28.8	0.60	0.96	0.31	0.16	9.4	24.9	33.6	21.7	8.0	2.0	0.4
3–	$CoCl_2$	14.4	0.97	0.00	0.02	0.03	25.8	48.4	25.1	0.5	0.2	—	—
	$NiCl_2$	18.9	0.52	0.23	0.13	0.45	23.7	37.0	26.1	10.1	2.6	0.4	—
	$PtO_2 \cdot 2H_2O$	27.5	0.95	0.43	0.45	0.05	9.2	26.5	35.1	22.1	6.1	0.7	0.2
4–	$CoCl_2$	17.7	1.07	0.03	—	0.21	21.2	42.5	29.7	5.9	0.5	0.1	—
	$NiCl_2$	23.6	0.53	0.26	—	0.86	15.6	32.0	30.8	15.9	4.8	0.8	0.1
	$PtO_2 \cdot 2H_2O$	24.7	1.15	0.40	—	0.18	10.7	31.9	36.3	16.8	3.7	0.5	0.1
			Self-activation										
2–	$PtO_2 \cdot 2H_2O$[b]	15.1	0.14	0.16	0.14	0.63	41.5	26.1	18.6	12.9	0.8	0.1	—
	K_2PdCl_4	27.5	0.65	0.10	0.29	0.91	12.1	24.6	32.1	21.9	7.9	1.2	0.1
	HCl	22.1	0.04	0.00	0.01	1.51	17.0	28.4	37.7	16.8	0.1	—	—
3–	$PtO_2 \cdot 2H_2O$	4.0	0.06	0.00	0.03	0.21	84.1	8.4	3.7	3.1	0.5	0.1	—
	K_2PdCl_4	28.6	1.06	0.18	0.06	0.71	11.9	21.8	35.0	20.0	8.6	2.4	0.3
4–	$PtO_2 \cdot 2H_2O$	37.6	0.93	0.00	—	1.70	6.0	12.7	24.4	31.3	20.4	5.0	0.2
	K_2PdCl_4	33.7	0.88	0.00	—	1.48	5.0	18.0	31.8	29.6	12.9	2.6	0.1
	Palladium-on-charcoal[113]	—	1.68	1.38	—	1.41	—	—	—	—	—	—	—

[a] Reaction conditions as in Table VI, equilibrium = 41.7% D.
[b] $RuCl_3$, $RhCl_3$, K_2PtCl_4, and H_2PtCl_6 gave results similar to HCl.

The exchange results with 4-picoline generally resemble the data for 2-picoline with some significant differences. Borohydride-reduced cobalt shows appreciable isotope scrambling at the methyl group. In this respect the reactivity of the methyl groups in the picolines is 4 ≫ 2 > 3. This may reflect the increasing participation of species (**25**) such as is

$$\underset{\underset{Co}{\downarrow}}{\underset{\ddot{N}}{\bigcirc}}\text{—CH}_3 \quad \rightleftharpoons \quad \underset{\underset{Co}{\downarrow}}{\underset{N}{\bigcirc}}\text{⋯CH}_2\text{⋯} \;+\; \underset{Co}{\overset{H}{|}} \tag{32}$$

(**25**)

depicted in Eq. (32) in the exchange mechanism with cobalt. The second important difference is the remarkable selectivity with self-activated platinum oxide and potassium chloropalladite, no isotope being incorporated in the β positions with either catalyst, whereas all positions exchange with palladium-on-charcoal.[113]

In general, self-activation of all picolines with platinum oxide yields a predominance of isotope incorporation in the methyl side chain. This orientation has been attributed to a site effect in that catalytic sites formed by alkyl group reduction of platinum oxide favor side-chain deuteration at the expense of ring exchange.

The remaining interesting feature of the picoline results is the direction of the compensation effect between the increased adsorption due to the nitrogen lone-pair interaction and the steric hindrance of the methyl group. The data from 3-picoline for the α exchange show that some deactivation of one position (presumably the 2-position, or ortho to the methyl) is observed, although the effect is not as pronounced as in the ortho positions of toluene. Thus, the compensation effect would appear to favor increased adsorption from the lone pair on the nitrogen and this is confirmed by the results of 4-picoline on borohydride-reduced platinum.

5. *Lutidines*

The patterns of exchange in the lutidines (Table X) are similar to those observed for pyridine and the picolines. With cobalt, little deuteration occurs when the methyl groups are substituted in the 2- and 6-positions, as in 2,6-lutidine. In 2,4-lutidine, the hydrogen atom in position 3 is flanked by two methyl groups, while the hydrogen at position 5 is ortho to a methyl group. Thus, very severe deactivation

[113] D. P. Biddiscombe, E. F. G. Herington, I. J. Lawrenson, and J. F. Martin, *J. Chem. Soc.*, 444 (1963).

TABLE X
CATALYTIC DEUTERATION OF LUTIDINES[a]

Lutidine	Catalyst from	Atom % deuterium	Average No. of deuterium atoms at			Me			Deuterium distribution						
			α	β	γ	(2,6)	(4,5)	d_0	d_1	d_2	d_3	d_4	d_5	d_6
			Sodium borohydride reduction											
2,4-	$CoCl_2$	8.8	0.56	0.04	—	0.05	0.14	37.8	47.5	12.7	1.8	0.2	—	—
	$NiCl_2$	13.7	0.43	0.10	—	0.54	0.15	28.0	36.3	24.1	8.9	2.2	0.4	0.1
	$PtO_2 \cdot xH_2O$	7.2	0.43	0.10	—	0.09	0.03	51.6	36.4	9.2	2.1	0.6	0.1	—
2,5-	$CoCl_2$	5.7	0.47	0.02	0.02	0.00	0.00	53.2	43.3	3.0	0.4	0.1	—	—
	$NiCl_2$	9.1	0.20	0.02	0.00	0.53	0.08	42.9	37.2	15.6	3.7	0.5	0.1	—
	$PtO_2 \cdot xH_2O$	24.5	0.42	0.40	0.39	1.02	0.00	14.9	21.9	23.7	20.7	12.5	5.0	1.3
2,6-	$CoCl_2$	1.3	—	—	—	—	—	89.3	10.1	0.5	0.1	—	—	—
	$NiCl_2$	3.1	—	—	—	—	—	75.5	21.1	3.0	0.3	—	—	—
	$PtO_2 \cdot xH_2O$	17.8	—	0.71	0.46	0.43	—	32.8	22.3	19.9	13.7	5.2	2.9	1.7
			Self-activation[b]											
2,4-	$PtO_2 \cdot xH_2O$	20.0	0.14	0.07	—	0.81	0.78	28.4	24.8	17.0	12.1	9.0	6.1	2.6
	K_2PdCl_4	29.8	0.45	0.05	—	1.14	1.05	6.1	15.2	24.3	26.1	18.4	7.9	2.0
2,5-	$PtO_2 \cdot xH_2O$	19.2	0.37	0.14	0.28	0.86	0.08	23.6	26.2	22.2	15.7	8.3	3.2	0.8
	K_2PdCl_4	3.3	—	—	—	—	—	80.6	12.8	4.1	1.7	0.5	0.2	0.1
2,6-	$PtO_2 \cdot xH_2O$	12.4	—	0.27	0.57	0.27	—	31.6	40.7	17.8	6.7	2.2	0.7	0.3
	K_2PdCl_4	3.1	—	—	—	—	—	74.4	21.3	3.0	0.3	—	—	—

[a] Reaction conditions as in Table VI; equilibrium = 35.7% D.
[b] Reaction conditions as in Table VIII.

of these positions is observed and the results are analogous to the alkylbenzenes (m-xylene and mesitylene).[61,62,98] Exchange with borohydride-reduced nickel shows that the reactivity of the methyl group is 2 > 4 > 3 which is the same order as with self-activated platinum oxide.

6. Effect of Polar Substituents on Pyridine Exchange

For the deuteration of heterocycles containing charge-transfer active substituents, the results for the substituted benzenes must be taken as a guide since little deuteration has been done with heterocyclic derivatives themselves. For halogen substituents, both rates of exchange and extent of ortho deactivation are $F > Cl > Br \gg I$. Of other common functional groups, carboxylic acids and ethers also give extensive ortho deactivation, while cyano, nitro, and acetyl poison ring exchange almost completely.[114] By contrast, amino and hydroxyl groups normally exhibit ortho activation on platinum.

7. Azines

Pyridazine (26), pyrimidine (27), and pyrazine (28) are all readily labeled by exchange on platinum (Table XI); however, s-triazine (29) decomposes to an insoluble product. Borohydride-reduced cobalt is

TABLE XI

ORIENTATION OF DEUTERIUM IN AZINES LABELED ON COBALT, NICKEL, AND PLATINUM CATALYSTS[a]

Azine[b]	Catalyst from	Atom % deuterium	Average No. of deuterium atoms at		Deuterium distribution				
			α	β	d_0	d_1	d_2	d_3	d_4
Pyridazine	$CoCl_2$	7.2	0.29	0.00	73.9	23.7	2.2	0.2	—
Pyridazine	$NiCl_2$	7.0	0.28	0.01	73.9	24.2	1.8	0.1	—
Pyridazine	PtO_2	49.1	0.98	0.98	8.2	25.7	34.8	24.0	1.3
Pyrimidine	PtO_2	25.3	—	—	33.6	39.1	20.5	6.1	0.7
Pyrazine	PtO_2	45.2	—	—	8.1	26.7	41.8	23.4	—

[a] Reaction conditions as in Table VI.
[b] s-Triazine decomposed to an insoluble product.

[114] J. L. Garnett, L. Henderson, and W. A. Sollich, Proc. Symp. Tritium Phys. Biol. Sci., 1961 2, 47 (1962).

again specific for the α position in pyridazine, whereas platinum oxide catalyzes exchange in all positions. Thus, the pattern of results is similar to those obtained with pyridine. The remaining important result is the increased toxicity (i.e., stronger adsorption) of the diazines compared

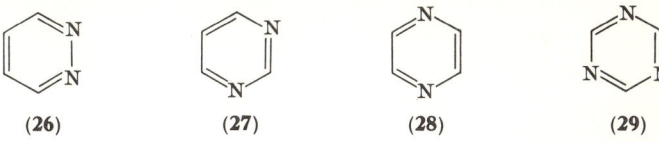

(26)   (27)   (28)   (29)

with pyridine as evidenced by the decrease in deuterium incorporation in benzene from mixed exchanges.[90]

8. *Thiophene and Furan*

Consistent with the fact that thiophene is a poison in metal-catalyzed hydrogenation reactions, thiophene is deuterated only slowly on the prereduced transition metals listed in Table XII, except for iridium. In both hydrogenation[115] and exchange[89] the poisoning has been attributed to the influence of the heteroatom in the adsorption process, presumably as species (30), or even of elemental sulfur as a consequence

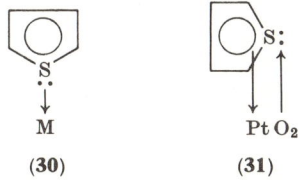

(30)   (31)

of C–S bond rupture. The best metal-catalyzed labeling system for thiophene is self-activated platinum oxide (self-activated K_2PdCl_4 and H_2PtCl_6 both give acid exchange). Self-activation produces a layer of active platinum atoms in the support of platinum oxide, exchange then occurring on this catalyst by the established mechanisms.[82, 101] Since thiophene exchanges only on self-activated, but not prereduced platinum, a plausible explanation for the process involves initial adsorption as species (31), the effect of the oxygen in the platinum oxide being sufficient to tilt the molecule so that the sulfur no longer touches the catalyst surface and thus prevents poisoning. In the subsequent exchange, the sulfur atom is still repelled from the self-activated surface by the oxygen in the bulk PtO_2; however, a compensation effect

[115] F. A. Matsen, A. C. Makrides, and N. Hackerman, *J. Chem. Phys.* 22, 1800 (1954).

TABLE XII

Deuteration of Thiophene and Furan on Borohydride-Reduced (A) and Self-Activated (B) Catalysts[a]

Compound	Catalyst activation	Catalyst from	Atom % deuterium	Average No. of deuterium atoms at		Deuterium distribution				
				α (2,5)	β (3,4)	d_0	d_1	d_2	d_3	d_4
Thiophene[b]	A	$(NH_4)_2IrCl_6$	18.8	0.59	0.16	50.4	30.3	14.4	3.7	1.2
	B	$RuO_2 \cdot H_2O$	5.6	0.25	0.00	78.3	16.4	4.0	1.1	0.2
	B	$IrO_2 \cdot 2H_2O$	6.0	0.16	0.08	85.7	5.9	4.8	2.1	6.5
	B	$PtO_2 \cdot 2H_2O$	19.5	0.64	0.14	55.3	20.3	17.3	4.8	2.3
	B	$CoCl_2$	1.7	—	—	93.5	6.2	0.3	—	—
	B	$NiCl_2$	4.1	0.16	0.00	86.4	11.6	1.4	0.5	0.1
	B	K_2PdCl_4	37.7	1.16	0.35	12.5	38.5	36.1	10.1	2.5
	B	H_2PtCl_6	43.3	1.05	0.68	11.1	32.1	34.3	17.6	4.9
	B	HCl	39.4	1.05	0.53	12.9	36.2	34.3	13.5	3.1
Furan[c]	A	$(NH_4)_2IrCl_6$	5.8	0.25	0.02	80.4	16.6	2.5	0.4	0.1
	B	$RuO_2 \cdot H_2O$	7.6	0.22	0.01	82.4	8.6	6.1	1.8	1.1
	B	$IrO_2 \cdot 2H_2O$	23.3	0.74	0.19	47.6	23.3	19.5	7.5	2.1
	B	$PtO_2 \cdot 2H_2O$	7.7	0.30	0.01	76.3	17.9	4.8	0.9	0.1
	B	$CoCl_2$	4.8	—	—	82.1	16.6	1.2	0.1	—
	B	$NiCl_2$	19.4	0.77	0.01	40.9	40.7	18.0	0.3	0.1

[a] Reaction conditions as in Table VI, except for organic reactant (5.0×10^{-2} mole).
[b] Borohydride reduction of $FeCl_3$, $CoCl_2$, $NiCl_2$, $RhCl_3$, K_2PdCl_4, and $PtO_2 \cdot 2H_2O$ and hydrogen reduction of $PtO_2 \cdot 2H_2O$ gave inactive catalysts for thiophene.
[c] Self-activation with Fe_2O_3, Co_2O_3, NiO, PdO, and Rh_2O_3 gave inactive catalysts. Self-activation with $FeCl_3$, $RuCl_3$, $RhCl_3$, K_2PdCl_4, $(NH_4)_2IrCl_6$, H_2PtCl_6, and HCl polymerized furan during exchange.

appears to occur since the α positions exchange more readily than the β positions. Thus, the lone pair on the sulfur, while not strong enough to inhibit exchange, accentuates the formation of the charge-transfer complex on the surface of the catalyst.

The explanation for the mechanism of exchange with thiophene is particularly important since a number of workers have found poor reproducibility with prereduced platinum, i.e., some prereduced catalysts actually catalyze exchange, whereas no deuteration is observed in duplicate samples.[105] The reason for this apparent anomaly is presumably that reduction of platinum oxide with hydrogen is very slow and inefficient. Unless *complete* reduction is achieved, some PtO or PtO_2 remains on which self-activation can readily occur.

The results in Table XII show that self-activated platinum oxide exchanges in both α and β positions of thiophene, whereas self-activated ruthenium oxide catalyzes deuteration almost exclusively in the α position. A similar type of selectivity has been observed for the exchange of thiophene on the disulfides of tungsten (both α and β deuteration) and molybdenum (predominantly α) using evaporated metal films.[116]

The labeling of furan (Table XII) is similar to thiophene, since the most active of the borohydride-reduced catalysts is iridium. However, self-activated iridium oxide is the most active of all for general deuteration, whereas nickel chloride is the most efficient and selective for the α position, only slight polymerization being observed. Acid catalysis should be used with caution for exchange labeling, since furan readily polymerizes above 30° in the presence of hydrochloric acid and chlorides.

Furan has also been labeled with heavy water on supported catalysts (chromium, zinc, and manganese oxides promoted with K_2CO_3) at a temperature of 350°.[117] Deuterated furan has also been obtained from the vapor phase decarbonylation of furfural over mixed oxide catalysts in the presence of heavy water. Both of these systems utilize extreme experimental conditions and the methods outlined in Table XII are to be preferred for preparative labeling.

9. *Miscellaneous Heterocyclic Derivatives*

The platinum-catalyzed tritiation of a representative number of miscellaneous heterocycles is shown in Table XIII. In addition to these compounds, cytidine, deoxycytidine hydrochloride, deoxyuridine,

---

[116] P. Kieran and C. Kemball, *J. Catal.* **4**, 394 (1965).
[117] A. Ya. Karmil'chik, S. Hillers, and M. V. Shimanskaya, *Latv. PSR Zinatnu Akad. Vestis, Kim. Ser.*, 328 (1966) [*Chem. Abstr.* **65**, 15175 (1966)].

TABLE XIII

Platinum-Catalyzed Tritiation of Miscellaneous Heterocycles[5,120]

Compound	Weight (gm)	Catalyst (gm)	Solvent[a]	Vol. (ml)	Specific activity (Ci/ml)	Temp. (°C)	Time (hr)	Specific activity (mCi/mM)
Adenine	0.3	0.25	Acetic	3	67	130	14	439
Adenosine	0.32	0.2	Acetic	4	63	110	8	680
Caffeine	0.5	0.2	Acetic	2	25	150	16	1280
Cytosine	0.5	0.2	Acetic	4	40	100	20	293
Deoxyadenosine	0.2	0.2	Acetic	3	67	100	6.5	627
Deoxyguanosine	0.5	0.2	Water	2	87	120	3	1000
Hypoxanthine	0.3	0.21	Acetic	3	50	150	15	630
5-Methylcytosine	0.35	0.2	Water	2	100	130	16	15600
Nicotine	0.5	0.2	Water	2	35	100	72	82
Nicotinic acid	0.5	0.2	Water	2	37.5	130	160	320
Thymidine	0.2	0.2	Water	1	2	100	24	14.9
Thymine	0.26	0.23	Acetic	2	100	170	15	8600
Uridine	0.5	0.2	Acetic	2	75	120	16	10080

[a] Acetic means 75% acetic acid/water.

guanine, guanosine, inosine, isonicotinic hydrazide, orotic acid, theophylline, and uracil have been tritiated to reasonable specific activities by this method.[5] Adenine[79] (32) was one of the first heterocycles labeled

(32)

in this manner; however, purines, pyrimidines, and nucleosides[118–121] have also been successfully tagged.

IV. Current Trends in Labeling

While research continues particularly concerning *heterogeneous* catalysis and, to a lesser extent, the Wilzbach method to improve the

---

[118] M. L. Eidinoff, J. E. Knoll, D. K. Fukushima, and T. F. Gallagher, *J. Amer. Chem. Soc.* **74**, 5280 (1952).

[119] J. H. Taylor, P. S. Woods, and W. L. Hughes, *Proc. Nat. Acad. Sci. U.S.* **43**, 122 (1957).

[120] W. G. Verly and G. Hunebelle, *Bull. Soc. Chim. Belg.* **66**, 640 (1957).

[121] A. Murray, D. F. Peterson, N. F. Hayes, and M. Magee, *Los Alamos Biol. Med. Res. Gp. Rep. Hlth Div.*, LAMS-2627, p. 51 (Jan.–June, 1961).

techniques and efficiency of labeling, other new methods which are of potential value have recently been discovered. These include homogeneous metal catalysis,[56] aspects of acid–base catalysis[122] and the most recently reported radiation-induced exchange with T_2O. These current developments in labeling will be discussed in this section with special reference to methods available for tagging heterocyclic compounds both generally and specifically.

A. Homogeneous Metal Catalysis

1. Conditions of Technique

This procedure constitutes the *homogeneous* analog of the previous (Section III) *heterogeneous* system. It consists of exchanging the hydrogen in isotopic water with the organic compound in the presence of *homogeneous* metal catalysts such as Pt^{II} salts. A typical reaction procedure utilizes acetic acid as a common solvent, isotopic water, sodium platinum(II) chloride as catalyst, the organic compound, and hydrochloric acid to stabilize the homogeneous catalyst. The components react in evacuated sealed tubes[56, 123–125] or are heated at a temperature within the range 25°–130°, this being determined by the isotope used and the enrichment required. The acidity of the solution is critical, since this should be high enough to stabilize the catalyst complex against reduction and, thus, prevent precipitation or "self-activation" of the platinum.[68] However, the acidity should not be too high, otherwise the metal-catalyzed process is inhibited.

2. Mechanisms of Homogeneous Exchange

Only preliminary studies of the homogeneous technique have been performed and the procedure has been used to label only relatively simple molecules. However, extrapolation to some heterocyclic systems has been achieved. In general, the orientation of isotope incorporated by both homogeneous and heterogeneous procedures is similar, ortho deactivation effects being observed for positions adjacent to bulky substituents such as alkyl groups.

Two mechanisms have been proposed for the exchange of ring hydrogens in heterocycles,[123–126] termed the homogeneous associative and

---

[122] A. R. Katritzky and C. D. Johnson, *Angew. Chem. Int. Ed. Engl.* **6**, 608 (1967).
[123] R. J. Hodges and J. L. Garnett, *J. Phys. Chem.* **72**, 1673 (1968).
[124] R. J. Hodges and J. L. Garnett, *J. Catal.* **13**, 83 (1969).
[125] R. J. Hodges and J. L. Garnett, *J. Phys. Chem.* **73**, 1525 (1969).
[126] J. L. Garnett, R. J. Hodges, and W. A. Sollich-Baumgartner, *Proc. 4th Int. Congr. Catal., 1968*, Paper 1, in press.

dissociative π-complex mechanisms [Eqs. (33) and (34)], and they are analogous to the corresponding heterogeneous mechanisms. If alkyl substituents are present in the heterocycle, homogeneous metal-catalyzed exchange may also be observed in such groups and π-allylic species have been proposed to account for such deuteration.[124]

3. *Exchange in Specific Heterocyclic Systems*

Mechanistically, homogeneous metal-catalyzed exchange in heterocycles[125, 127] is complicated by the effect of the lone-pair electrons from the heteroatom and also the presence of acid in the exchange medium. In the present preliminary homogeneous studies (Table XIV) only a

TABLE XIV

Homogeneous Metal-Catalyzed Deuteration of Heterocyclic Compounds[a]

Run No.	Compound	Atom % deuterium	Deuterium distribution				
			d_0	d_1	d_2	d_3	d_4
1	Pyridine	0.1	99.5	0.5	—	—	—
2	Benzo[h]quinoline	0.12	98.9	1.1	—	—	—
3	Benzo[b]quinoline	0.0	100.0	—	—	—	—
4	Furan	33.8	19.2	40.3	27.8	12.1	0.75
5	Furan[b]	30.5	15.6	46.3	37.8	—	—
6	Thiophene	47.9	10.5	25.3	30.4	28.9	4.7

[a] Exchange conditions involved Na_2PtCl_4 (0.02 M), HCl (0.005 M), organic reactant (0.4 gm), in CH_3COOD (2.4 gm) and D_2O (0.46 gm) at 100° for 5 hours.
[b] No Pt^{II} present, i.e., acid catalyzed only.

small degree of exchange ($<1\%$) has been observed in unsubstituted pyridine. Thus, pyridine may form a strong σ complex with platinum via its nitrogen atom and this would inhibit π-complex formation or, alternatively, the pyridine is protonated as the pyridinium ion, which with its positive charge may not form the necessary π complex for exchange.

Thiophene and furan both are active to electrophilic attack in the α positions and readily undergo acid exchange in those positions. For thiophene the ratio of rates of acid exchange[128–130] at H-2 and H-3 is

[127] J. L. Garnett and J. C. West, unpublished data.
[128] H. Schreiner, *Monatsh.* **82**, 702 (1951).
[129] K. Halvarson and L. Melander, *Arkiv Kemi* **8**, 29 (1955).
[130] B. Ostman and S. Olsson, *Arkiv Kemi* **15**, 275 (1959).

approximately 1000; thus, most of the d_3 and d_4 contribution to the low voltage mass spectrum of thiophene in run 6 (Table XIV) is homogeneous metal-catalyzed. This has been proved by inhibitor experiments.[127] In a similar manner, the d_3 and d_4 component of the furan low

$$\left[\begin{array}{c}\text{Cl} \\ |\text{II} \\ \text{Cl—Pt} \\ | \\ \text{Cl} \quad \text{N}\end{array}\right]^{-} \underset{-\text{H}^+}{\overset{+\text{D}^+}{\rightleftharpoons}} \left[\begin{array}{c}\text{Cl} \\ |\text{II} \\ \text{Cl—Pt} \\ | \\ \text{Cl} \quad \text{N} \quad \text{D}\end{array}\right] \quad (33)$$

$$\left[\begin{array}{c}\text{Cl} \\ |\text{II} \\ \text{Cl—Pt—Cl} \\ | \\ \text{Cl}\end{array}\right]^{2-} + \bigcirc_{\text{N}} \rightleftharpoons \left[\begin{array}{c}\text{Cl} \\ \text{Cl} |\text{II} \\ \text{Cl—Pt} \\ | \\ \text{Cl} \quad \text{N}\end{array}\right]^{2-} \rightleftharpoons \left[\begin{array}{c}\text{Cl} \\ |\text{II} \\ \text{Cl—Pt} \\ | \\ \text{Cl} \quad \text{N}\end{array}\right] + \text{Cl}^{-}$$

$$\rightleftharpoons \left[\begin{array}{c}\text{Cl Cl} \\ |\text{IV} \\ \text{Cl—Pt} \\ \text{H} | \quad \text{N} \\ \text{Cl}\end{array}\right]^{2-} \underset{-\text{D}^+\text{Cl}^-}{\overset{-\text{H}^+\text{Cl}^-}{\rightleftharpoons}} \left[\begin{array}{c}\text{Cl} \\ |\text{II} \\ \text{Cl—Pt} \\ | \\ \text{Cl} \quad \text{N}\end{array}\right]^{2-} \quad (34)$$

voltage mass spectrum (run 4 vs. run 5) has been shown to be formed by a metal-catalyzed process.[127] In the present circumstances, it is possible to separate metal-catalyzed exchange from acid exchange; however, for general labeling purposes the simultaneous use of both techniques is valuable.

Attempts have been made to reduce the poisoning effect of the nitrogen lone pair in pyridine by complexing with cobalt as a cocatalyst; however, no significant increase in deuteration rate was observed under these conditions. Similar efforts to reduce the lone-pair effect of heterocyclic nitrogen atoms involved exchanging heterocyclic molecules where steric hindrance may occur. Thus, in the benzo[h]quinolines and benzo-[b]quinolines (33) and (34) some exchange is observed with the more

(33)   (34)   (35)

sterically hindered **33**. This is analogous to exchange in phenanthrene[125] (**35**) where the formation of a σ-Pt complex in the 4-position is more sterically hindered than in the 1-position.

It is thus clear that homogeneous metal catalysis is satisfactory for tritium-labeling benzo[*h*]quinoline, but not for deuteration, whereas the homogeneous method is very satisfactory for the deuteration and/or tritiation of furan and thiophene. This homogeneous technique is still in the preliminary stages of development and it may be possible in future work to overcome the problem associated with the toxicity of the lone pair in nitrogen heterocycles. Thus, a discovery of this nature would be valuable since the homogeneous metal-catalyzed technique possesses several advantages when compared with the corresponding heterogeneous process which have been summarized elsewhere.[126]

B. Acid–Base Catalysis

Both acid and base catalysis have been used extensively to catalyze exchange in aromatic, and to a lesser extent, heterocyclic molecules. In acid exchange, the most widely used catalysts are sulfuric acid,[122, 129, 131] phosphoric acid,[132] trifluoroacetic acid[5, 133] perchloric acid,[134] aluminum chloride,[135] and the phosphoric acid–boron trifluoride complex.[132] These reactions constitute the simplest electrophilic substitution. The mechanism for such substitution in benzenoid compounds is now comparatively well understood;[122] however, the problem of heteroaromatic electrophilic substitution is still being clarified and has led to renewed interest in acid-catalyzed exchange in heterocyclic compounds.[122]

1. *Acid Exchange in Specific Heterocyclic Compounds*

The sulfuric acid system has been the most extensively studied and will be used to demonstrate the application of the method in determining the reactivity of heterocyclic compounds to deuterium and/or tritium labeling. A summary of representative data with the other acidic catalysts for labeling heterocyclic compounds will complete this section.

a. *Pyridines*. Even under the most vigorous conditions at temperatures of approximately 200°, pyridine and 4-picoline exchange slowly in the ring positions, whereas 2,6-di- and 2,4,6-trimethylpyridine labeled more

[131] P. Avinur and A. Nir, *Bull. Res. Counc. Isr. Sect. A* **7**, 74 (1958).
[132] P. M. Yavorsky and E. Gorin, *J. Amer. Chem. Soc.* **84**, 1071 (1962).
[133] B. Aliprandi and F. Cacace, *Ann. Chim.* (Rome) **51**, 397 (1961).
[134] B. Aliprandi and F. Cacace, *Ann. Chim.* (Rome) **49**, 2011 (1959).
[135] C. Mantescu, A. Genunche, and A. T. Balaban, *J. Label. Compounds* **1**, 178 (1965).

readily, but in the 3- and 5-positions only. 1,2,4,6-Tetramethylpyridinium ion (36) reacts slightly faster than the 2,4,6-trimethylpyridine (37) and again exchanges only in the ring positions.[122,136]

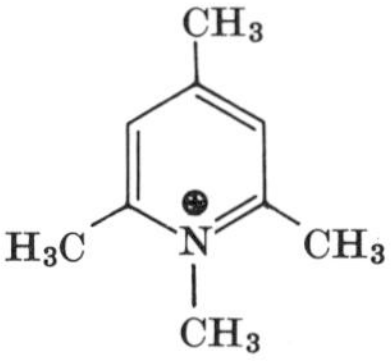

(36) 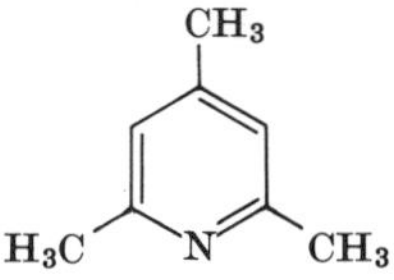   (37)

b. *Pyridine N-Oxides.* Hydrogen exchange in 2,4,6-trimethylpyridine N-oxide proceeds smoothly in the 3- and 5-positions at 200° without decomposition, whereas the 2- and 4-hydrogen atoms are exchanged in 3,5-dimethylpyridine N-oxide (38).[122,137] The labeling of the 1-methoxy-

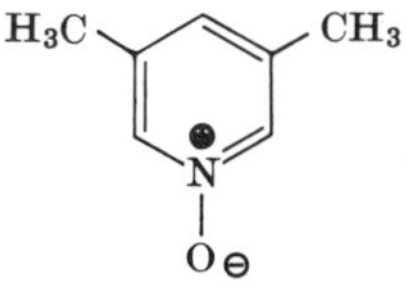

(38)

2,4,6-trimethylpyridinium ion and 2,6-dimethylpyridine N-oxide is difficult because of concurrent decomposition. With care, exchange can be shown to occur in the 3-position with the latter compound. Preliminary studies[122] on 3-hydroxypyridine show that exchange does occur; however, whether this incorporation is at the 2- or 6-position has not yet been clarified. 3,5-Dimethoxypyridine N-oxide exchanges in the 2- and 6-positions.

c. *Pyridones.* 4-Pyridone (39) and 1-methyl-4-pyridone (40) exchange in the 3- and 5-positions readily,[138] while 5-methyl-2-pyridone (41) and 3-methyl-2-pyridone (42) are deuterated in the 3- and 5-positions,

(39) 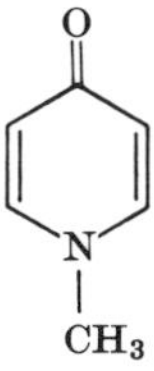   (40) 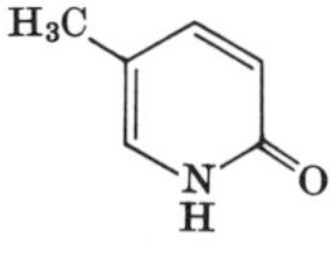   (41)

[136] A. R. Katritzky and B. J. Ridgewell, *J. Chem. Soc.*, 3753 (1963).
[137] A. R. Katritzky, B. J. Ridgewell, and A. M. White, *Chem. Ind.* (London), 1576 (1964).
[138] P. Bellingham, C. D. Johnson, and A. R. Katritzky, *Chem. Ind.* (London), 1384 (1965).

respectively. It has also been shown that 4-quinolone at low acidities ($H_0 < -7$) exchanges as the neutral species (**43**) at the 3-position, whereas on increasing the acidity, reaction via the conjugate acid occurs successively in the 3-, 6-, 8-, and 5-positions (**44**).[139]

(**42**)   (**43**)   (**44**)

d. *Thiophene.* In the acid-catalyzed exchange of thiophene[128–130] Halvarson and Melander,[129] using 59.4% sulfuric acid at 25°, reported a value of 955 ± 140 for relative rates of tritium exchange at H-2 and H-3, and Ostman and Olsson,[130] with 57% sulfuric acid also at 25°, found relative exchange rates at these positions of 1045 ± 61 for deuterium and 911 ± 60 for tritium.

2. *Miscellaneous Acid Exchange Reactions*

A number of isolated acid-induced exchange systems have been reported and are of use in labeling heterocycles. For example, pyridine with DCl in a sealed tube at 218° exchanges only in the 2- and 6-positions.[140] Of the remaining acid reagents, the two most promising are aluminum chloride,[135] which forms the Bronsted acid $H^+(AlCl_3OH)^-$, and the phosphoric acid–boron trifluoride complex.[132] Using the former reagent to tritiate a series of heterocycles, the results (Table XV) reflect the order of decreasing reactivity toward electrophilic substitution, i.e., 2,4,6-collidine > 2,6-lutidine > pyridine. Balaban and co-workers have also tritiated a series of pyrimidine derivatives with the same acid.[135]

With the phosphoric acid–boron trifluoride reagent,[132] exchange in heterocycles (Table XVI) occurs at room temperature with both deuterium and tritium.

More complicated heterocyclic compounds such as imidazolo[1,2-*a*]-pyridine (**45**), imidazolo[1,2-*a*]pyrimidine (**46**), and 1,2,4-triazolo[1,5-*a*]-pyrimidine (**47**) have been deuterated[141] by 3.0 M D_2SO_4 at 100°. In **45** and **46** exchange occurred at position 3, whereas in **47** deuterium substituted at position 6. Thus, as a general rule, in acid-catalyzed

---

[139] G. P. Bean, A. R. Katritzky, and A. Marzec, *Bull. Acad. Pol. Sci., Ser. Sci. Chim.* **16**, 453 (1968) [*Chem. Abstr.*, **70**, 76963 (1969)].
[140] J. A. Zoltewicz and C. L. Smith, *J. Amer. Chem. Soc.* **89**, 3358 (1967).
[141] W. A. Pandler and L. S. Helmick, *Chem. Commun.*, 377 (1967).

TABLE XV
LABELING OF HETEROCYCLES WITH TRITIATED $H(AlCl_3OH)$[a]

Compound	Reaction time (hr)	Specific activity (d.p.s./mole)
2,4,6-Collidine	3	7.5×10^6
2,4,6-Collidine	7	1×10^7
2,6-Di-t-butyl-4-picoline	3	2.5×10^7
2,6-Di-t-butyl-4-picoline	7	4×10^7
2,6-Lutidine	7	4.8×10^6
Pyridine	12	5×10^5

[a] All reactions involved organic compound (0.1 mole), with methylene dichloride (20 ml; except for pyridine, 25 ml), tritiated water (0.05 mole) at specific activity of 3.96×10^9 d.p.s./mole containing aluminum chloride (0.1 mole), heated to 41°–42°.

TABLE XVI
LABELING OF HETEROCYCLES WITH $TH_2PO_4 \cdot BF_3$[a]

Compound	Reaction time (hr)	Reagent (μCi/gm)	Product (μCi/gm)	Theoretical (μCi/gm)
Pyridine	22	40.7	0.07	24.6
Pyridine hydrochloride	22	40.7	0.9	26.1
Pyridine N-oxide	45.5	40.7	29.8	22.3
Quinoline	22	41.4	0	47.2
Thiophene	20	63.3[b]	35	34

[a] All reactions performed at 23°, except pyridine N-oxide (70°).
[b] TH_2PO_4 only.

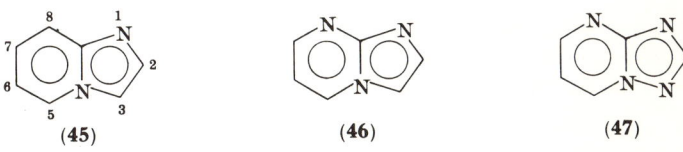

(45)   (46)   (47)

exchange with heterocyclic molecules isotope incorporation occurs predominantly at those positions which also undergo electrophilic substitution most readily. Some of the limitations of this rule have already been discussed.[122]

3. Base-Catalyzed Exchange

Shatenshtein[142] detected no difference in exchange reactivity of five hydrogen atoms in pyridine when deuterated in liquid ND_3 and KNH_2 at 25°. Pyridine when heated with NaOH in D_2O in a monel bomb exchanged in all positions.[140] At low base concentration, rates of exchange were greatest at position 2, but at high base concentrations position 4 was the most active. Relative rates of exchange under these latter conditions in unprotonated pyridine at positions 4, 3, and 2 were 3.0, 2.3, and 1.0, respectively. The mechanism of deuteration presumably involves removal of a proton to give the pyridyl anion, which then abstracts a deuteron from the solvent. The above results[140] are to be compared with the earlier work of Kawazoe and co-workers,[143] who found only α exchange in pyridine in strongly basic deuterium oxide at 220°.

Of the substituted pyridines, the halogenated derivatives have been the most intensively studied.[144,145] Treatment of 3,5-dichloropyridine N-oxide at 74° with 0.1 N NaOD led to exchange in three positions of the molecule, whereas with 3-chloropyridine N-oxide relative rates of exchange were position 2 > 6 > 4 > 5. In 1-methyl-4-pyridone, 1,3,5-trimethyl-4-pyridone, and 3,5-dibromo-1-methyl-4-pyridone, deuteration in basic D_2O at 100° gives 2- and 6-substitution.[146] With the polyazaindenes (45)–(47) already discussed in the acid exchange section,[141] base-catalyzed deuteration occurs in the positions indicated: **45** 3 and 5; **46** 2, 3, 5, and 6; and **47** 2, 5, 6, and 7. In other isolated heterocycles some selectivity is observed in base-catalyzed exchange, e.g., certain imidazoles,[147] thiazole,[148] isothiazole,[148] benzothiazole,[149] and benzoxazole.[149]

C. Radiation-Induced Exchange with Tritium Oxide

The most recent labeling technique applicable only to tritium has been the use of tritium oxide as its own radiation source for radiation-

---

[142] A. I. Shatenshtein, "Isotopic Exchange and the Replacement of Hydrogen in Organic Compounds." Consultants Bureau, New York, 1962.

[143] Y. Kawazoe, M. Ohnishi, and Y. Yoshioka, *Chem. Pharm. Bull.* (Tokyo) **12**, 1386 (1964).

[144] R. A. Abramovitch, G. M. Singer, and A. R. Vinutha, *Chem. Commun.*, 55 (1967).

[145] J. A. Zoltewicz and G. M. Kauffman, *Tetrahedron Lett.*, 337 (1967).

[146] P. Beak and J. Bonham, *J. Amer. Chem. Soc.* **87**, 3365 (1965).

[147] T. M. Harris and J. C. Randall, *Chem. Ind.* (London), 1728 (1965).

[148] R. A. Olofson, J. M. Landesberg, K. N. Houk, and J. S. Michelman, *J. Amer. Chem. Soc.* **88**, 4265 (1966).

[149] M. Fox, F. Taddei, and P. E. Todesco, *Proc. Sci. Fac. Chim. Ind. Bologna* **23**, 225 (1965).

induced exchange with organic compounds. Rate of incorporation of isotope is slow and exposure times of approximately one month are required to obtain mCi/mg levels of specific activity in the organic compound with water of 5 Ci/ml. Preliminary studies only have been reported for ten compounds including pyridine. The results suggest that radiation-induced degradation does not interfere with the labeling process as it does for Wilzbach gas exchange. Gold and co-workers[150] originally found that radiation-induced exchange could occur between tritium oxide and toluene; however, the potential of the technique as a labeling tool for a variety of compounds has only recently been discovered.[151] At this time, the method is very promising as a general tritiation procedure of the future, especially for heterocycles.

D. COMPARISON OF LABELING TECHNIQUES

The radiation-induced procedures (Section II), of which the Wilzbach method is the most useful, are limited to tritium labeling only. The predominant advantage of the method is its simplicity, since any compound containing hydrogen will exchange under these conditions. The main difficulty is to obtain satisfactory *radiochemical* purity in the final product, since many competing decomposition and hydrogenation reactions occur simultaneously leading to a significant proportion of the activity being lost in these waste products. For example, some of the side reactions which have been observed in gas exposure labeling are fragmentation, addition of fragments to the compound exposed, polymerization, replacement of substituents, addition of tritium at points of unsaturation, isomerization, and racemization. Products formed as a result of fragmentation, polymerization, or replacement of a substituent usually have properties which are significantly different from those of the parent compound and can be readily removed. Addition of tritium to unsaturated molecules is a more difficult problem and actually precludes the labeling of many compounds such as unbranched olefins. The final limitation of this labeling method is the relatively low levels of specific activity which can be incorporated when compared with other procedures such as catalytic exchange with isotopic water. However, if all other labeling methods fail, tritium exchange can be achieved by this technique, and provided that radiochemical purification of the product is obtained, the method will always yield the required tritiated compound.

By comparison, catalytic exchange procedures (Sections III and IV) are applicable to both deuterium and tritium labeling and are the most

[150] J. R. Adsetts and V. Gold, *Chem. Commun.*, 915 (1968).
[151] D. Fong, J. L. Garnett, and M. Long, *J. Label. Compounds* (in press),

useful general methods for this purpose. For tritiation, high radiochemical yields and relatively high specific activities (100 mCi/mM) are easily obtained. The advantages of the catalytic procedures have already

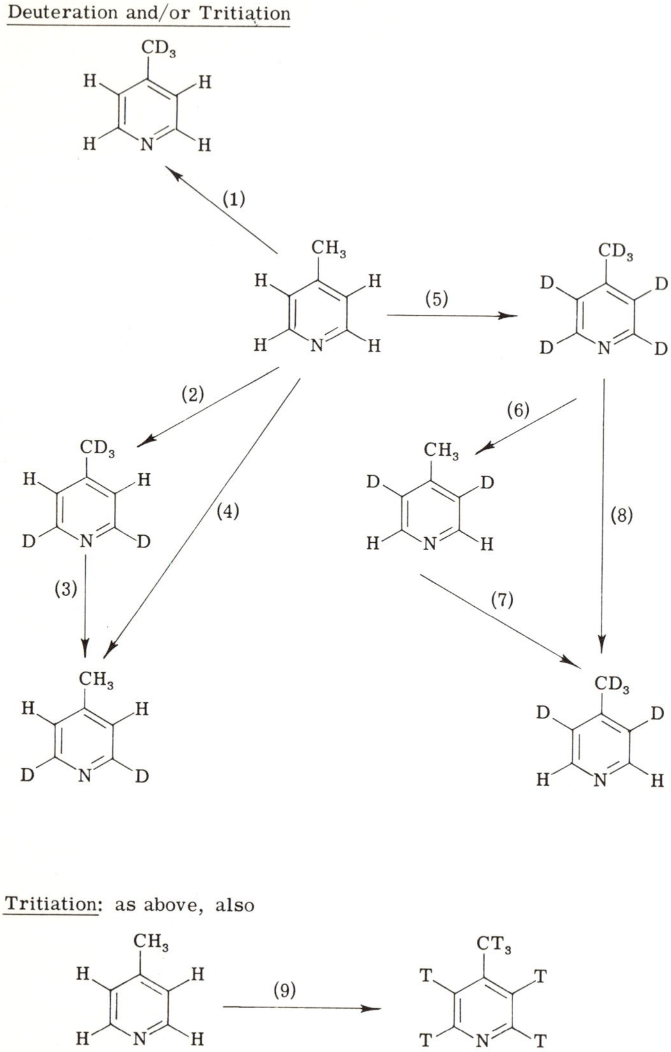

Fig. 1. Various possibilities for labeling 4-picoline. (1) Dilute DCl at 100°C, (2) PtO_2 (self-activated) + D_2O, (3) dilute HCl at 100°C, (4) $CoCl_2$ + $NaBH_4$ + D_2O, (5) PtO_2 or $NiCl_2$ + $NaBH_4$ + D_2O, (6) PtO_2 (self-activated) + H_2O, (7) dilute DCl at 100°C, (8) $CoCl_2$ + $NaBH_4$ + H_2O, and (9) Wilzbach or T_2O. (1), (3), and (7): from self-activation of aqueous solutions of H_2PtCl_6.

been summarized briefly.[64] For heterocyclic systems, acid–base catalysis usually requires rigorous conditions for labeling the simple unsubstituted members of the series, whereas for heterogeneous metal catalysis much lower temperatures are usually necessary to achieve satisfactory tagging. In those substituted heterocyclic compounds where exchange (particularly acid–base catalysis) occurs under relatively mild conditions, care should be exercised in the subsequent use of the labeled material, since many reaction media in both chemical and biochemical systems are sufficiently acidic or basic to catalyze back exchange during the course of a particular experiment.

Homogeneous metal catalysis is a promising recent development, although its application to heterocyclic systems has been retarded because of the difficulty of overcoming strong irreversible complex formation between the catalyst and the heterocyclic compound to be labeled. Future developments should certainly overcome this problem and the method will be a valuable addition to the others discussed in Sections III and IV. As a consequence of procedures available, it is now possible to use a number of methods successively to achieve specificity in labeling. Some of the possibilities are demonstrated in Fig. 1 using 4-picoline as representative model compound. Many more variations are, of course, feasible depending on the structure of the particular heterocyclic molecule. The limitations of such procedures will essentially depend on the magnitude of the isotopic scrambling during each step, and experimental conditions may have to be carefully controlled to eliminate this problem.

Acknowledgments

Acknowledgment is given to the Australian Research Grants Committee, the Australian Atomic Energy Commission, and the Australian Institute of Nuclear Science and Engineering for the support of graduate students, who have worked with the present authors in the field of isotopic hydrogen labeling of heterocyclic systems.

The Chemistry of 1-Pyrindines

FILLMORE FREEMAN

Department of Chemistry, California State University, Long Beach, California

I. Introduction	187
II. Preparation and Reactions	189
A. Unsubstituted 1H-1-, 5H-1-, and 7H-1-Pyrindines	189
B. Substituted 1H-1-, 5H-1-, and 7H-1-Pyrindines	195
C. 6,7-Dihydro-5H-1-pyrindines	202
D. Hydroxy- and Oxo-6,7-dihydro-5H-1-pyrindine Derivatives	212
E. 5H-1-pyrindine-5,7(6H)-diones	217
F. Hexahydro- and Octahydro-1-pyrindines	222
G. Hexahydro- and Octahydro-1-pyrindones	225
III. Biomedical Applications	229

I. Introduction

Although the chemistry of the polycyclic nitrogen heterocycles containing one heteronitrogen atom has been the subject of many comprehensive reviews in recent years, the chemistry of pyrindines has received no such attention.[1a, 1b] This review will systematically survey the chemical literature, through 1969, concerning the chemistry of the derivatives of 1-pyrindine (**1**)–(**9**), 6,7-dihydro-5H-1-pyrindine (**10**), hydroxy- and oxo-6,7-dihydro-5H-1-pyrindine, 5H-1-pyrindine-5,7(6H)-dione (**11**), and hexahydro- and octahydro-1-pyrindine. Some medically important 1-pyrindine derivatives, such as the octahydro- and decahydrobenzo[a]cyclopenta[f]quinolizines (**12**) and (**13**) will also be discussed.

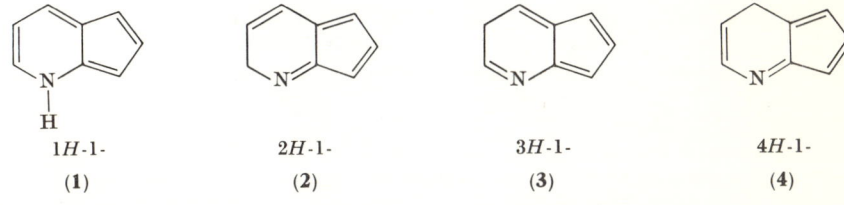

1H-1-	2H-1-	3H-1-	4H-1-
(**1**)	(**2**)	(**3**)	(**4**)

[1a] R. C. Elderfield and E. T. Losin, *in* "Heterocyclic Compounds" (R. C. Elderfield, ed.), Vol. III, Chapter 4. Wiley, New York, 1952.

[1b] It appears that most modern books on heterocyclic chemistry have little, if any, discussion of the chemistry of 1- and 2-pyrindines.

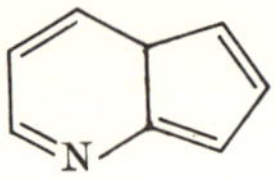

4aH-1-	7aH-1-	5H-1-	6H-1-
(5)	(6)	(7)	(8)

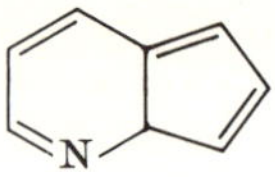

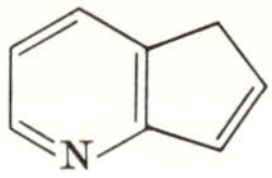

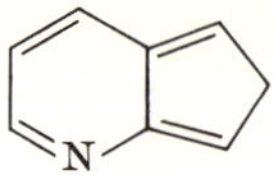

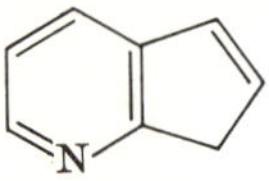

7H-1-		
(9)	(10)	(11)

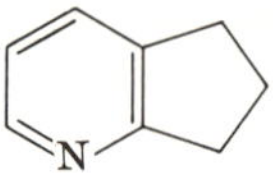

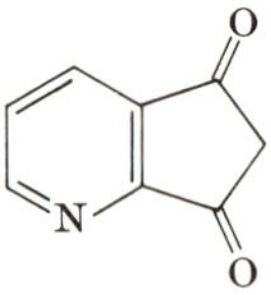

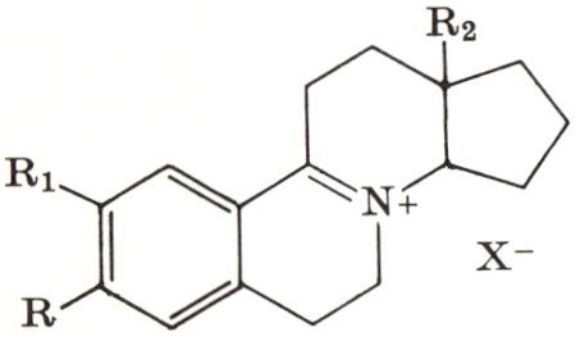

(12) 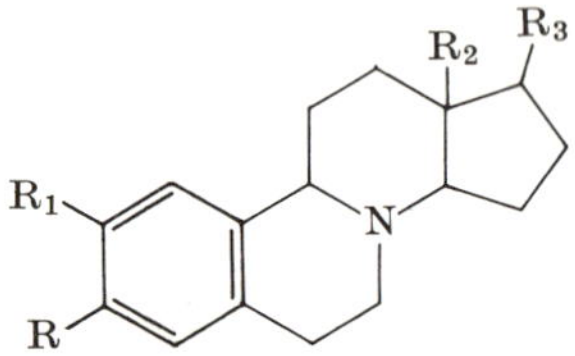   (13)

The *Chemical Abstracts* nomenclature and numbering system will be used throughout the discussion.

In recent years, much work has been directed toward the synthesis of unsubstituted and substituted pyrindines.[1-18] The tautomeric mixture of **1**, **7**, and **9** contains the only unsubstituted 1-pyrindine prepared to date. Demonstrations of aromatic character in azulene have

---

[2] T. Eguchi, *Bull. Chem. Soc. Jap.* **3**, 239 (1928).
[3] W. C. Thompson and J. R. Bailey, *J. Amer. Chem. Soc.* **53**, 1002 (1931).
[4] W. C. Thompson, *J. Amer. Chem. Soc.* **53**, 3160 (1931).
[5] H. L. Lochte and A. G. Pittman, *J. Org. Chem.* **25**, 1492 (1960).
[6] H. L. Lochte and A. G. Pittman, *J. Amer. Chem. Soc.* **82**, 469 (1960).
[7] P. Arnall, *J. Chem. Soc.*, 4040 (1954).
[8] C. F. Bradenburg and D. R. Latham, *J. Chem. Eng. Data* **13**, 391 (1968).
[9] M. M. Robison, *J. Amer. Chem. Soc.* **80**, 6254 (1958).
[10] V. Prelog and S. Szpilfogel, *Helv. Chim. Acta* **28**, 1684 (1945).
[11] A. G. Anderson, W. F. Harrison, R. G. Anderson, and A. G. Osborne, *J. Amer. Chem. Soc.* **81**, 1255 (1959).
[12] C. B. Reese, *J. Amer. Chem. Soc.* **84**, 3979 (1962).
[13] M. M. Robison and B. L. Robison, *J. Amer. Chem. Soc.* **77**, 6554 (1955).
[14] H. L. Ammon, Ph.D. Thesis, University of Washington, Seattle, Washington, 1963; *Diss. Abstr.* **24**, 1393 (1963).
[15] A. G. Anderson and H. L. Ammon, *Tetrahedron Lett.*, 2579 (1966).
[16] A. G. Anderson and H. L. Ammon, *Tetrahedron* **23**, 3601 (1967).
[17] H. L. Ammon and L. H. Jensen, *J. Amer. Chem. Soc.* **88**, 681 (1966).
[18] W. E. Hahn and J. Epsztajn, *Rocz. Chem.* **38**, 989 (1964).

Sec. II.A.] THE CHEMISTRY OF 1-PYRINDINES 189

stimulated study of analogous structures in which heteroatoms have replaced one or more carbon atoms in either the five- or seven-membered ring. These compounds may be classified as π-equivalent (same number of π electrons as ring atoms) or π-excessive analogs of azulene (more π electrons than ring atoms).[19-23] The parent pseudoazulene of the 1-pyrindine group is the π-excessive 1H-1-pyrindine (**1**). Also, preparation of the iso-π-electronic heteroanalogs of azulene, 2-phenyl- and

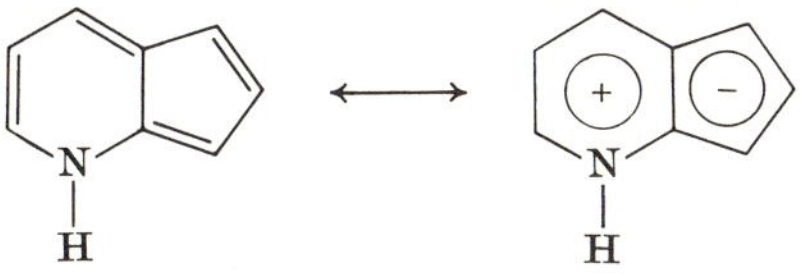

2-methyl-2-pyrindine (**14a**) and (**14b**)[24-26] and an examination of their

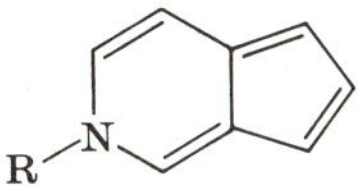

(**14a**) R = C$_6$H$_5$
(**14b**) R = CH$_3$

spectral properties lend further support to the concept of aromatic character in the 1-pyrindine and 2-pyrindine ring systems.

II. Preparation and Reactions

A. Unsubstituted 1H-1-, 5H-1-, and 7H-1-Pyrindines

Although 6,7-dihydro-5H-1-pyrindine (**10**) and its derivatives have been known for years,[1-8] the first unsubstituted 5H-1-pyrindine (**7**) was not prepared until 1958.[9] Attempts to dehydrogenate **10** with selenium at 400° or with palladium–charcoal at 350° were unsuccessful.[10]

---

[19] W. F. Harrison, Ph.D. Thesis, University of Washington, Seattle, Washington, 1960.
[20] A. Albert, "Heterocyclic Chemistry, An Introduction." Athlone Press, London, 1968.
[21] R. Mayer, *Angew. Chem.* **69**, 481 (1957).
[22] R. Mayer, J. Franke, V. Horak, I. Hanker, and R. Zahradnik, *Tetrahedron Lett.*, 289 (1961).
[23] V. W. Armit and R. Robinson, *J. Chem. Soc.*, 827 (1922).
[24] A. G. Anderson, W. F. Harrison, and R. G. Anderson, *J. Amer. Chem. Soc.* **85**, 3448 (1963).
[25] A. G. Anderson and W. F. Harrison, *J. Amer. Chem. Soc.* **86**, 708 (1964).
[26] A. G. Anderson and D. M. Forkey, *J. Amer. Chem. Soc.* **91**, 924 (1969).

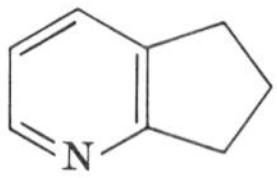

(10)

However, Robison,[9] starting with **10** and using the thermal rearrangement of the pyrindane 1-oxide (**15**), prepared 5H-1-pyrindine (**7**) according to Scheme 1. It was suggested that the reaction product

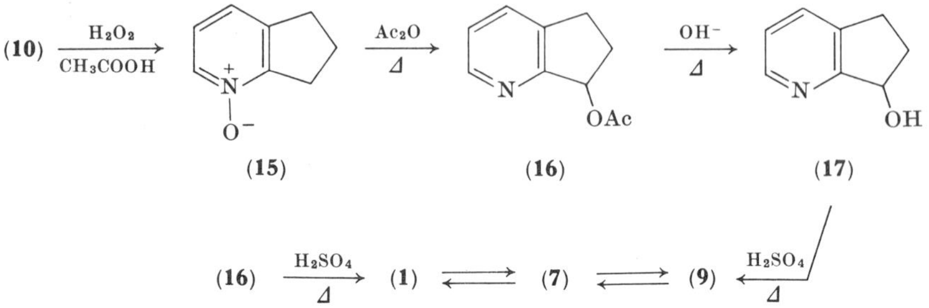

(10) $\xrightarrow[\text{CH}_3\text{COOH}]{\text{H}_2\text{O}_2}$ (15) $\xrightarrow[\Delta]{\text{Ac}_2\text{O}}$ (16) $\xrightarrow[\Delta]{\text{OH}^-}$ (17)

(16) $\xrightarrow[\Delta]{\text{H}_2\text{SO}_4}$ (1) $\rightleftharpoons$ (7) $\rightleftharpoons$ (9) $\xleftarrow[\Delta]{\text{H}_2\text{SO}_4}$

SCHEME 1

consisted predominantly of a mixture of the 5H- and 7H-tautomers, and it was postulated that the orange color of the freshly distilled product was due to the presence of the pseudoazulene (**1**) (1H-tautomer).[11, 16]

Although a plausible mechanism for the rearrangement of **15** to **16** is shown below,[27a–29b] an alternative free radical[30a–30d] or cyclic intramolecular mechanism is also reasonable.[29b]

In an attempt to demonstrate the presence of **1** in the tautomeric mixture Reese[12] repeated Robison's synthesis, then treated the product(s) with a large excess of methyl iodide, and suggested that the

[27a] V. J. Traynelis and R. F. Martello, *J. Amer. Chem. Soc.* **82**, 2744 (1960).
[27b] V. Boekelheide and W. J. Linn, *J. Amer. Chem. Soc.* **76**, 1286 (1954).
[27c] J. A. Berson and R. Cohen, *J. Amer. Chem. Soc.* **77**, 1281 (1955).
[27d] O. H. Bullitt and J. T. Maynard, *J. Amer. Chem. Soc.* **76**, 1370 (1954).
[28a] S. Oae, S. Tamagaki, T. Negoro, K. Ogino, and S. Kozuka, *Tetrahedron Lett.*, 917 (1968).
[28b] C. W. Muth, R. S. Darlak, M. L. DeMatte, and G. F. Chovanec, *J. Org. Chem.* **33**, 2762 (1968).
[29a] S. Tamagaki, S. Kazuka, and S. Oae, *Tetrahedron Lett.*, 4765 (1968).
[29b] An excellent discussion of heteroaromatic N-oxides has appeared recently. A. R. Katritzky and J. M. Lagowski, "Chemistry of the Heterocyclic N-oxides." Academic Press, New York, 1971.
[30a] A. Alkaitis and M. Calvin, *Chem. Commun.*, 292 (1968).
[30b] N. Hata, E. Okutsu and F. Tanaka, *Bull. Chem. Soc. Jap.* **71**, 1769 (1968).
[30c] P. L. Kumler and O. Buchardt, *Chem. Commun.*, 1321 (1968).
[30d] O. Buchardt and B. Jensen, *Tetrahedron Lett.*, 5233 (1967).

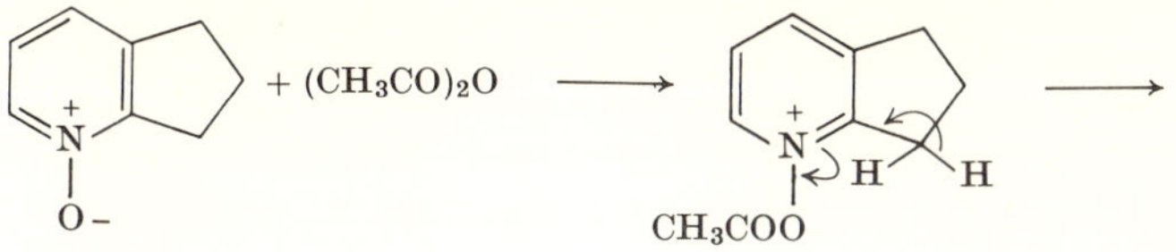

resulting product contained 1-methyl-5H-1-azonianindene iodide (**18a**). Subsequent treatment of the product with alkali gave a dark orange oil, presumably **19** which resisted purification.[12] However Anderson and

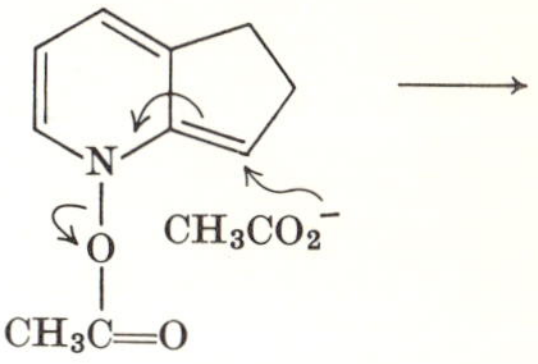

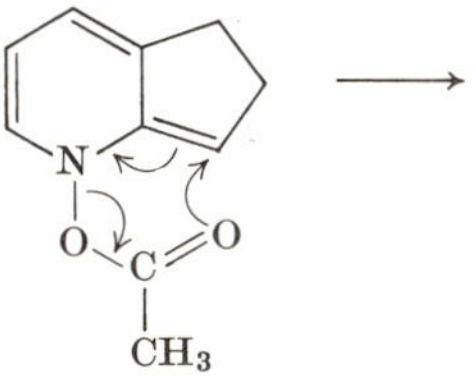

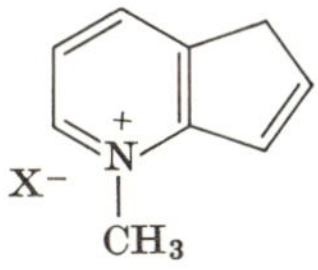

(**18a**) X = I
(**18b**) X = Cl
(**18c**) X = p-CH$_3$C$_6$H$_4$SO$_3^-$

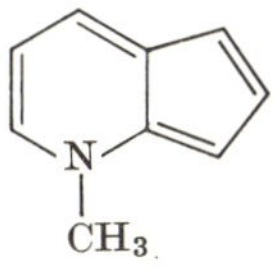

(**19**)

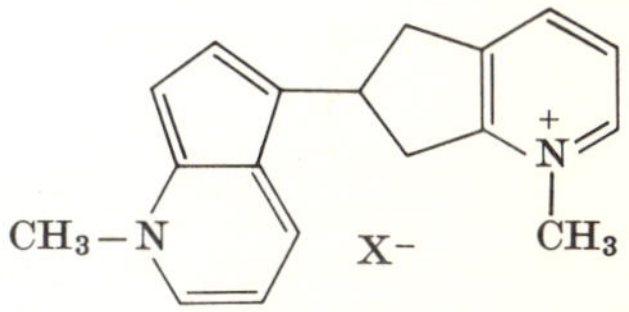

(**20a**) X = I
(**20b**) X = Cl

Ammon[16] were able to obtain **19** in sufficient purity for characterization from **18b**,[14-16] and thus confirmed the postulate[11] about the color of 1H-1-pyrindine (**1**).[12] Analysis of the visible absorption spectrum of the tautomeric mixture showed that approximately 0.1% of **1** was in tautomeric equilibrium with **7** and **9**.[12,16]

Although Anderson and Ammon[14-16] confirmed the postulate that the presence of (ca. 0.1%) the 1H-tautomer (**1**) is responsible for the orange-red color in the tautomeric mixture, their subsequent experiments on the quaternization of **7** with methyl iodide followed by extractive basification gave a 78% yield of **20a** and a small amount of **19**. Under similar conditions **18b** gave a 40% yield of **19** and no **20b**. The crystal and molecular structure of 1-methyl-6[5-(1-methyl-1H-1-pyrindinyl)]-1-azoniaindan iodide (**20a**) has been solved by X-ray diffraction.[17]

More recently, **7** and 2-phenyl-5H-1-pyrindine (**22**) have been prepared by the dehydration of the corresponding 2-substituted 7-hydroxy-6,7-dihydro-5H-1-pyrindines (**21**) in excellent yields.[18] Similarly,

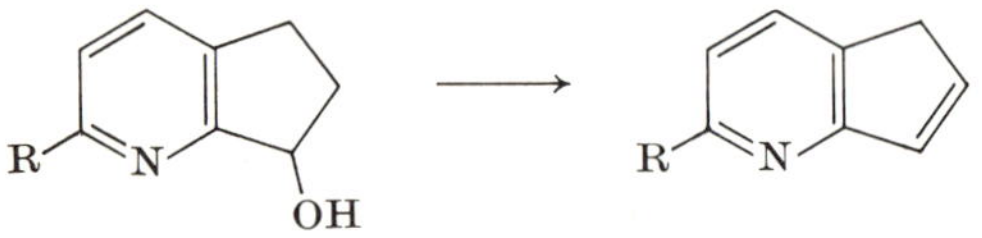

(**21**)

(**7**) R = H (76%)
(**22**) R = C$_6$H$_5$ (71%)

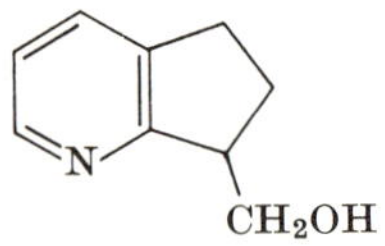

(**23**) CH$_2$OH

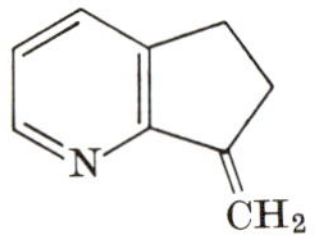

(**24**) CH$_2$

dehydration of **23** gives 6,7-dihydro-7-methylene-5H-1-pyrindine (**24**) in 81% yield.

1. Molecular Orbital Calculations

During the past few years several groups of workers have performed a variety of quantum mechanical calculations on π-electronic analogs of the nonalternant aromatic hydrocarbon azulene, and N-heteroaromatic systems.[31a–36] Raimondi and Favini[31a] (using the semiempirical

[31a] M. Raimondi and G. Favini, *Gazz. Chem. Ital.* **98**, 433 (1968).
[31b] S. F. Mason, *J. Chem. Soc.*, 3999 (1963).
[31c] E. M. Evleth, J. A. Berson, and S. L. Manatt, *J. Amer. Chem. Soc.* **87**, 2908 (1965).
[31d] G. Bergson and A. M. Weidler, *Acta Chem. Scand.* **16**, 2464 (1962).
[32] R. Borsdorf, *J. Prakt. Chem.* **4**, 211 (1966).
[33] J. Feitelson, *J. Chem. Phys.* **43**, 2511 (1965).
[34] W. Moffitt, *J. Chem. Phys.* **22**, 320 (1954).
[35] M. Los and W. H. Stafford, *J. Chem. Soc.*, 1680 (1959).
[36] W. Treibs and J. Beger, *Ann. Chem.* **652**, 192 (1962).

Sec. II.A.]   THE CHEMISTRY OF 1-PYRINDINES   193

ASMO-CI method of Pariser and Parr) and Borsdorf[32] (using LCAO-HMO) calculated the electron densities and the π-bond orders for the five 1-pyrindines (**1**), (**19**), (**25**)–(**27**). Comparison of the data with azulene

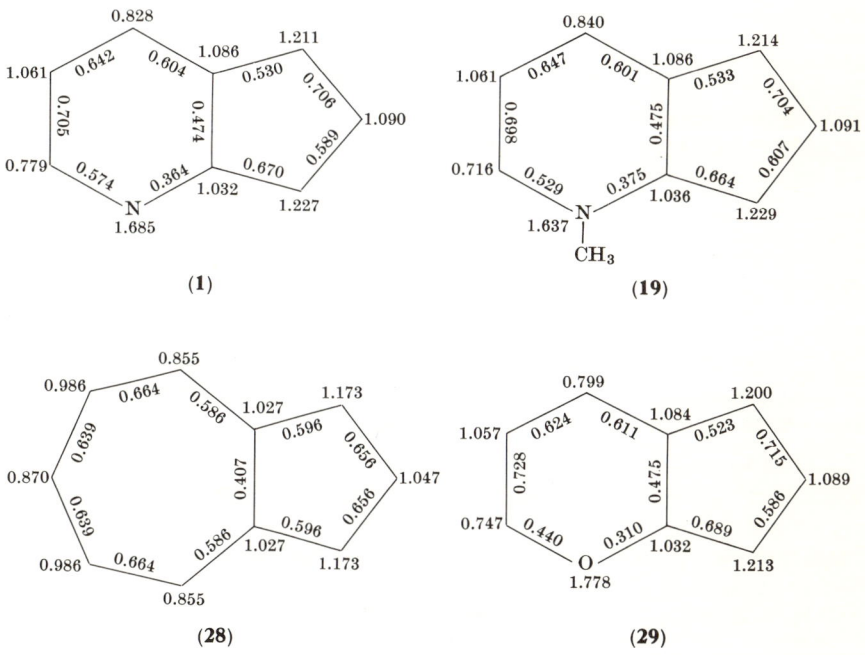

(**28**) and oxalene (**29**) revealed an azulenelike electron excess in the five membered ring and a deficiency in the six-membered ring. All the carbon atoms in **1** (except position 7a) and in **19** showed a greater electron density than those in **29**. Calculation of the localization energy and free valences

for **1** and **19** showed that electrophilic substitution should occur most readily at positions 5 and 7, and that the greatest reactivity for nucleophilic substitution should occur at carbon atoms 2 and 4. For radical substitution reactions, position 6 should be most reactive. The delocalization energies (β units) and $N \to V_1$ transitions[37, 38] of **1**, **19**, and **25–29** are compared in Table I.

TABLE I

DELOCALIZATION ENERGIES AND $N \to V_1$ TRANSITIONS FOR SOME 1-PYRINDINES[32]

Compound	Delocalization energy (β)	$N \to V_1$ transitions[37, 38]
(**1**)	2.79	0.907
(**19**)	2.96	0.901
(**25**)	4.74	0.814
(**26**)	4.72	0.730
(**27**)	6.54	0.729
(**28**)[39]	3.36	0.877
(**29**)[40]	2.66	0.880

Consideration of the $N \to V_1$ transitions for **7** and **19** suggest that alkyl groups in positions 2, 4, or 6 should show a hyposochromic shift, whereas alkyl groups in positions 1, 3, 5, or 7 should lead to a bathochromic displacement. The bathochromic shift with **19** has been demonstrated.[31a–31c]

2. *Spectra*

The similar ultraviolet spectra of **7**, **18a**, **18b**, and **19** are listed in Table II. The ultraviolet and visible spectra of **19** are very similar to those of analogous heteroanalogs of **28**, and confirm the presence of **1** in **7**.[12, 15, 16]

The infrared spectra of **7**,[15, 18] **19**,[15] **22**,[18] and **24**[18] have been reported. An infrared spectrum (neat) of **7** shows a broad band (at 3408 cm^{-1}) which decreases appreciably in carbon tetrachloride solution. This band has been attributed to the ν NH of the 1H-tautomer (**1**).[15]

[37] R. S. Mulliken, *J. Chem. Phys.* **8**, 234 (1940); **23**, 1997 (1955).
[38] $N \to V$ transitions are those in which the principal quantum number of the electron is unchanged and involve the "valence shell" and the ground or "normal state." V_1, V_2, V_3, V_4, etc., are excited states of successively higher energy.[39]
[39] A. Streitwieser, "Molecular Orbital Theory for Organic Chemists," p. 244. Wiley, New York, 1961.
[40] R. Borsdorf, *Z. Chem.* **4**, 422 (1964).

The proton magnetic resonance (PMR) spectrum of **7** is very complex and is compatible with the postulate that 1-pyrindine is composed predominantly of a mixture of **7**, **9**, and a small amount of **1**.[15, 16] The complex PMR spectra of **19** and **20** have also been recorded.[15, 16]

TABLE II

ULTRAVIOLET SPECTRA OF SOME 1-PYRINDINES

Compound	Solvent	λ_{max} mμ (log ϵ)	Ref.
(7)	Neat	466 (0.68)	16
(7)	Cyclohexane	247 (3.91), 276 (3.68), 281 (3.74), 285 (3.73), 291 (3.7)	16
(7)	Methanol	244 (3.87), 278 (3.73, sh),[a] 283 (3.78), 292 (3.68)	16
(7)	0.005 N HI in methanol	256 (3.45, sh), 2.95 (4.08)	16
(18a)	Acetonitrile	250 (4.52), 272 (4.11, sh), 300 (3.78)	16
(18a)	0.018 N HI	276 (4.08), 301 (3.77)	16
(18b)	Water; 0.01 N HCl	233 (4.07), 275 (4.1), 302 (3.79)	16
(18b)	0.1 N NaOH	220 (3.53), 256 (4.06), 456 (2.84)	12
(19)	Cyclohexane	260 (4.38), 268 (4.24, sh), 315 (3.79, sh), 321 (3.92), 328 (3.87), 335 (3.91), 459 (2.9, sh)	12, 16
(19)	0.01 N NaOH	256 (4.42), 320 (3.92), 325 (3.89, sh), 331 (3.83), 462 (3.02)	16

[a] sh = Shoulder

B. SUBSTITUTED 1H-1-, 5H-1-, AND 7H-1-PYRINDINES

Very few examples of substituted derivatives of **1**, **7**, and **9** have been reported.[41a-42] Two examples (**19**) and (**20**) of **1** are described above. The thermal condensation of o-aminobenzaldehydes and o-aminoacetophenones with substituted cyclopentanones yields derivatives of the cyclopentaquinoline system (**30**).[41b] Dehydrogenation of **31** with

---

[41a] The fused ring derivative, 1H-cyclohepta[d,e]pyridine is also known: A. G. Anderson and L. L. Replogle, *J. Amer. Chem. Soc.* **83**, 3333 (1961).
[41b] W. Treibs and G. Kempter, *Chem. Ber.* **92**, 601 (1959).
[42] H. Junek and R. J. Schaur, *Monatsh. Chem.* **99**, 89 (1968).

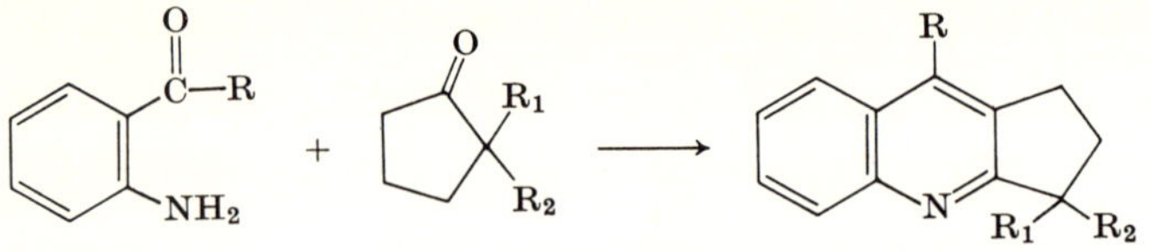

(30a) R = CH$_3$; R$_1$ = H; R$_2$ = C$_6$H$_5$
(30b) R = CH$_3$; R$_1$R$_2$ = CHC$_6$H$_5$
(30c) R = R$_1$ = H; R$_2$ = CH$_2$C$_6$H$_5$
(30d) R = H; R$_1$R$_2$ = CHC$_6$H$_5$
(30e) R = CH$_3$; R$_1$ = R$_2$ = H

palladium–charcoal in xylene gives the pseudoazulene (32).[41a, 41b]

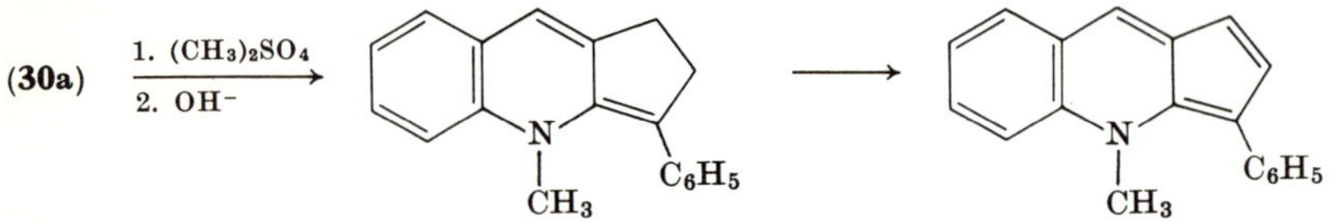

Recently, Junek and Schaur[42] have shown that malononitrile (33)[43] condenses with 2,5-hexanedione and 3-methylhexane-2,5-dione to

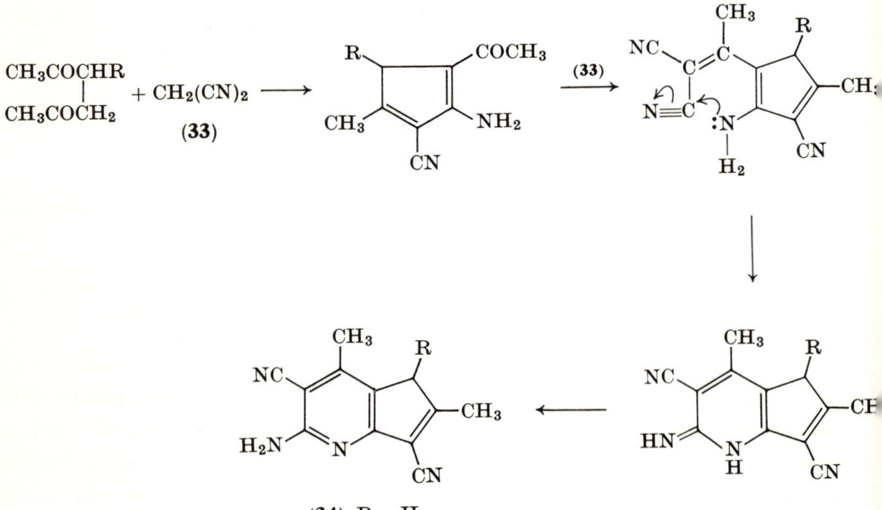

(34) R = H
(35) R = CH$_3$

SCHEME 2

[43] F. Freeman, *Chem. Rev.* **69**, 591 (1969).

give 2-amino-3,7-dicyano-4,6-dimethyl-5H-1-pyrindine (**34**) and 2-amino-3,7-dicyano-4,5,6-trimethyl-5H-1-pyrindine (**35**), respectively. A possible mechanism for the condensation is shown in Scheme 2.[44]

Hydrolytic decarboxylation[42,43] of **34** with sulfuric acid gives the amide (**36**), and treatment of **34** with acetyl chloride in dioxane results in a milder hydrolysis to give **37**. Compound **34** condenses with aromatic aldehydes and nitroso compounds at the 5-position to give useful benzylidene dye derivatives (**39**)–(**43**) in 70–80% yields.

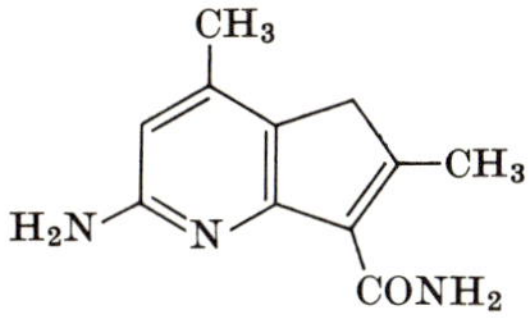

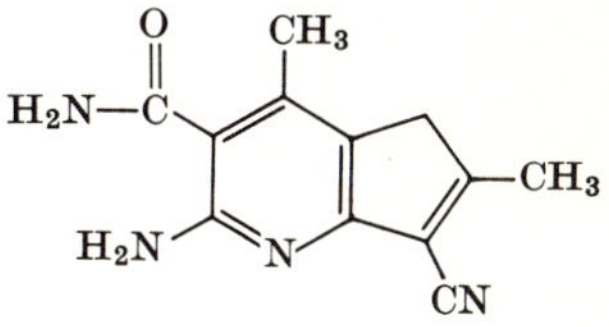

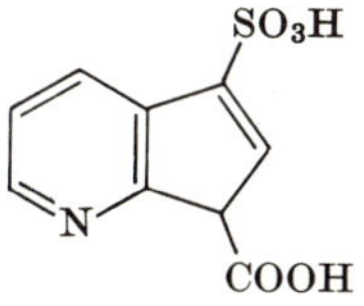

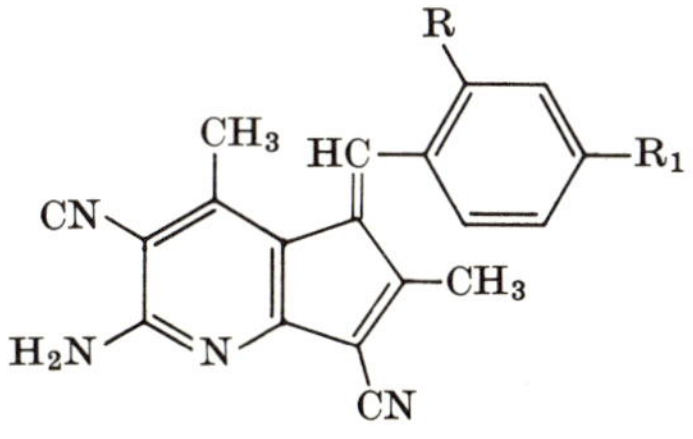

(**39**) R = R₁ = H
(**40**) R = H; R₁ = N(CH₃)₂
(**41**) R = Cl; R₁ = H

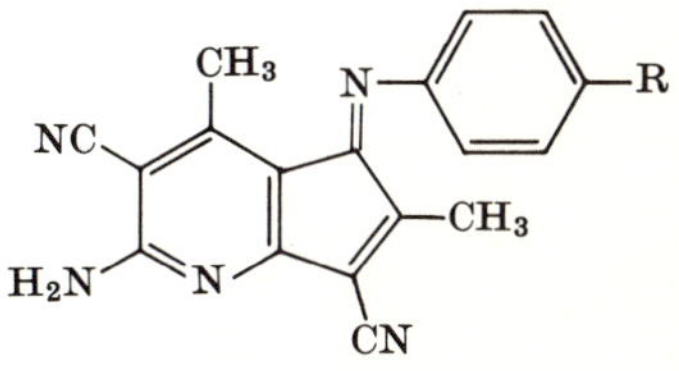

(**42**) R = H
(**43**) R = N(CH₃)₂

The preparation of the first 7H-1-pyrindine (**9**) derivative, which involved a five-step synthesis of the proposed 5-sulfo-7-carboxy-7H-1-pyrindine (**38**) from 8-hydroxyquinoline in a yield of less than 3%, was

[44] F. Freeman and T. I. Ito, *J. Org. Chem.* **34**, 3670 (1969).

reported in 1955.[41-48] The tautomeric structure is uncertain. More recently, however, several 6-benzamido-7,7-bis(substituted phenyl)-7H-1-pyrindine derivatives (**45**)[47] have been prepared by cyclization of the products obtained from the reaction of Grignard reagents with the unsaturated azlactone (**46**) which is derived from pyridine-3-carboxaldehyde (**44**) and hippuric acid. A suggested reaction route is given in Scheme 3.[47] It is also seen from Table III that yields are lower when the 7H-1-pyrindine derivatives (**45**) are formed from the corresponding N-oxide.[47]

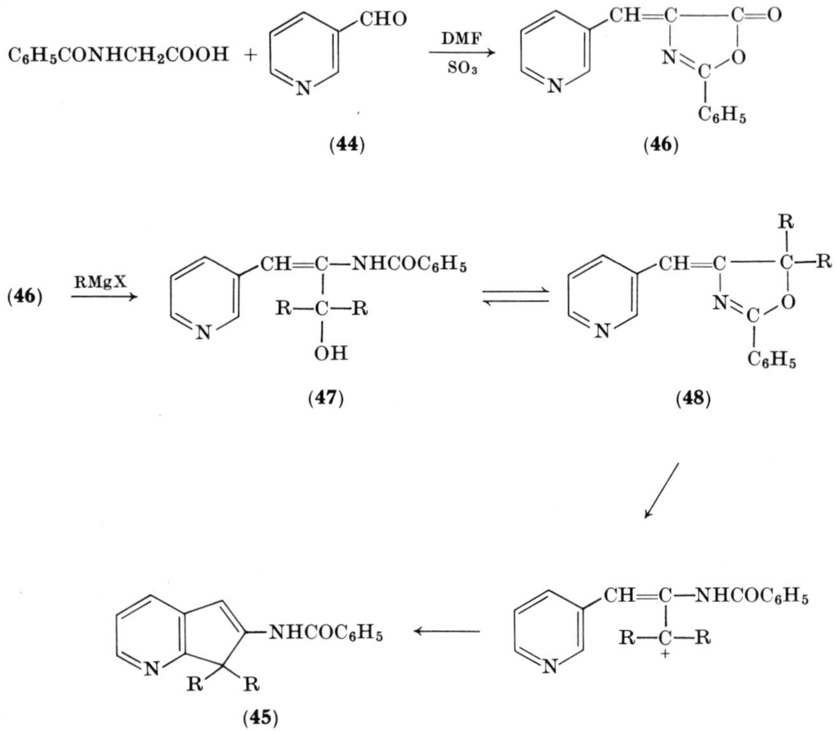

Scheme 3

[45] O. Süs, K. Moller, R. Dietrich, H. Eberhardt, M. Glos, M. Grundkotter, H. Hoffman, and H. Schäfer, *Ann. Chem.* **593**, 91 (1955).
[46] K. Matsumara, *J. Amer. Chem. Soc.* **49**, 813 (1927).
[47] G. Slater and A. W. Somerville, *Tetrahedron* **22**, 35 (1966).
[48] O. Süs, German Patent 939,327 (1956); *Chem. Abstr.* **52**, 14707 (1958).

TABLE III

PREPARATION AND PROPERTIES OF 6-BENZAMIDO-7,7-BIS-
(SUBSTITUTED PHENYL)-$7H$-1-PYRINDINES[47]

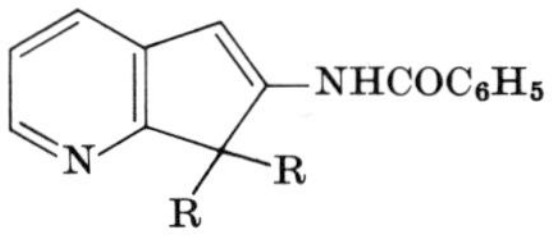

R	Starting material	Yield (%)	M.p. (°C)	λ_{max}, mμ (log ϵ)
C_6H_5	Carbinol (**47**)	88	217	236 (4.39)
C_6H_5	N-oxide (**49**)	60	218	
o-C_6H_4Cl	(**47**)	68	176	235 (4.43)
o-C_6H_4Cl	(**49**)	41	176	
p-C_6H_4Cl	(**47**)	78	179	235 (4.47)
p-C_6H_4Cl	(**49**)	42	179	
o-$C_6H_4OCH_3$	(**47**)	72	214	223 (4.52)
o-$C_6H_4OCH_3$	(**49**)	60	214	
p-$C_6H_4OCH_3$	Oxazoline (**48**)	51	182	225 (4.49)
CH_3	(**48**)	60	216	225 (4.38)

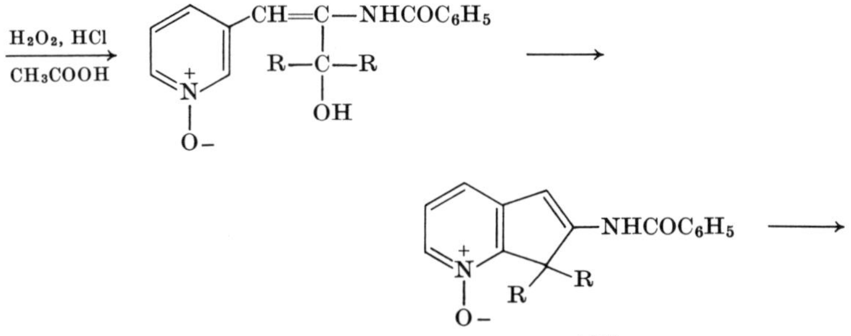

One obtains **50** from the condensation of 4-hydroxycoumarin and 2-pyridinecarboxaldehydes.[49]

Dehydration of the symmetrical ditertiary glycol (**51**) with polyphosphoric acid above 200° gives, alongside the butadiene (**52**), 6-

[49] F. Litvan and W. G. Stell, *Helv. Chim. Acta* **42**, 878 (1959).

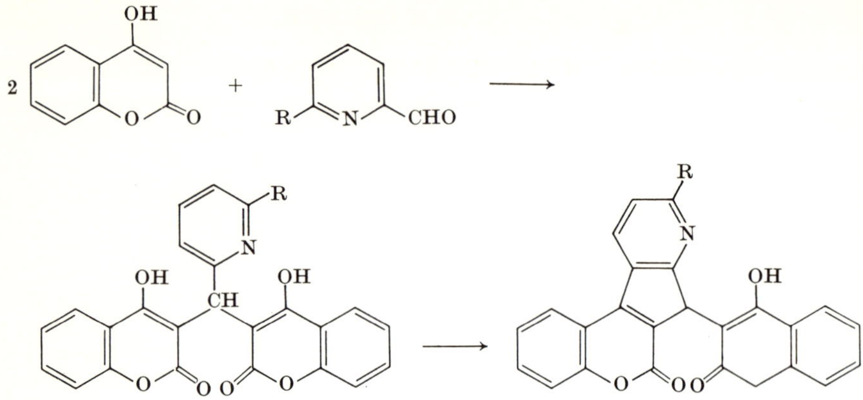

(3-pyridyl)-5-methyl-7H-1-pyrindine (**53**). The formation of **53**, which possesses endocrine activity, is favored at higher temperatures.[50, 51]

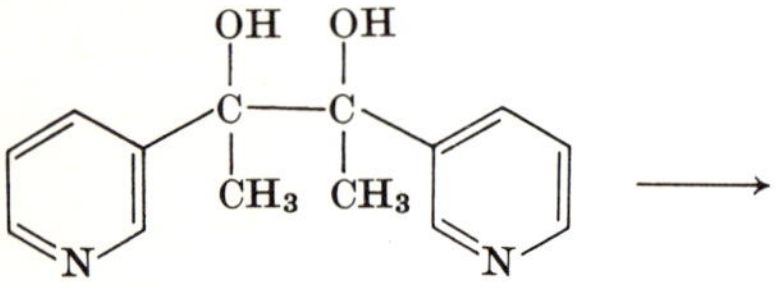

In their work on nonsteroidal antiinflammatory agents, Greenwald and Shen[52] prepared the previously unknown azabenzofulvene (**54**) by condensing **55** or **56** with p-chlorobenzaldehyde. Dehydration of **61**, which can be prepared from **57** via **58**, **59**, and **60**,[53] gives **55b** and **56b** in

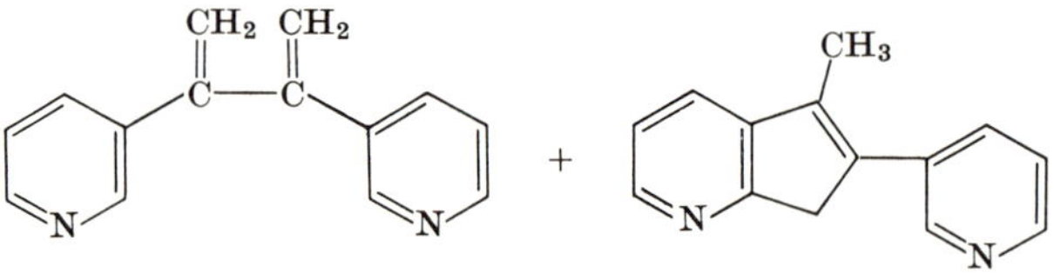

[50] W. L. Bencze and M. J. Allen, *J. Amer. Chem. Soc.* **81**, 4015 (1959).
[51] W. L. Bencze, C. A. Burckhardt, and W. L. Yost, *J. Org. Chem.* **27**, 2865 (1962).
[52] R. B. Greenwald and T. Y. Shen, *J. Heterocycl. Chem.* **7**, 683 (1970).
[53] F. Ramirez and A. P. Paul, *J. Amer. Chem. Soc.* **77**, 1035 (1955).

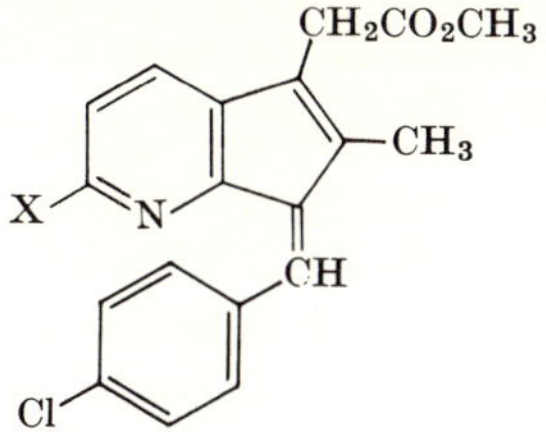

(54a) X = H	(55a) X = H	(56a) X = H
(54b) X = Cl	(55b) X = Cl	(56b) X = Cl

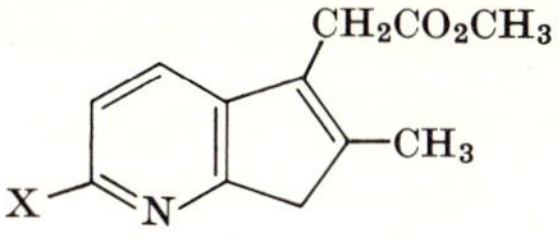

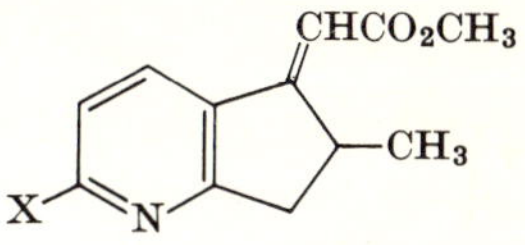

a ratio of 1:4. **55a** and **56a** are prepared by dehydration of **62**.

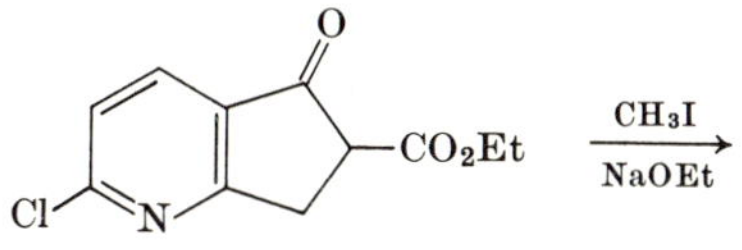

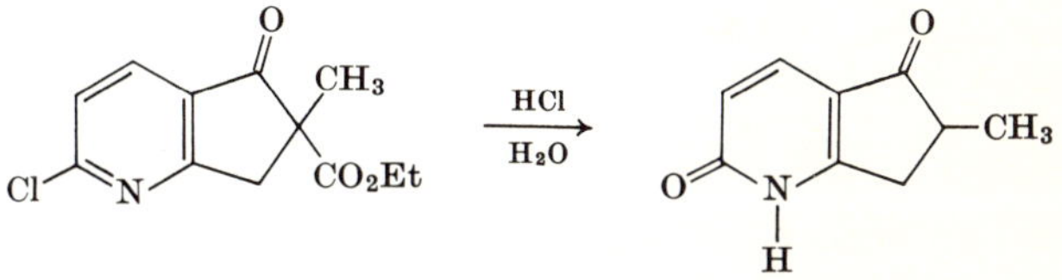

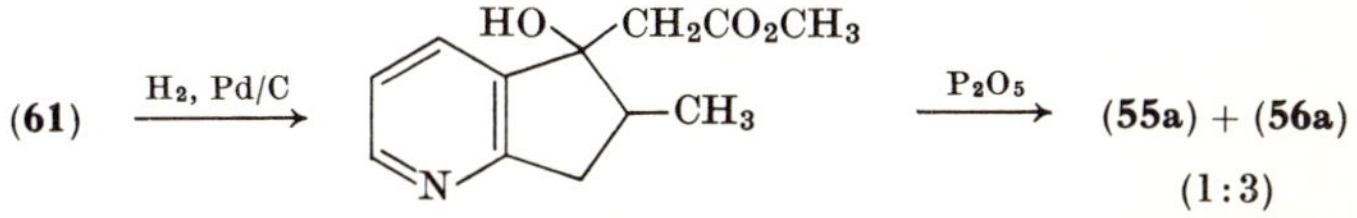

The interesting PMR spectra of **54** showed an unusual downfield shift of nearly 1 ppm of the protons meta to the chlorine. This phenomenon could arise from long-range interaction with the unshared pair of electrons on nitrogen.[52, 54, 55]

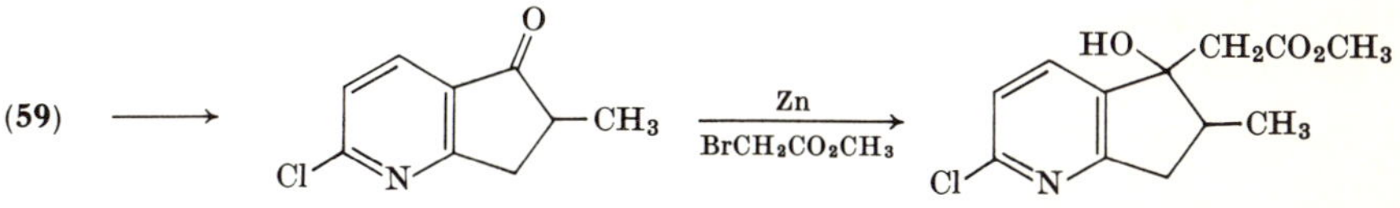

---

[54] R. B. Greenwald and E. C. Taylor, *J. Amer. Chem. Soc.* **90**, 5272 (1968).
[55] C. V. Fritchie and J. L. Wells, *Chem. Commun.*, 917 (1968).

C. 6,7-Dihydro-5H-1-pyrindines

Derivatives of 6,7-dihydro-5H-1-pyrindine (10) comprise the largest group of 1-pyrindines. Studies of the basic nitrogen compounds in the 120°–375° petroleum distillate fractions via chemical reactivity and infrared, mass, ultraviolet, and proton magnetic resonance spectrometry have led to the separation and identification of many derivatives of **10**.[2, 3, 5–8, 56–66]

Although these nitrogen compounds are only present in trace amounts, they cause serious problems (catalyst poisoning, polymerization, etc.) in the processing of petroleum. Some of the derivatives of **10** which have been identified and characterized are 2-methyl- and 5-methyl-6,7-dihydro-5H-1-pyrindine (**63**)[5, 6, 56] and (**64**)[5] and 2,4,7a,8,8-pentamethylcyclopenta[f]pyrindan (**65**).[8] Compound **63**, which is found

(63)   (64)   (65)

in California petroleum,[6] is identical with the unambiguously synthesized compound.[6, 67] Basu[67a, 67b] first reported the preparation of **63** by hydrolytic decarboxylation of the product obtained from the condensation of hydroxymethylenecyclopentanone and ethyl β-aminocrotonate. This condensation reaction provides a general procedure for the preparation of several derivatives (**66**) of **63**.

[56] H. Suzumura, *Bull. Chem. Soc. Jap.* **34**, 1097 (1962).
[57] H. Suzumura, *Bull Chem. Soc. Jap.* **34**, 1846 (1962).
[58] S. Tanaka and H. Arakawa, *Bunseki Kagaku* **6**, 281 (1957); *Chem. Abstr.* **52**, 11662 (1958).
[59] W. Funasaka and T. Kojima, *Bunseki Kagaku* **9**, 741 (1960); *Chem. Abstr.* **56**, 9393 (1962).
[60] S. Tanaka and H. Arakawa, *Bunseki Kagaku* **5**, 513 (1956); *Chem. Abstr.* **51**, 17609 (1957).
[61] C. Karr and T-C. L. Chang, *J. Inst. Fuel* **31**, 522 (1958).
[62] G. E. Mapstone, *J. Proc. Roy. Soc. N. S. Wales* **82**, 79 (1948); *Chem. Abstr.* **44**, 7517 (1950).
[63] F. Runge, J. Freytag, and J. Kolbe, *Chem. Ber.* **87**, 873 (1954).
[64] A. C. Nixon and R. E. Thorpe, *J. Chem. Eng. Data* **7**, 429 (1962).
[65] H. V. Drushel and A. L. Sommers, *Anal. Chem.* **38**, 19 (1966).
[66] P. Arnall, *J. Chem. Soc.*, 1702 (1958).
[67a] U. Basu, *Ann. Chem.* **530**, 131 (1937).
[67b] U. Basu, *Sci. Cult.* **2**, 466 (1937); *Chem. Abstr.* **31**, 3919 (1939).

(63)

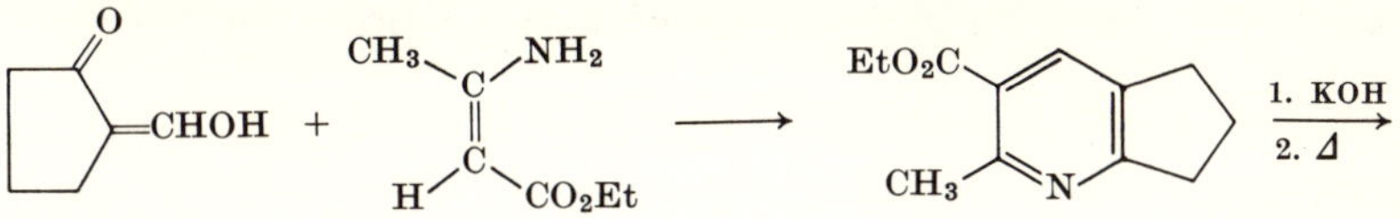

In order to demonstrate the absence of 3-methyl- and 4-methyl-6,7-dihydro-5*H*-1-pyrindine (**67**) and (**68**) in California petroleum, Lochte and Pittman[6] prepared both compounds from **69** as indicated. In both

(**66a**) X = COR
(**66b**) X = CN

(**67**)

(**68**)

(**69**) (**70**)

(**71**)
(**67**)

(**69**)
(**68**)

(**72**) (**73**)

preparations the first step involves addition of the α,β-unsaturated nitriles to the pyrrolidine enamine of cyclopentanone (**69**) to give the respective ketonitrile (**70**) or (**72**) in about 33% yield. Reductive cyclization of (**70**) and (**72**) gives the octahydro-5*H*-1-pyrindines (**71**) and (**73**), which can be dehydrogenated in the vapor phase. Colonge and co-workers[68a] have also prepared **10** and **68** in 20–30% yield by treatment of the corresponding

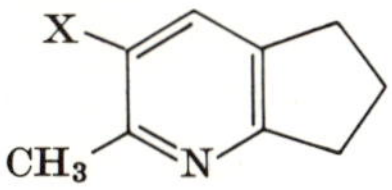

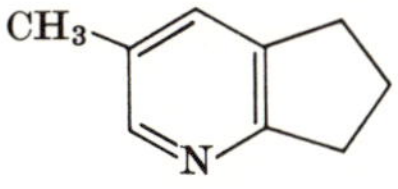

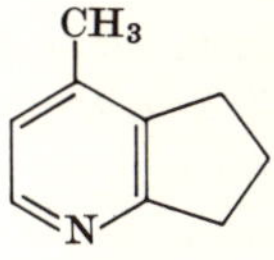

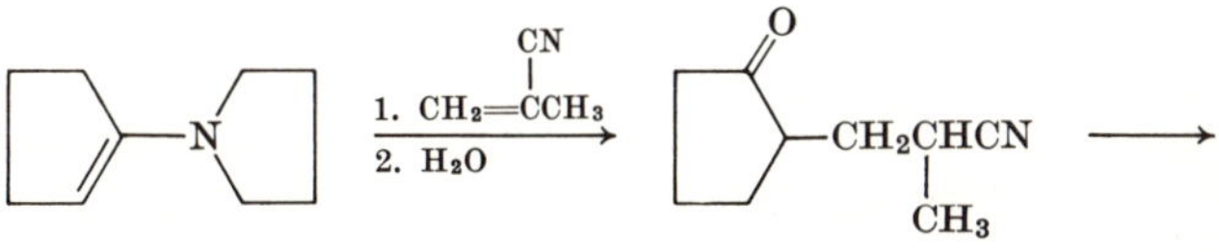

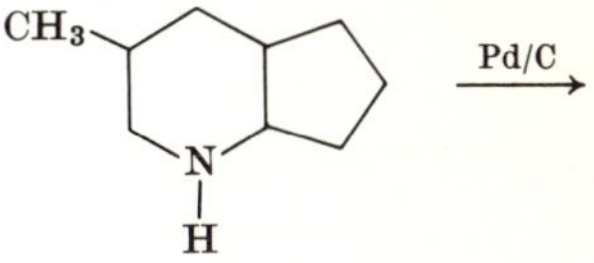

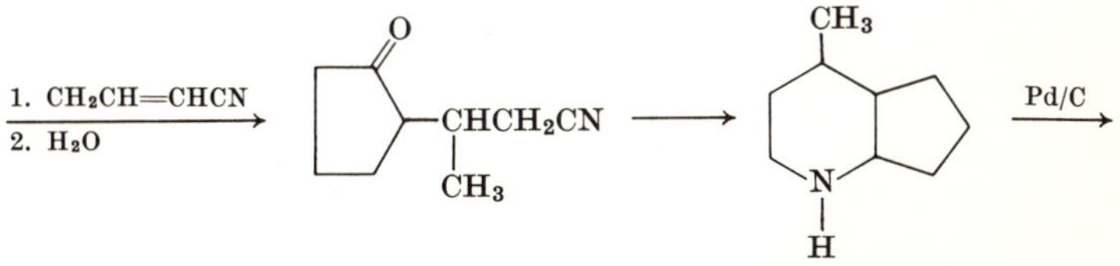

68a J. Colonge, J. Dreux, and M. Thiers, *Bull. Soc. Chim. Fr.*, 1459 (1959).

δ-oxoaldehydes (**74**) with hydroxylamine hydrochloride and then with sodium hydroxide solution.

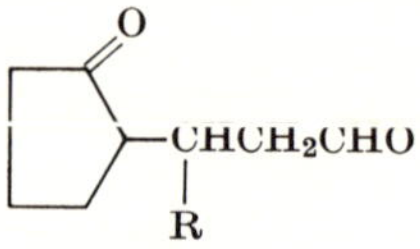

R = H, CH₃
(**74**)

Several investigators have shown that the ultraviolet spectra of **63**, **67**, and **68** are similar to 2,3,4-trimethylpyridine.[6, 66]

5-Methyl-6,7-dihydro-5H-1-pyrindine (**64**) has also been identified as a component of California petroleum.[5] The isomeric 5- and 6-methyl-6,7-dihydro-5H-1-pyrindines (**64**, **75**) were prepared as described above from the respective enamines (**76**) and (**77**). The position of the methyl

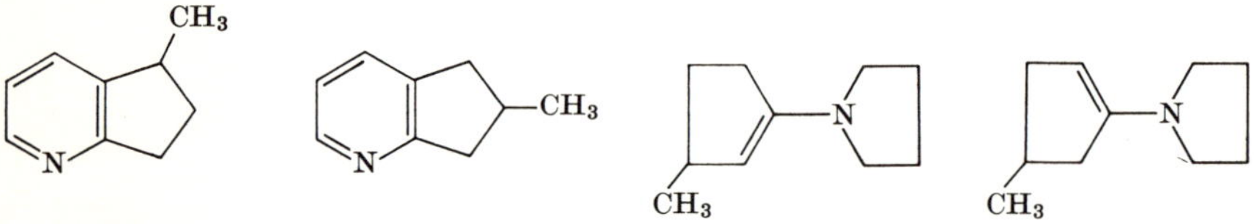

(**64**)   (**75**)   (**76**)   (**77**)

group in **64** was demonstrated by its alternate synthesis from **78**.

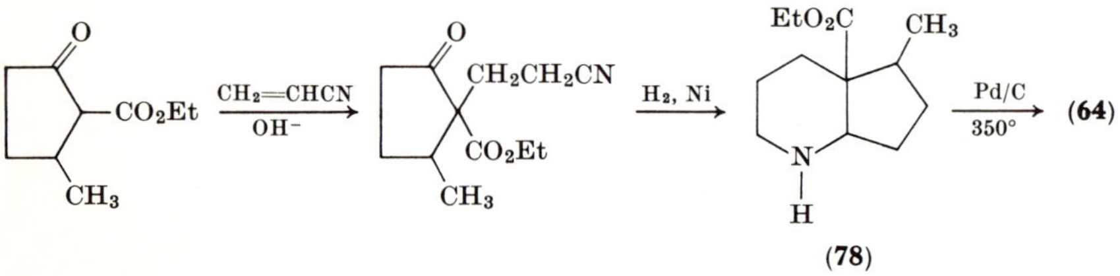

(**78**)

Table IV shows a comparison of the physical properties of some derivatives of **10**.

A variety of condensation reactions involving cyclopentanone and its derivatives has been used to prepare a large number of substituted 6,7-dihydro-5H-1-pyrindines. Compound **10** was first prepared[68b] starting with the condensation of 2-hydroxymethylenecyclopentanone (**79**)[68a–69] and cyanoacetamide (Scheme 4). The following mechanism

[68b] E. Godar and R. P. Mariella, *J. Amer. Chem. Soc.* **79**, 1402 (1957).
[69] C. Ainsworth, *Org. Syn. Coll. Vol.* **4**, 536 (1963).

TABLE IV

Physical Properties of Some Derivatives of
6,7-Dihydro-5H-1-pyrindine

Compound	Boiling point (°C)	Refractive index	Derivative	Melting point (°C)	Ref.
(10)	199.5	1.5380^{25}	Picrolonate	238.8	57, 66
(63)	(208–209/755 mm)	$1.5316^{24.5}$	Picrate	151–152	6, 57, 67a–68b
(64)	(207–208/746 mm)	—	Picrate	147–149	5
(67)	(225/756 mm)	—	Picrate	204–206	6, 56
(68)	(226/752 mm)	$1.5360^{24.5}$	Picrate	169–170	6

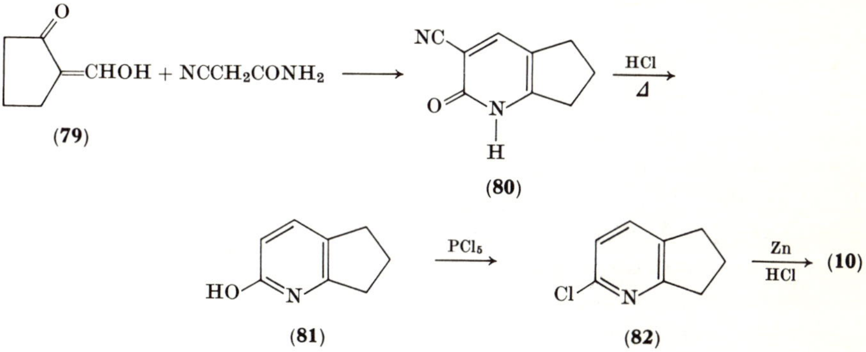

Scheme 4

appears to be plausible for the formation of **80** (Scheme 5).[44, 70]

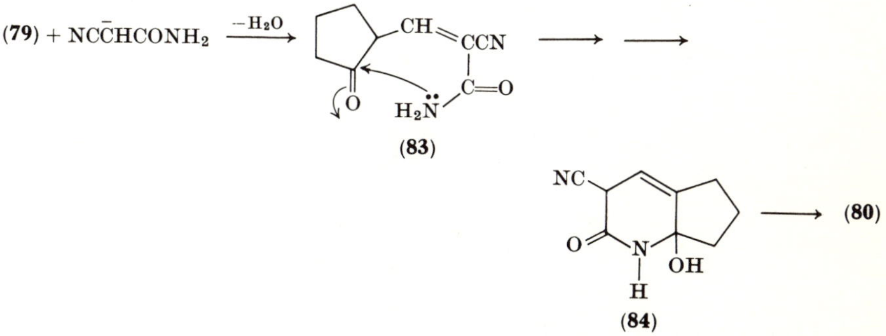

Scheme 5

[70] F. Freeman, D. K. Farquhar, and R. L. Walker, *J. Org. Chem.* **33**, 3648 (1968).

The first step presumably involves attack of the cyanoacetamide anion at the acyclic carbonyl carbon atom to give the condensation product (83). Nucleophilic attack of the unshared pair of electrons on nitrogen at the ring carbonyl carbon atom leads to 84, which eliminates water to give 80. Initial condensation at the ketone carbonyl would lead to the 5H-2-pyrindine derivative (85).[44, 70, 71]

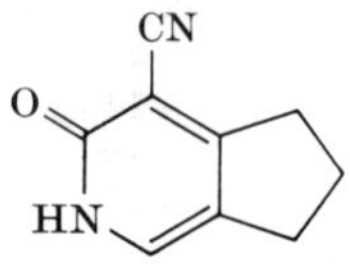

(85)

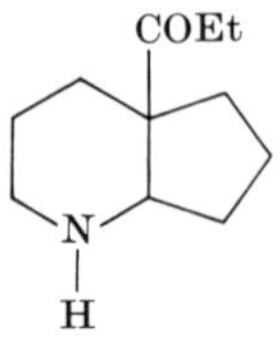

(86)

Lochte and Pittman[6] have prepared 10 by monocyanoethylating 69 with acrylonitrile, followed by reductive ring closure and catalytic dehydrogenation. It has also been shown[5] that passage of 86 over 30% Pd/C at 350° gives 10 in approximately 31% yield. Condensation of allyl alcohol, cyclopentanone, and ammonia in the presence of $Cd_3(PO_4)_2$-activated earth catalyst also gives 10 in 31% yield.[72, 73]

2-Aminocyclopentenecarbonitrile, which is prepared by Thorpe cyclization of adiponitrile, reacts with acid anhydrides to give 2-(acylamino)cyclopentenecarbonitriles (87).[74] Excellent yields of 4-amino-2-hydroxy-6,7-dihydro-5H-1-pyrindines (88–91) are obtained when derivatives of 87 are treated with sodamide in liquid ammonia

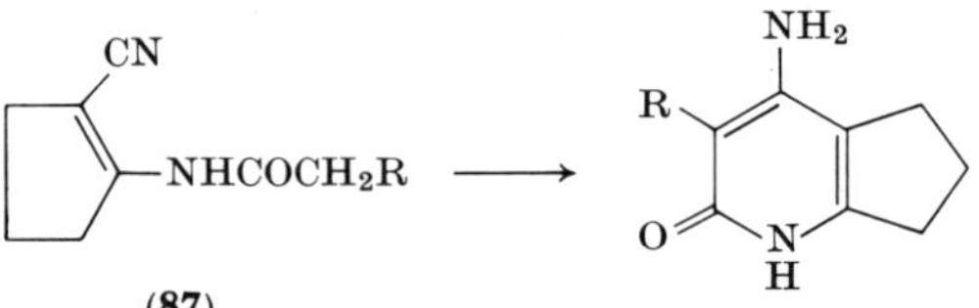

(87)

(88) R = H (97%)
(89) R = $COCH_3$ (85%)
(90) R = Cl (92%)
(91) R = CH_3 (97%)

[71] A. Lapworth and R. H. F. Manske, *J. Chem. Soc.*, 2533 (1928); 1976 (1930).
[72] T. Ishiguro, Y. Morita, and K. Ikushima, *Yakugaku Zasshi* 78, 216 (1958); *Chem. Abstr.* 52, 11847 (1958).
[73] T. Ishiguro, Japanese Patent 8319, 1959, *Chem. Abstr.* 54, 17425 (1960).
[74] H. E. Schroeder and G. W. Rigby, *J. Amer. Chem. Soc.* 71, 2205 (1949).

at −33°. Compound **88** almost certainly exists in the pyridone structure shown. Diazotization of **88** with sodium nitrite in cold concentrated sulfuric acid gives 2,4-dihydroxy-6,7-dihydro-5H-1-pyrindine (**92**).

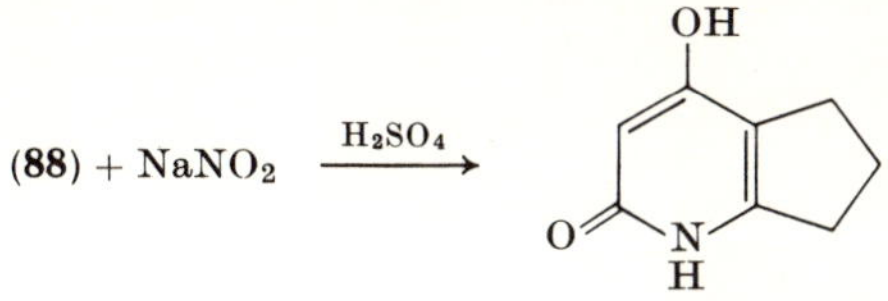

$$(88) + \text{NaNO}_2 \xrightarrow{\text{H}_2\text{SO}_4} (92)$$

Compound **92** has also been prepared in good yield by Prelog and Szpilfogel[10] as shown below. Robison[9] chlorodehydroxylated **92** to the

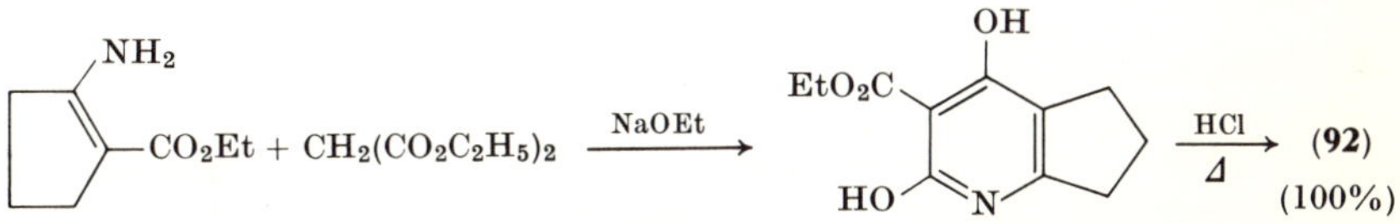

2,4-dichloro derivative (**93**) with phenylphosphonic dichloride. Reduction of **93** with 5% Pd/C gives **10** in 86% yield.[9] Phenylphosphonic dichloride is a better reagent than phosphorus oxychloride for reactions of this type.[9, 70, 75]

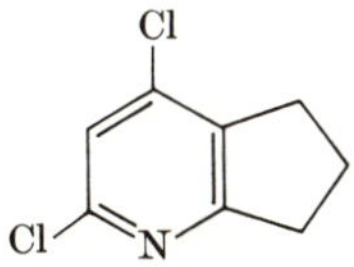

The versatile chemical behavior of **88** is shown in the reactions of Scheme 6. No attempt was made to establish the proposed structure (**101**).

Malonamide amidine hydrochloride (**102**) and the sodium derivative of **79** give 2-amino-3-carboxamido-6,7-dihydro-5H-1-pyrindine (**103**).[76] Compound **103a** was converted into **103b–e** via conventional reactions.

---

[75] M. M. Robison, *J. Amer. Chem. Soc.* **80**, 5481 (1958).
[76] A. Dornow and E. Neuse, *Arch. Pharm.* (*Weinheim*) **287**, 361 (1954).

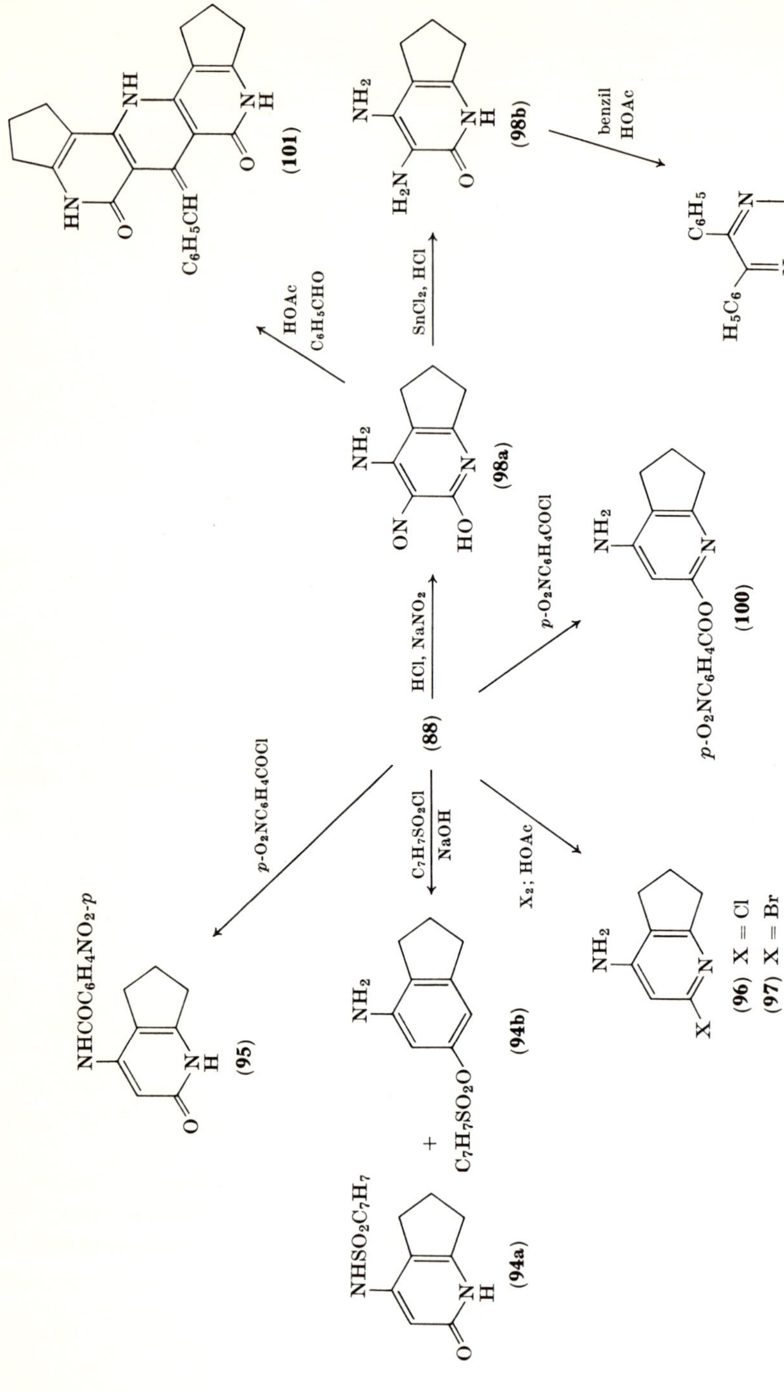

SCHEME 6

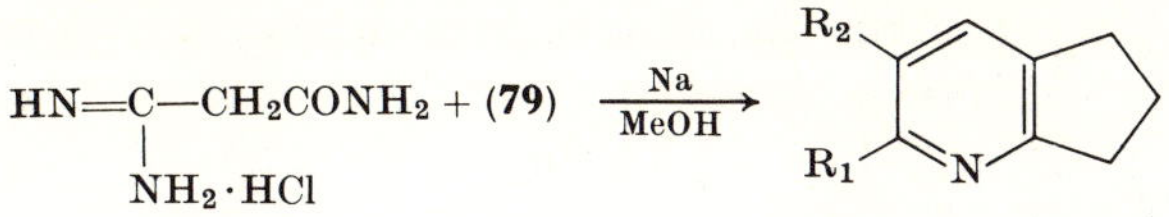

$HN=C-CH_2CONH_2$ + (**79**) $\xrightarrow{\text{Na}}{\text{MeOH}}$
|
$NH_2 \cdot HCl$
(**102**)

(**103a**) $R_1 = NH_2$, $R_2 = CONH_2$
(**103b**) $R_1 = NH_2$, $R_2 = COOH$
(**103c**) $R_1 = OH$, $R_2 = CONH_2$
(**103d**) $R_1 = OH$, $R_2 = H$
(**103e**) $R_1 = OH$, $R_2 = COOH$

Chalcones add to cyclopentanone in the presence of a secondary amine to yield 1,5-diketones which react with hydroxylamine to give the 2,4-diaryl-6,7-dihydro-5H-1-pyrindines (**104**) and (**105**).[77a,b] Similarly,

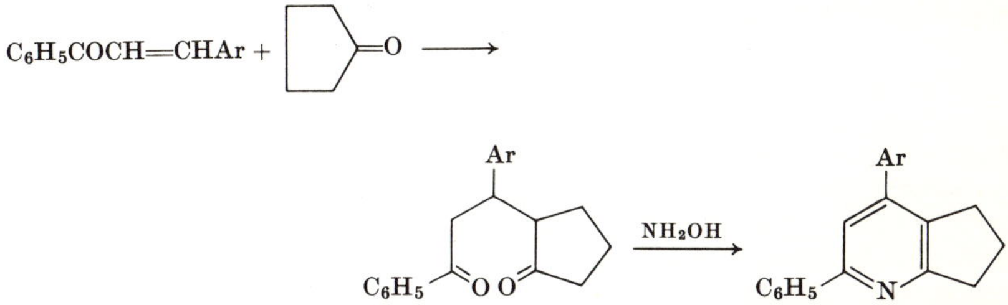

$C_6H_5COCH=CHAr$ + [cyclopentanone] $\longrightarrow$

$\xrightarrow{NH_2OH}$

(**104**) Ar = p-$H_3CC_6H_4$
(**105**) Ar = 3,4-$CH_2OC_6H_3$

compound **106** is prepared in 28% yield as indicated below. The styryl derivative (**106**) reacts with m-nitrobenzaldehyde in acetic anhydride to give **107**. Under similar conditions **63** reacts with benzaldehyde and

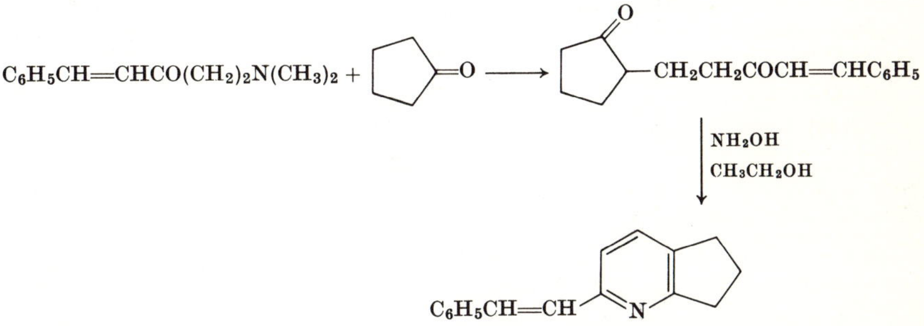

$C_6H_5CH=CHCO(CH_2)_2N(CH_3)_2$ + [cyclopentanone] $\longrightarrow$ [cyclopentanone]-$CH_2CH_2COCH=CHC_6H_5$

$\downarrow$ NH_2OH, CH_3CH_2OH

$C_6H_5CH=CH$-[pyrindine]
(**106**)

[77a] C. Striegler, *J. Prakt. Chem.* **86**, 241 (1912).
[77b] H. Stobbe, *J. Prakt. Chem.* **86**, 211 (1912), and preceding refs.

m-nitrobenzaldehyde to give the benzylidene derivatives (**108**) and (**109**), respectively.[78] Condensation at the 7-position of **63** suggests that the methylene hydrogens in the alicyclic ring are more reactive than the

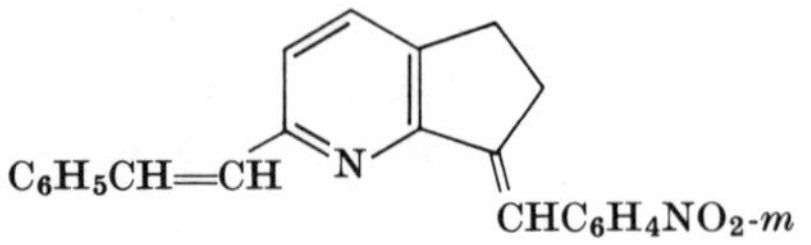

(**107**) $C_6H_5CH=CH$ ... $CHC_6H_4NO_2$-*m*

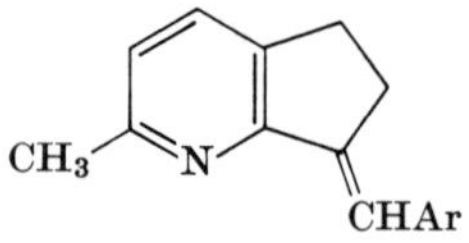

(**108**) Ar = C_6H_5
(**109**) Ar = *m*-$O_2NC_6H_4$

hydrogens on the 2-methyl carbon. Further condensation of **109** with *m*-nitrobenzaldehyde yields **107**. In these reactions **63** is similar to **110** and unlike **111**.[78]

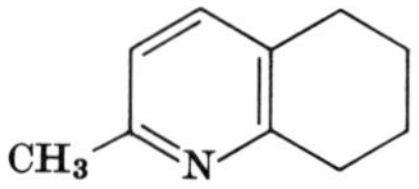

(**110**)

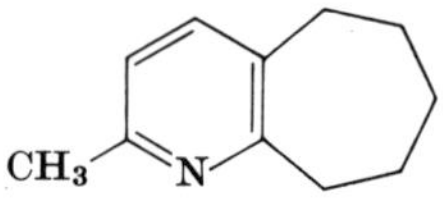

(**111**)

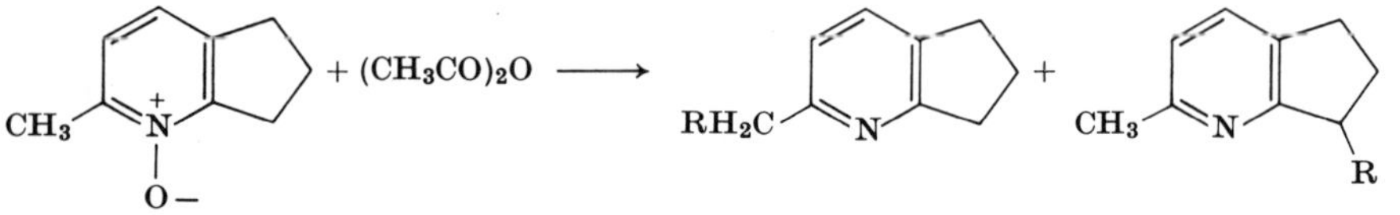

(**112**) + $(CH_3CO)_2O$ ⟶ (**113**) RH_2C + (**114**) CH_3 , R

(**113**) R = OH or CH_3CO_2
(**114**) R = OH or CH_3CO_2

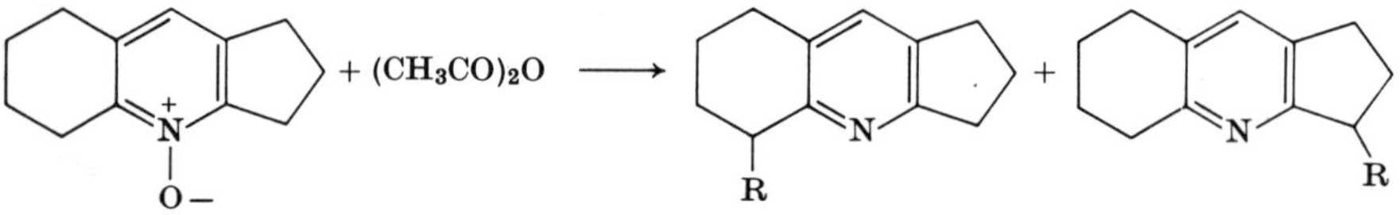

(**115**) + $(CH_3CO)_2O$ ⟶ (**116**) + (**117**)

(**116**) R = OH or CH_3CO_2
(**117**) R = OH or CH_3CO_2

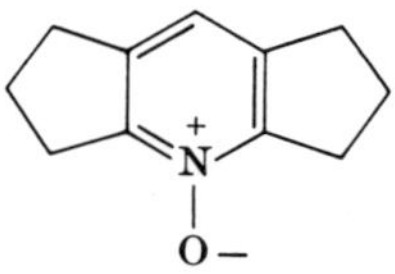

(**118**)

[78] J. Epsztajn, W. E. Hahn, and B. K. Tosik, *Rocz. Chem.* **43**, 807 (1969).

The rearrangement products of the N-oxides (**112**) and (**115**)[79] provide another comparison of the relative reactivities of the 2-methyl group and the 7-methylene hydrogens.[28a, 28b, 29b] In contrast to these results, the symmetrical N-oxide (**118**) yielded only acetoxyl derivatives.

Many substituted 6,7-dihydro-5H-1-pyrindines (**119**) and (**120**)[80, 81] have been found to be useful intermediates in the dye, photographic, and pharmaceutical industries.

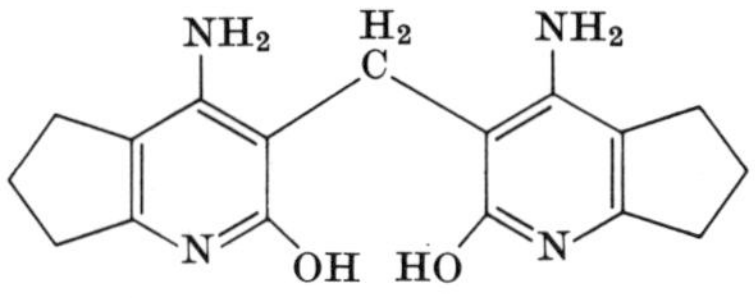

(**119**)

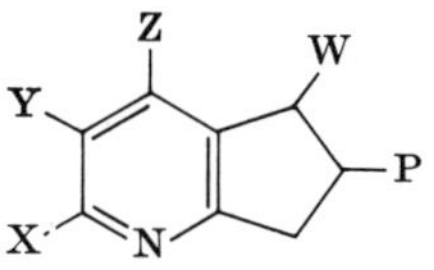

(**120**)

X = CH_3, OH or OAc
Y = H or Ac
Z = H, NH_2, or $CONH_2$
W = H or CH_3
P = H or Ph

A general synthetic scheme, starting with 3-cyano-6,7-dihydro-5H-1-pyrind-2-one (**121**) and using standard reactions, for the preparation of a wide variety of 6,7-dihydro-5H-1-pyrindine derivatives, has been reported by Godar and Mariella.[68a, 68b, 82, 83] It was also found[82]

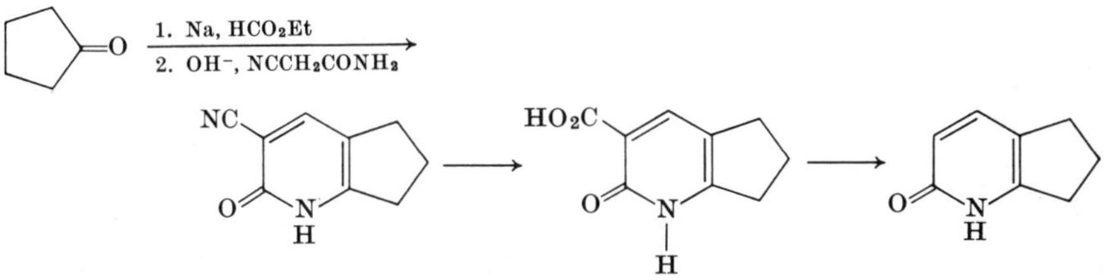

1. Na, HCO_2Et
2. OH^-, $NCCH_2CONH_2$

(**121**) (**122**) (**123**)

that it was possible selectively to reduce cyanochloropyrindines to primary amines in acidic ethanolic medium. In slightly basic or neutral solution, the halogen is rapidly removed.

[79] W. E. Hahn, J. Epsztajn, B. Olejniczak, and S. Stasiak, *Rocz. Chem.* **40**, 149 (1966).
[80] B. Reichert and A. Lechner, *Arzneim-Forsch.* **15**, 36 (1965); *Chem. Abstr.* **62**, 10403 (1965).
[81] G. W. Rigby, U.S. Patent 233349 (1944); *Chem. Abstr.* **38**, 2508 (1944).
[82] E. M. Godar and R. P. Mariella, *J. Org. Chem.* **25**, 557 (1960).
[83] E. M. Godar and R. P. Mariella, *Appl. Spectrosc.* **15**, 29 (1961).

Hydroxymethylation occurs when derivatives of **10** are heated in a sealed tube with paraformaldehyde.[84] The structures of the products

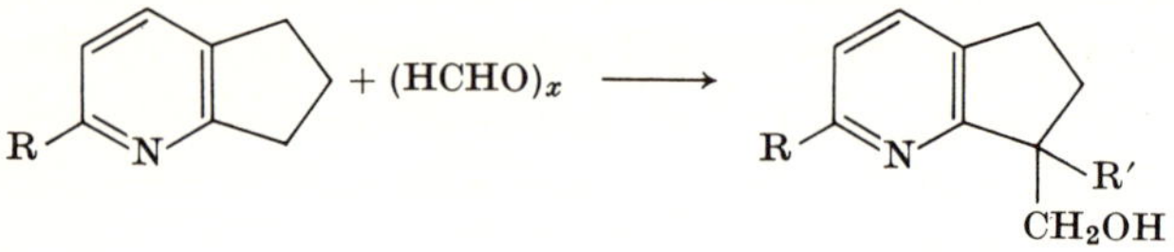

R = H or Ph
R' = H or CH$_2$OH

(124)

were confirmed by infrared spectroscopy. The N-oxide of **10** and its 2-phenyl derivative yield acetoxyl derivatives when heated with acetic anhydride. Hydrolysis of the acetates affords the corresponding alcohol.[85]

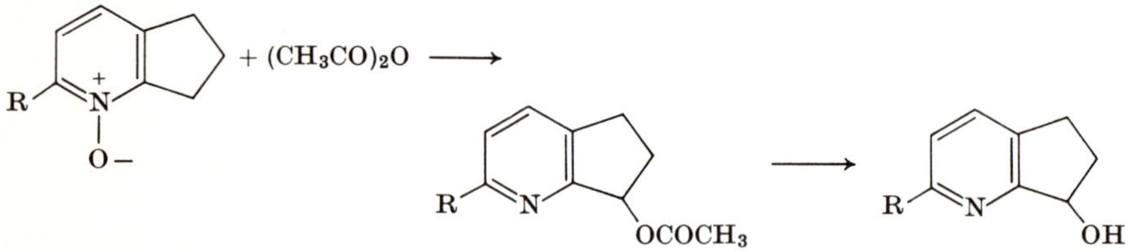

(125)
R = H (71%)
R = C$_6$H$_5$ (84%)

(126)
R = H or C$_6$H$_5$

D. Hydroxy- and Oxo-6,7-dihydro-5H-1-pyrindine Derivatives

The reaction of ethyl acetoacetate with cyclopentanone in the presence of excess ammonium acetate gives 2-hydroxy-4-methyl-6,7-dihydro-5H-1-pyrindine (**127**) in 23.5% yield.[20, 86] The spectral evidence available is not conclusive, but by anology with simpler systems, it seems certain that the pyridone form (**127b**) is predominant.

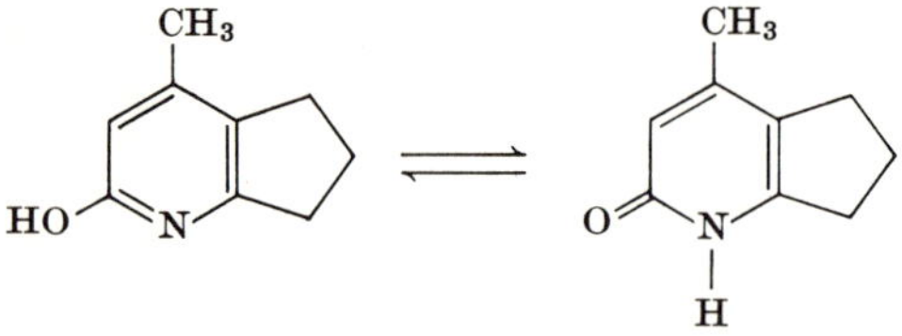

(127a)   **(127b)**

[84] W. E. Hahn and J. Epsztajn, *Rocz. Chem.* **37**, 395 (1963).
[85] W. E. Hahn and J. Epsztajn, *Rocz. Chem.* **37**, 403 (1963).
[86] A. Sakurai and H. Midorikawa, *Bull. Chem. Soc. Jap.* **41**, 165 (1968).

An interesting preparation of **128** involves the cycloaddition of carbon suboxide to the anil of cyclopentanone (**129**).[87] Compound **129** also reacts with monosubstituted malonyl chlorides to give N-aryl-2-oxo-3-alkyl-4-hydroxy-6,7-dihydro-5H-1-pyrindines (**131**) in fair yields.[88] Similarly, the oxime ether of cyclopentanone (**132**) reacts to give the cyclic hydroxamic acid derivative (**133**),[89] and benzylmalonyl chloride reacts with the N,N-dimethylhydrazone of cyclopentanone (**134a**) to give **134b** in 90% yield.[90]

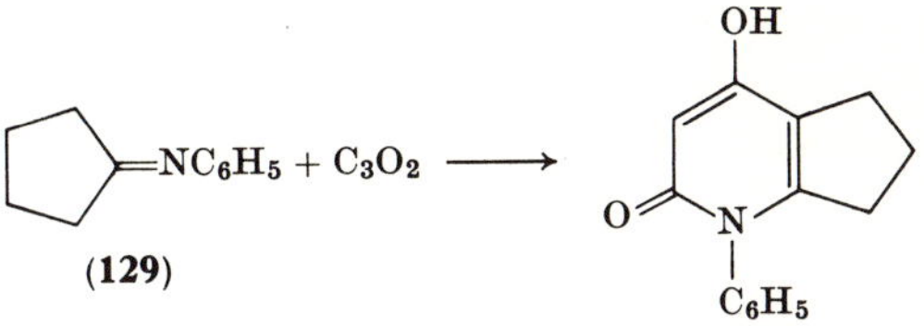

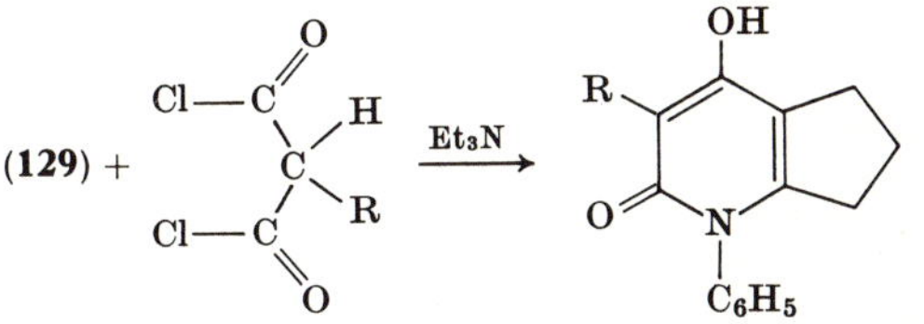

(131a) R = i-C$_3$H$_7$
(131b) R = C$_2$H$_5$
(131c) R = CH$_3$

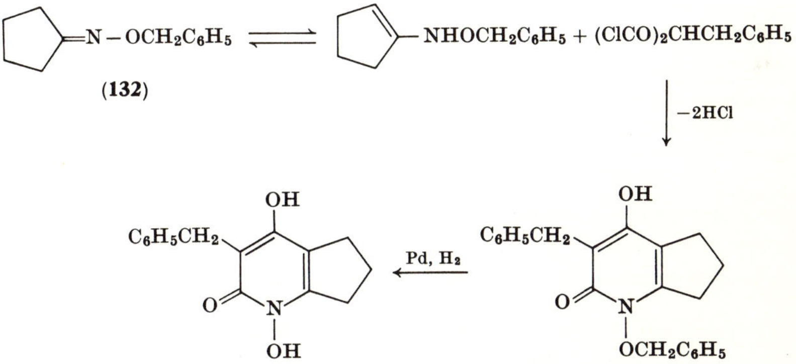

[87] E. Ziegler, F. Hradetzky, and M. Eder, *Monatsh. Chem.* **97**, 1394 (1966).
[88] E. Ziegler, G. Kleineberg, and K. Belegrates, *Monatsh. Chem.* **98**, 76 (1967).
[89] E. Ziegler and K. Belegrates, *Monatsh. Chem.* **99**, 1454 (1968).
[90] E. Ziegler and K. Belegrates, *Monatsh. Chem.* **99**, 1460 (1968).

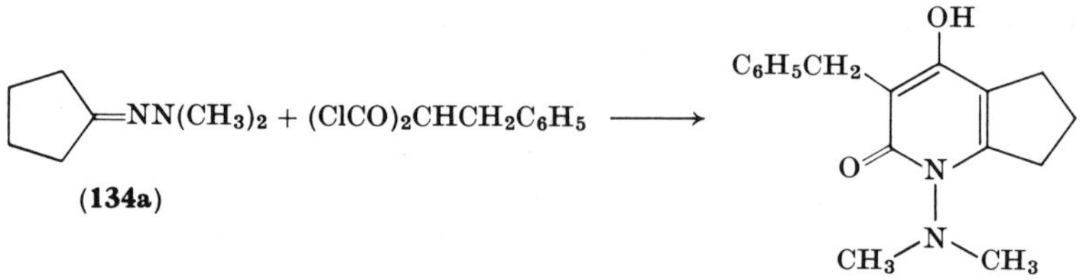

(Structures 134a, 134b shown)

Trichloroacrylonitrile condenses with β-oxocarboxylic ester (**135**) to give **136**, which undergoes ring closure to **137** on treatment with 95% sulfuric acid.[91] Although acrylonitrile generally reacts with ketones in the presence of an amino acid catalyst to afford the α-monocyanoethyl

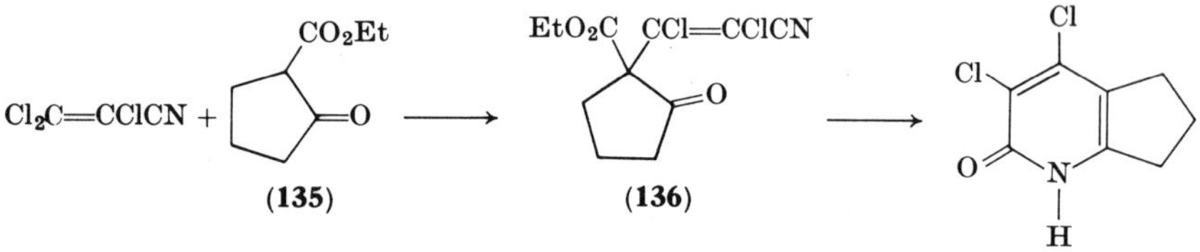

(Structures 135, 136, 137 shown)

ketones, it has recently been found[92-96] that γ,δ-unsaturated lactams are also formed in good yields when the reaction is carried out above 200° in the presence of a small amount of water. The reaction involves the

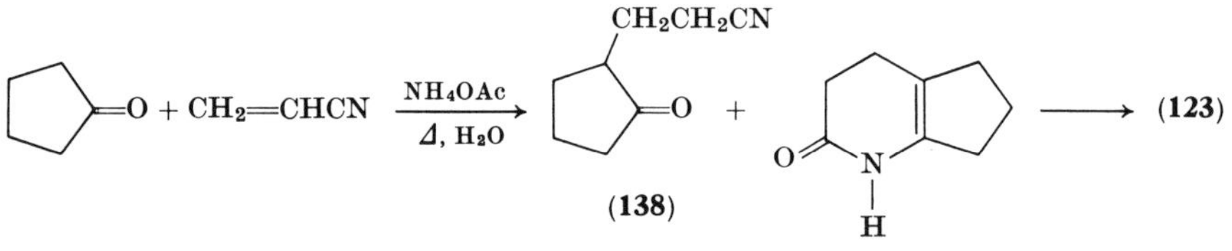

(Structures 138, 139, 123 shown)

formation of the α-monocyanomethyl ketone (**138**) which is hydrolyzed to the corresponding amide. Cyclodehydration of the amide gives 1,2,3,4,5,6-hexahydro-7H-1-pyrind-2-one (**139**).[92] Similar results are

[91] K. Grohe and A. Roedeg, *Chem. Ber.* **100**, 2953 (1967).
[92] J. J. Vill, T. R. Steadman, and J. J. Godfrey, *J. Org. Chem.* **29**, 2780 (1964).
[93] A. Vigier and J. Dreux, *Bull. Soc. Chem. Fr.* **10**, 2293 (1963).
[94] O. Y. Magidson, *Zh. Obsch. Khim.* **33**, 2173 (1963); *Chem. Abstr.* **59**, 13942 (1963).
[95] I. N. Nazarov, G. A. Shvekhgeimer, and V. A. Rudenko, *J. Gen. Chem. USSR* **24**, 325 (1954).
[96] J. A. Adamcik and R. J. Flores, *J. Org. Chem.* **29**, 572 (1964).

obtained in two steps under milder conditions.[97] For example, treatment of **138** with cold 96% sulfuric acid gave the corresponding amide which after heating at 170°–180° for 1 hour afforded (**139**). Aromatization of **139** to **123** occurs in cold 96% sulfuric acid.[97]

The infrared spectrum of **139** exhibited bands at 3225, 3125 (NH, NH associated), and 1667 cm^{-1} (C=O), and the ultraviolet maximum in ethyl alcohol appeared at 246 mµ ($\epsilon = 5029$).[97]

The action of mineral acids on 2-(2'-oxocyclohexyl)methyl-6-chloronicotinic acid (**140**) yields 6-(4'-carboxy)butyl-2-hydroxy-5-oxo-6,7-dihydro-5H-1-pyrindine (**141**).[98a–98c] Chemical and spectra data, as well as an unequivocal synthesis starting with ethyl 2-bromomethyl-

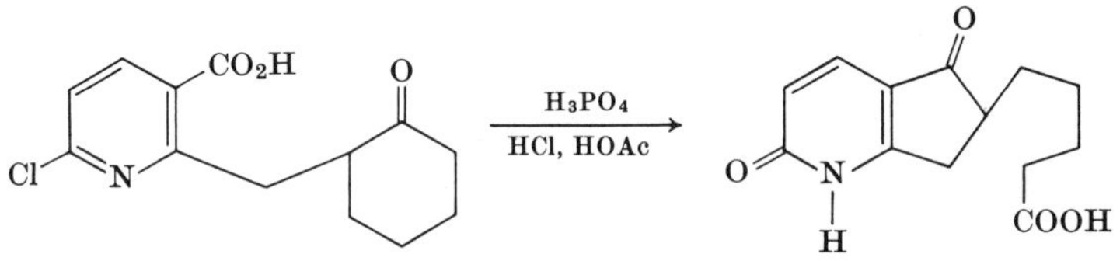

6-chloronicotinate, were used to help elucidate the structure of **141**. The route to **141** is probably (via **142**) as shown. Esterification of **141** with alcoholic hydrochloric acid gave the corresponding methyl and

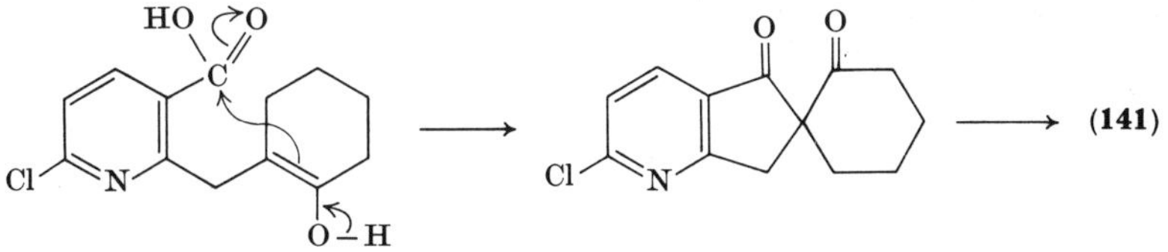

ethyl esters, and treatment of the sodium salt with methyl iodide or the reaction of **141** with diazomethane gives **143**. Ramirez and Paul[98a] also

---

[97] A. I. Meyers and G. Garcia-Muñoz, *J. Org. Chem.* **29**, 1435 (1964).
[98a] F. Ramirez and A. P. Paul, *J. Amer. Chem. Soc.* **77**, 1035 (1955).
[98b] This example of N-methylation of an α-pyridone derivative with diazomethane is in contrast to the report that treatment of the tautomeric mixture of 2-pyridone and 2-hydroxypyridine gives only 2-methoxypyridine.[98c]
[98c] H. S. Mosher, *in* "Heterocyclic Compounds" (R. C. Elderfield, ed.), Vol. I, p. 435. Wiley, New York, 1952.

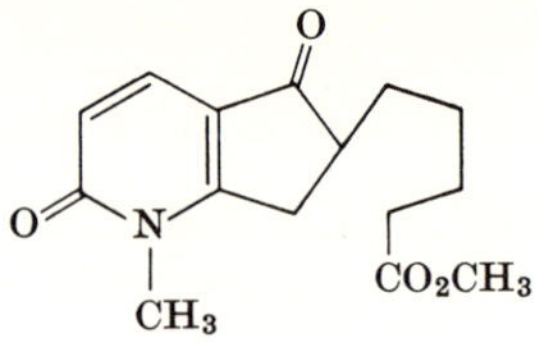

(143)

prepared the compounds **144–148** for spectral studies.

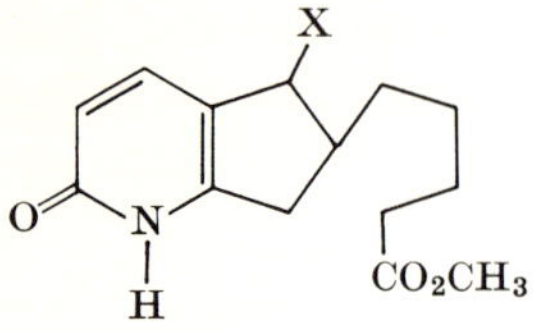

(**144**)
X = OH or Cl

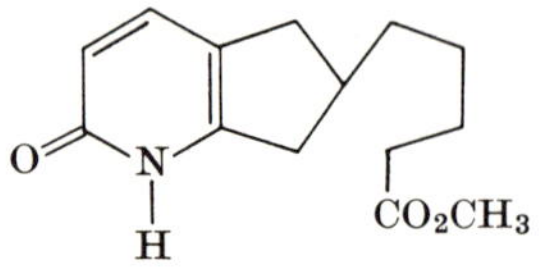

(**145**)

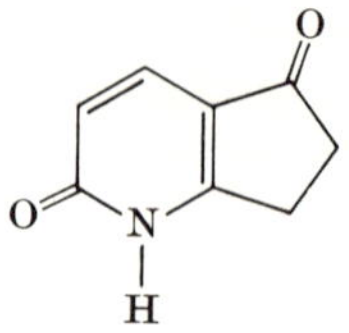

(**146**)

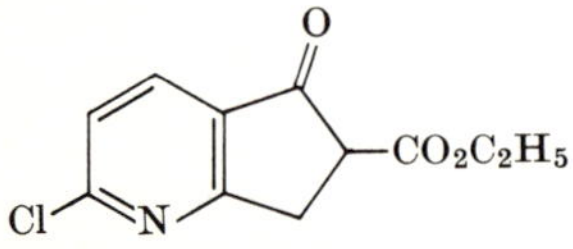

(**147**)

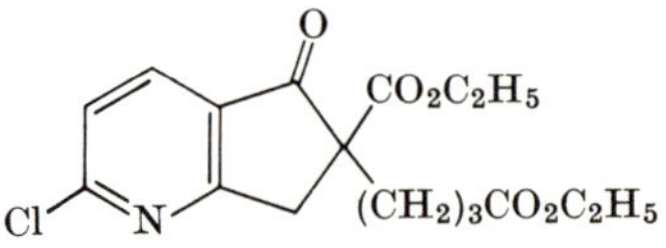

(**148**)

Another novel synthesis involves the condensation of cyclopentanone with α-cyanocinnamic acid (**149**) to give **150** and **151**.[99] Presumably a

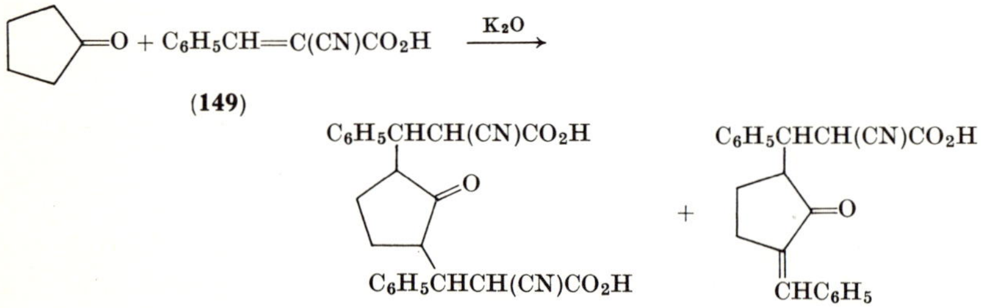

(**149**)

(**151**)

(**150**)

double Michael addition of α-cyanocinnamic acid to cyclopentanone gives **151** which, in the alkaline solution, undergoes partial degradation by a retro-Michael reaction to give **150**. Acid hydrolysis of **150** gives **152**.

[99] J. P. Schneider and P. Cordier, *C. R. Acad. Sci.* **265**, 638 (1967).

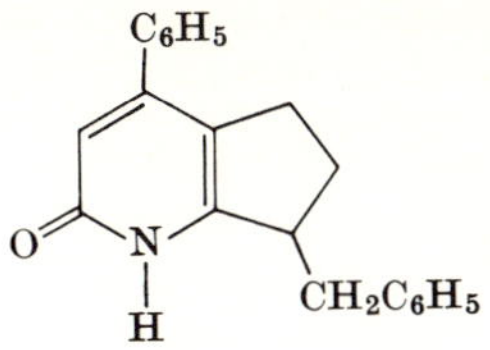

(152)

E. 5H-1-PYRINDINE-5,7(6H)-DIONES

5H-1-Pyrindine-5,7(6H)-dione (11) and 5H-2-pyrindine-5,7-(6H)-dione (153) are also described in *Chemical Abstracts* as 4-azaindan-1,3-dione and 5-azaindan-1,3-dione, respectively.[100–108]

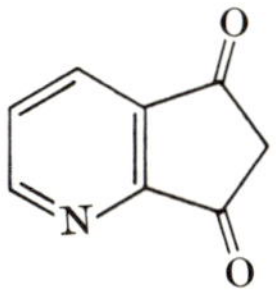

(11) 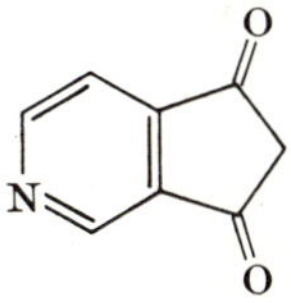   (153)

Derivatives of 11 are obtained by condensing the lactone of 2-(hydroxymethyl)nicotinic acid and aromatic aldehydes in the presence

[100] I. V. Turovskii, J. Linabergs, and O. Y. Neiland, *Khim. Geterotsikl. Soedin.*, 158 (1967); *Chem. Abstr.* **67**, 5963 (1967).
[101] L. E. Neilands and G. Y. Vanags, *Khim. Geterotsikl. Soedin.*, 114 (1967); *Chem. Abstr.* **67**, 64269 (1967).
[102] V. Grinsteins, A. Veveris, and L. Neilands, *Latv. PSR Zinat. Akad. Vestis, Kim Ser.*, 212 (1966); *Chem. Abstr.* **65**, 10557 (1966).
[103] L. Neilands and G. Y. Vanags, *Latv. PSR Zinat. Akad. Vestis, Kim Ser.*, 74 (1963); *Chem. Abstr.* **60**, 4102 (1964).
[104a] L. Neilands, A. Karklins, A. Veiss, and G. Vanags, *Latv. PSR Zinat. Akad. Vestis. Kim. Ser.*, 7 (1964); *Chem. Abstr.* **61**, 3699 (1964).
[104b] A. Veiss, A. Karklins, and I. Tale, *Khim. G. S., Akad. Nauk., Latv. SSR*, 205 (1962); *Chem. Abstr.* **59**, 1096, 14582 (1962).
[104c] A. Karklins and A. Veiss, *Khim. G.S., Akad. Nauk., Latv. SSR*, 124 (1963); *Chem. Abstr.* **60**, 28 (1964).
[105] L. Neilands and G. Vanags, *Khim. G.S., Akad. Nauk., Latv. SSR*, 93 (1965); *Chem. Abstr.* **63**, 3785 (1965).
[106] J. Pelcere, A. Karklins, O. Neilands, and A. Veiss, *Latv. PSR Zinat. Akad. Vestis, Kim. Ser.* 524 (1969); *Chem. Abstr.* **72**, 62126 (1970).
[107] J. Freimanis, E. Silins, O. Neilands, and L. Taure, *Latv. PSR Zinat. Akad. Vestis, Fiz. Teh. Zinat. Ser.* 28 (1967); *Chem. Abstr.* **69**, 47810 (1968).
[108] L. Neilands, *Khim. Geterotsikl. Soedin.*, 647 (1970); *Chem. Abstr.* **73**, 77013 (1970).

of sodium methoxide.[101] An alternative procedure using quinolinic acid and arylacetic acids gives lower yields.[101] Treatment of ice-cold

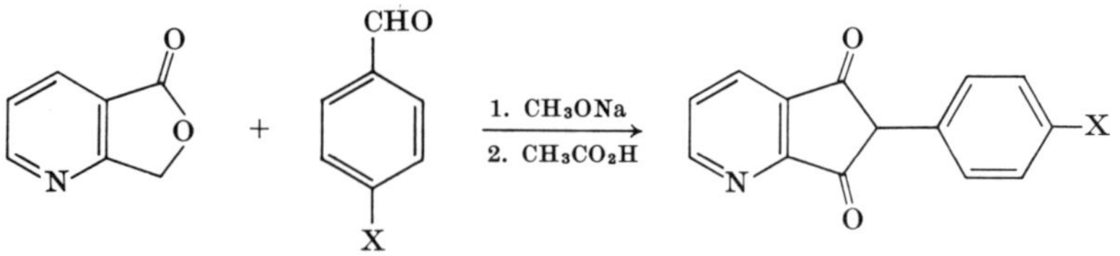

(154a) X = H (47%)
(154b) X = N(CH$_3$)$_2$ (34%)
(154c) X = N(CH$_3$)$_2 \cdot$ HCl
(154d) X = NO$_2$ (58%)
(154e) X = NO$_2 \cdot$ HCl

suspensions of **154a, b, e** in diethyl ether and dimethylformamide with diazomethane afforded the respective N-methylbetaines.[101, 108]

Alkaline treatment of **155** affords 5-carboxy-5-hydroxy-6.6-dichloro-5H-1-pyrind-7-one (**156**) and **157–158**.[2, 109, 110] A plausible mechanism involves an initial attack of hydroxide ion at the 6-carbonyl carbon of **155**, followed by ring contraction (Scheme 7). Other mechanisms are also reasonable.

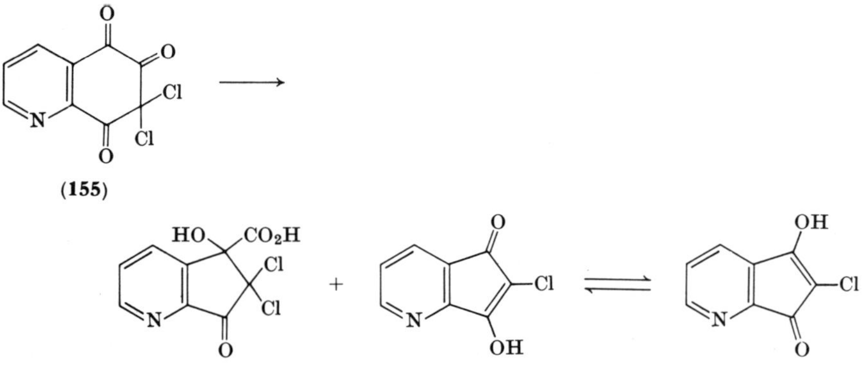

SCHEME 7

6-Carbomethoxy-5H-1-pyrindine-5,7(6H)-dione (**159**), which can exist in the two tautomeric hydroxyketone forms (**160**) and (**161**) and

[109] T. Zincke and H. Müller, *Ann. Chem.* **264**, 201 (1891).
[110] T. Zincke, *Ann. Chem.* **290**, 321 (1896).

the NH form (**162**), is prepared in near quantitative yield from methyl acetate and dimethyl quinolinate.[103, 111] Treatment of **159** with diazomethane or its sodium salt with dimethyl sulfate gives the N-methyl-

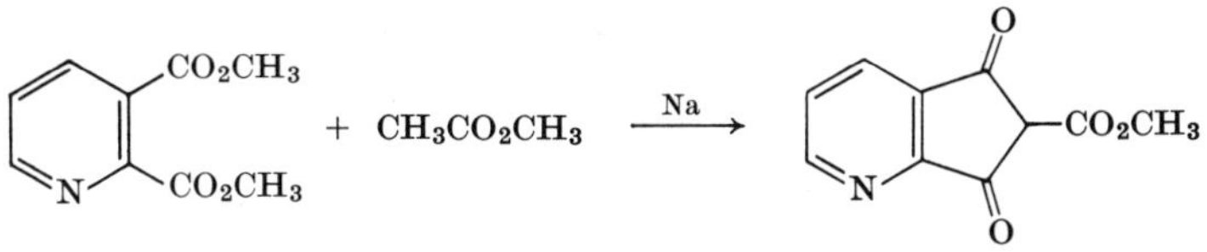

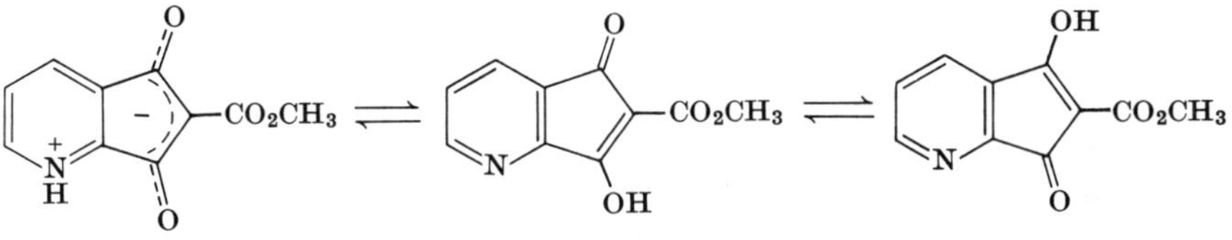

betaine (**163**) in 75 and 59% yields, respectively.[103]

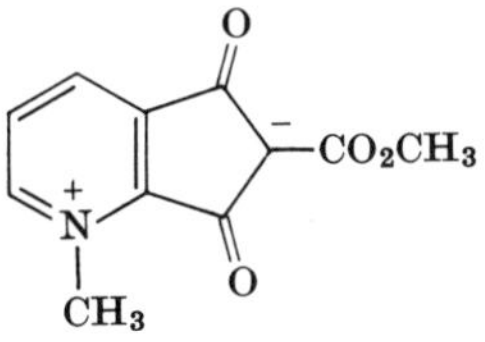

In a similar reaction, dimethyl quinolinate reacts with methyl ketones to give 6-acyl- and 6-aroyl-5H-1-pyrindine-5,7(6H)-diones.[112] Another interesting reaction involves condensation of imines having at least one α-methylene group with dimethylmalonyl chloride in the presence of triethylamine to give substituted pyrindinediones.[113] For example, condensation with N-cyclopentylidenepropylamine gives 1-propyl-3,3-dimethyl-6,7-dihydro-5H-1-pyrindine-2,4(1H,3H)-dione (**164**) in 76% yield.

[111] K. Bittner, *Chem. Ber.* **35**, 1411 (1902).
[112] T. El-Zimaity, *Diss. Abstr.* **24**, 2094 (1964).
[113] J. C. Martin, K. C. Brannock, and R. H. Meen, *J. Org. Chem.* **31**, 2966 (1966).

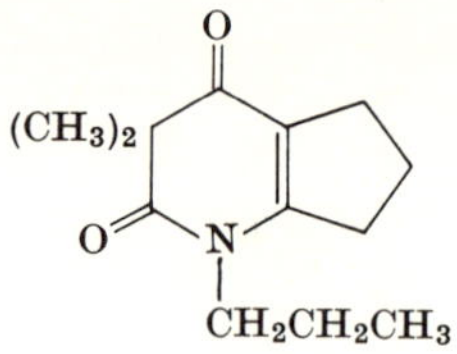

(164)

In acidic medium **154a** exists as a mixture of the dipolar ion [(**165**), pK = 3.94] and the N-protonated enol [(**166**), pK = −0.51]. At pH values greater than 3, an equilibrium is established between **165** and

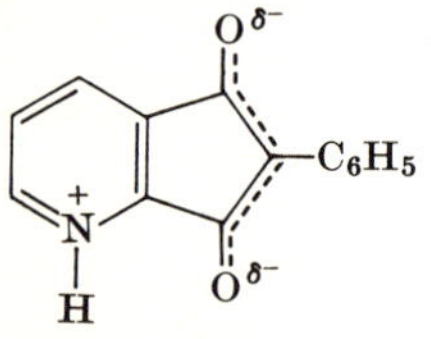

(165) 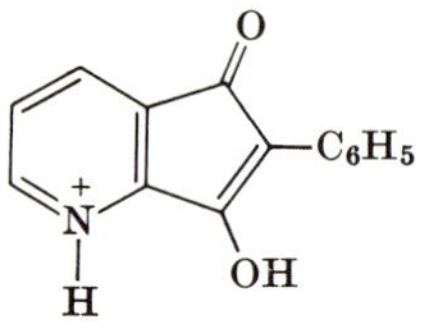   **(166)** 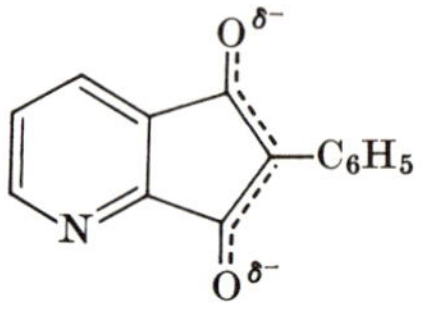   **(167)**

167,[100, 103] and no evidence for the presence of **168** was found. Similarly, **169** exists as the dipolar ion (**170**) in neutral and alkaline solution, and as the enolic cation (**171**) in hydrochloric acid.

Presumably substitution reactions and opening of the five-membered ring occur in the coulometric titration of **154a, 162, 163,** and **172** with electrolytically generated bromine and chlorine in 33–75% acetic acid and in 50% ethyl alcohol–sulfuric acid solutions.[104a] Substitution values greater than theoretical and a comparison of similar reactions with 1,3-indandione derivatives[104a, 104b] suggest that the negative inductive effect of the nitrogen atom favors ring opening. Consistent with this argument is the report that passage of chlorine through an aqueous solution of **162** gives 3-carbomethoxydichloroacetylpicolinic acid (**173**).[104a]

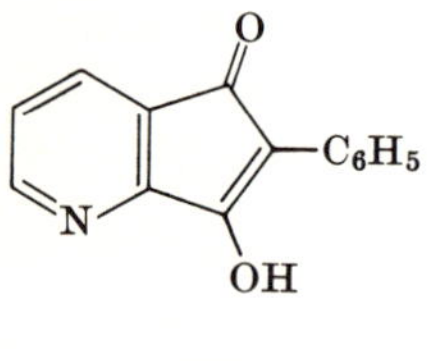

(168) 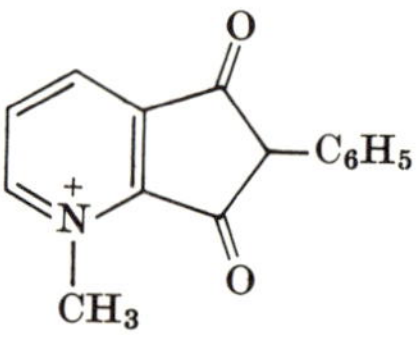   **(169)**

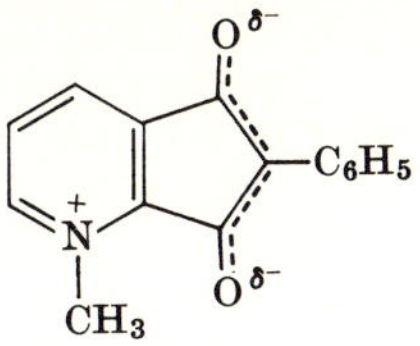

Sec. II.E.]   THE CHEMISTRY OF 1-PYRINDINES   221

(170) 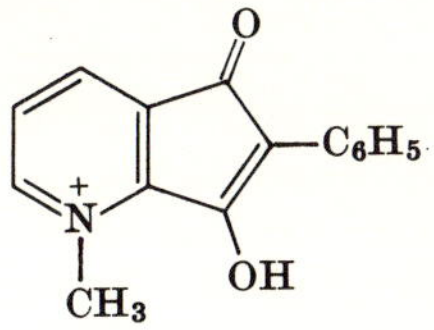   (171)

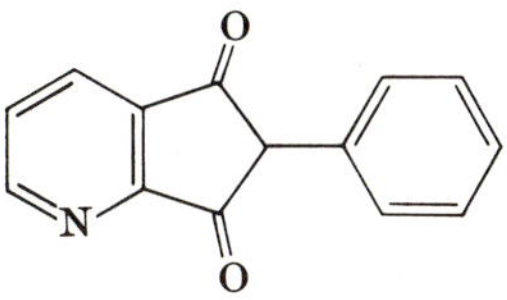

(154a) 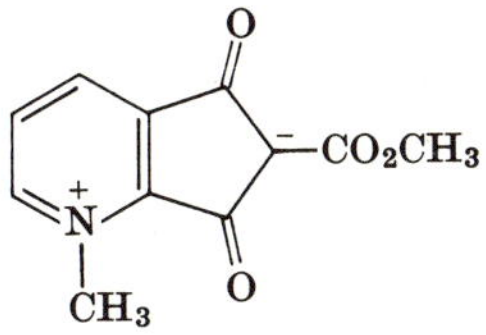   (163) 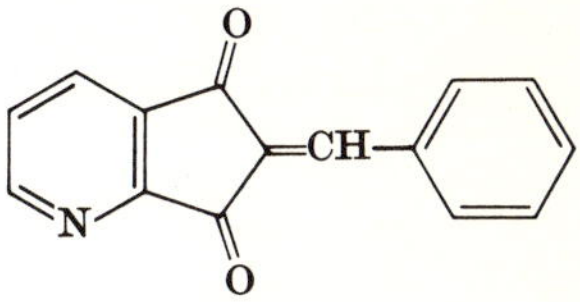   (172)

The carbonyl groups in the derivatives of **159** undergo mono- and diaddition reactions. For example, mono- and diguanylhydrazones are prepared by heating **159** with aminoguanidine salts.[102] Treatment of **174**

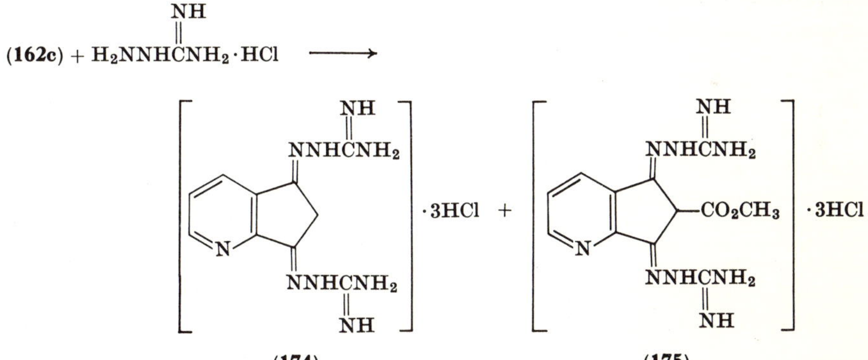

(162c) + H$_2$NNHCNH$_2$·HCl $\longrightarrow$

(174)   (175)

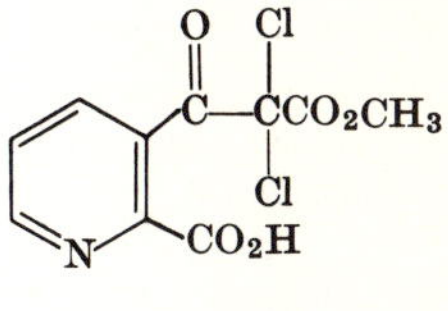

(173)

with thiosemicarbazide and concentrated hydrochloric acid gives the guanylhydrazone-thiosemicarbazone (176) and the dithiosemicarbazone (177).[102] Thiosemicarbazide also reacts with 159 to give the monothiosemicarbazone.

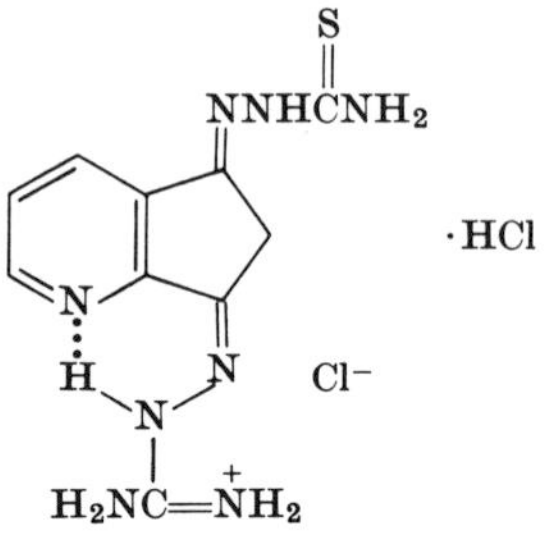

(176) 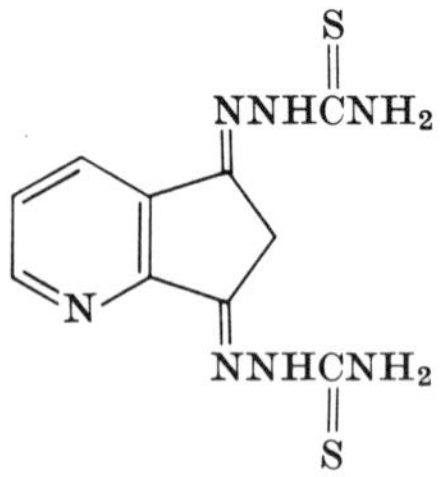   (177)

Although some guanylhydrazone-thiosemicarbazones showed high antitubercular activity, 176 did not. 6-Acyl- and 6-aroyl-5H-1-pyrindine-5,7(6H)-diones also react with hydrazine.[112]

F. Hexahydro- and Octahydro-1-pyrindines

The reaction of acrylonitrile and substituted cyclopentanones (cyanoethylation) is convenient for the preparation of suitable precursors to reduced heterocycles of known stereochemistry.[95, 114–117] For example, $\Delta^{1(7a)}$-hexahydro-1-pyrindine (180) can conveniently be prepared in 73% yield by the hydrolytic ring closure of 179[115a] or in good yield from the ketal (182).[114] Reduction of 180 with sodium and ethanol gives pure trans-octahydro-1-pyrindine (183),[114] and the cis isomer (184) or a mixture of 183 and 184, under hydrogen with a variety of catalysts and solvents.[114, 117] Compound 184 is also formed in good yield from 186.[118] The yield is lower (15%) when benzoyl-cis-octahydro-1-pyrindine (187) is refluxed in dioxane saturated with dry hydrogen chloride for 40 minutes.[118, 119] In contrast, benzoyl-trans-

[114] T. Henshall and E. W. Parnell, *J. Chem. Soc.*, 661 (1962).
[115a] E. F. Godefroi and L. H. Simanji, *J. Org. Chem.* **27**, 3882 (1962).
[115b] K. Schofield and R. J. Wells, *J. Chem. Soc. C*, 621 (1967).
[116] L. A. Cohen and B. Witkop, *J. Amer. Chem. Soc.* **77**, 6595 (1955).
[117] E. A. Mistryukov, *Izv. Akad. Nauk SSSR, Ser. Khim.* 2001 (1965); *Chem. Abstr.* **64**, 6610 (1966).
[118] E. A. Mistryukov, *Izv. Akad. Nauk SSSR, Ser Khim.* 1249 (1964); *Chem. Abstr.* **61**, 11962 (1964).
[119] E. A. Mistryukov, *Izv. Akad. Nauk SSSR, Ser. Khim.* 2006 (1965); *Chem. Abstr.* **64**, 6459 (1966).

(178) (179) (180)

(180)

(181) (182)

(183)

(184)

(184)

(185) (186)

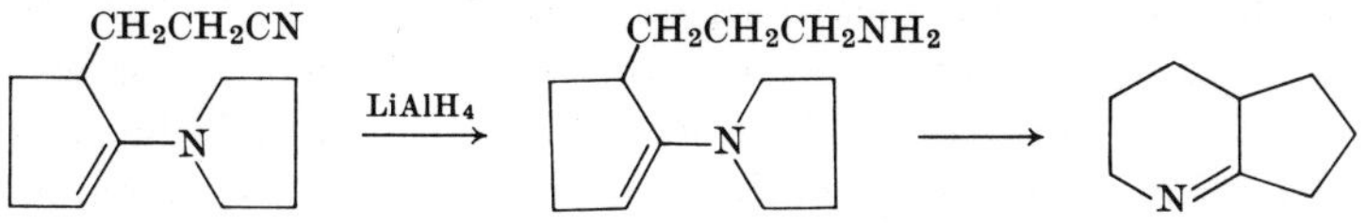

octahydro-1-pyrindine (**188**) gives a 96% yield of **183**·HCl under similar conditions. Presumably the close proximity of the carbonyl oxygen of the amide group and the hydrogen atom at C–8 enables **188**

(187)

(188)

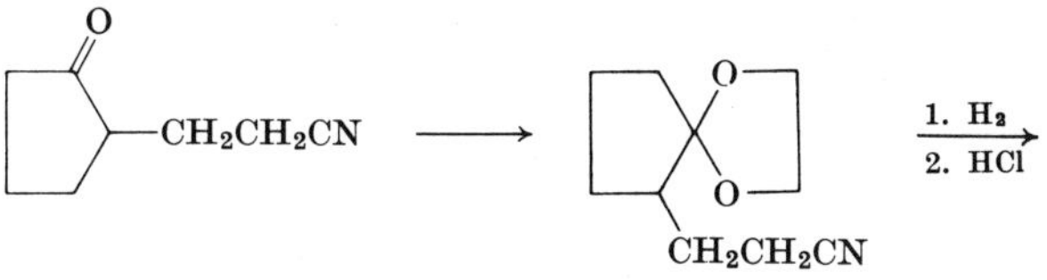

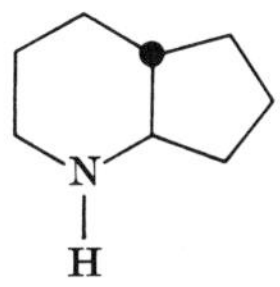

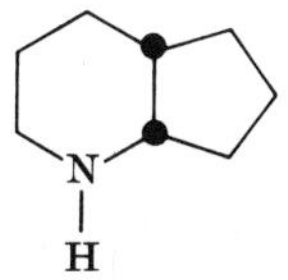

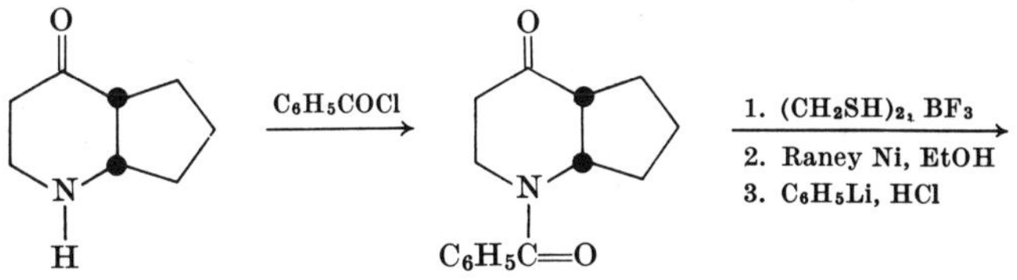

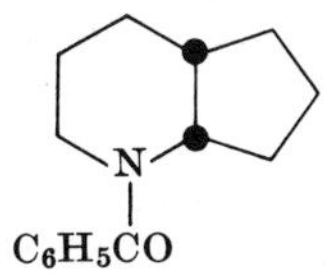

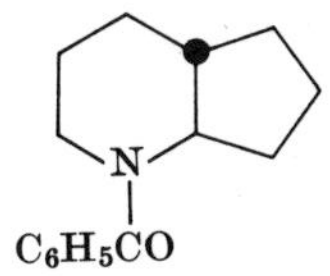

to react faster. This proximity is avoided in **187** since the C–N bond is in the axial conformation.[119]

The basic properties of octahydro-1-pyrindines[93, 96, 99–103, 120–125] and the chromatographs of **184** on aluminum oxide have been reported.[105, 126]

Reaction of the lactones (**189**) and (**191**) with ammonia has also been shown to be a convenient method for introducing nitrogen into the ring system.[127–129]

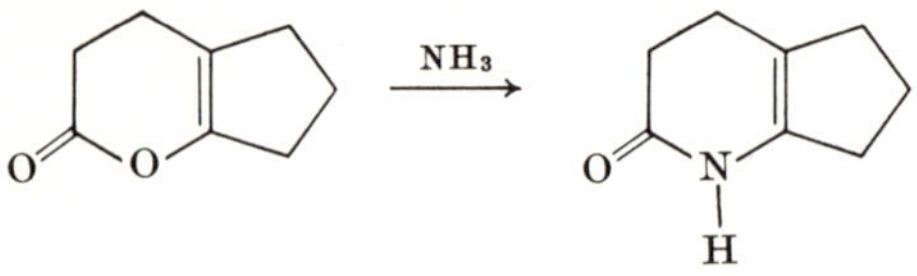

(**189**) → NH₃ → (**190**)

(**191**) → NH₃, 250° → (**192**) → LiAlH₄ → (**193**)

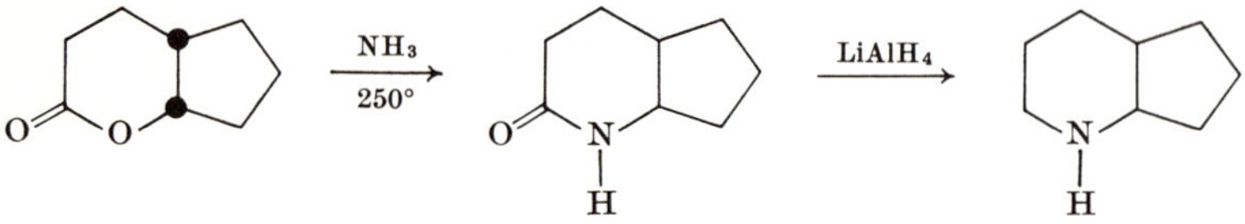

N-Methyl-3-bromopropylamine hydrobromide reacts with 1-(1-cyclopentenyl)piperidine to afford a mixture of **194** and **195** in an overall yield of 74%.[130, 131] Curiously, 3-bromopropylamine hydrobromide

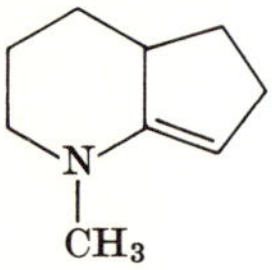

(**194**) 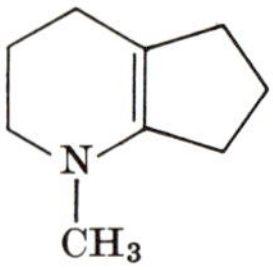   (**195**) 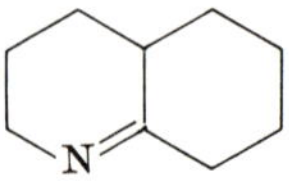   (**196**)

[120] N. F. Albertson, U.S. Patent 2,585,210 (1952); *Chem. Abstr.* **46**, 9617 (1952).
[121] N. F. Albertson, *J. Amer. Chem. Soc.* **72**, 2594 (1950).
[122] I. N. Nazurov, G. A. Shuekhgeimer, and V. A. Rudenko, *Zh. Obshch. Khim.* **24**, 319 (1954); *Chem. Abstr.* **49**, 4651 (1955).
[123] H. Henecka and L. Schultz, German Patent 904,532 (1954); *Chem. Abstr.* **52**, 9222 (1958).
[124] R. Longeraz, A. Vigier, and J. Dreux, *C. R. Acad. Sci.* **253**, 1810 (1951).
[125] E. A. Mistryukov and N. I. Aronova, *Izv. Akad. Nauk SSSR, Ser. Khim.* 789 (1967); *Chem. Abstr.* **68**, 1158 (1968).
[126] E. A. Mistryukov, *J. Chromatogr.* **9**, 311 (1962).
[127] S. V. Kessar, A. Kumar, and A. L. Rampal, *J. Indian Chem. Soc.* **40**, 655 (1963).
[128] N. P. Shusherina, R. Y. Levina, and N. D. Dmitrieva, *Vestn. Mosk. Univ., Ser. Mat., Mekh., Astron., Fiz., Khim.* **13**, 191 (1958); *Chem. Abstr.* **53**, 12287 (1959).
[129] N. K. Chaudhuri and P. C. Mukharji, *J. Indian Chem. Soc.* **33**, 155 (1956).
[130] R. F. Parcell, *J. Amer. Chem. Soc.* **81**, 2596 (1959).
[131] R. F. Parcell and F. P. Hauck, *J. Org. Chem.* **28**, 3468 (1963).

reacts with 1-(1-cyclohexenyl)piperidine to give **196**, but with 1-(1-cyclopentyl)piperidine it does not give **180**.

G. Hexahydro- and Octahydro-1-pyrindones

Dev[132] reported that the reaction of 2-(cyclopentan-2-one)butyrate and ammonium acetate gives the 2H-1-pyrindin-2-one (**197**). However, it is difficult to reconcile the reactants and the suggested product (**197**) since a reasonable mechanism, which involves conversion of the ester to an amide followed by cyclization and dehydration, would predict **198** as a possible product. Alternatively, if the reactant ester was 3-(cyclopentan-2-one)butyrate, the same mechanism would predict **197** as the product.[44]

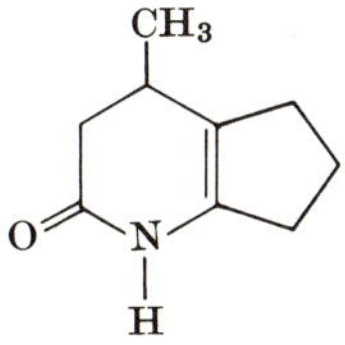

(**197**) 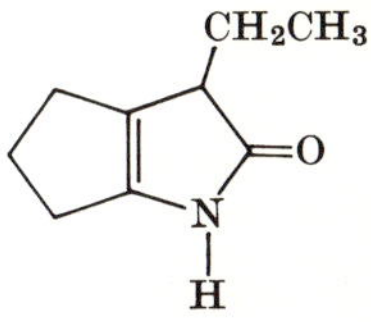   (**198**)

A one-step method for the preparation of 2H-1-pyrindin-2-one derivatives involves a novel modification of the well-known amine-catalyzed condensation of acrylonitrile and ketones. In this procedure the reactants are treated with aqueous ammonium hydroxide solutions.[92, 95, 97, 114–117, 133, 134] Similarly, heating crude 2-diethylamino-ethyl 1-cyclopentenyl ketone (**199**) with 36% aqueous methylamine or 25% ammonium hydroxide in dioxane gives 1-methylperhydro-4-

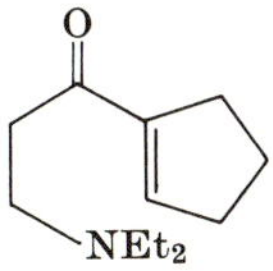

(**199**) 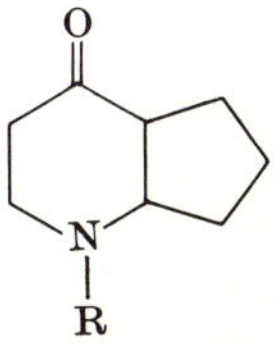   (**200**) R = CH$_3$
           (**201**) R = H
           (**202**) R = C(CH$_3$)$_2$C≡CH

---

[132] S. Dev, *J. Indian Chem. Soc.* **30**, 443 (1953).
[133] J. J. Vill, U.S. Patent 3,267,111 (1966); *Chem. Abstr.* **65**, 70108 (1966).
[134] W. I. Taylor and M. M. Robison, U.S. Patent 3,210,357 (1965); *Chem. Abstr.* **65**, 2235 (1966).

pyrindone (**200**) or perhydro-4-pyrindone (**201**).[135] When **199** is heated with 3-amino-3-methyl-1-butyne in aqueous hydrochloric acid N-(3-methyl-1-butyn-3-yl)perhydro-4-pyrindone (**202**) is formed in 20% yield.[136]

A 74.3% mixture of α- and β-2-methyl-4-oxooctahydro-1-pyrindine (**203**) and (**204**) is obtained from the condensation of propenyl 1-cyclopentenyl ketone and aqueous ammonia.[137] Chromatography on aluminum oxide (2:1 diethyl ether–petroleum ether) gave 40% **203** and 42%

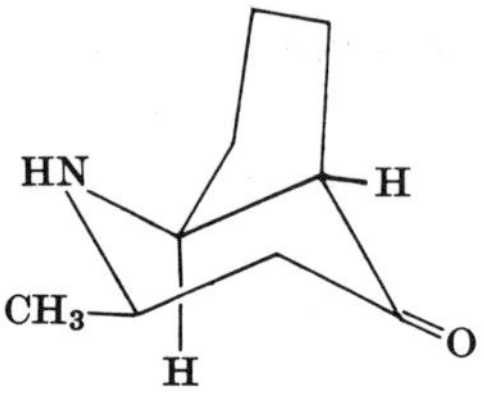

(**203**) 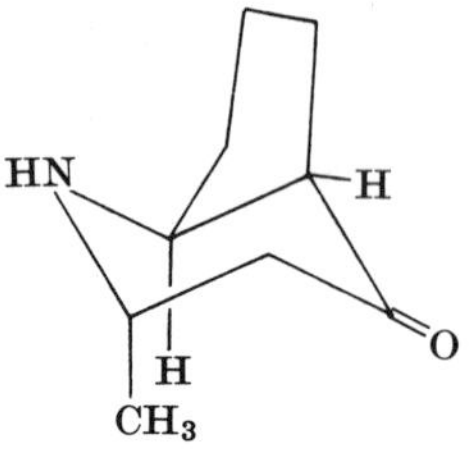   (**204**)

204. The small differences in the dipole moments of **203** (2.90 D) and **204** (3.16 D) suggest that both isomers have the same ring conformation.[138]

Condensation of cyclopentanone with the amino ester (**205**) gives 1,3-dimethyl-1,2,3,5,6,7-hexahydro-4H-1-pyrindin-4-one (**206**), in 38% yield, which can be reduced to the octahydro derivative with lithium aluminum hydride.[139] A variety of substituted 4-pyrindones (**203**, **204**, **207–209**) is obtained by heating 2-methoxypropyl 1-cyclopenten-1-yl

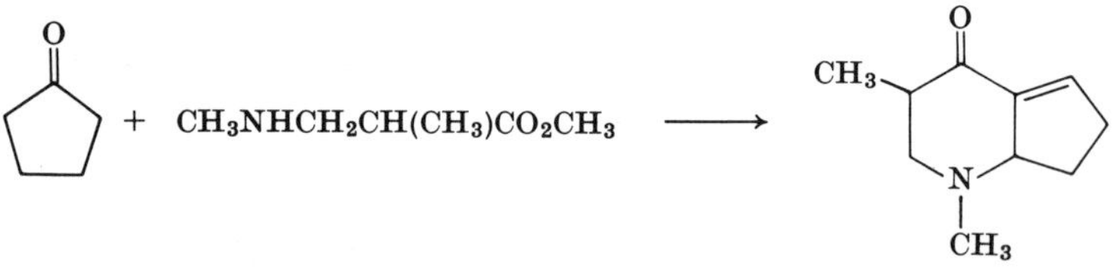

+ $CH_3NHCH_2CH(CH_3)CO_2CH_3$ ⟶

(**205**)   (**206**)

---

[135] E. A. Mistryukov, *Izv. Akad. Nauk SSSR, Otd. Khim. Nauk*, 584 (1958); *Chem. Abstr.* **52**, 20159 (1958).

[136] E. A. Mistryukov, N. I. Aronova, and V. F. Kucherov, *Izv. Akad. Nauk SSSR Ser. Khim.* 512 (1964); *Chem. Abstr.* **60**, 15825 (1964).

[137] Zh. I. Isen, B. T. Sydykov, and D. V. Sokolov, *Izv. Akad. Nauk SSSR Ser. Khim.* **16**, 60 (1966); *Chem. Abstr.* **66**, 5167 (1967).

[138] I. Yu. Kokoreva, *Zh. Strukt. Khim.* **5**, 314 (1964); *Chem. Abstr.* **61**, 2573 (1964).

[139] Z. Horii, C. Iwata, and Y. Tamura, *Chem. Pharm. Bull.* **10**, 940 (1962); *Chem. Abstr.* **59**, 3890 (1963).

ketone or allyl 2,4-dimethyl-1-cyclopenten-1-yl ketone with aqueous methanolic methylamine or ammonium hydroxide.[140]

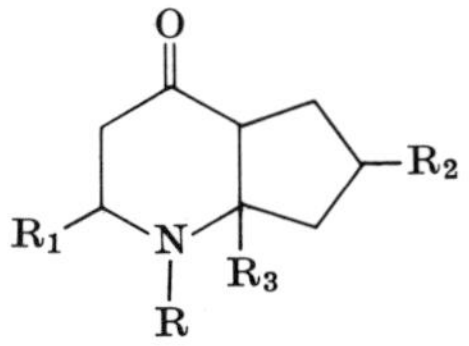

(203, 204) $R = R_2 = R_3 = H; R_1 = CH_3$
(207) $R = R_1 = CH_3; R_2 = R_3 = H$
(208) $R = R_1 = R_2 = R_3 = CH_3$
(209) $R = H; R_1 = R_2 = R_3 = CH_3$

Meyers and co-workers[141] have recently reported an improved method for obtaining enaminoketones from β-aminonitriles instead of β-amino esters (Scheme 8). The enamines, which are easily obtained from β-aminopropionitriles, yield the amino dihydropyrindinium salt (210) on warming with anhydrous magnesium perchlorate or magnesium iodide in benzene or toluene. It was proposed that the role of magnesium

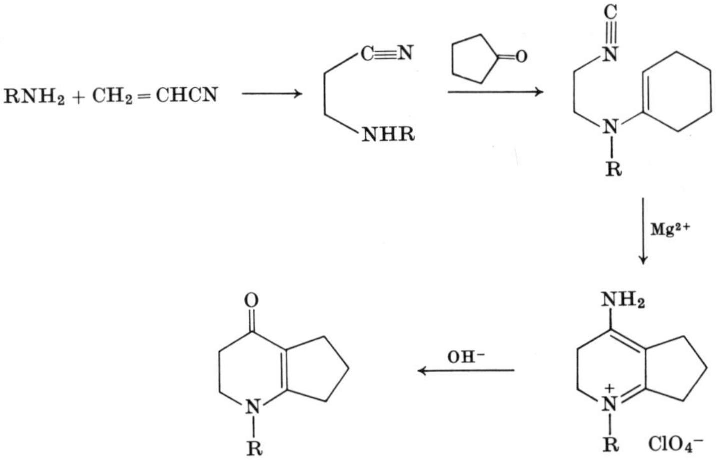

(211) $R = CH_3$ (92%)
(212) $R = (CH_2)_2N(C_2H_5)_2$ (85%)

(210a) $R = CH_3$ (74%)
(210b) $R = (CH_2)_2N(C_2H_5)_2$ (50%)

SCHEME 8

[140] I. N. Nazarov, L. I. Ukhova, and V. A. Rudenko, *Izv. Akad. Nauk SSSR, Otd. Khim. Nauk.* 498 (1953); *Chem. Abstr.* **48**, 9371 (1948).
[141] A. I. Meyers, A. H. Reine, J. C. Sircar, K. B. Rao, S. Singh, H. Widmann, and M. Fitzpatrick, *J. Heterocycl. Chem.* **5**, 151 (1968).

ion in the cyclization is somewhat similar to the reaction of Grignard reagents and nitriles.[141]

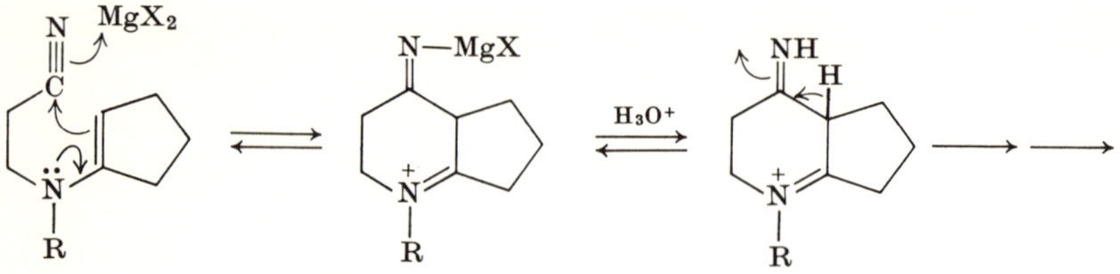

The carbonyl carbon atom in the octahydro-1-pyrindones undergoes typical carbonyl addition reactions.[142-147] For example, the cyanohydrin (213) is obtained from the corresponding oxo compound. Alcoholysis of 213 gives the ester.[142] Phenyllithium adds smoothly to 1,2-dimethyl-4-oxo-perhydro-1-pyrindine (207) to give 214.[143-145]

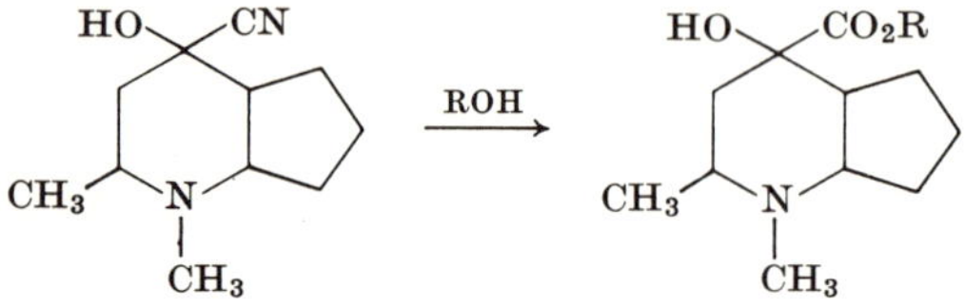

The reduction of 4-oxo-*cis*-octahydro-4H-1-pyrindine with eight different reducing agents gave varying amounts of axial (30–90%) and

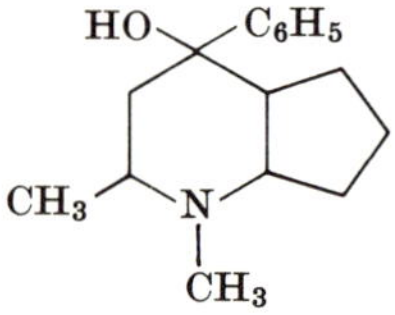

---

[142] I. N. Nazarov and N. I. Shvetsov, *Zh. Obshch. Khim.* **26**, 3170, 3181 (1956); *Chem. Abstr.* **51**, 8742 (1957).

[143] I. N. Nazarov, L. I. Ukhova, and V. A. Rudenko, *Izv. Akad. Nauk SSSR, Otd. Khim. Nauk*, 730 (1953); *Chem. Abstr.* **48**, 12746 (1954).

[144] E. A. Mistryukov, *Izv. Akad. Nauk SSSR, Otd. Khim. Nauk*, 623 (1961); *Chem. Abstr.* **55**, 23521 (1961).

[145] E. A. Mistryukov and V. F. Kucherov, *Izv. Akad. Nauk SSSR, Otd. Khim. Nauk*, 627 (1961); *Chem. Abstr.* **55**, 27305 (1961).

[146] E. A. Mistryukov, *Izv. Akad. Nauk SSSR, Ser. Khim.*, 2155 (1965); *Chem. Abstr.* **64**, 9678 (1966).

[147] E. A. Mistryukov and N. I. Aronova, *Izv. Akad. Nauk SSSR. Ser. Khim.*, 143 (1967); *Chem. Abstr.* **66**, 9769 (1967).

equatorial epimers (10–70%) of the corresponding alcohols.[146] Substituent effects on the configuration of octahydro-1-pyrindin-4-ols and cis-octahydro-1-pyrindin-4-ones have also been studied.[147]

The physical properties of a wide range of reduced 1-pyrindin-2- and 4-ones are given in Refs. 135–137, 140, 141, and 148–150.

III. Biomedical Applications

A wide variety of compounds containing the basic 1-pyrindine skeleton have found applications in the biomedical field.[50–52, 102] Compounds 52 and 53 described above, possess valuable biological activities similar to mepyrapone which has been found to act as a specific 11-β-hydroxylase inhibitor in the biosynthesis of corticoid hormones in man as well as in animals.[51] Antishock activity has been found in a series of octahydro- and decahydrobenzo[a]cyclopenta[f]quinolizenes (12) and (13).[151a] The cardiovascular and potent antishock properties of 2,3,3a,5,6,11,12,12a-octahydro-8-hydroxy-1H-benzo[a]cyclopenta[f]-quinolizinium bromide (12a),[151b] as well as its positive inotropic effect on the cat papillary muscle preparation,[151c] have been reported. Methods of synthesis for a series of these compounds and tests of their relative antishock activity have been described.[151a] Although some compounds were active, comparison of these results pointed up the lack of structure-activity relationships in the seven compounds tested.

(12a)

---

[148] S. A. Vartanyan and G. A. Chukhadzhyan, *Izv. Akad. Nauk. Arm. SSR, Khim. Nauk* **15**, 53 (1962); *Chem. Abstr.* **59**, 3873 (1963).
[149] D. Libermann, M. Moyeux, A. Rouaix, N. Rist, and F. Grumbach, *C. R. Acad. Sci.* **244**, 402 (1957).
[150] H. Person and A. Foucaud, *C. R. Acad. Sci.* **265**, 1007 (1967).
[151a] R. E. Brown, D. M. Lustgarten, R. J. Stanaback, M. W. Osborne, and R. I. Meltzer, *J. Med. Chem.* **7**, 232 (1964).
[151b] M. W. Osborne, M. M. Winbury, and W. M. Govier, *Federation Proc.*, **22**, 308 (1963).
[151c] R. L. Detar, G. C. Boxhill, and M. M. Winbury, *Amer. Soc. Pharmacol. Exp. Therapeut. Pharmacol. Meeting, San Francisco, Calif.*, 1963. Paper 034.

Quinolizines similar to **13a** and **13b** have been screened in the treatment of circulatory collapse, endocrine disorders, and cardiovascular ailments.[152-154] Many of these compounds, which are prepared by condensation of substituted phenethylamines with a suitable oxo or oxo

(**13c**) $R_1 = OH$, $R_2 = \alpha\text{-}CH_3CO$
(**13d**) $R_1 = OH$, $R_2 = \beta\text{-}CH_3O$
(**13e**) $R_1 = H$, $R_2 = \beta\text{-}CH_3CO$

(**13a**) $R = COCH_2OCOCH_3$
(**13b**) $R = CO_2CH_2CH_3$

ester, also showed some steroidal activity.[154-156] When tested for progestational activity, **13c** and **13d** were inactive. However, structurally similar **13e** gave full progestational responses at a dose of 50 mg/kg/day.[159]

Some 1-pyrindine derivatives have been found to show analgesic and anesthetic activity.[157-159] 1-Pyrindine derivatives have been screened for their effect as antimetabolites on the growth of *Entamoeba histolytica*,[160] and other pyrindine derivatives have been studied as possible tuberculostatic[102, 108, 161] and antimalarial agents.[162]

Recently, the absolute structure of the optically active antiviral antibiotic from *Streptomyces abikoensis*, abikoviromycin (**215**), was

[152] I. N. Nazarov, *Izb. Tr. I. N. Nazarov. Akad. Nauk, SSSR*, 558 (1961); *Chem. Abstr.* **56**, 8682 (1962).
[153] M. Protiva, V. Mychajlyszya, and J. O. Vilek, *Chem. Listy* **49**, 1045 (1955); *Chem. Abstr.* **50**, 3476 (1956).
[154] S. Sugasawa and S. Saito, Japanese Patent 5532 (1958); *Chem. Abstr.* **53**, 18061 (1959).
[155] M. Nakamura, *Exp. Cell. Res.* **25**, 648 (1961); *Chem. Abstr.* **57**, 15610 (1962).
[156] R. I. Meltzer and R. E. Brown, U.S. Patent 3,341,544 (1967); *Chem. Abstr.* **68**, 87202 (1968).
[157] D. Libermann, N. Rist, F. Grumbach, and S. Cals, *Bull. Soc. Chem. Fr.* **39**, 1195 (1957).
[158] R. I. Meltzer and R. E. Brown, Belgian Patent 642,060 (1964); *Chem. Abstr.* **63**, 5714 (1965).
[159] R. I. Meltzer and R. E. Brown, French Patent 1,401,060 (1965); *Chem. Abstr.* **63**, 13147 (1965).
[160] R. E. Brown, D. M. Lustgarten, R. J. Stanaback, and R. I. Meltzer, *J. Med. Chem.* **10**, 451 (1967).
[161] A. I. Meyers, J. Schneller, and N. K. Ralhan, *J. Org. Chem.* **28**, 2944 (1963).
[162] M. J. S. Dewar, *J. Chem. Soc.*, 615 (1944).

reported.[163] Spectral data and chemical reactivity (Scheme 9) support the structure (4S,4aR)-5-ethylidene-2,3-dihydro-5H-1-pyrindine 4,4a-oxide (**215**).

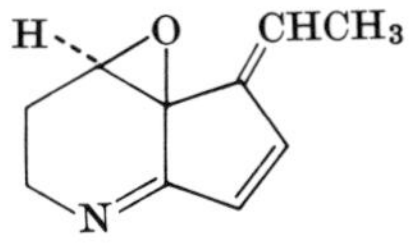

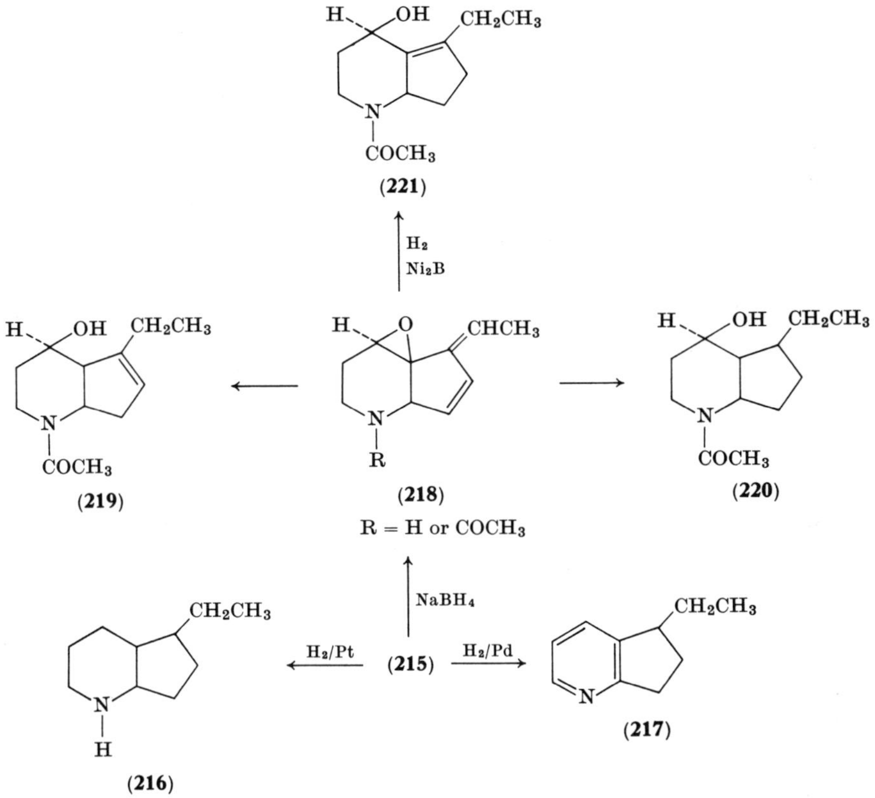

SCHEME 9

[163] A. I. Gurevoch, M. N. Kolosov, V. G. Korobko, and V. V. Onoprienko, *Tetrahedron Lett.*, 2209 (1968).

3-Oxo-2,3-dihydrobenz[d]isothiazole-1,1-dioxide (Saccharin) and Derivatives

H. HETTLER

*Department of Chemistry, Max-Planck-Institute for Experimental Medicine, Göttingen, Germany**

I. Introduction	234
II. Nomenclature	234
III. Structures and Physical Properties	235
A. Crystal Structure	235
B. NMR Spectra	237
C. IR Spectra and Molecular Association	238
D. Mass Spectra	239
IV. The 3-Oxo-2,3-dihydrobenz[d]isothiazole-1,1-dioxide System	239
A. Preparation	239
B. Properties	242
C. Physiological Effects	243
D. Substitution at the Five-Membered Ring	244
E. Reactions with Cleavage of the Five-Membered Ring	255
V. The 3-Thioxo-2,3-dihydrobenz[d]isothiazole-1,1-dioxide System	260
VI. 3-Alkoxy and 3-Aryloxybenz[d]isothiazole-1,1-dioxides	262
A. Preparation and Properties	262
B. Rearrangement	264
C. Elimination	265
VII. 3-Aminobenz[d]isothiazole-1,1-dioxide and Its Derivatives	265
A. Preparation and Properties	265
B. Reactions	267
VIII. 3-Imino-2,3-dihydrobenz[d]isothiazole-1,1-dioxide and Its Derivatives	269
IX. Miscellaneous Compounds	270
X. (Benz[d]isothiazolyl-3-oximino)alkane-1,1-dioxides	271
XI. 3-Chlorobenz[d]isothiazole-1,1-dioxide	273
XII. Elimination Reactions with 3-Substituted Benz[d]isothiazole-1,1-dioxides	274

---

* *Present Address*: 7858 Weil/Rhein, Hauptstr. 163, West Germany.

I. Introduction

The present review deals with 3-oxo-2,3-dihydrobenz[d]isothiazole-1,1-dioxide (**1**)[1] (saccharin) and in particular with such derivatives thereof which are formally on the same level of oxidation, e.g., 3-thioxo-2,3-dihydrobenz[d]isothiazole-1,1-dioxide (**2**) (thiosaccharin) and the so-called "pseudosaccharins" (**3**)–(**6**)[2,3] which correspond to the lactim form of **1**. The topic has been thoroughly reviewed by Bambas in 1952.[4]

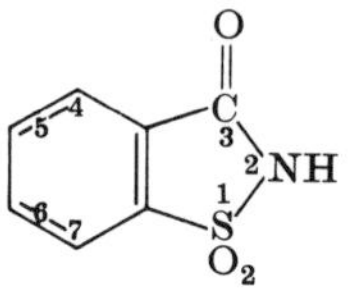

(**1**) 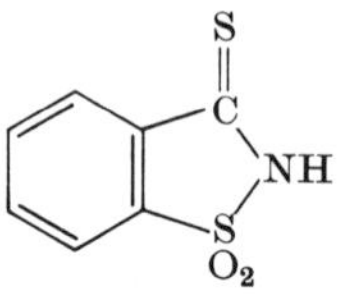   (**2**)

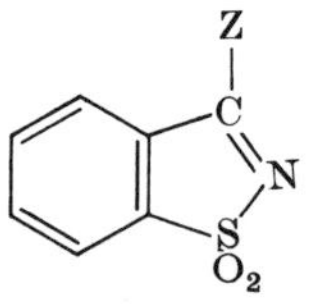

(3) Z = alkoxy, aryloxy
(3a) Z = alkoxy
(4) Z = alkylthio
(5) Z = amino substituent
(5a) Z = amino
(5b) Z = dialkylamino
(6) Z = chloro

The objective here is to report on the more recent developments in this area, focusing on preparative aspects.

II. Nomenclature

The history of saccharin (**1**)[5] since its discovery by Remsen and Fahlberg in 1879,[6,7] is reflected in a host of names; e.g., 2,3-dihydro-3-oxo-benzisosulfonazole; 1,2-benzoisothiazolin-3-one-1,1-dioxide; o-sulfobenzimide; o-benzosulfimide; benzoic sulfimide; benzoic sulfonimide; benzoic sulfinide; 1-benzosulfonazol-2(1)-one; besides clearly

---

[1] Nomenclature is in accordance with IUPAC rules.
[2] R. Rambaud, in "Traité de chimie organique" (V. Grignard, ed.), Vol. 21, pp. 548–557. Masson, Paris, 1953.
[3] J. A. Jesurun, Ber. **26**, 2286 (1893).
[4] L. L. Bambas, in "The Chemistry of Heterocyclic Compounds" (A. Weissberger, ed.) Vol. 4, pp. 297–353. Wiley (Interscience), New York, 1952.
[5] For an account see H. J. Jungel, Deut. Lebensm. Rundsch. **47**, 7 (1951).
[6] I. Remsen and C. Fahlberg, Amer. Chem. J. **1**, 426 (1879–1880).
[7] I. Remsen and C. Fahlberg, Ber. **12**, 469 (1879).

trivial names like gluside, garantose, saccharinol, saccharinose, saccharol, saxin, and sykose. For the purpose of instant identification the terms "saccharin" and "pseudosaccharins" for **1** and **3–6** respectively, will be used throughout this review as has become standing practice.[2] 3-Oxo-2,3-dihydrobenz[*d*]isothiazole-1,1-dioxide is considered to be the systematic name.[1]

III. Structures and Physical Properties

A. Crystal Structure

The dominating feature common to compounds (**1**)–(**6**) is the strongly electron withdrawing SO_2 group.[8] Two representative crystal structures have been determined by X-ray crystallography: (i) The structure of saccharin (**1**),[9,10] a fairly strong acid ($pK_a = 1.30$)[11] with a tendency to dimerize, and (ii) the structure of 3-diisopropylaminobenz[*d*] isothiazole-1,1-dioxide (**7**)[12] a neutral "pseudosaccharin" derivative with considerable steric crowding.

The most remarkable feature of the molecular configuration of **1** is the narrow C–S–N angle of $92.2°$[10] $(92.7°)$[9] in the five-membered ring, as a compromise between ring strain and angular distortion. The molecule as a whole (apart from the S-oxide oxygens) is practically planar. The crystal is built from centrosymmetric dimers, with N–H · · · O=C hydrogen bonds.

Hydrogen bonding may account[9] for the high m.p. of **1** (228°) as compared to *N*-methyl saccharin (m.p. 131°) (**8**); however, equally high melting points of 3-dialkylaminobenz[*d*]isothiazole-1,1-dioxides[13] (**5b**) suggest the influence of other factors (perhaps a dipolar structure combined with close packing) as well.

The molecular structure of **7** reflects a compromise between maximum resonance, which would require an all-planar geometry and steric hindrance from the isopropyl substituents. The result is unusual. Not only has the benzene ring been squeezed by the "pressure" of bulky substituents R_1 and R_2, but its plane has been forced to form an angle of

---

[8] K. F. Reid, "Properties and Reactions of Bonds in Organic Molecules," p. 414. Longmans, Green, New York, 1968.

[9] J. C. J. Bart, *J. Chem. Soc. B*, 376 (1968).

[10] Y. Okaya, *Acta Crystallogr. Sect. B* **25**, 2257 (1969).

[11] I. H. Pitman, H.-S. Dawn, T. Higuchi, and A. A. Hussain, *J. Chem. Soc. B*, 1230 (1969).

[12] W. Saenger and H. Hettler, *Chem. Ber.* **102**, 1468 (1969).

[13] H. Hettler, *Z. Anal. Chem.* **220**, 9 (1966).

6° 4′ with the plane of the five-membered ring. Substituents R_1 and R_2 are coplanar with nitrogen, but their axis is turned out of the isothiazole plane by 14° 4′.[12] Resonance states like **7a**,[12] which include the sulfoxide

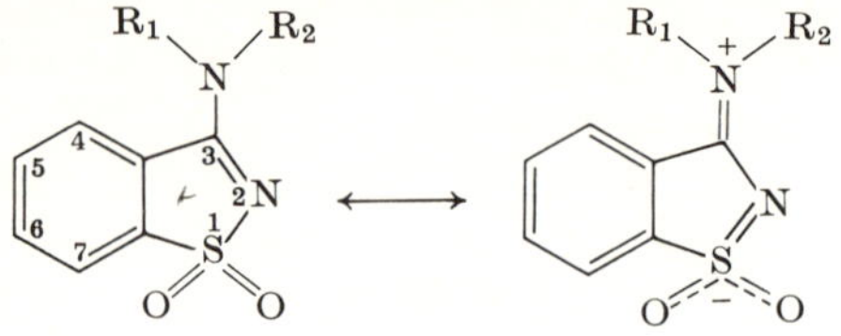

(7) $R_1, R_2 = CH(CH_3)_2$ (7a)

TABLE I

COMPARATIVE BOND LENGTHS AND ANGLE DATA[a]

Bond length	1^9 Å	1^{10} Å	7^{12} Å	Angle[b]	1^9	1^{10}	7^{12}
C_1–C_2	1.382	1.394	1.391	C_1–C_2–C_3	121.0	120.8	121.9
C_2–C_3	1.391	1.384	1.374	C_2–C_3–C_4	121.7	121.8	119.7
C_3–C_4	1.365	1.389	1.376	C_3–C_4–C_5	116.7	116.0	118.4
C_4–C_5	1.383	1.388	1.379	C_4–C_5–C_6	122.1	123.1	123.4
C_5–C_6	1.385	1.369	1.388	C_5–C_6–C_1	120.8	120.6	117.4
C_6–C_1	1.368	1.382	1.395	C_6–C_1–C_2	117.7	117.7	119.2
C_6–C_7	1.480	1.474	1.520	C_5–C_6–C_7	112.6	112.9	108.5
C_7–N_1	1.375	1.369	1.320	C_6–C_7–N_1	109.8	109.6	114.7
C_5–S	1.761	1.758	1.753	C_7–N_1–S	115.0	115.1	111.8
S–N_1	1.663	1.663	1.616	N_1–S–C_5	92.7	92.2	96.5
S–O_1	1.427	1.429	1.436	O_1–S–O_2	117.4	117.7	116.0
S–O_2	1.428	1.409	1.435	N_1–S–O_2	110.3	110.2	110.5
C_7–N_2	—	—	1.333	C_5–S–O_1	112.8	112.8	109.0
C_7–O_3	1.220	1.214	—	O_3–C_7–N_1	123.7	123.9	119.6

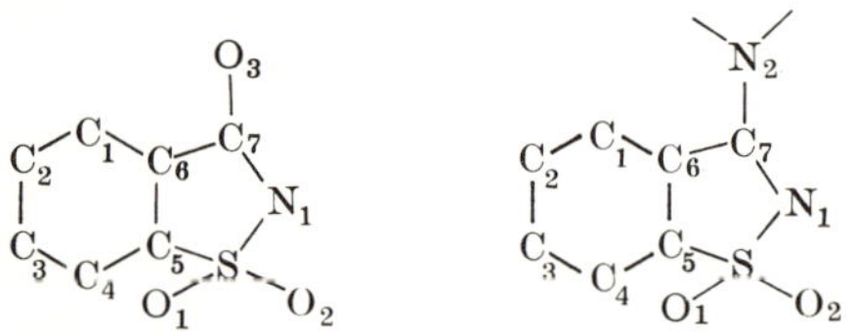

[a] The numbering suggested[9,12] has been used for comparing crystallographic data.
[b] The center atom denotes the vertex.

group, should contribute even more to the ground state of sterically less hindered N-substituted 3-aminobenz[d]isothiazole-1,1-dioxides.

Since R_1 and R_2 in **7a** are no longer equivalent with respect to their chemical environment as a result of restricted rotation, different sets of protons for R_1 and R_2 are observed in the NMR.[14]

A few conclusions have been drawn from structural work which may bear on reactivity:

The five-membered ring in **1** is strained. Both electron-withdrawing groups, the C=O and the SO_2, compete for the pair of electrons at the nitrogen in **1**,[12] whereas in **7** through resonance with **7a** a high degree of stabilization is achieved.

B. NMR Spectra

NMR studies showed that barriers to rotation exist for 3-dialkylaminobenz[d]isothiazole-1,1-dioxides; their numerical values, coalescence temperatures, etc., are, however, dependent on the solvent system.[14] From frequency shifts in the infrared it is known that vibrations of S–O bonds[15] in sulfonamides, particularly those attributed to antisymmetric (ν_3) vibrations[16] are susceptible to solvent influences.

Neither monosubstituted 3-aminobenz[d]isothiazole-1,1-dioxides (**9**) nor 3-alkoxy or aryloxy derivatives (**3**),[17] over a temperature range of 150°, gave any indication of a separation of ^{1}H NMR signals,[14, 18] due to type **9b** (E-form)[19] structures. Spin coupling of the N–H protons with the protons of substituents R (R = Alkyl) excludes an appreciable proportion of 3-alkyl imino structures (**9c**)[14, 18] in the equilibrium.

From the results of ^{1}H NMR studies[14] and X-ray analysis of **7**[12] the conclusion was drawn that the ground state of substituted 3-aminobenz[d]isothiazole-1,1-dioxides is best represented by structures (**7a**) and (**9a**).

Only if steric hindrance is introduced, as in disubstituted 3-amino derivatives, is transition to free rotation shifted into a temperature

---

[14] U. Krüger and H. Hettler, *Ber. Bunsenges. Phys. Chem.* **73**, 15 (1969).
[15] H. H. Szmant, in "Sulfur in Organic and Inorganic Chemistry" (A. Senning, ed.), pp. 107–152. Dekker, New York, 1971.
[16] J. N. Baxter, J. Cymerman-Craig, and J. B. Willis, *J. Chem. Soc.*, 669 (1955).
[17] H. Hettler, unpublished results (1967–1969), partly in collaboration with H. H. Wilkening.
[18] H. Böshagen, W. Geiger, and H. Medenwald, *Chem. Ber.* **103**, 3166 (1970).
[19] J. E. Blackwood, C. L. Gladys, K. L. Loening, A. E. Petrarca, and J. E. Rush, *J. Amer. Chem. Soc.* **90**, 509 (1968).

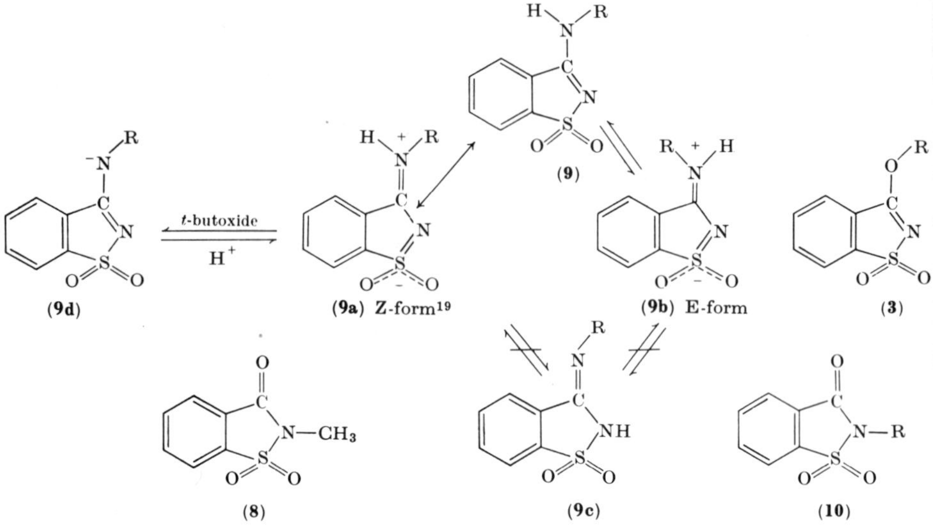

range where it can be observed in the NMR. The tautomeric structure **9c** appears in mass spectrometric fragmentation[20] of **9**.

C. IR Spectra and Molecular Association

Analysis of the infrared spectrum of **1**[16, 21] showed an antisymmetric SO_2 frequency somewhat higher than in linear sulfonamides, probably due to ring strain.[16] IR spectra of metal salts of **1** in solution have also been studied.[22]

The principal IR absorptions of **3**,[20, 23] **5**, and **9**[13, 18] (which compounds have been recommended for the identification of alcohols[24, 25] and primary or secondary amines[13]) and of 2-substituted saccharins (**10**)[20, 23] recorded under identical conditions appear as sets of relatively constant frequencies. For structures with an acidic hydrogen, e.g., compounds (**1**)[16] (**2**), (**9**), self-association and complex formation is expected, but the dipolar 3-dialkylamino derivatives (**5b**) also tend to associate in various solvents.[14] Saccharin (**1**) has been reported to form 1:1 complexes with purines, e.g., theophylline and caffeine, and with amides and

---

[20] H. Hettler, H. M. Schiebel, and H. Budzikiewicz, *Org. Mass Spectrom.* **2**, 1117 (1969).
[21] D. G. O'Sullivan, *Spectrochim. Acta* **16**, 762 (1960).
[22] Yu. N. Sheinker and Yu. Pomeranzew, *Zh. Fiz. Khim.* **33**, 1819 (1959); *Chem. Zentr.*, 8198 (1962).
[23] K. Abe, *J. Pharm. Soc. Jap.* **75**, 159 (1955); *Chem. Zentr.*, 1194 (1962).
[24] J. R. Meadow and E. E. Reid, *J. Amer. Chem. Soc.* **65**, 457 (1943).
[25] H. Böhme and H. Opfer, *Z. Anal. Chem.* **139**, 255 (1953).

phenols in aqueous solution.[26] Solvent influence on the complexing of **1** with N,N-dimethylacetamide has been estimated.[27]

D. MASS SPECTRA

The mass spectrometric fragmentation of selected compounds representing most of what is covered in this review has been analyzed recently.[20] Apart from typical fragmentations (loss of SO_2; rearrangement followed by loss of CO, etc.) the pattern is frequently dominated by the fragmentation of substituents. The MacLafferty rearrangement of 3-substituted benz[d]isothiazole-1,1-dioxides with a proton in the appropriate position resembles the thermal fragmentation[28] (cf. Section XII).

IV. The 3-Oxo-2,3-dihydrobenz[d]isothiazole-1,1-dioxide System

A. PREPARATION

Within five years after its discovery in 1879,[6,7] saccharin (**1**) became an industrial product as the first noncarbohydrate sweetener,[5] and it is still holding an important position in the market.[29-31] Substantial quantities are used as additives in the electroplating industry.[32] Numerous synthetic approaches are known.[4, 29-31] More recent synthesis of **1** and derivatives substituted in the phenyl ring are based essentially on the following procedures.

(i) Industrial processes follow Fahlberg's original preparation,[7] i.e., ring closure of a benzene sulfonamide (or precursor) bearing a carboxy group (or a suitable derivative thereof) in the ortho position.[29, 30] One

---

[26] J. R. Marvel and A. P. Lemberger, *J. Amer. Pharm. Ass. Sci. Ed.* **49**, 417 (1960).
[27] J. R. Marvel and A. P. Lemberger, *J. Amer. Pharm. Assoc. Sci. Ed.* **49**, 420 (1960).
[28] H. Hettler, *Tetrahedron Lett.*, 6031 (1966).
[29] W. Foerst (ed.), "Ullmann's Encyclopädie der technischen Chemie," Vol. 16, pp. 478–486. Urban & Schwarzenberg, Munich, 1965.
[30] H. F. Mark (ed.), "Kirk–Othmer Encyclopedia of Chemical Technology," Vol. 19, pp. 595–607. Wiley (Interscience), New York, 1969.
[31] E. Profft, *Chem. Tech.* (Berlin) **3**, 362 (1951).
[32] W. Foerst (ed.), "Ullmann's Encyclopädie der technischen Chemie," Vol. 7, p. 832. Urban & Schwarzenberg, Munich, 1956.

commercial process uses the reaction of *o*-methoxycarbonylphenylsulfonyl chloride (obtained from anthranilic acid) with ammonia.[33] Ordinarily the key intermediate is *o*-tolyl sulfonamide (**11**) from chlorosulfonation of toluene and subsequent aminolysis. The oxidative step **11** → *o*-carboxybenzene sulfonamide (**12**) followed by ring closure to **1**, has been achieved with a variety of oxidizing agents,[2,4,29-31] potassium permanganate,[6,7] chromic acid,[34-38] and electrolytic oxidation,[39,40] and a combination of chemical and electrolytic methods.[38]

^{14}C-Labeled saccharin (**1a**) has been employed in metabolic studies.[41]

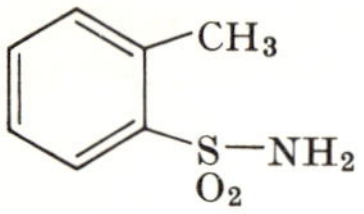

(**11**) 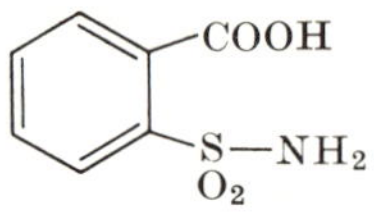 (**12**) 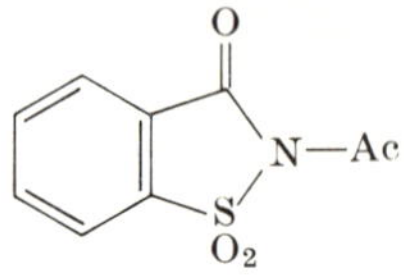 (**13**) Ac = Acyl

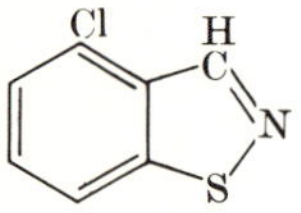

(**14**) 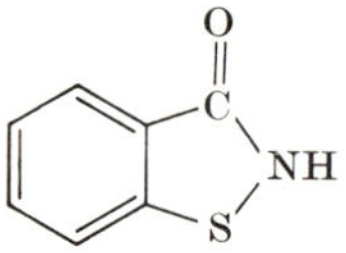 (**15**) 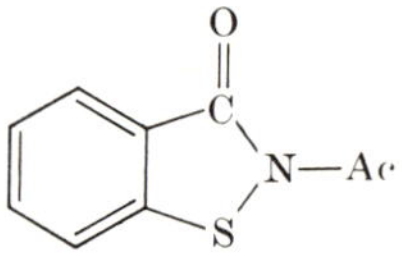 (**16**) Ac = Acyl

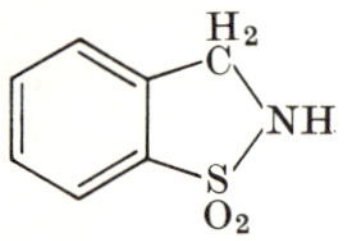

(**17**) 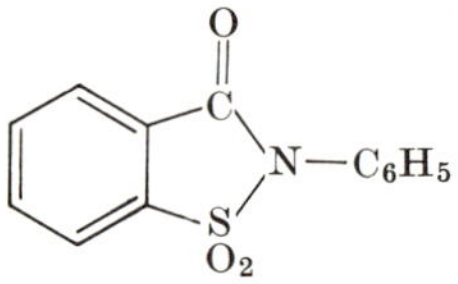 (**18**)

[33] O. F. Senn, U.S. Patent 2,667,503; *Chem. Abstr.* **49**, 3527e (1955); *Chem. Eng.* **61** (7), 128 (1954).
[34] Société Chimique des Usines du Rhöne, British Patent 153,520 (1920); *Chem. Zentr.* **II**, 369 (1921).
[35] J. Bebie, U.S. Patent 1,366,349 (1920); *Chem. Abstr.* **15**, 1030 (1921).
[36] Société Chimique des Usines du Rhöne, German Patent 347,140 (1922); *Chem. Zentr.* **II**, 1137 (1922).
[37] J. W. Orelup, U.S. Patent 1,601,505 (1926); *Chem. Abstr.* **20**, 3696 (1926).
[38] J. Mizuguchi, *J. Chem. Soc. Jap., Ind. Chem. Sect.* **54**, 291 (1951); *Chem. Zentr.*, 3357 (1954).
[39] Farbwerke Hoechst, Dutch Patent 80,092 (1948); *Chem. Zentr.*, 6588 (1953).
[40] J. Eisenbrand, *Z. Elektrochem. Angew. Phys. Chem.* **54**, 314 (1950); *Chem. Zentr.*, 3357 (1954).
[41] R. M. Pitkin, D. W. Anderson, W. A. Reynolds, and L. J. Filer, *Proc. Soc. Exp. Biol. Med.* **137** (3), 803 (1971).

Controlled hydrolysis of 3-substituted benz[d]isothiazole-1,1-dioxides like (**2**), (**3**), (**4**), (**6**), etc., or 2-acylsaccharins (**13**)[42] is a way of returning to the saccharin system. 3-^{18}O-Labeled saccharin has thus been prepared from **6** and $H_2^{18}O$.[43]

(ii) Oxidation of certain benz[d]isothiazole derivatives, e.g., 4-chlorobenz[d]isothiazole (**14**) with hydrogen peroxide in glacial acetic acid.[44,45] It may be noted that permanganate oxidation of 5-aminobenz[d]-isothiazole attacks the benzene ring with formation of isothiazole dicarboxylic acid.[46,47]

(iii) Oxidation of 3-oxo-2,3-dihydrobenz[d]isothiazoles (**15**) with potassium permanganate[48] or hydrogen peroxide.[49-51] Following Reissert's procedure[49] a number of 2-alkyl and aryl derivatives of **15** have been prepared and in some instances oxidized to 2-substituted saccharins (**10**)[52] with hydrogen peroxide in glacial acetic acid. Under similar conditions 2-acyl-3-oxo-2,3-dihydrobenz[d]isothiazoles (**16**) afford saccharin (**1**).[53]

(iv) Oxidation of 2,3-dihydrobenz[d]isothiazole-1,1-dioxides (**17**).[54] Besides **1**, a number of saccharins substituted in the phenyl ring have been prepared by permanganate oxidation of the corresponding compounds of type **17**.[55]

(v) Ortho-metalation of *N*-methyl and *N*-phenyl benzene sulfonamides by excess *n*-butyl lithium followed by carbonation and ring closure[56] affords 2-methylsaccharin (**8**) and 2-phenylsaccharin (**18**). Functional groups sensitive to *n*-butyl lithium cannot be employed.[57]

[42] L. E. Hart, E. W. McClelland, and F. S. Fowkes, *J. Chem. Soc.*, 2114 (1938).
[43] H. Hettler and H. Neygenfind, *Chem. Ber.* **103**, 1397 (1970).
[44] F. Becke and H. Hagen, *Ann. Chem.* **729**, 146 (1969).
[45] F. Becke, *Int. J. Sulfur Chem. B* (*Quart. Rep.*) **6**, 77 (1971).
[46] A. Adams and R. Slack, *Chem. Ind.* (*London*), 1232 (1956).
[47] A. Adams and R. Slack, *J. Chem. Soc.*, 3061 (1959).
[48] M. McKibben and E. W. McClelland, *J. Chem. Soc.* **123**, 170 (1923).
[49] A. Reissert and E. Manns, *Chem. Ber.* **61**, 1308 (1928).
[50] E. W. McClelland and A. J. Gait, *J. Chem. Soc.*, 921 (1926).
[51] R. G. Bartlett, L. E. Hart, and E. W. McClelland, *J. Chem. Soc.*, 760 (1939).
[52] F. Gialdi, R. Ponci, and A. Baruffini, *Farmaco-Ed-Sci.* **16**, 509 (1961); *Chem. Zentr.*, 7-1363 (1966).
[53] F. Gialdi, R. Ponci, A. Baruffini, and P. Borgna, *Farmaço-Ed-Sci.* **19**, 76 (1964); *Chem. Zentr.*, 19-1216 (1965).
[54] L. L. Bambas, "The Chemistry of Heterocyclic Compounds" (A. Weissberger, ed.), Vol. 4, pp. 278–296. Wiley (Interscience), New York, 1952.
[55] Y. Nitta, M. Shindo, T. Takasu, and C. Isono, *J. Pharm. Soc. Jap.* **84**, 493 (1964); *Chem. Zentr.*, 47-0981 (1967).
[56] H. Watanabe, R. L. Gay, and C. R. Hauser, *J. Org. Chem.* **33**, 900 (1968).
[57] L. Lombardino, *J. Org. Chem.* **36**, 1843 (1971).

Hauser's approach was extended to cover **1** and 3-oxo-2,3-dihydrobenz-[*d*]isothiazole-1,1-dioxides with suitable substituents in the phenyl ring by employing *N-t*-butylarylsulfonamides.[57] The substituted 2-(*N-t*-butylsulfamoyl)benzoic acids are cyclized and dealkylated in one step by polyphosphoric acid.[57]

Among the many derivatives of **1** substituted in the phenyl ring[4] those with substituents in the 6-position predominate. Practically all procedures start with the desired substitution pattern, as under conditions of electrophilic or radical substitution the five-membered ring will invariably be affected. More recently the interest in pharmaceutical use and in potential thermostable polymers[58] promoted synthetic work on such compounds as 4-amino-,[59] 4-acetamido-,[59] 5-amino-,[58,60] 5-acetamido-,[60-62] 6-amino-,[58] 6-acetamido-,[63] 6-bromo-,[64,65] 4-chloro-,[66,67] 5-chloro-,[57,68] 6-chloro-,[64,65] 5-fluoro-,[57] 5-methyl-,[57] 6-methyl-,[69,70] 5-methoxy-,[57,71] 4-nitro-,[59] 5-nitro-,[58,60,72] 6-nitro-,[58,63,72,73] and 6-sulfamoylsaccharin[74,75] and also 6-sulfamoylsaccharin with a halogen, alkyl, alkoxy, nitro, or amino group in the 5-position.[76]

B. Properties

Saccharin (**1**) is a fairly strong acid ($pK_a = 1.30$[11] earlier work reported $pK_a = 1.61$[77]) which remains stable in aqueous buffered solutions of pH 3.3, 7.0, and 9.0 at temperatures of up to 150° for 1 hr.[78]

---

[58] G. F. D'Alelio, W. A. Fessler, and D. M. Feigl, *J. Macromol. Sci., Chem.* **3**, 941 (1969).
[59] G. H. Hamor, *J. Amer. Pharm. Ass. Sci.* **49**, 280 (1960).
[60] E. Warren and G. H. Hamor, *J. Pharm. Sci.* **50**, 625 (1961).
[61] O. G. Backeberg and J. L. C. Marais, *J. Chem. Soc.*, 78 (1943).
[62] L. Szabo, *Bull. Soc. Chim. Fr.*, 771 (1953).
[63] B. Loev and M. Kormendy, *J. Org. Chem.* **27**, 2177 (1962).
[64] G. H. Hamor and B. L. Reavlin, *J. Pharm. Sci.* **56**, 134 (1967).
[65] R. de Roode, *Amer. Chem. J.* **8**, 167 (1886).
[66] G. H. Hamor and N. Farraj, *J. Pharm. Sci.* **54**, 1265 (1965).
[67] W. Davies, *J. Chem. Soc.* **119**, 876 (1921).
[68] J. H. Gerver, *Verh. Akad Wetensch. Amsterdam* **30**, 236 (1922); *Chem. Abstr.* **16**, 1572 (1922).
[69] R. Fischer and H. Humi, *Arzneim-Forsch.* **14**, 1301 (1964).
[70] W. W. Randall, *Amer. Chem. J.* **13**, 256 (1891).
[71] R. D. Haworth and A. Lapworth, *J. Chem. Soc.* **125**, 1299 (1924).
[72] W. Davies and Q. N. Porter, *J. Chem. Soc.*, 826 (1957).
[73] N. C. Rose, *J. Heterocycl. Chem.* **6**, 749 (1969).
[74] G. H. Hamor, *J. Pharm. Sci.* **51**, 1109 (1962).
[75] W. Herzog, *Z. Angew. Chem.* **39**, 728 (1926); *Chem. Abstr.* **20**, 3450 (1926).
[76] F. C. Novello, U.S. Patent 2,957,883 (1957); *Chem. Abstr.* **55**, 6632b (1961).
[77] I. M. Kolthoff, *Rec. Trav. Chim.* **44**, 629 (1925).
[78] O. De Garmo, G. Ashworth, C. Eaker, and R. J. Munch, *J. Amer. Pharm. Ass. Sci. Ed.* **41** (1), 17 (1952).

Saccharin (**1**), thiosaccharin (**2**), 3-aminobenz[*d*]isothiazole-1,1-dioxide, (**5a**) and monosubstituted derivatives (**9**) have in common that their respective anions (**19**), (**2a**), and (**9d**) are ambident[79] (ambifunctional[80]). The saccharin anion (**19**) may react via the nitrogen or the

(**19**)

oxygen in a nucleophilic reaction depending on such factors as temperature, the electrophile, the cation, and the solvent.[79, 80] Of the two electrophilic centers the carbonyl group is the site of attack preferred over the sulfoxide.

C. Physiological Effects

Saccharin (**1**) and its sodium salt (**19a**) have found extensive use as nonnutritive sweetening agents.[81]

Based on dilution in distilled water to threshold sweetness **1** is about 300 times as sweet as sucrose.[30] Results with about 80 saccharin derivatives indicate that substitution in the 2- or 3-position gives tasteless compounds.[82] Exceptions are Mannich bases (**20**) and 3-oxo-2-hydroxymethyl-2,3-dihydrobenz[*d*]isothiazole-1,1-dioxide (**21**) which possess a sweet taste.[83] A structural unit $O=\overset{|}{C}-\overset{|}{N}H-\overset{|}{S}O_2$ (or perhaps $HO-\overset{|}{C}=\overset{|}{N}-\overset{|}{S}O_2$) incorporated in an appropriate ring system might be held responsible for the sweet taste of **1** and the related 3,4-dihydro-1,2,3-oxathiazin-4-one-2,2-dioxides.[84] Shallenberger and Acree have stated that a system $A(H) \xrightarrow{2.5-4\text{Å}} B$ where A and B are electronegative atoms, is a prerequisite for sweet taste.[85]

Exposure of rats to X-rays induced aversion to the saccharin sodium salt (**19a**).[86, 87]

[79] N. Kornblum, R. A. Smiley, R. K. Blackwood, and D. C. Iffland, *J. Amer. Chem. Soc.* **77**, 6269 (1955).
[80] R. Gompper, *Angew. Chem.* **76**, 412 (1964).
[81] R. Froelich, *Pharm. Praxis*, 38 (1961).
[82] G. H. Hamor, *Science* **134**, 1416 (1961).
[83] H. Zinner, U. Zelck, and G. Rembarz, *J. Prakt. Chem.* [4] **8**, 150 (1959).
[84] K. Clauss and H. Jensen, *Deut. Offenlegungsschr.* 2,001,017 (1971).
[85] R. S. Shallenberger and T. E. Acree, *Nature (London)* **216**, 480 (1967).
[86] R. M. Bradley and C. M. Mistretta, *J. Comp. Physiol. Psychol.* **75**, 186 (1971).
[87] T. E. Chaddock, *Experientia* **26**, 164 (1970).

Many feeding experiments with mammals have proved that neither **1** nor any of its degradation products is retained in the organism in appreciable quantities.[41, 88]

In 1957 Allen *et al.* first produced evidence that saccharin when implanted into the bladder of mice might produce urinary bladder carcinomas.[89] Further studies have confirmed those findings.[90-93] However, no evidence that saccharin when used as a substitute for sugar has detrimental effects[29, 30, 88] on test animals[94] or humans[88, 95] has yet been produced.

D. Substitution at the Five-Membered Ring

1. *Nucleophilic Reactivity*

A major distinction for nucleophilic reactions with ambient anions is whether they proceed with kinetic or thermodynamic control.[80] N-Substituted saccharins (**10**) should be thermodynamically more stable because of amide character than the isomeric "pseudosaccharin" (**3**) of imidate structure. In fact **3** may be rearranged thermally to **10** in an irreversible reaction.[96] The threshold for thermodynamic control appears to be lowered for electrophiles with multiple bonds, e.g., formaldehyde, reactive derivatives of carboxylic acids, but also quaternary salts of N-heterocyclic compounds.[80] It will be seen that in those cases substitution indeed occurs at the nitrogen, not necessarily through thermodynamic control.

2. *Alkylation with Alkyl Halides*

N-Alkyl- (**10a**) and *N*-aryl-saccharins can be synthesized with *o*-chlorocarbonyl phenylsulfonyl chloride and alkylamines or aryl-

---

[88] G. Bungard, *Deut. Apoth.* **23**, 97 (1971).
[89] M. J. Allen, E. Boyland, C. E. Dukes, E. S. Horning, and J. G. Watson, *Brit. J. Cancer* **11**, 212 (1957).
[90] J. M. Price, C. G. Biava, B. L. Oser, E. E. Vogin, J. Steinfeld, and H. L. Ley, *Science* **167**, 1131 (1970).
[91] G. T. Bryan and E. Ertürk, *Science* **167**, 996 (1970).
[92] G. T. Bryan, E. Ertürk, and O. Yoshida, *Science* **168**, 1238 (1971).
[93] G. T. Bryan and O. Yoshida, *Arch. Environ. Health* **23**, 6 (1971).
[94] O. G. Fitzhugh, A. A. Nelson, and J. P. Frawley, *J. Amer. Pharm. Ass.* **40**, 538 (1951).
[95] F. Bär, *Umschau*, 218 (1970).
[96] H. Hettler, *Tetrahedron Lett.*, 1793 (1968).

amines,[97-99] however, not in a straightforward reaction.[100, 101] The alkyl residue may also be introduced by reacting the saccharin anion (**19**) or derivatives substituted in the phenyl ring with alkylating agents. The results of alkylation reactions are consistent with the postulate that alkylation at the more electronegative atom of an ambident ion generally increases with increasing S_N1 character of the reaction.[78] Alkyl halides ordinarily react at the less electronegative nitrogen atom to form **10a**, except when the conditions favor a S_N1 reaction.

(**22**)   (**10a**)   (**19a**) M = Na
                               (**19b**) M = K
                               (**19c**) M = Ag

(**3a**) R = Alkyl

The reaction of **19a** on heating with alkyl bromides and iodides first reported by Fahlberg[102] and widely used since[55, 103-106] appears to be rather sluggish in protic solvents, e.g., diethylene glycol.[105] Chlorinated reagents seemed to react readily only when the halogen is activated by a neighboring carbonyl[107] or carboxy group[107, 108] or in an allylic or benzylic[55, 105, 109] position. (See, however, Ref. 110.)

Rice and Pettit[111-113] greatly improved the procedure by employing

[97] I. Remsen and C. E. Coates, *Amer. Chem. J.* **17**, 309 (1895).
[98] I. Remsen and E. P. Kohler, *Amer. Chem. J.* **17**, 330 (1895).
[99] I. Remsen and F. E. Clark, *Amer. Chem. J.* **30**, 278 (1903).
[100] W. E. Henderson, *Amer. Chem. J.* **25**, 1 (1901).
[101] W. M. Blanchard, *Amer. Chem. J.* **30**, 485 (1903).
[102] C. Fahlberg and A. List, *Chem. Ber.* **20**, 1596 (1887).
[103] H. Eckenroth and G. Koerppen, *Chem. Ber.* **30**, 1265 (1897).
[104] T. Sachs, E. von Wolff, and A. Ludwig, *Chem. Ber.* **37**, 3252 (1904).
[105] L. L. Merritt, S. Levey, and H. B. Cutter, *J. Amer. Chem. Soc.* **61**, 15 (1939).
[106] G. H. Hamor and T. O. Soine, *J. Amer. Pharm. Ass., Sci. Ed.* **43**, 120 (1954).
[107] H. Eckenroth and K. Klein, *Chem. Ber.* **29**, 329 (1896).
[108] G. R. Pettit and R. E. Kadunce, *J. Org. Chem.* **27**, 4566 (1962).
[109] E. Profft and L. Lemke, *J. Prakt. Chem.* [4] **13**, 253 (1961).
[110] W. H. Arnold and N. E. Searle, U.S. Patent 2,462,835 (1949); *Chem. Abstr.* **43**, 4421 (1949).
[111] H. L. Rice and G. R. Pettit, *J. Amer. Chem. Soc.* **76**, 302 (1954).
[112] L. M. Rice, C. H. Grogan, and E. E. Reid, *J. Amer. Chem. Soc.* **75**, 4304 (1953).
[113] E. E. Reid, L. M. Rice, and C. H. Grogan, *J. Amer. Chem. Soc.* **77**, 5628 (1955).

dimethylformamide[23] (DMF) as solvent. Moreover, addition of sodium iodide enhances the reaction of alkyl chlorides.[105, 111]

Alternatively, p-tosylates in DMF may be used for preparing **10a**, particularly with a sugar (e.g., protected 6-O-p-tosyl-D-glucofuranose[114]) as starting material. Sugar derivatives (**10a**; R = tetraacetylglycosidyl,[115] 2,3,4-triacetyl-β-methylglucosid-6-yl[116]) have been obtained from the corresponding iodo-sugars and **19c** and **19a** respectively. The anhydrous sodium salt **19a** is conveniently obtained from **1** and sodium ethoxide in ethanol.[117] For alkylation reactions, preparation of **19a** from sodium hydride and **1** in hexamethylphosphotriamide (HMTP) has been recommended in view of the well-recognized power of HMTP to solvate alkali cations.[118] Fusion of alkali salts **19a**, **19b** with suitable halides, e.g., dialkylaminoalkyl chlorides[106, 119] also yields N-alkyl derivatives (**10a**).

Some reactions deserve discussion under mechanistic aspects; N-methylsaccharin (**8**) is obtained in about 95% yield when **19a** is heated with methyl iodide in DMF for 3–4 hr[120] or when kept for several days at room temperature.[17] When isopropyl and sec-butyl halides are used instead,[120] 3-isopropoxy- and 3-sec-butoxybenz[d]isothiazole-1,1-dioxide [**3a**; R = –OCH(CH$_3$)$_2$, –OCH(CH$_3$)(C$_2$H$_5$)] are the main products. Similarly, the silver cations in **19c** promote o-alkylation[121] presumably through attack by silver on the halogen of the alkyl halide.[78] In both cases the contribution of alkyl carbonium ion to the transition state is enhanced and therefore alkylation takes place at the more electronegative oxygen.[78]

Apart from activating properties, dimethylformamide does not shield **19** by selective solvation to the same extent as protic solvents do.[80]

3. *Other Alkylating Agents*

Reaction of saccharin salts with p-tosyl esters of alcohols provides a very satisfactory method for preparing N-alkylsaccharins (**10a**),[114]

---

[114] E. M. Acton, J. E. Christensen, H. Stone, and L. Goodman, *Experientia* **24**, 998 (1968).
[115] K. Josephson, *Chem. Ber.* **60**, 1822 (1927).
[116] E. Hardegger and O. Jucker, *Helv. Chim. Acta* **32**, 1158 (1949).
[117] E. Stephen and H. Stephen, *J. Chem. Soc.* 492 (1957).
[118] H. Normant and Th. Cuvigny, *Bull. Soc. Chim. Fr.*, 1866 (1965).
[119] J. Braun, K. Heider, and E. Müller, *Chem. Ber.* **51**, 273 (1918).
[120] K. Abe, *J. Pharm. Soc. Jap.* **75**, 153 (1955); *Chem. Zentr.*, 1193 (1962).
[121] G. Heller, A. Buchwald, R. Fuchs, W. Kleinicke, and J. Kloss, *J. Prakt. Chem.* [2] **111**, 1 (1925).

e.g., N-methylsaccharin (**8**),[122, 123] N-ethylsaccharin (**22**),[122, 123] and others.[124] Likewise **8** and **22** result from the reaction of dimethyl sulfate[55] and diethyl sulfate, respectively, on **19a**.[125]

The N-alkylsaccharins **8** and **22** are also produced when **1** is boiled with the corresponding trialkyl phosphites under nitrogen for 48 hr.[126] 2-Trimethylammonio-pyrimidines react with hard bases[127] to yield new 2-substituted pyrimidines with loss of trimethylamine.[128] However, in formamide at 120° 2-trimethylammonio-pyrimidines act as methylating agents toward **19**.[128]

The reaction of saccharin (**1**) with diazomethane[121] shows that N- versus O-methylation can depend on parameters such as temperature, solvent, etc.[121, 129] Arndt has also found[130, 131] that in the presence of a high concentration of diazomethane, methylation takes place at the nitrogen, whereas on slow addition of diazomethane in ether oxygen is methylated.

Once again the individual influences may be summarized by stating that N-methylsaccharin (**8**) is formed in an S_N2-transition state, whereas 3-methoxybenz[d]isothiazole-1,1-dioxide (**3a**; $R = CH_3$) results from an S_N1-type reaction.[132] Polar solvents tend to favor S_N1 reactions.[132] Rate constants for the reaction of **1** with diphenyldiazomethane have been determined.[133]

4. *Other Methods of Introducing Substituents at the Nitrogen*

Direct phenylation of saccharin has been effected in 75% yield. Diphenyliodonium bromide reacts with (**19a**) via nucleophilic displacement to form **18**.[134]

---

[122] O. J. Magidson and S. W. Gorbatschow, *Chem. Ber.* **56**, 1810 (1923).
[123] B. J. Frydman and A. Troparesky, *An. Asoc. Quim. Argent.* **45**, 79 (1957); *Chem. Zentr.*, 2802 (1959).
[124] K. Abe, *J. Pharm. Soc. Jap.* **75**, 164 (1955); *Chem. Zentr.*, 1588 (1962).
[125] J. R. Meadow and J. C. Cavagnol, *J. Org. Chem.* **16**, 1582 (1951).
[126] A. M. Roe and J. B. Harbridge, *Chem. Ind. (London)*, 182 (1965).
[127] R. G. Pearson, *Surv. Prog. Chem.* **5**, 1 (1969).
[128] W. Klötzer, *Monatsh. Chem.* **87**, 536 (1956).
[129] E. Ayça, *Rev. Fac. Sci. Forest. Univ. Instanbul, Ser. C* **22**, 383 (1957); *Chem. Abstr.* **53**, 11346e (1959).
[130] F. Arndt, *Angew. Chem.* **61**, 397 (1949).
[131] F. Arndt, in "Organic Analysis," Vol. I, pp. 197–241. Wiley (Interscience), New York, 1953.
[132] R. Gompper, *Chem. Ber.* **93**, 187 (1960).
[133] J. Hine and W. C. Bailey, *J. Amer. Chem. Soc.* **81**, 2075 (1959).
[134] F. M. Beringer, A. Brierley, M. Drexler, E. M. Gindler, and C. C. Lumpkin, *J. Amer. Chem. Soc.* **75**, 2708 (1953).

Hauser's metalation procedure[56] leads to formation of N-alkylsaccharin (**10a**) and N-phenylsaccharin (**18**). Acid catalyzed hydrolysis of saccharin anils (**23**) is an alternative route to compounds (**10a**) and (**18**).[135]

(**23**)   (**24**)   (**25**)

Furthermore, oxidation of N-alkyl/aryl derivatives of 3-oxo-2,3-dihydrobenz[d]isothiazole has been used to prepare N-substituted saccharins.[52]

3-Oxo-2-vinyl-2,3-dihydrobenz[d]isothiazole-1,1-dioxide (N-vinylsaccharin) (**24**) has been synthesized in various ways, e.g., by transvinylation of vinyl acetate with saccharin (**1**) in the presence of mercuric sulfate,[136] by addition of **1** to acetylene with zinc acetate as a catalyst,[136] or by ester pyrrolysis of N-(acetoxyethyl)saccharin.[137] **19b** reacts with ethylene chlorohydrin at 140°–150° to yield N-(2-hydroxyethyl)-saccharin (**25**). The compound is esterified with acetic anhydride and pyrolyzed at 460°–470° to give (**24**).[137]

25 is also obtained from hydroxyethylation of **1** with ethylene carbonate.[138] N-Allylsaccharin is accessible not only from reaction with allyl halides[120] but also from the addition of allyl alcohol to **1** catalyzed by mercuric acetate and sulfuric acid.[139]

5. Addition Reactions

3-Oxo-2-hydroxymethyl-2,3-dihydrobenz[d]isothiazole-1,1-dioxide (**21**) (N-hydroxymethylsaccharin, m.p. 128°),[83, 140] is formed in the reaction of saccharin (**1**) with an excess of formaldehyde in ethanol.[83] Earlier work[141, 142] on **21** reported a melting point of 225°[141] indicating

---

[135] H. Watanabe, C.-L. Mao, and C. R. Hauser, *J. Org. Chem.* **34**, 1786 (1969).
[136] H. Hopff, U. Wyss, and H. Lyssi, *Helv. Chim. Acta* **43**, 135 (1960).
[137] K. Kato and S Wada, *J. Chem. Soc. Jap., Pure Chem. Sect.* **83**, 501 (1962); *Chem. Zentr.*, 29-0923 (1966).
[138] K. Yanagi and S. Akiyoshi, *J. Org. Chem.* **24**, 1122 (1959).
[139] H. Hopff and H. Lyssi, *Helv. Chim. Acta* **46**, 1052 (1963).
[140] H. Böhme and F. Eiden, *Arch. Pharm.* (*Weinheim*) **292**, 642 (1959).
[141] C. Maselli, *Gazz. Chim. Ital.* **30**, 31 (1900); *Chem. Zentr.* **II**, 71-629 (1900).
[142] G. Parmeggiani, *Boll. Chim. Farm.* **47**, 37 (1908); *Chem. Zentr.* **I**, 1389 (1908).

that formation of **21** is reversed on heating. Under Mannich conditions, i.e., in the presence of formaldehyde and secondary amines, one obtains Mannich bases (**20**; R, R' = alkyl).[83]

<chemical structures>
(20) saccharin-N-CH₂NRR'
(21) saccharin-N-CH₂OH
(26) saccharin-N-CH₂Cl

(27) ortho-disubstituted benzene with CH₂OH and SO₂NHCH₃
</chemical structures>

21 is converted into the N-chloromethyl derivative (**26**) with thionyl chloride[83] or with phosphorus pentachloride in ice-cold ether.[140] Besides **26**, the N-bromomethyl and the N-iodomethyl derivatives[143] have been prepared. The identity of **21** has been established by lithium aluminum hydride reduction of **21** and of N-methylsaccharin (**8**) both of which gave the same product (**27**).[83] Reaction of **26** with *sec* amines affords **20**.[83] With mercaptides, e.g., sodium benzyl mercaptide, one obtains the N-benzyl mercaptomethyl derivative (**28**), and with potassium thiocyanate the corresponding N-thiocyanatomethyl compound.[144] Moreover **26** has been reacted with a salt, sodium *trans*-chrysanthemate, to give the ester.[145]

21 adds on to acrylonitrile under acid catalysis to yield (N-saccharinylmethyl)acrylamide.[146]

Condensation of **21** with 2-thio-4-methyluracil in the 5-position has been reported.[147] The structural assignment of the product was based on analogies.[147, 148] Reaction of **19** with 1-chloropropane-2,3-diol followed by periodate cleavage affords saccharin-N-acetaldehyde.[149]

[143] Chien-Pen Lo, U.S. Patent 3,002,884 (1961); *Chem. Zentr.*, 15-2011 (1964).
[144] Chien-Pen Lo, U.S. Patent 2,949,399 (1960); *Chem. Abstr.* **55**, 7435 (1961).
[145] J. Martel, French Patent 1,451,417 (1966); *Chem. Abstr.* **66**, 115839k (1967).
[146] D. T. Mowry, U.S. Patent, 2,529,455 (1950); *Chem. Zentr.* **II**, 2109 (1951).
[147] L. Monti and G. Franchi, *Gazz. Chim. Ital.* **81**, 191 (1951); *Chem. Zentr.*, 1650 (1952).
[148] H. Hellmann, *in* "Neuere Methoden der praeparativen organischen Chemie" (W. Foerst, ed.), Vol. II, p. 195. Chemie, Heidelberg, 1960.
[149] F. Toffoli, F. de Martilis, and M. Dolci, *Farmaco Ed. Sci.*, 516 (1956); *Chem. Zentr.*, 1577 (1958).

Saccharin **1** adds on to activated double bonds. In a base-catalyzed reaction with methyl vinyl ketone, N-(3-ketobutyl)saccharin[150] is formed. Acid catalyzed addition of **1** to 2,3-dihydropyran occurs as expected, again at the nitrogen.[151, 152]

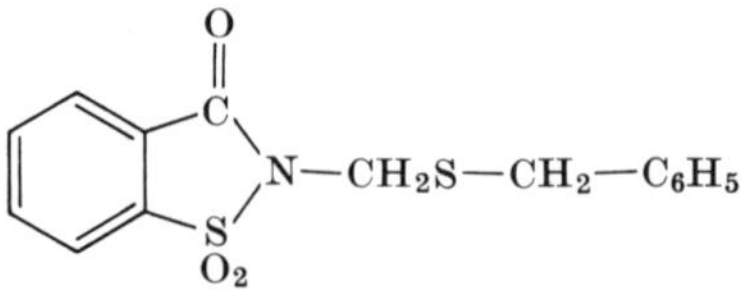

$$\underset{(28)}{\text{saccharin-}N\text{—CH}_2\text{S—CH}_2\text{—C}_6\text{H}_5}$$

6. Acylation

Treatment of ambident saccharin anion (**19**) with acylating agents yields 3-oxo-2-acyl-2,3-dihydrobenz[d]isothiazole-1,1-dioxides (N-acylsaccharins). Many examples are known: (**13**; Ac = acetyl[153]) to octadedecanoyl[117, 122] benzoyl,[103] subst. benzoyl,[117] carbamyl,[154, 155] phenylsulfonyl,[42] trichlormethylsulfenyl,[156] nitro,[157] etc. There is one claimed exception[42] to give the alternative product (**29**) of acylimidate structure; if (**29**) is formed it is liable to undergo spontaneous Mumm rearrangement[158, 159] to **13**.

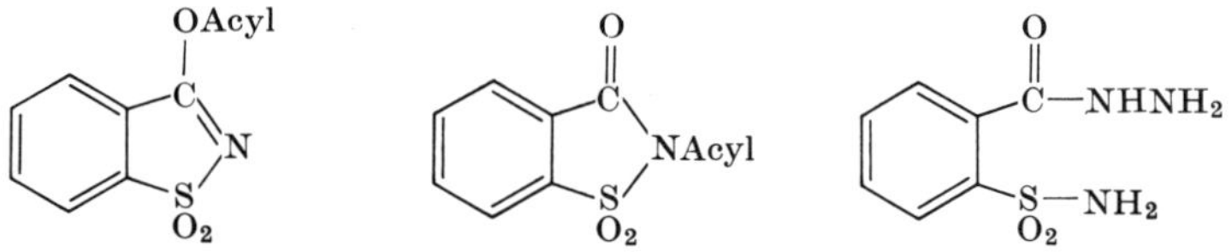

(**29**)　　(**13**)　　(**30**)

[150] H. Arai, S. Shima, and N. Murata, *J. Chem. Soc. Jap. Ind. Chem. Sect.* **62**, 82 (1959); *Chem. Zentr.* 818 (1960).
[151] A. J. Speziale, K. W. Ratts, and G. J. Marco, *J. Org. Chem.* **26**, 4311 (1961).
[152] A. J. Speziale and G. J. Marco, U.S. Patent 3,145,213 (1964); *Chem. Zentr.*, 51-1861 (1965).
[153] H. Eckenroth and G. Koerppen, *Chem. Ber.* **29**, 1048 (1896).
[154] I. Shioyama, S. Mine, and K. Murata, *Deut. Offenlegungsschr.* 1,953,422 (1970); *Chem. Abstr.* **73**, 366 (1970).
[155] S. J. Mehta and G. H. Hamor, *J. Pharm. Sci.* **50**, 672 (1961).
[156] A. R. Kittleson, *Science* **115**, 84 (1952).
[157] J. Runge and W. Treibs, *J. Prakt. Chem.* [4] **15** (287), 223 (1962).
[158] O. Mumm and H. Hesse, *Chem. Ber.* **65**, 1192 (1932).
[159] W. Schulenburg and S. Archer, "Organic Reactions," (A. C. Cope, ed.), Vol. 14, pp. 1–15. Wiley, New York, 1965.

Conventional procedures: (i) reaction of a saccharin salt, e.g., **19a**, **19b**, **19c**[42, 121] with an acid chloride[42, 103, 117, 121, 122] or anhydride.[153] Ring closure and acetylation are achieved in one step when o-sulfamoylbenzoic acid (**12**)[160] or o-sulfamoylbenzohydrazide (**30**)[161] are treated with acetic anhydride.

(ii) Dicyclohexylcarbodiimide (DCC) for condensing **1** with a carboxylic acid. Since no activated acyl derivative is required, the method has been used for preparing derivatives of N-carbobenzoxyaminoacids (**31**)[162–165] or oligopeptides, although with some racemization.[162] Losses of DCC occur from a relatively slow side reaction with **1**.[162]

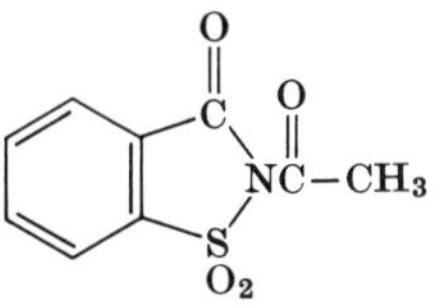

(33) 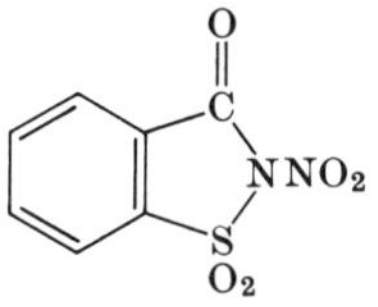   (34) 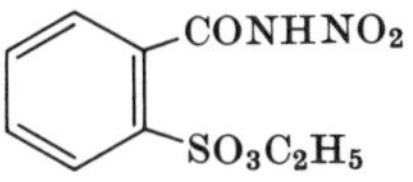   (35)

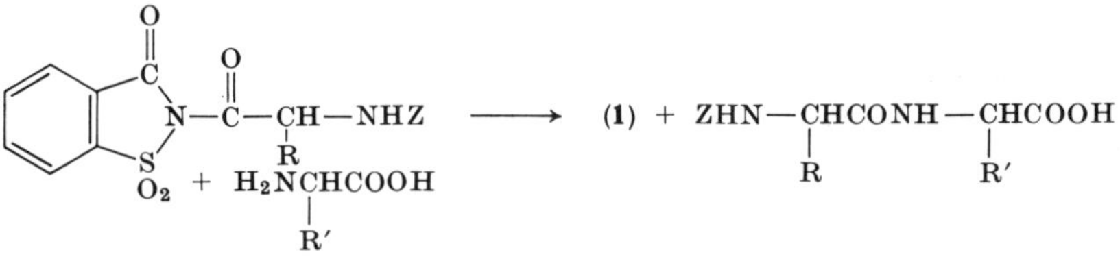

(31)   (38)

Acids or salts are utilized directly in Stephen's method:[166] (iii) reaction with 3-chlorobenz[d]isothiazole-1,1-dioxide (**6**) (pseudosaccharin chloride[3, 166]). The initial step is presumably formation of **29**, followed by Mumm rearrangement. Frequently one obtains pseudosaccharin anhydride (**32**)[25, 162] as a by-product. From the reactions with silver acetate and below 3° only N-acetylsaccharin (**33**) besides **32** was isolated.[167] In this context reexamination of the reported 3-O-benzenesulfonylbenz[d] isothiazole-1,1-dioxide[42] and the supposed 3-O-benzoyl

[160] C. W. Whitehead, J. T. Traverso, J. F. Bell, and P. W. Willard, *J. Med. Chem.* **10**, 844 (1967).
[161] C. W. Whitehead and J. T. Traverso, *J. Org. Chem.* **25**, 413 (1960).
[162] F. Micheel and M. Lorenz, *Tetrahedron Lett.*, 2119 (1963).
[163] F. Micheel and M. Lorenz, *Ann. Chem.* **698**, 242 (1966).
[164] F. Micheel and M. Lorenz, *Deut. Auslegeschr.* 1,217,392 (1964); *Chem. Zentr.*, 17-1797 (1967).
[165] F. Micheel and M. Lorenz, *Deut. Auslegeschr.* 1,225,654, 1,225,655 (1965); *Chem. Zentr.*, 35-2421 (1967).
[166] E. Stephen and H. Stephen, *J. Chem. Soc.*, 490 (1957).
[167] H. Hettler, *Tetrahedron Lett.*, 4049 (1966).

derivative[121] would be of interest. Saccharin, like other heterocyclic imides and imino compounds, has been reacted with aromatic isocyanates[168] for identification.

Ketene reacts with **1** to form *N*-acetylsaccharin (**33**).[169] Interestingly, one obtains *N*-nitrosaccharin (**34**) from the exposure of **1** to dinitrogen pentoxide in nitromethane.[157] In contrast to the usual ring opening of type (**10**) compounds, **34** is cleaved with S–N fission on solvolysis with ethanol[157] to form (**35**). **1** reacts with an excess of oxalyl chloride

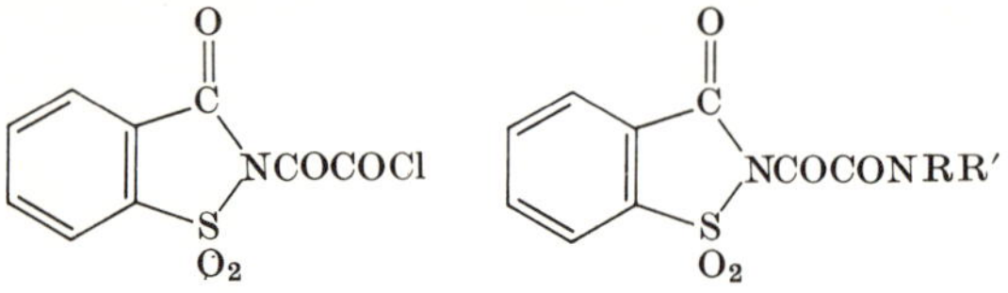

(36)   (37)

to form **36**,[17] which behaves as a bifunctional reagent. With less than molar quantities of amine or aminoacid ester, **37** is obtained, which on nucleophilic attack reacts with displacement of **19**.

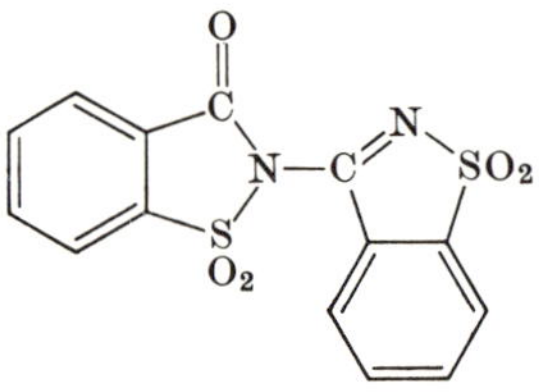

(32)

N-Acylsaccharins (**13**) possess a certain potential as acylating agents. They will acylate amines, but will react with water or alcohols only when acid or base is present.[167] The method was used to acylate α-aminopenicillanic acid.[170] Micheel[162–165] has based a peptide (**38**) synthesis on the acyl transfer from **31** [Z = carbobenzoxy, obtained through reaction with DCC or with pseudosaccharin chloride (**6**) or with thionyl chloride and imidazole] to amino acids. Pseudosaccharin anhydride **32**[3, 166] is the product of a condensation between **6** and **1**, mostly from hydrolysis of **6**. Formation of **32** tends to occur in nonprotic solvents with base catalysis, even when practical precautions are taken to exclude moisture. Water and protic solvents seem to shield the anion **19** and prevent attack on **6**.

[168] R. A. Henry and W. M. Dehn, *J. Amer. Chem. Soc.* **71**, 2297 (1949).
[169] R. E. Dunbar and W. S. Swenson, *J. Org. Chem.* **23**, 1793 (1958).
[170] K. Bauer, *Deut. Auslegeschr.* 1,745,619 (1969); *Chem. Abstr.* **72**, P 3483g (1970).

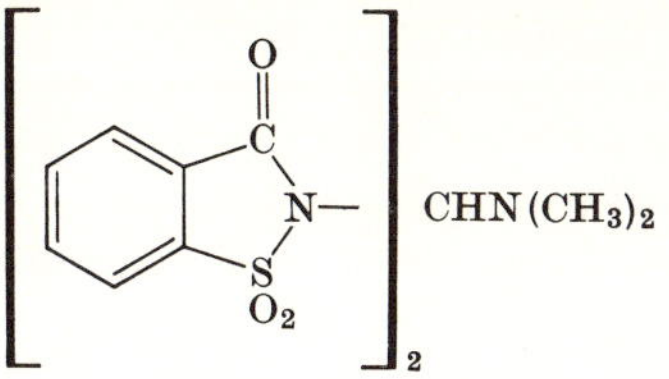

[structure of (39): saccharin N– attached, with CHN(CH₃)₂, subscript 2]

(Chloromethylene)dimethylammonium chloride[171] combines with two molecules of **19** to form the triaminomethane derivative (**39**).[172]

7. 3-Oxo-2-halo-2,3-dihydrobenz[d]isothiazole-1,1-dioxides

Free halogen reacts with the saccharin anion (**19**). The standard procedure for preparing 3-oxo-2-chloro-2,3-dihydrobenz[d]isothiazole-1,1-dioxide (N-chlorosaccharin) (**40**) is still passing of chlorine gas through a cold aqueous solution of **19a**.[173, 174]

N-Bromosaccharin (**41**), first obtained from the reaction of bromine with **19c** in carbon tetrachloride[175] has since been made from **1** in basic solution.[176] N-Halosaccharins are sources of positive halogen.[177]

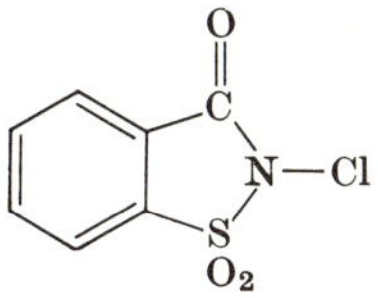

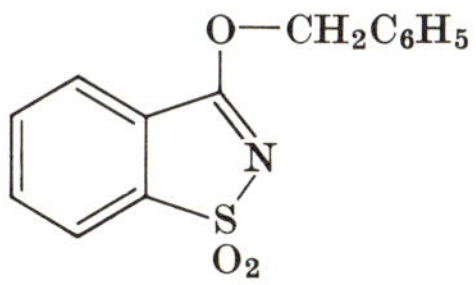

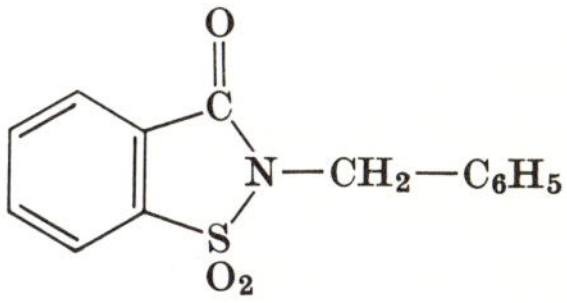

(**40**) N—Cl   (**41**) N—Br

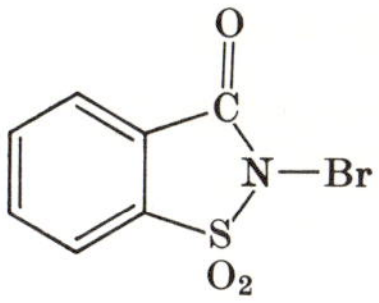

(**42**) O—CH₂C₆H₅   (**43**) N—CH₂—C₆H₅

---

[171] H. H. Bosshard, R. Mory, M. Schmid, and H. Zollinger, *Helv. Chim. Acta* **42**, 1653 (1959).
[172] H. Böhme and F. Soldan, *Chem. Ber.* **94**, 3109 (1961).
[173] F. D. Chattaway, *J. Chem. Soc.* **87**, 1882 (1905).
[174] H.-S. Dawn, I. H. Pitman, T. Higuchi, and S. Young, *J. Pharm. Sci.* **59**, 955 (1970).
[175] K. Ziegler, A. Späth, E. Schaaf, W. Schumann, and E. Winkelmann, *Ann. Chem.* **551**, 80 (1942).
[176] M. E. Fondovila and P. M. R. Schröder, *Rev. Fac. Cien. Quim. Univ. Nac. La Plata* **33**, 33 (1961); *Chem. Zentr.*, 14-0994 (1966).
[177] P. Fresenius, *Angew. Chem.* **64**, 470 (1952).

On account of its low chlorine potential[178] ($pK_{cp} = 4.85$ at $25°$)[11] **40** is a stronger chlorinating reagent in water than either chloramine-T or N-chlorosuccinimide.[174] Methanolysis of **40** yields **1** and methyl hypochlorite.[174] An earlier report[122] and references[179] on formation of N-methylsaccharin (**8**) in this reaction should be revised.

The kinetics of the addition of **40** to cyclohexene[175] in which N-(2-chlorocyclohexyl) saccharin is formed, besides 3-chlorocyclohexene were found to be first order in both reactants.[180]

8. *Free Radical Reactions*

Competition between O- and N-substitution is also found in homolytic reactions with saccharin (**1**). In the presence of *tert*-butyl peroxide and traces of copper, toluene, and p-xylene on heating yield O-benzylsaccharin (**42**) and O-p-methylbenzylsaccharin, respectively, besides N-benzylsaccharin (**43**) and N-p-methylbenzylsaccharin in a 4:1 ratio in each case.[181] N-Cyclohexenylsaccharin is obtained from cyclohexene under similar conditions.[181,182]

9. *Reactions with Metalorganic Compounds*

Treatment of saccharin (**1**) with an organolithium reagent (in tetrahydrofuran at $-78°$) gives only the 3-substituted benz[d]isothiazole-1,1-dioxide (**44**).[183] When **1** is treated with either an organolithium or a Grignard reagent at $0°$ to $80°$ **44** (R = alkyl, aryl) is obtained besides the o-sulfamylphenylcarbinol (**45a**) in varying yields depending on the temperature.[183] Earlier work on the Grignard reaction with (**1**)[184,185] should be reexamined.

45b is obtained from the reaction of a ketone RRCO with N-O-dilithiated phenylsulfonamide.[56] **45b** affords **46** on thermal cyclodehydration.[56]

45c is known to result from the reaction of N-methylsaccharin (**8**) and N-ethylsaccharin (**22**) with Grignard reagents.[186] 3-Phenylbenz[d]-

---

[178] T. Higuchi, A. A. Hussain, and I. H. Pitman, *J. Chem. Soc. B*, 626 (1969).
[179] S. S. Verma and R. C. Srivastava, *Indian J. Chem.* **4**, 445 (1966).
[180] H.-S. Dawn, I. H. Pitman, and T. Higuchi, *J. Pharm. Sci.* **59**, 429 (1970).
[181] A. Fono, *Chem. Ind. (London)* 414 (1958).
[182] M. S. Kharasch and A. Fono, *J. Org. Chem.* **23**, 325 (1958).
[183] R. A. Abramovitch, B. Purtshert, E. M. Smith, and P. C. Srinivasan, *Abstr. 23rd Southeastern Reg. Meeting, Amer. Chem. Soc. Abstr.*, 175 (1971).
[184] B. Oddo and Q. Mingoia, *Gazz. Chim. Ital.* **57**, 465 (1927); *Chem. Abstr.* **21**, 3202 (1927).
[185] A. Mustafa and M. K. Hilmy, *J. Chem. Soc.*, 1339 (1952).
[186] F. Sachs and A. Ludwig, *Chem. Ber.* **37**, 385 (1904).

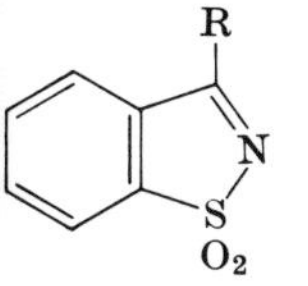

SACCHARIN AND ITS DERIVATIVES

(44)

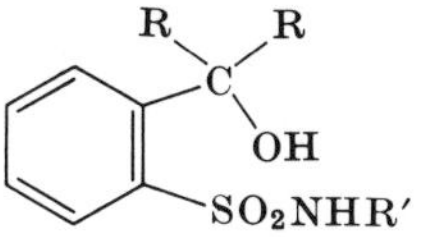

(45a) R = alkyl, aryl; R′ = H
(45b) R, R′ = alkyl, aryl
(45c) R = alkyl; R′ = CH_3, C_2H_5

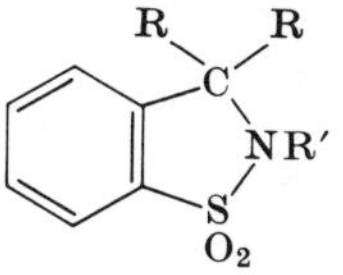

(46)

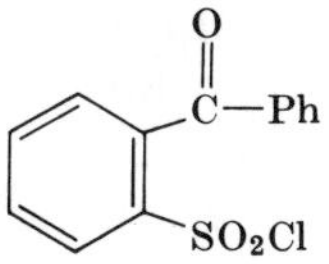

(47)

isothiazole-1,1-dioxide (**44**, R = C_6H_5) and derivatives are also obtained from the reaction of **47** with ammonia.[187]

E. Reactions with Cleavage of the Five-Membered Ring

Ring cleavage of the five-membered ring in (**1**) and N-substituted derivatives (**10**), (**18**) follows the same pattern (Scheme I).

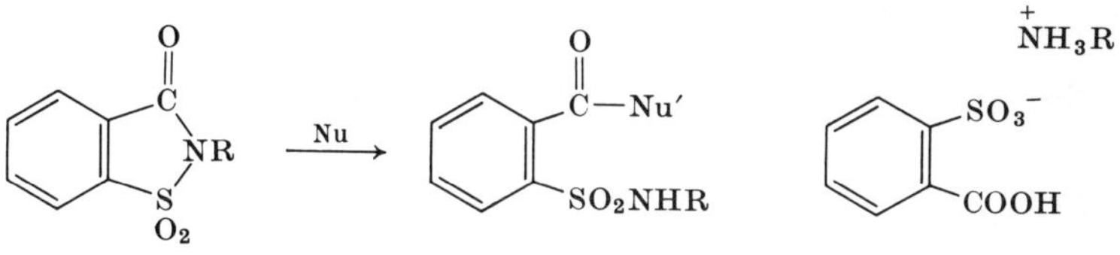

(1) R = H
(19a) R = Na
(10) R = alkyl
(18) R = phenyl

(48) R = H
(49) R = Na
(50) R = alkyl
(51) R = phenyl

(52a) R = H, alkyl
(52b) R = CH_3

Scheme 1

Attack of a nucleophilic reagent (Nu) ordinarily brings about C–N fission and produces sulfonamides Nu = OH, giving **49**,[188] **50**,[113] **51**;[134] Nu = NH_2R', R′ = alkyl,[161] aryl,[161,189] giving **48** and **50**, Nu′ = NHR′; Nu = N_2H_4, giving **48** and **50**, Nu′ = $NHNH_2$.[161] Prolonged heating of **48** (Nu′ = NH-aryl) favors ring closure with formation of the 3-arylaminobenz[d]isothiazole-1,1-dioxide.[189]

[187] J. B. Wright, *J. Heterocycl. Chem.* **5**, 453 (1968).
[188] C. Fahlberg and A. List, *Chem. Ber.* **21**, 242 (1888).
[189] A. Mannessier-Mameli, *Gazz. Chim. Ital.* **65**, 51 (1935); *Chem. Abstr.* **29**, 3997 (1935).

1 is hydrolyzed to **52a** under strongly acidic conditions and on heating in water.[7,190,191] Acid-catalyzed alcoholysis yields **48** (Nu' = OMe,[3,192] OEt,[192] O-alkyl[193]), and similar substituted sulfamoylbenzoic esters, e.g., the 6-halo,[64,66] 4 (or 6)-nitro,[194] and 4-amino derivatives[194] of **48** are formed.

Acid hydrolysis of N-methylsaccharin (**8**) in which **52b** is formed has been carried out under reflux with concentrated acid.[125] The instability of N-(dialkylaminoalkyl) derivatives of 6-nitrosaccharin in alcohol may be the result of alcoholysis.[106]

The product of the condensation of N-chloromethylsaccharin (**26**) with the sodium salt of 2-methylmalonic ester, on heating with alcoholic potassium hydroxide, is ring opened and monodecarboxylated.[140]

Synthetic aspects have promoted work on the degradation of **10**. K. Abe has utilized the following reaction sequence for the preparation of primary (**53**) and secondary amines (**54**),[23,120,124,195-197] in particular those in which the alkyl substituents R and R' are different. **10** is

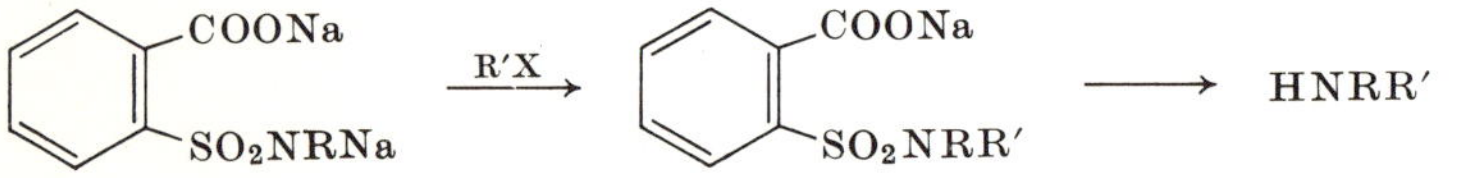

COONa
SO$_2$NRNa
(**50a**)

$\xrightarrow{R'X}$

COONa
SO$_2$NRR'
(**55**)

$\longrightarrow$ HNRR'

(**53**) R' = H
(**54**)

N(CH$_2$)n—X
(with fused benzisothiazole carbonyl/SO$_2$ ring)

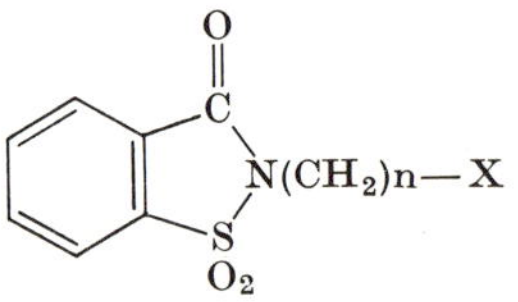

(**56**) X = Halogen
(**57**) X = NRR'

---

[190] H. D. Richmond and C. A. Hill, *J. Soc. Chem. Ind.* **37**, 246 (1918); *Chem. Abstr.* **12**, 2215 (1918).
[191] H. D. Richmond and C. A. Hill, *J. Soc. Chem. Ind.* **38**, 8T (1919); *Chem. Abstr.* **13**, 979 (1919).
[192] B. Loev and M. Kormendy, *J. Org. Chem.* **27**, 1703 (1962).
[193] G. H. Hamor, U.S. Patent 2,946,815 (1959); *Chem. Zentr.*, 11426 (1961).
[194] G. H. Hamor and M. Janfaza, *J. Pharm. Sci.* **52**, 102 (1963).
[195] K. Abe, *J. Pharm. Soc. Jap.* **75**, 168 (1956); *Chem. Zentr.*, 8572 (1962).
[196] K. Abe, Japanese Patent Appl. 9681 (1958); *Chem. Zentr.*, 6240 (1961)
[197] K. D. Gundermann, *Ann. Chem.* **588**, 167 (1954).

hydrolyzed with 10% alkali hydroxide to **50a** or alternatively under acidic conditions to primary amines (**53**)[197] directly. Alkylation of the dialkali salt **50a** with a reagent R'X (R'X = alkyl halide,[120] dimethylsulfate[196, 198] etc.) yields *N*-dialkylsulfonamides (**55**). Finally, the amine (**54**) is liberated by heating **55** with 20% HCl at 130°–140°.[23] Hydrolysis of the sulfonamide group in **55** is clearly aided by the *o*-carboxy function[195] and is retarded by increase of branching in R.[23, 195] *N*-Phenacylsaccharin (**10c**; R = –CH$_2$COC$_6$H$_5$)[107] and derivatives after reduction served as starting material for phenylalkanolamines,[196, 198] e.g., ephedrine and pseudoephedrine. Further amines of biological and pharmaceutical interest like kynurenamine,[199] *S*-benzylcysteine, and *S*-benzylisocysteine[197] have been prepared.

N-Halogenoalkyl saccharins (**56**)[103, 153, 200] are key intermediates for the preparation of symmetrically substituted ethylenediamines, ethanolamines, and homologs[120] through direct exchange with an amine (**57**) or with acetate, in the latter case followed by Abe's procedure. (**57**) is also obtained from alkylation of **1** with alkyl or dialkylaminoalkyl halides.[106, 119].

At higher temperatures, compounds **56** condense with another molecule of **1**,[103, 113, 120, 200, 201] or with potassium phthalimide[120, 202] to form **58a** or **58b**, respectively, and with *N*-hydroxyphthalimide.[202]

(**58a**) Z = SO$_2$
(**58b**) Z = CO

58a and **58b** serve as a source for symm. *N,N*-dialkylpolymethylene diamines and mono-*N*-alkylated polymethylene diamines, respectively, when Abe's procedure is applied.[120] Pettit has shown that desulfurization

---

[198] K. Abe, S. Yamamoto, and K. Matsui, *J. Pharm. Soc. Jap.* **76**, 1058 (1956); *Chem. Zentr.* 21518 (1963).
[199] Y. Jo, *J. Jap. Biochem. Soc.* **27**, 670 (1956); *Chem. Zentr.*, 13136 (1959).
[200] J. D. Commerford and H. B. Donahoe, *J. Org. Chem.* **21**, 583 (1956).
[201] G. H. Hamor and J. M. Balikian, *J. Amer. Pharm. Ass. Sci. Ed.* **49**, 283 (1960).
[202] G. H. Hamor and F. Rubessa, *Farmaco, Ed. Sci.* **25**, 36 (1970); *Chem. Abstr.* **72**, 386 (1970).

with Raney nickel[204] is an alternative method for degrading N-substituted saccharins (**10**).[108] Depending on the quality of the catalyst and the temperature one obtains N-substituted cyclohexylcarboxamides, or under very mild conditions N-substituted benzamides.[108] Reduction of *N*-alkylsaccharins with lithium aluminum hydride affords *o*-hydroxymethyl-*N*-alkylbenzenesulfonamides like **27**,[83, 112] whereas reduction of **1** yields 1,2-benzisothiazoline-1,1-dioxide.[203]

D'Alelio *et al.* have been working on thermostable polymers[58, 205-208] starting from 5- or preferably 6-aminosaccharin[208] and from "metabisaccharin" (**59a**)[205] with bifunctional aromatic amines through soluble hemipolymer stages in melts and in solution.

(**59a**) Y = CO, Z = SO₂
(**59b**) Y = SO₂, Z = CO

The ring closed polymers seem to contain "pseudosaccharin" units and sulfur-free, possibly indazole units besides imide groups.[208] **59a** should be regarded as the only well-established bisaccharin,[205] since earlier claims[209, 210] to **59b** and the "ortho-bisaccharin" isomer[211, 212] could not be confirmed.[205, 212]

[203] S. J. Childress and T. Baum, U.S. Patent 3,164,602 (1962); *Chem. Abstr.* **62**, P 9140h (1965).
[204] G. R. Pettit and E. E. van Tamelen, *in* "Organic Reactions," (A. C. Cope, ed.), Vol. 12, pp. 401, 412. Wiley, New York, 1962.
[205] G. F. D'Alelio, D. M. Feigl, W. A. Fessler, Y. Giza, and A. Chang, *J. Macromol. Sci., Chem.* **3**, 927 (1969).
[206] G. F. D'Alelio, Y. Giza, and D. M. Feigl, *J. Macromol. Sci., Chem.* **3**, 1105 (1969).
[207] G. F. D'Alelio, W. A. Fessler, Y. Giza, D. M. Feigl, A. Chang, and M. Saha, *J. Macromol. Sci., Chem.* **4**, 159 (1970).
[208] G. F. D'Alelio, W. A. Fessler, Y. Giza, D. M. Feigl, A. Chang, and S. Saha, *J. Macromol. Sci., Chem.* **5**, 383 (1971).
[209] A. F. Holleman, *Rec. Trav. Chim.* **42**, 839 (1923).
[210] A. F. Hollemann and H. J. Choufoer, *Versl. Koninklijee Akad. Wetensch.*, Amsterdam **33**, 307 (1924); *Proc. Acad. Sci. Amsterdam* **27**, 353 (1925); *Chem. Zentr.* II, 632 (1924).
[211] R. Wischin, *Chem. Ber.* **23**, 3113 (1890).
[212] A. F. Hollemann and H. J. Coufoer, *Rec. Trav. Chim.* **48**, 1075 (1929).

1. Ring Expansion

N-Substituted saccharins with a β-carbonyl group undergo base-catalyzed ring expansion to give benzothiazines. Compounds forming more stable carbanions, e.g., **60a**, yield 1,3-benzothiazine (**61**).[213,214] The β-keto compounds (**60b**)[215,216] and (**60c**)[196,216] by contrast afford 1,2-benzothiazines **63**. Very likely the reaction proceeds via alcoholytic ring opening and subsequent Dieckmann ring closure. Thus, **60b** with one equivalent of sodium ethoxide gives **62**, which could be condensed to **63a**

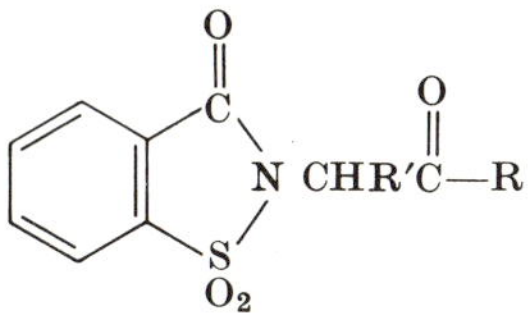

(**60a**) R′ = C_6H_5, R = OC_2H_5
(**60b**) R′ = H, R = CH_3
(**60c**) R′ = H,CH_3, R = C_6H_5
(**60d**) R′ = H, R = $(CH_2)_3Cl$

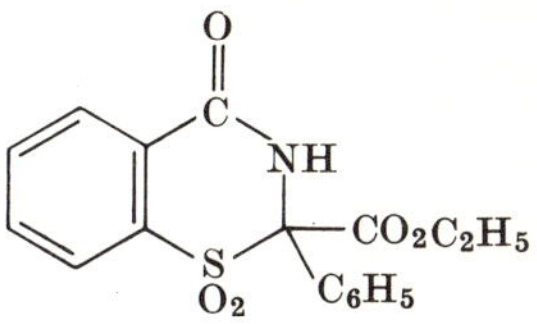

(**61**)

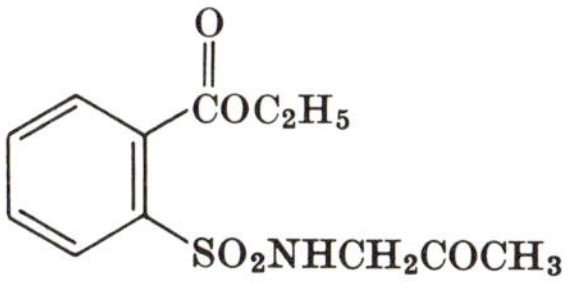

(**62**)

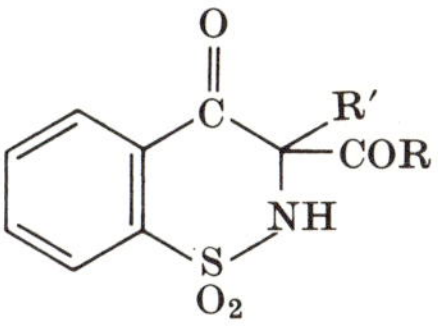

(**63**)

(R′ = H, R = CH_3) with two more equivalents of base.[215] When R in compounds of type (**60**) contains a halogen in the ω-position (**60d**),[217] a tricyclic system including an enol ether (e.g., 2,3-dihydro-6H-oxepino-[c][1,2]-benzothiazine-5(4H)-one-7,7-dioxide)[218] is formed.

---

[213] H. Zinnes, R. A. Comes, and J. Shavel, *J. Org. Chem.* **29**, 2068 (1964).
[214] C. Rasmussen, U.S. Patent 3,501,466 (1970); *Chem. Abstr.* **72**, 121,562e (1970).
[215] H. Zinnes, R. A. Comes, F. R. Zuleski, A. N. Caro, and J. Shavel, *J. Org. Chem.* **30**, 2241 (1965).
[216] A. Kraaijeveld and A. M. Akkerman, S. African Patent 646,150 (1964); *Chem. Abstr.* **64**, 12,689d (1966).
[217] H. Hart and G. Levitt, *J. Org. Chem.* **24**, 1261 (1959).
[218] H. Zinnes, R. A. Comes, and J. Shavel, *J. Med. Chem.* **10**, 223 (1967).

2. Reactions of N-Acyl Saccharins

N-Acyl saccharins (13) in part resemble N-alkyl saccharins in their reactions. From the hydrolysis of 3-oxo-2-acetyl-6-chloro-2,3-dihydrobenz[d]isothiazole-1,1-dioxide (64) with a base/alcohol mixture one obtains N-acetyl-4-chloro-2-sulfamoylbenzoic acid (65). On heating with base the acetyl group is hydrolyzed off.[160]

Alkaline cleavage of N-benzenesulfonylsaccharin is reported to occur at the ring S–N bond.[50]

Treatment of 64 with anhydrous hydrazine produces 3,6-dihydrazinobenz[d]isothiazole-1,1-dioxide, with alcoholic hydrazine, however, some splitting of the acetyl group together with some ring cleavage occurs with formation of the hydrazine salt of 6-chlorosaccharin and 4-chloro-2-sulfamoylbenzohydrazide.[160]

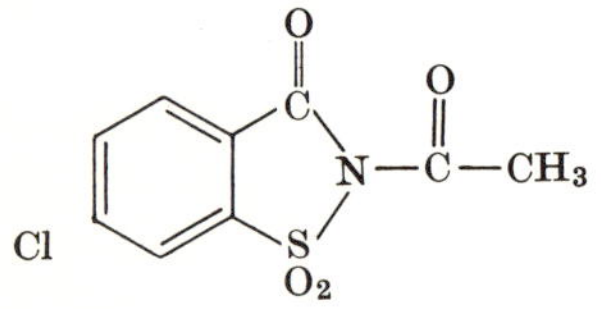

Heating 64 with concentrated sulfuric acid affords 65 besides 6-chlorosaccharin.[160] The acylating properties of 13 have been discussed in Section IV, D, 6.

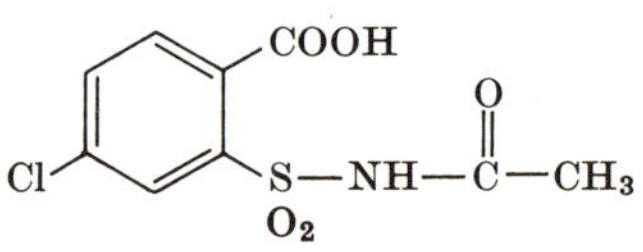

V. The 3-Thioxo-2,3-dihydrobenz[d]isothiazole-1,1-dioxide System

3-Thioxo-2,3-dihydrobenz[d]isothiazole-1,1-dioxide (thiosaccharin) (2) is obtained from the reaction of phosphorus pentasulfide on saccharin (1)[125, 160, 219, 220] or on 2-sulfamoylbenzoic acid (12) and its substituted derivatives.[160] Above about 140° the components react exothermically, and for the preparation of derivatives of (2) substituted in the phenyl ring a diluent like sand or tetralin is employed.[160] Formation of 2 is usually accompanied by small amounts of 1,2-benzodithiole-3-thione[219-221] when the temperature exceeds 160°. 2 is supposed to be an

---

[219] A. Mannessier, *Gazz. Chim. Ital.* **45**, I, 540 (1915); *Chem. Abstr.* **10**, 1174 (1916).
[220] A. Mannessier, *Gazz. Chim. Ital.* **46**, I, 231 (1916); *Chem. Abstr.* **10**, 2892 (1916).
[221] L. Legrand, Y. Mollier, and N. Lozac'h, *Bull. Soc. Chim. Fr.* 327 (1953).

intermediate in this side reaction.[226] (Likewise loss of SO_2 is encountered in the preparation of o-chlorobenzonitrile[222] by heating **1** with phosphorus pentachloride above 230°.[3] Heating **1** with highly concentrated sulfuric acid, however, yields benzoic-o-sulfonic acid anhydride.[223]) Indicative for the C=S group as in **2** is an absorption at about 5000 Å which imparts a yellow color.[224] Due to this fairly reactive group, **2** is relatively unstable. The reactions of the thiocarbonyl group partly differ from, and partly resemble those of a carbonyl group.

(i) Nucleophilic attack on the C=S group does not require catalysis.[224] With water and dilute alkali carbonate **2** is hydrolyzed to saccharin (**1**). Treatment with acid or with base yield **48** (Nu' = OH) and **52a**, respectively.[219]

(ii) Replacement by a nucleophile H_2NX, e.g., an amine (X = H, aryl),[189, 225, 226] hydroxylamine (X = OH),[227] hydrazine (X = NH_2),[228] NHC_6H_5,[229] semicarbazide (X = $NHCONH_2$)[230] makes compounds of type (**66**) and/or their corresponding tautomers available.

(iii) The thiocarbonyl group in **2** displays the expected nucleophilic activity.[224] With alkylating agents, e.g., methyl bromide or iodide[160] or dimethyl or diethyl sulfate[125] only 3-alkylthiobenz[d]isothiazole-1,1-dioxides (**4**) were isolated.

(**66**) (**67**) (**68**)

[222] I. Remsen and A. R. L. Dohme, *Amer. Chem. J.* **11**, 321 (1889).
[223] L. H. Perrier, French Patent 1,187,090 (1959); *Chem. Zentr.* 9598 (1961).
[224] K. F. Reid, "Properties and Reactions of Bonds in Organic Molecules," p. 422. Longmans, Green, New York, 1968.
[225] A. Mannessier-Mameli, *Gazz. Chim. Ital.* **71**, 3 (1941); *Chem. Abstr.* **36**, 1028 (1942).
[226] A. Mannessier-Mameli, *Gazz. Chim. Ital.* **65**, 69 (1935); *Chem. Abstr.* **29**, 3997 (1935).
[227] A. Mannessier-Mameli, *Gazz. Chim. Ital.* **62**, 1067 (1932); *Chem. Abstr.* **27**, 2682 (1933).
[228] A. Mannessier-Mameli, *Gazz. Chim. Ital.* **71**, 18 (1941); *Chem. Abstr.* **36**, 1029 (1942).
[229] A. Mannessier-Mameli, *Gazz. Chim. Ital.* **71**, 596 (1941); *Chem. Abstr.* **37**, 100 (1943).
[230] A. Mannessier-Mameli, *Gazz. Chim. Ital.* **71**, 25 (1941); *Chem. Abstr.* **36**, 1030 (1942).

Originally, the product from the reaction with dimethyl sulfate had been assigned the structure of N-methylthiosaccharin (**67**).[219] Meadow and Cavagnol prepared **67** from N-methylsaccharin (**8**) and phosphorus pentasulfide in analogy with **2**.[125]

Acid hydrolysis of **67** back to **8** seems to require rather drastic conditions[125] in keeping with the soft character[127] of **67** as a base. Eventually **8** is hydrolyzed to the methylammonium salt of o-sulfobenzoic acid (**52b**).[125] 3-Alkylthiosaccharins (**4**) are hydrolyzed with acid to **52b** and the thiol somewhat more readily.[125]

The most convenient method for preparing **4** is, however, the reaction of 3-chlorobenz[d]isothiazole-1,1-dioxide (**6**) with thiols or alkali thiolates.[231] All attempts to rearrange **4** to N-alkylthiosaccharins have failed.[17, 96] Hydrazinolysis of **4** as well as **3** has been used for preparing 3-hydrazinobenz[d]isothiazole-1,1-dioxide (**68**).[160]

VI. 3-Alkoxy and 3-Aryloxybenz[*d*]isothiazole-1,1-dioxides

A. Preparation and Properties

Generally, 3-alkoxy- and 3-aryloxybenz[d]isothiazole-1,1-dioxides, "pseudosaccharin ethers," (**3**) are obtained by reacting 3-chlorobenz-[d]isothiazole-1,1-dioxide (**6**) with the corresponding aliphatic or aromatic alcohols.[3, 20, 24, 25, 120, 232, 233]

Such derivatives are usually crystalline and have been recommended for the characterization of alcohols and phenols[24, 25, 234] and for protecting alcoholic functions.[96, 167] Compounds of type **3** exhibit a characteristic IR pattern, which aids identification.[20, 23]

In earlier procedures **6** was refluxed with the respective alcohol.[3, 24, 235] For reacting complex systems like steroid alcohols[17, 20] or nucleosides,[17] pyridine together with an inert solvent was employed.[25] **32** is a common by-product under such conditions,[25] but is separable from **3** on account of its poor solubility.

To a limited extent 3-alkoxybenz[d]isothiazole-1,1-dioxides (**3a**) are available from direct alkylation of **1** with isopropyl and s-butyl halides[120] and in the presence of silver ion,[121] presumably in an S_N1 reaction (cf. Section IV, D, 2). 3-Methoxybenz[d]isothiazole-1,1-dioxide was

---

[231] J. R. Meadow and J. C. Cavagnol, *J. Org. Chem.* **17**, 488 (1952).
[232] C. H. Grogan, E. E. Reid, and L. M. Rice, *J. Org. Chem.* **20**, 1425 (1955).
[233] C. H. Grogan and L. M. Rice, U.S. Patent 2,751,392 (1953); *Chem. Zentr.*, 10822 (1957).
[234] C. D. Hodgman, "Tables for Identification of Organic Compounds," Chemical Rubber Publ., Cleveland, Ohio, 1960.
[235] C. Maselli, *Gazz. Chim. Ital.* **30**, 529 (1901); *Chem. Zentr.* **I**, 517 (1901).

produced on a preparative scale in a somewhat unexpected reaction by passing chlorine gas through a solution of 3-methylmercaptobenz[d]-isothiazole-1,1-dioxide in methanol.[160]

The parent alcohol is conveniently liberated from **3** on treatment with gaseous ammonia[3, 96] or a primary or secondary amine.[13] Since the resulting 3-aminobenz[d]isothiazole-1,1-dioxides (**5**) are sparingly soluble in most solvents, separation from the alcohol offers no problem.

Optically active alcohols like cholest-5-en-3β-ol or 2-(d)-octanol when recovered in this way show no loss of optical activity.[17] There is such a strong tendency toward formation of the 3-aminobenz[d]isothiazole-1,1-dioxide system (**5**), that **3** reacts with aminolysis in a not too dilute aqueous solution of the amine.[13] **3** reacts just as readily in hydrazinolysis with formation of 3-hydrazinobenz[d]isothiazole-1,1-dioxide (**68**).[160]

TABLE II

HYDROLYSIS OF 3-ALKOXYBENZ[d]ISOTHIAZOLE-1,1-DIOXIDES

Compound	Alkyl residue	Experimental conditions	Time	Result	Reference
69	Ethyl	Water at 100°		No hydrolysis	3
69	Ethyl	Water at 150° (sealed tube)		Hydrolysis to **1** and further to **48a** (Nu′ = OH)	3
70	Methyl	0.1 N NaOH/dioxan (1:1) at room temperature	18 hr	About 12% hydrolysis to **1**	17
71	5′-Thymidyl	1 N NaOH/dioxan (1:10) at room temperature	10 min	Quantitative hydrolysis to **1** and thymidine	236
69	Ethyl	1 N HCl/dioxan at 78°	2 hr	No appreciable hydrolysis	17
72	n-Hexadecyl	30% HCl at reflux	4 hr	Hydrolysis to **52a**	125

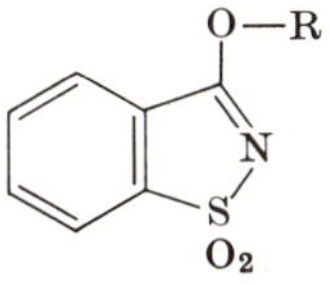

(**69**) R = C_2H_5
(**70**) R = CH_3
(**71**) R = 5′-Thymidyl
(**72**) R = n-Hexadecyl

[236] B. Jastorff, Göttingen, unpublished results (1968).

Results of hydrolysis (cf. Table II) indicate that compounds (**3**) are relatively stable in acid, but will hydrolyze more readily with alkali.

B. Rearrangement

3-Alkoxybenz[*d*]isothiazole-1,1-dioxides (**3a**) rearrange to *N*-alkylsaccharins (**10a**) in an apparently uncatalyzed thermal rearrangement[20, 96] as well as by lithium iodide catalysis.[96] The known rearrangement of alkyl imidates to *N*-alkylamides[237] (Lander rearrangement[159, 238] frequently treated as a special case of a Chapman–Mumm rearrangement[239]) by and large requires catalysis, in particular by alkyl halides.

The "critical" temperature for the thermal rearrangement of **3a** under various conditions (in the melt,[96] in quinoline[20] or naphthalene[17]) appears to be in the vicinity of 180°. The 3-methoxy derivative (**70**) isomerizes readily with lithium iodide (in methyl ethyl ketone at 64°).[17, 96] When ethyl iodide is added to this reaction, *N*-methylsaccharin (**8**) and *N*-ethylsaccharin (**22**) are formed together.[96] Lithium iodide catalysis is particularly useful in rearranging nucleoside derivatives, e.g., **71**.[17] The 3-benzyloxy derivative (**73**) reacts with debenzylation.[17] Even more complex systems like derivatives of steroid alcohols[17, 20] are isomerized on heating in reasonable yields. When the optically active 3-(2-(*d*)octyloxy)benz[*d*]isothiazole-1,1-dioxide was thermally re-

(**73**) (**74**)

arranged, the resulting *N*-(2-octyl) saccharin turned out to be optically inactive.[17]

Reaction with **6** followed by rearrangement to **10a** and subsequent cleavage using Abe's[195, 196, 198] or Pettit's[108] procedure (cf. Section IV, E) may be used for converting alcohols into primary or secondary amines.

When tetraalkylammonium picrate was allowed to react with **6**, *N*-picrylsaccharin[103] was isolated, among other products.[96] The tendency toward a Mumm rearrangement is enhanced as the acid character of substituents R in **3** increases (cf. Section IV, D, 6).

[237] C. G. McCarty, in "The Chemistry of the Carbon-Nitrogen Bond" (S. Patai, ed.), p. 443. Wiley (Interscience), New York, 1970.
[238] G. D. Lander, *J. Chem. Soc.* **83**, 406 (1903).
[239] R. Roger and D. G. Neilson, *Chem. Rev.* **61**, 179 (1961).

All attempts to isomerize 3-alkyl thiobenz[d]isothiazole-1,1-dioxides (4) in a similar fashion have failed.[17, 96] This is in agreement with Chapman's findings that sulfur analogs of aryl imidates show little tendency to rearrange.[240]

C. Elimination

The thermal energy which brings about alkylimidate → N-alkyl amide rearrangement is also sufficient to cause pyrrolytic cis-elimination.[241] Unlike ordinary alkyl imidates[242] 3-alkoxy compounds (3a) are liable to isomerize even if requirements for elimination are met.

Elimination reactions with saccharin (1) as a leaving group fit into a more general fragmentation scheme (cf. Section XII). Ideally the proton is abstracted from a benzylic carbon atom in β-position to the exocyclic oxygen. Thus styrene is formed in preparative pyrolysis of 3-(β-phenylethoxy)benz[d]isothiazole-1,1-dioxide (74)[28] as well as in mass spectrometric fragmentation.[20] Derivatives of terpene alcohols[17, 96] and those with bulky substituents, e.g., 3-neopentyloxybenz[d]isothiazole-1,1-dioxide tend to decompose on heating.[17]

In another interesting type of elimination, whereby the saccharin anion (19) acts as a leaving group, styrene is obtained from alkali fusion of N-(2'-phenylethyl)saccharin.[243]

VII. 3-Aminobenz[d]isothiazole-1,1-dioxide and Its Derivatives

A. Preparation and Properties

3-Aminobenz[d]isothiazole-1,1-dioxide (5) and its N-substituted derivatives are remarkably stable, due to resonance (cf. Section III, A). Hence it is not surprising that 5a is found in a variety of pyrolytic reactions, e.g., starting from: (i) the ammonium salts of saccharin and thiosaccharin,[244, 245] or from 1 or 2 in the presence of NH_3 donors,[244]

---

[240] A. W. Chapman, *J. Chem. Soc.*, 2296 (1926).
[241] W. Schulenburg and S. Archer, in "Organic Reactions," (A. C. Cope, ed.), Vol. 14, pp. 26, 27. Wiley, New York, 1965.
[242] N. P. Marullo, C. D. Smith, and J. F. Terapane, *Tetrahedron Lett.*, 6279 (1966).
[243] R. J. Baumgarten and P. L. De Christopher, *Tetrahedron Lett.*, 3027 (1967).
[244] E. Mameli and A. Mannessier-Mameli, *Gazz. Chim. Ital.* 70, 855 (1940); *Chem. Abstr.* 36, 1026 (1942).
[245] I. M. Kogan and V. M. Dziomko, *J. Gen. Chem. USSR* 23, 1234 (1953); *Chem. Abstr.* 47, 12280d (1953).

(ii) from saccharin oxime[244, 246] (besides **1** and nitric acid), (iii) from 3-phenylhydrazinobenz[d]isothiazole-1,1-dioxide.[229] The thioketo group in **2** reacts in part with aminolysis on boiling with aqueous ammonia.[225] Moreover 3-arylamino compounds (**9**; R = aryl) are produced when one heats 2-N-arylcarbamyl-phenylsulfonamides (**75**).[189, 208] Similarly, treating related compounds like o-cyanophenylsulfonyl chloride[3, 247] with ammonia or melting of o-cyanophenylsulfonamide leads to **5a**.[247]

Practical syntheses of **5a** or N-substituted 3-aminobenz[d]isothiazole-1,1-dioxides are mostly based on the following procedures: (iv) Refluxing saccharin with an excess of a monoalkyl or aryl amine above 130°, presumably via ring opening, by analogy with **75**.[161, 189, 248]

<chemical structure>

(**75**) R = aryl

(v) Reacting thiosaccharin with an aliphatic or aromatic amine[189, 225] (cf. Section V).

(vi) Aminolysis of 3-alkoxy and 3-aryloxy derivatives (**3**)[3, 13] (cf. Section VI, A).

(vii) Reaction of 3-chlorobenz[d]isothiazole-1,1-dioxide (**6**) with ammonia,[3] and primary or secondary amines.[3, 13, 161, 166, 232, 248, 249]

The reaction has been suggested for analytical characterization of primary and secondary amines.[13] It may be carried out in the presence of water since **6** reacts much faster with an amine than with water.[13] Moreover, preparation of derivatives of pyrazole,[250] hydrazine[13] (see [251]), arylsulfonamides[252] and ureas[253] illustrates the tendency of **6** to react with N–H groups.

A couple of more recent examples are of theoretical rather than preparative interest, as the yields are generally low.

[246] A. Mannessier-Mameli, *Gazz. Chim. Ital.* **65**, 77 (1935); *Chem. Abstr.* **29**, 3999 (1935).
[247] H. Bradshaw, *Amer. Chem. J.* **35**, 335 (1906).
[248] R. P. Singh, *J. Indian Chem. Soc.* **36**, 479 (1959); *Chem. Abstr.* **54**, 9890h (1960).
[249] W. M. McLamore and G. D. Laubach, *C. R. Congr. Int. Chim. Ind., 31st, Liege* (1958); *Chem. Abstr.* **54**, 5960 (1960).
[250] J. T. Traverso, C. W. Whitehead, J. F. Bell, H. E. Boaz, and P. W. Willard, *J. Med. Chem.* **10**, 840 (1967).
[251] E. Schrader, *J. Prakt. Chem.* [2] **95**, 312 (1917).
[252] A. Klages, E. Sturm, and J. Weniger, *J. Prakt. Chem.* [2] **116**, 163 (1927).
[253] H. Hettler, *Tetrahedron Lett.*, 1791 (1968).

(viii) Reaction of **19a** with dialkylcarbamoyl chloride, which yields 3-dialkylaminobenz[*d*]-isothiazole-1,1-dioxides (**5b**) besides 2-(dialkylcarbamoyl)saccharin (**76**).[155]

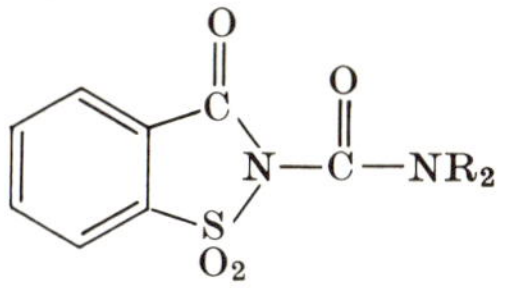

$$\text{(76)}$$

(ix) Oxidation of 3-alkylaminobenz[*d*]isothiazoles.[18]

Structural aspects have been discussed in Section III,A. Derivatives with an NH function, e.g., **5a** and **9**, are weak acids,[244] the latter with pK_a values of about 12.5 (in 66% DMF).[161] Concentrated alkali is needed to dissolve **5**.[244] The silver salt of **5a** precipitates readily.[244] For preparing the anhydrous potassium salt of **9** one simply evaporates to dryness with potassium *tert*-butoxide in *tert*-butanol.[17, 254] Compounds (**9**) form addition complexes with mercury acetate.[247]

B. Reactions

N-Substituted derivatives (**5b**) and (**9**) are moderately stable toward alkaline hydrolysis, and are remarkably stable toward acid.[13]

Alkaline hydrolysis of **5a** appears to pass through **1**,[244] leading to the same final product (*o*-carboxybenzenesulfonamide). The monomethyl and ethyl derivatives (**9**; R = Me, Et) hydrolyze slowly to **1** in dilute alkali,[13] 3-(*o*-carboxyanilino)benz[*d*]isothiazole-1,1-dioxide (**77**) and functional derivatives form a tetracyclic system (**78**) on treatment with acetic anhydride.[166] The amide of **77** is converted by dilute alkali into **79**.[166]

Reduction of **9** with sodium borohydride/aluminum chloride has been reported.[13]

The potassium salt of **9** reacts with alkyl iodide in dimethylformamide.[17] Alkylated products proved to be identical with those obtained from **6** and the corresponding secondary amine.[17]

High yields of primary and secondary amines have been reported in the alkali fusion of **9** and **5b**, respectively.[17, 28] Alkylation of any 3-monoalkyl- or 3-aryl-aminobenz[*d*]isothiazole-1,1-dioxide (**9**) followed

---

[254] H. Neygenfind, Thesis, Braunschweig Technical University, 1969.

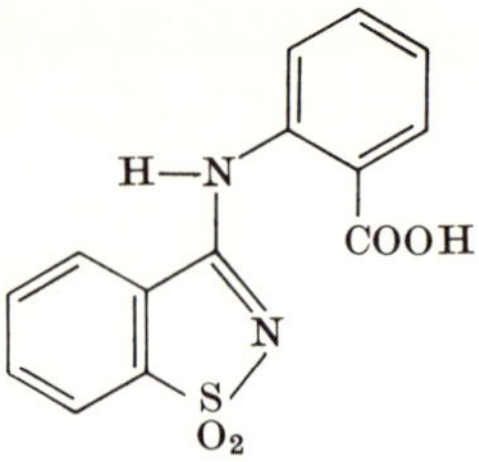

(77)

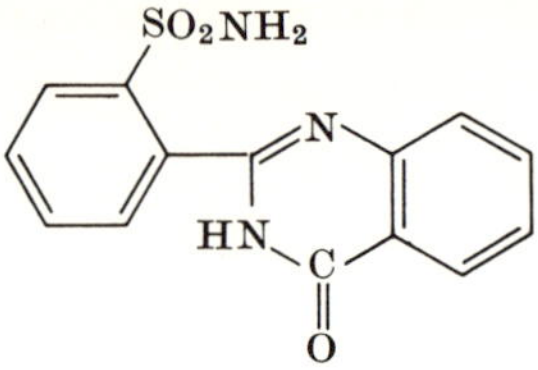

(79)

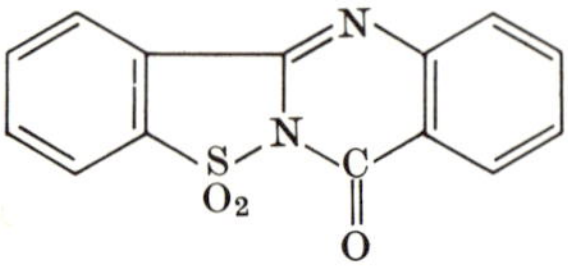

(78)

by alkali fusion is a way of preparing unsymmetrical secondary amines. 5a has been acylated in the 3-position using acetic anhydride.[246]

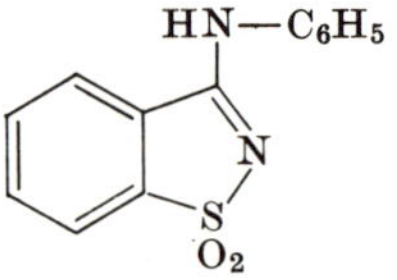

(80)

The potassium salt was employed in acylation of 3-anilinobenz[d]-isothiazole-1,1-dioxide (80) with benzoyl chloride.[225]

1. *3-Hydrazinobenz[d]isothiazole-1,1-dioxide*

3-Hydrazinobenz[d]isothiazole-1,1-dioxide (68) is prepared in analogy to 3-amino compounds according to procedures (iv)–(viii) by reacting 1,[228, 251] 2,[228] 3,[160, 251] and 6[13] with hydrazine. In addition, hydrazinolysis of 3-alkylmercapto compounds (4)[160] and ring closure of o-carbethoxyphenylsulfonamide with hydrazine[251] yield 68. A pK_a value of 7.8 has been determined for 68.[160] A large number of hydrazone,[160, 251] carbamide, and sulfamide derivatives of 68 and derivatives substituted in the phenyl ring have been prepared[160] and evaluated for pharmaceutical activity.[255]

Condensation of 68 with aldehydes and ketones is conveniently carried out in methanol with trifluoroacetic acid as a catalyst[17] or in refluxing

[255] P. W. Willard, C. W. Whitehead, and J. J. Traverso, *J. Med. Chem.* **10**, 849 (1967).

cellosolve.²⁵⁰ When 1,3-diketones, keto esters, ketoacetaldehyde acetals, diacetals, and unsaturated ketones are reacted with **68** in cellosolve, 3-pyrazolyl- and 3-pyrazolinylbenz[*d*]isothiazole-1,1-dioxides (**81**) are formed.²⁵⁰

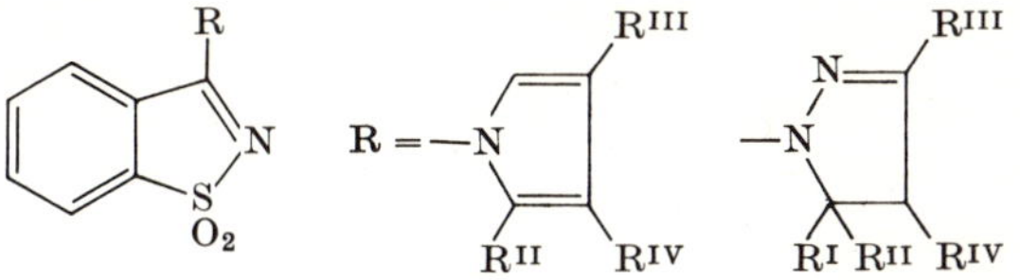

(**81**)

R^I = OH, methyl R^{III} = H, alkyl
R^{II} = H, alkyl, phenyl R^{IV} = H, alkyl

Traverso *et al.*, in their analysis of NMR spectra, showed that pyrazole and pyrazoline rings in **81** are coplanar with the benzisothiazole moiety,²⁵⁰ probably due to resonance structures of type **7a**.

VIII. 3-Imino-2,3-dihydrobenz[*d*]isothiazole-1,1-dioxide and Its Derivatives

True 3-imino-2,3-dihydrobenz[*d*]isothiazole-1,1-dioxides (**82**) are capable of existence provided they bear an alkyl or aryl substituent in 2-position as in saccharin anils (**23**). The free imino compound (**82a**; R = Me, R' = H) is obtained from isomerization of *o*-cyano-*N*-methyl-phenylsulfonamide in the presence of methylamine,²⁵⁶ or more generally preparation of **82** may start from *o*-cyanophenylsulfonyl chloride together with an excess of amine.³,²⁵⁶ A straightforward approach uses cyclization of an *o*-sulfamylbenzamide monosubstituted at each amide function, like *o*-(*N*-phenylsulfamyl)benzanilide (**83**)²⁵⁷ with phosphorus pentoxide, phosphorus pentachloride, or preferably phosphorus oxychloride.¹³⁵,²⁵⁷ Similarly, 2-chlorocarbonylbenzene sulfonyl chloride

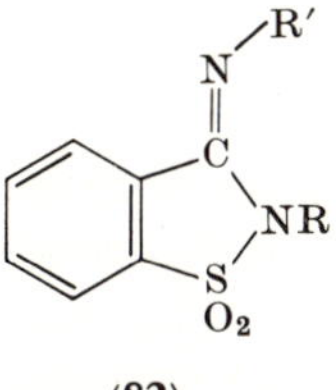

(**82**)   (**83**) 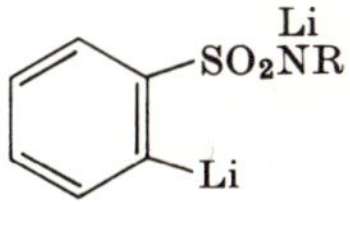   (**85**; R = Me, Ph) 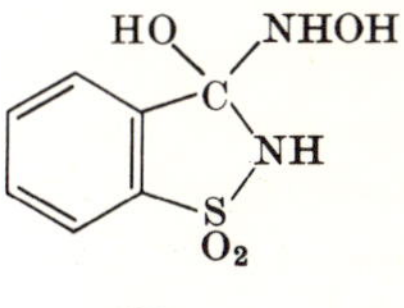   (**87**)
(**84**; R = R' = Ph)

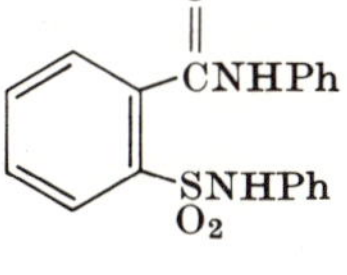

---

[256] G. Cignarella and U. Teotino, *J. Amer. Chem. Soc.* **82**, 1594 (1960).
[257] I. Remsen and J. H. Hunter, *Amer. Chem. J.* **18**, 809 (1896).

when reacted with an excess of aniline yielded 2-phenyl-3-phenylimino-2,3-dihydrobenz[d]isothiazole-1,1-dioxide (84) besides N-phenylsaccharin (18).[99, 100] In the course of his studies on the Beckmann rearrangement, Hauser found that o-(N-methylsulfamyl)benzophenone oxime with phosphorus pentachloride affords N-phenylsaccharin anil (84).[135]

Acid hydrolysis of 82 gives 2-substituted saccharins (10),[135, 257] heating with alcoholic sodium hydroxide or glacial acetic acid yields the ring open diamide.

As 83 has become more readily available from the reaction of the versatile intermediate (85) with phenyl isocyanate, saccharin anils (23) may serve as intermediates for the preparation of 2-substituted saccharins (10).[135]

IX. Miscellaneous Compounds

1. *3-Azidobenz[d]isothiazole-1,1-dioxide*

3-Azidobenz[d]isothiazole-1,1-dioxide (86) is obtained from exchange

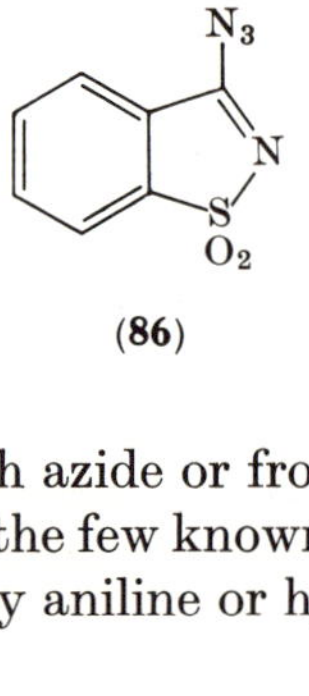

(86) 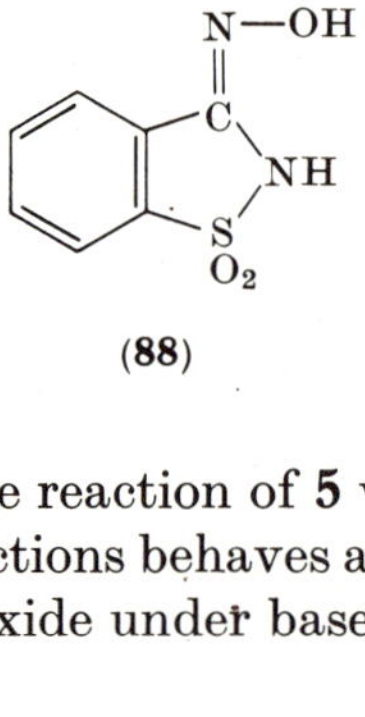   (88)

of chloride in 6 with azide or from the reaction of 5 with nitric acid.[251] The azide group in the few known reactions behaves as a pseudohalogen, i.e., it is replaced by aniline or hydroxide under base catalysis.[251]

2. *3-Hydroxy-3-hydroxylamino-2,3-dihydrobenz[d]isothiazole-1,1-dioxide*

Saccharin (1) combines with hydroxylamine to give derivative 3-hydroxy-3-hydroxylamino (87) and not 3-oximino-2,3-dihydrobenz[d]isothiazole (88) which was obtained from 2 under similar conditions.[227, 258] The addition is reversed and 87 converted to 1 by heating or treatment with concentrated sulfuric acid.[227]

[258] A. Gaudiano, F. Toffoli, and L. Benedetto, *Rend. Ist. Super. Sanita* **20**, 70 (1957); *Chem. Abstr.* **54**, 2311 (1960).

X. (Benz[d]isothiazolyl-3-oximino)alkane-1,1-dioxides

Oximes react very smoothly with the chloro compound **6** to form crystalline (benz [d] isothiazolyl - 3 - oximino) alkane - 1, 1 - dioxides ("pseudosaccharin oxime ethers") (**89**; R = alkyl, aryl, R' = H, alkyl, aryl).[20, 28, 43, 254] Since **6** reacts preferably with the oxime, the reaction is carried out in the presence of water.[43]

The parent oximes are conveniently liberated from **89** using gaseous ammonia in dioxane[43] in analogy to **3**[3, 96] (cf. Sections VI, A and VII, A) the saccharin residue proves to be a practical leaving group. In rearrangement reactions the saccharin anion may recombine with a positive center via the oxygen or the nitrogen atom.[43]

1. Ketoxime Derivatives

Ketoxime derivatives (**89a**; R, R' = alkyl, aryl) undergo the usual acid-catalyzed Beckmann rearrangement with great ease.[254, 259]

In addition, heating of **89a** in pyridine/water gives rise to a Beckmann rearrangement,[254] whereas heating in dry pyridine causes rearrangement to **90**, possibly through an ion pair mechanism and reentry of the saccharin anion via the nitrogen.[43, 254]

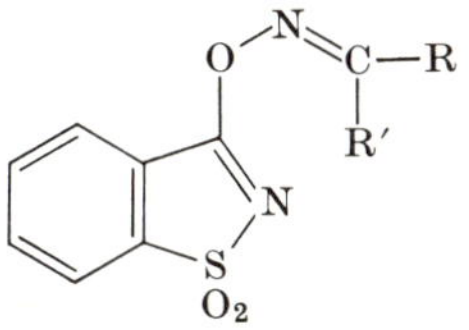

(**89**)
(**91**; R = R' = Ph)

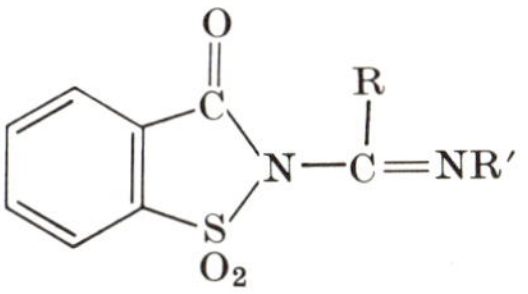

(**90**)

On melting the benzophenonoxime pseudosaccharin ether, (**91**) undergoes what appears to be a two-stage rearrangement to (**94**)[259] with (**92**) and possibly (**93**) as intermediate and transition state,[260] respectively.

The acetophenonoxime pseudosaccharin ether affords 3-anilino-benz[d]isothiazole-1,1-dioxide (**80**) and apparently ketene under comparable conditions.[254, 259] A similar mechanism involving **95** may be operative in the formation of **5b** from **19** and dialkylcarbamoyl

---

[259] H. Neygenfind and H. Hettler, *Tetrahedron Lett.*, 5509 (1968).
[260] K. B. Wiberg and B. I. Rowland, *J. Amer. Chem. Soc.* **77**, 2205 (1955).

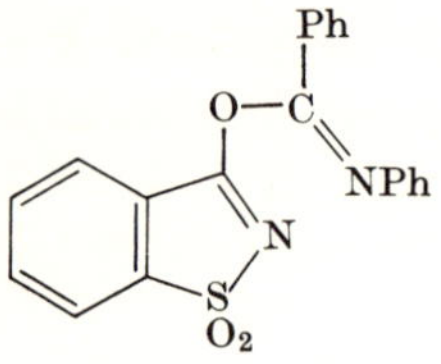

chloride[155] (cf. Section VII, A). Saccharin oxime reacts with **6** like a ketoxime.[43]

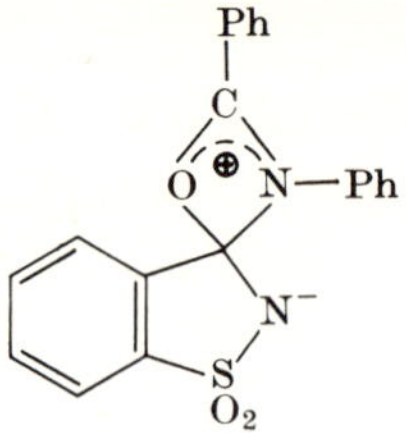

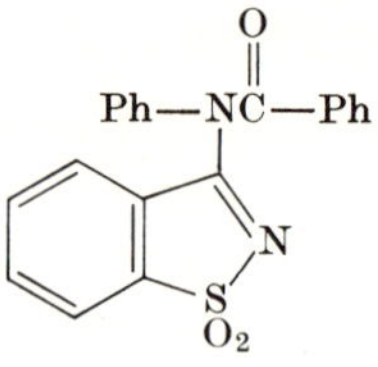

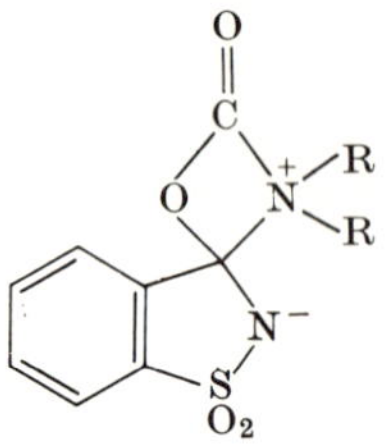

2. Aldoxime Derivatives

Preparation of *syn*-aldoxime derivatives (**89**; R = alkyl, aryl, R' = H) is much the same as that of ketoximes except that the *syn*-pseudosaccharin aldoxime ethers are more unstable.[43] *Syn*-Benzaldoxime derivatives (**96**; R = alkyl, nitro, halogen, etc.) not only undergo the classical Beckmann fragmentation to nitriles on heating or induced by light, but they are also capable of forming formanilides in a regular Beckmann rearrangement.[43]

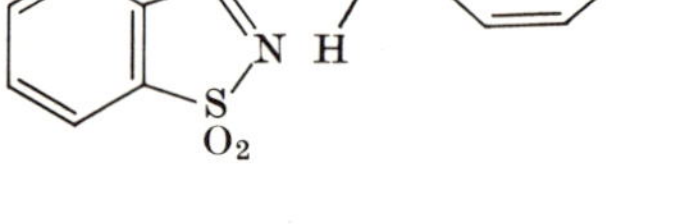

Electron donating substituents R and the presence of orthosubstituents facilitate the rearrangement to formanilides.[43] The *p*-dialkylaminobenzaldoxime derivatives (**96**; R = *p*-NMe$_2$, NEt$_2$) can isomerize to an analog of **90**.[43] Beckmann fragmentation of aldoxime derivatives (optionally without isolation) has been suggested for the synthesis of nitriles.[28, 43] The reaction is first order, the rate constant being dependent on the solvent.[43]

o-Hydroxy-*syn*-benzaldoxime derivatives react smoothly with base to give benzisoxazoles, with elimination of saccharin anions.[254]

XI. 3-Chlorobenz[d]isothiazole-1,1-dioxide

3-Chlorobenz[d]isothiazole-1,1-dioxide ("pseudosaccharin chloride") (**6**)[3, 24, 25, 166, 251, 261, 262] displays the reactivity of a cyclic imidoyl chloride[263] resembling very much carboxylic acid halides. In previous sections preparation of **3, 4, 5, 13, 68, 86, 89** from **6** has been mentioned. Derivatives of **6** substituted in the phenyl ring have been described.[250] Interestingly, Meadow observed[231] that crude **6** and material that contained phosphorus pentachloride reacted with thiols more readily than the pure compound. In the reaction of **6** with aromatic sulfonamides, aluminum chloride had been added for activation.[252]

Jesurun's original procedure,[3] i.e., heating **1** with phosphorus pentachloride at 180° has been modified in detail.[166, 232] Whether **6** is formed directly or whether o-cyanophenylsulfonyl chloride, the main product at low temperature, acts as precursor has not been clearly established.

According to Maselli[235] **6** is also formed when **1** is allowed to stand in a saturated solution of chlorine in water. Lately more complex reactions have been reported for **6**. It reacts with pyridine N-oxide to give N-2-pyridyl saccharin[264] presumably via an intermediate (**97**).

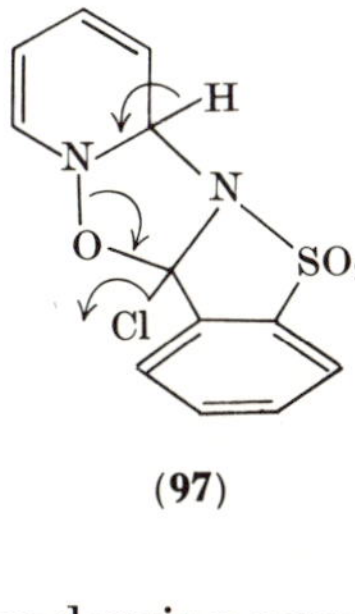

(**97**)

Apart from its use as a condensing agent in peptide synthesis[162-165] (cf. Section IV, D, 6) **6** has been applied to esterification of carboxylic

---

[261] P. Fritsch, *Ber.* **29**, 2290 (1896).
[262] A. J. Walker and E. Smith, *J. Chem. Soc.* **89**, 350 (1906).
[263] H. Ulrich, "The Chemistry of Imidoyl Halides," Plenum, New York, 1968.
[264] R. A. Abramovitch and G. M. Singer, *J. Amer. Chem. Soc.* **91**, 5672 (1969).

acids.[167] When, however, benzyl alcohols are employed under the same conditions, benzyl chlorides rather than carboxylic esters are obtained.[167]

6 reacts with dimethyl sulfoxide (DMSO) in an exothermic reaction.[167] Although no detailed study is available, formation of thioethers[167] (chloromethyl methyl sulfide, dimethyl sulfide), besides 1, is expected in accordance with the reactions given by phenylbenzimidoyl chloride.[265] When esterification with 6 as an activating agent started from the sodium salt of a carboxylic acid in DMSO, N-methylmercaptomethylene saccharin (98) was isolated, besides the ester.[20, 167] The reaction may

(98) (99) (100)

proceed via an imidoyloxysulfonium salt (99) followed by ylid formation (100). The reaction resembles self-alkylation of certain phenols with DCC in DMSO.[266, 267]

XII. Elimination Reactions with 3-Substituted Benz[d]isothiazole-1,1-dioxides

Saccharin (1) and 3-aminobenz[d]isothiazole-1,1-dioxide (5) are frequently encountered fragments in mass spectrometric fragmentation of 3-substituted benz[d]isothiazoles.[20] Both compounds may act as a leaving group in pyrrolytic elimination reactions[28] which lead to formation of multiple bonds. The following formal scheme for a hetero atom analog of a retro-ene reaction[268] (based on the widely adopted six-membered ring transition state for pyrrolytic cis-eliminations[269, 270]) has been put forward (Table III).

Elimination types (i)–(vi) have been verified experimentally, (i) and (ii) on a preparative scale. Type (i) is not restricted to the given example:[28] on the other hand, O → N migration (cf. Sections VI,B and C) may

[265] D. Martin and A. Weise, Ann. Chem. 702, 86 (1967).
[266] M. G. Burdon and J. G. Moffat, J. Amer. Chem. Soc. 89, 4725 (1967).
[267] J. P. Marino, K. E. Pfitzner, and R. A. Olofson, Tetrahedron 27, 4181 (1971).
[268] H. M. R. Hoffmann, Ann. Rep. Progr. Chem. 63, 333 (1967).
[269] C. D. Hurd and F. H. Blunck, J. Amer. Chem. Soc. 60, 2419 (1938).
[270] G. G. Smith and F. W. Kelly, Progr. Phys. Org. Chem. 8, 75–234 (1971).

TABLE III

Scheme for Eliminations in Some 3-Substituted
Benz[d]-isothiazole-1,1-dioxides

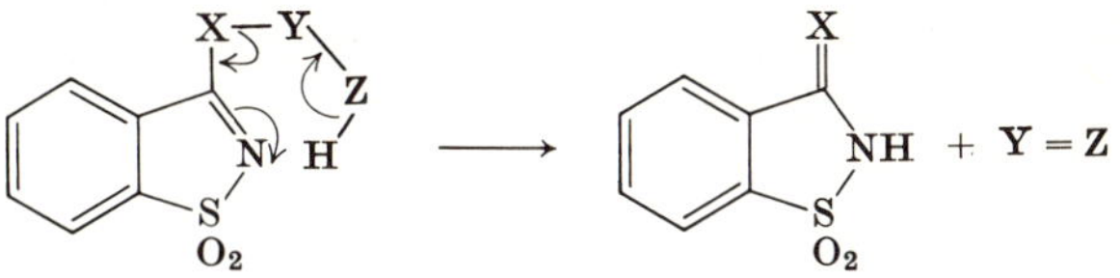

Type[a]	X	Y–Z	Y=Z	R
i[28]	O	CH$_2$–CHPh	H$_2$C=CHPh	—
ii[28,43]	O	N=CR	N≡CR	aryl (alkyl)
iii[28]	NH	N=CR	N≡CR	aryl
iv[254]	NH	C(=O)–CH$_3$	O=C=CH$_2$	—
v[28]	NR	C(=O)–NR	O=C=NR	aryl, alkyl
vi[28]	NR	C(=S)–NR	S=C=NR	cyclohexyl

compete.[96] Side reactions also take place together with type (iii) eliminations.[28] Formation of N-aryl-N'-(3-benz[d]isothiazolyl)carbodiimide-S,S-dioxide (101) from 6 and aryl ureas in a solvent formally fits into the scheme, although the mechanism is apparently not straightforward.[253]

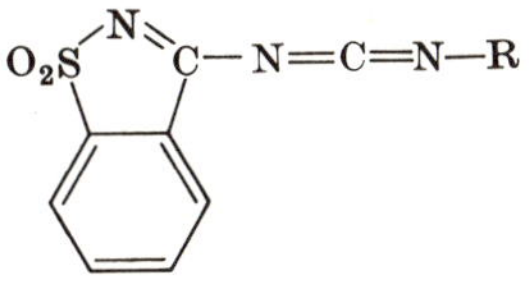

(101) R = aryl

Reactions (i)–(vi) (Table III) have in common that they proceed at relatively moderate temperatures (ca. 150°–300°) as compared with ester[271] or amide[272] pyrolyses.

[271] G. L. O'Connor and H. R. Nace, J. Amer. Chem. Soc. **74**, 5454 (1952).
[272] J. F. Bieron and F. J. Dinan, in "The Chemistry of the Amide Bond" (S. Patai, ed.), p. 279. Wiley (Interscience), New York, 1970.

An alternative elimination in solution, i.e., nitrile formation from the 2,4-dimethoxy-*syn*-benzaldoxime pseudosaccharyl ether was found to be first order.[43] The modest influence of the solvent on the rate of reaction was interpreted[254] in terms of its stabilizing effect on an all planar conformation.[14] If the mechanism is similar to type (ii) a coplanar transition state would be suited for an intramolecular elimination reaction.

Applications of Nuclear Magnetic Resonance Spectroscopy to Heterocyclic Chemistry: Indole and Its Derivatives

SHIVAYOGI P. HIREMATH* and RAMACHANDRA S. HOSMANE

Department of Chemistry, Karnatak University, Dharwar-3, India

I.	Introduction	278
II.	The Spectrum of Indole	279
III.	Elucidation of α and β Substitutions	282
	A. Solvent Effect	282
	B. Concentration Effect	283
	C. Temperature Effect	284
IV.	Solvent and Concentration Dependence of the 7-Proton Resonance	284
V.	Detection of α, β, and N–H Signals	285
VI.	Benzenoid Signals	286
	A. Appearance of Benzenoid Signals	286
	B. Assignment of Benzenoid Signals and Elucidation of Benzenoid Substitutions	290
VII.	Substituted Indoles	291
	A. Indoles Substituted in the Benzenoid Ring Only	292
	B. 1-Substituted Indoles with or without Substituents in the Benzenoid Ring	292
	C. 2-Substituted Indoles with Substituents in the Benzenoid Ring	293
	D. 3-Substituted Indoles with Substituents in the Benzenoid Ring	299
	E. Indoles with 1,2-, 1,3-, and 2,3-Substitution with Substituents also in the Benzenoid Ring	304
	F. 1,2,3-Trisubstituted Indoles with Substituents in the Benzenoid Ring	305
VIII.	Other Applications	318
	A. Protonation of Indoles	318
	B. Stereochemistry of the Side-Chain Substituents	319
	C. Mechanism of the Fischer Indole Synthesis	319
IX.	Conclusions	322

---

* *Present address*: Department of Chemistry, Karnatak University Postgraduate Centre, Gulbarga, India.

I. Introduction

Nuclear magnetic resonance (NMR) spectroscopy is fast becoming a unique tool to elucidate the structural features of even complex organic molecules. Although the technique has been widely applied since 1945, it is only comparatively recently that there have been systematic investigations of heterocyclic compounds. This paper aims principally at providing an account of the application of this technique to the chemistry of indole and its derivatives, which has been the major field of research in this laboratory[1-11] for several years. Considerable theoretical background is, of course, necessary if the discussions are to be understood effectively, and a short review[12] on the theory[13-22] of NMR may

---

[1] S. P. Hiremath and S. Siddappa, *J. Karnatak Univ.* **6**, 1 (1961).
[2] S. P. Hiremath and S. Siddappa, *J. Indian Chem. Soc.* **40**, 935 (1963).
[3] S. P. Hiremath and S. Siddappa, *J. Indian Chem. Soc.* **41**, 357 (1964).
[4] S. P. Hiremath and S. Siddappa, *J. Med. Chem.* **8**, 142 (1965).
[5] S. P. Hiremath, S. Siddappa, and M. Sirsi, *Arch. Pharm.* (*Weinheim*) **298**, 363 (1965).
[6] S. P. Hiremath and S. Siddappa, *J. Karnatak Univ.* **9**, 13 (1965).
[7] S. P. Hiremath and S. Siddappa, *J. Indian Chem. Soc.* **42**, 836 (1965).
[8] S. P. Hiremath and S. Siddappa, *J. Karnatak Univ.* **11**, 28 (1966).
[9] L. D. Basangoudar and S. Siddappa, *J. Chem. Soc. C*, 2599 (1967).
[10] S. N. Betkerur and S. Siddappa, *J. Chem. Soc. C*, 296 (1967).
[11] S. N. Betkerur and S. Siddappa, *J. Chem. Soc. C*, 1795 (1968).
[12] S. P. Hiremath and R. S. Hosmane, *J. Karnatak Univ.* **14**, 202 (1969).
[13] J. C. D. Brand, "Applications of Spectroscopy to Organic Chemistry." Oldbourne Press, London, 1965.
[14] D. Chapman and P. D. Magnus, "Introduction to Practical High Resolution Nuclear Magnetic Resonance Spectroscopy," p. 98. Academic Press, New York, 1966.
[15] J. R. Dyer, "Applications of Absorption Spectroscopy of Organic Compounds," p. 122. Prentice-Hall, Englewood Cliffs, New Jersey, 1965.
[16] L. M. Jackman, "Applications of Nuclear Magnetic Resonance Spectroscopy in Organic Chemistry," p. 24. Pergamon Press, London, 1962.
[17] J. A. Elvidge, *in* "Nuclear Magnetic Resonance for Organic Chemists" (D. W. Mathieson, ed.), p. 197. Academic Press, New York, 1967.
[18] J. D. Roberts, "Nuclear Magnetic Resonance," pp. 22–25. McGraw-Hill, New York, 1959.
[19] R. M. Silverstein and G. C. Bassler, "Spectrometric Identification of Organic Compounds," p. 127. Wiley, New York, 1968.
[20] R. F. M. White, *in* "Physical Methods in Heterocyclic Chemistry" (A. R. Katritzky, ed.), Vol. II, Chapter 9, p. 113. Academic Press, New York, 1963.
[21] J. A. Pople, W. G. Schneider, and H. J. Bernstein, "High Resolution Nuclear Magnetic Resonance," p. 424. McGraw-Hill, New York, 1959.
[22] J. W. Emsley, J. Feeney, and L. H. Sutcliffe, "High Resolution Nuclear Magnetic Resonance Spectroscopy," Vol. 1. Pergamon, Oxford, 1965.

aid the reader. Literature citations are taken predominantly from the references available from *Chemical Abstracts* under the title "NMR of Indoles."

II. The Spectrum of Indole

The α, β-, and N–H protons of indole (**1**) in carbon tetrachloride appear as triplets[23] (Fig. 4A, p. 286), readily separable from each other and from the aromatic protons. The triplet structure shows that, like pyrrole,[24–26]

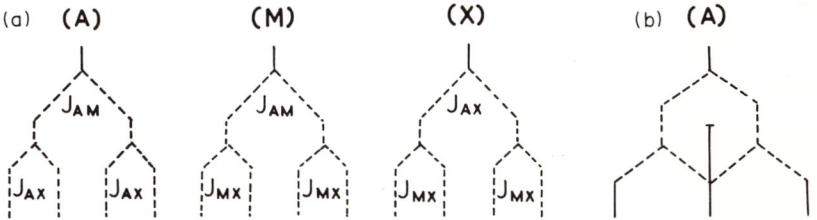

(**1**)

the heterocyclic ring protons are coupled not only with each other but also with the N–H proton. The 1-, 2-, and 3-protons of indole seem to form an ABC system which closely approaches the first-order AMX pattern.[19] The pattern AMX contains 12 lines of equal intensity, grouped in three well separated symmetrical quartets (Fig. 1A), when $J_{AM} \neq J_{AX} \neq J_{MX}$. But, when $J_{AM} = J_{AX} = J_{MX}$, each signal appears as a 1:2:1 triplet[17] (Fig. 1B). In indole, as $J_{1,2} \approx J_{2,3} \approx J_{1,3}$, a perfect 1:2:1 triplet is not observed; instead, the intensities of the signals are distorted.

The signal due to the 1-proton is broad owing to its coupling with the ^{14}N nucleus (spin = 1) which has a short relaxation time presumably by

FIG. 1. (a) AMX pattern, when $J_{AM} \neq J_{AX} \neq J_{MX}$. (b) "A" resonance of AMX pattern, when $J_{AM} = J_{AX} = J_{MX}$.

[23] L. A. Cohen, J. W. Daly, H. Kny, and B. Witkop, *J. Amer. Chem. Soc.* **82**, 2184 (1960).
[24] J. D. Roberts, *J. Amer. Chem. Soc.* **78**, 4495 (1956).
[25] H. Fukui, S. Shimokawa, and J. Sohma, *Mol. Phys.* **18**, 217 (1970).
[26] R. J. Abraham and H. J. Bernstein, *Can. J. Chem.* **37**, 1056 (1959).

TABLE I

CHEMICAL SHIFTS AND COUPLING CONSTANTS OF THE HETEROCYCLIC RING OF INDOLE[a]

Signals	Positions of resonance signals in τ (ppm)						Overall $\varDelta\tau$	Coupling constants, J in Hz
	CCl_4	$CDCl_3$	CH_3COCH_3	$(CH_3)_2SO$	Dioxane	CS_2	$(C_2H_5)_3N$		
α	3.46 (t)	3.26 (t)	<2.75	<2.71	<2.89	3.45 (t)	<3.12	0.76	$J_{1,2}=2.4; J_{1,3}=2.1;$ $J_{2,3}=3.3; J_{3,7}=0.7$
β	3.66 (t)	3.59 (t)	3.52 (m)	3.53 (m)	3.52 (m)	3.71 (t)	3.57 (m)	0.19	
N–H	(3.00)	(2.9)	0.00	−1.01	+0.48	(3.1)	−0.08	4.11	

[a] t, Triplet, m, multiplet. Figures in parentheses are tentative assignments. < indicates that actual values are certainly less than the values given.

virtue of its quadrupole moment.[27] The signals due to the α- and β-protons are clearly visible as two separate triplets (Fig. 2A), the β signal appearing at higher field than the α, and the difference in chemical shifts for the two signals lies between 0.2 and 0.8 ppm. Table I[28] lists the positions of resonance signals (measured on the τ scale using TMS as an internal reference standard) for the α, β, and N–H signals of indole in different solvents.

The fact that the 2- and 3-protons are coupled to the N–H proton is further proved by the appearance of a pair of doublets when the 1-proton is deuterated (Fig. 2B).

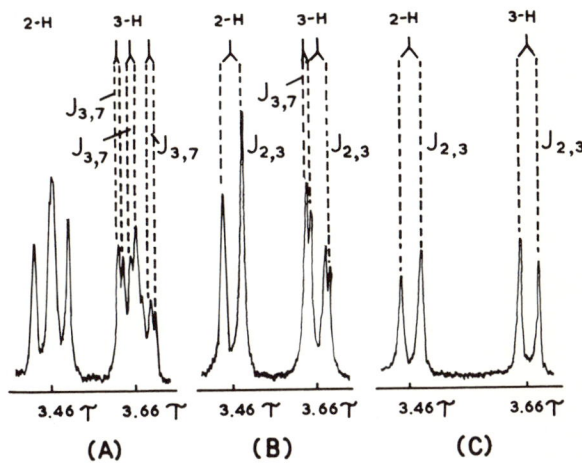

FIG. 2. The 2- and 3-proton signals of (A) indole, (B) N-deuterated indole, and (C) 7-methyl-[N-D]indole in CCl₄ at 60 MHz. (From Elvidge and Foster,[29] by permission of the Chemical Society.)

It can be seen that each line of the higher field signal (due to the 3-proton, in Fig. 2A and B), is further split.[29] This extra splitting is discernible also in the spectra of 4-, 5-, and 6-methylindoles,[29] but the spectrum of 7-methyl-1-D-indole (Fig. 2C) appears as a simple AB quartet, showing long-range 3,7-coupling[30–32] to be responsible for the extra splitting in the earlier cases. Thus, the multiplet patterns are

[27] J. A. Pople, *Mol. Phys.* **1**, 168 (1958).
[28] R. V. Jardine and R. K. Brown, *Can. J. Chem.* **41**, 2067 (1963).
[29] J. A. Elvidge and R. G. Foster, *J. Chem. Soc.*, 981 (1964).
[30] S. Sternhell, *Rev. Pure Appl. Chem.* **14**, 33 (1964).
[31] C. N. Banwell and N. Sheppard, *Discuss. Faraday Soc.* **34**, 120 (1962).
[32] R. A. Hoffman, *Ark. Kemi* **17**, 1 (1961).

first-order and provide the 2,3-, 1,2-, and 3,7-coupling constants directly. J values for indole are listed in Table I.

III. Elucidation of α and β Substitutions

In simple indoles, the signal of the β-proton appears at higher field than that of the α-proton. But, when the α-proton is substituted by an electronegative atom (such as sulfur or halogen) or an electronegative group[28] (acetyl, carbethoxy, etc.), the signal for the remaining β-proton, normally found well isolated at about 3.5 to 3.9τ, occurs considerably farther downfield (ca. 2.8–3.2τ) in the region of the α-proton. Thus, an ambiguity arises as to whether the α or β position has been substituted. Similarly, electron-donating groups in the β position shift the signal of α-proton upfield to where the β-proton absorbs.

The ambiguity, however, is resolved by the following effects.

A. Solvent Effect

One of the factors which affect the chemical shifts of protons is the solvent.[18, 21] Marked differences in chemical shift of the signal for the α-proton in indole and substituted indoles occur when spectra are obtained in nonpolar and polar solvents. In contrast, using polar and nonpolar solvents, only small differences are observed in the chemical shift of the signal for the β-proton in indoles.[28, 33] Hence, solvent shift, or lack of it, may be used to determine whether there is α or β substitution.

From the data in Table I, it is clear that the position of the β-proton signal is nearly the same ($\Delta\tau = 0.19$) even in more polar solvents like deuteriochloroform, acetone, and dimethyl sulfoxide, whereas there is a significant decrease in shielding for the α-proton ($\Delta\tau = 0.76$) in polar solvents. The greater solvent dependence of the α-proton is therefore a useful diagnostic criterion of α or β substitution.

Proton shifts at C-4 are practically unaffected by solvent.[34] The signal position of the N–H proton (Table I) in indole is influenced even more strongly ($\Delta\tau = 4.11$) by the solvent than is the α-proton.[28, 35] It was recently found[33] that the resonance of the 7-proton (see Section IV)

---

[33] M. G. Reinecke, H. W. Johnson, and J. F. Sebastian, *J. Amer. Chem. Soc.* **91**, 3817 (1969).

[34] J. F. K. Wilshire, *Aust. J. Chem.* **20**, 359 (1967).

[35] J. L. Mateos, E. Diaz, and R. Cetina, *Bol. Inst. Quim. Univ. Nacl. Auton. Mex.* **14**, 61 (1962); *Chem. Abstr.* **58**, 13328 (1963).

also undergoes remarkable downfield shifts in more polar solvents. In general, polar solvents and those which associate with indole (e.g., ethers) cause a shift to lower field, in the order dimethyl sulfoxide > triethylamine > acetone > dioxane > $CDCl_3$ > CCl_4 > CS_2.

B. Concentration Effect

The effect of concentration[33, 36] on the signal positions of the 2- and 3-protons of indole in carbon tetrachloride and tetrahydrofuran is illustrated in Fig. 3A and B.

The 3-proton resonance remains essentially constant, and that of the 2-proton moves to higher field with increasing indole concentration in

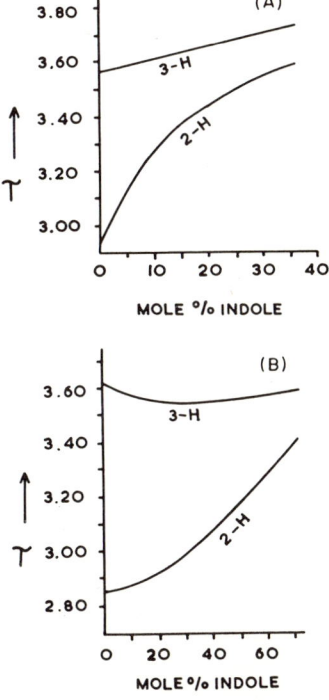

Fig. 3. (A) Chemical shifts of the 2- and 3-protons of indole, as a function of concentration in carbon tetrachloride. (From Reinecke et al.,[33] by permission of the American Chemical Society.) (B) Chemical shifts of the 2- and 3-protons of indole, as a function of concentration in tetrahydrofuran. (From Reinecke et al.,[33] by permission of the American Chemical Society.)

[36] M. G. Reinecke, H. W. Johnson, and J. F. Sebastian, Chem. Ind. (London), 151 (1964).

carbon tetrachloride. In relatively concentrated solutions (>25%) of indole in tetrahydrofuran both the 2- and, to a lesser extent, the 3-proton signal move to higher field with increasing concentration.

C. Temperature Effect

It has been found[33] that the 2-proton resonance in indole undergoes a shift to high field when the temperature is decreased (Table II).

TABLE II

Temperature Dependence of the 2- and 3-Proton Signals of Indole (10%) in Carbon Tetrachloride

Temperature (°C, ±2)	τ (±0.02)	
	2-H	3-H
32	3.25	3.58
14	3.29	3.59
0	3.37	3.60

It is evident that the effects of temperature and concentration might be useful in distinguishing the 2- and 3-protons of certain indoles, and in any case must be taken into consideration when structural interpretations based on NMR are presented. The coupling constant J also provides a useful means of distinguishing between α and β substitutions. The distinction between 2- and 3-chloroindole[37] was made on the basis that $J_{1,2} = 2.5$ Hz for 3-chloroindole, in contrast to 2-chloroindole, for which $J_{1,3}$ was only 2.0 Hz, the coupling constant being expected to decrease[16] with increase in the number of bonds separating the two protons.

IV. Solvent and Concentration Dependence of the 7-Proton Resonance

Recent studies[33] of solvent and concentration effects on the NMR of indoles have established that, in addition to the 2-proton, the 7-proton resonance in certain substituted indoles also undergoes downfield shifts in more polar and more dilute solutions. The chemical shifts of protons of indole and several methylindoles in two solvents, carbon tetrachloride and tetrahydrofuran, are listed in Ref. 33. Proton chemical shifts of the

[37] J. C. Powers, *J. Org. Chem.* **31**, 2627 (1966).

same compounds in carbon tetrachloride at 10 mole% concentration and infinite dilution, are given in the same paper.[33] The data substantiated the selective solvent and concentration dependence of the 2- and 7-proton resonances and also revealed a similar effect for the resonances of the 2- and 7-methyl groups.

For all the indoles investigated, except 1-methylindole, the concentration dependence curves for the 2- and 7-protons or 2- and 7-methyl groups were essentially similar in shape, but different from those of the protons or methyl groups at other positions. Thus, the 2- and 7-protons distinguish themselves from the remaining protons of the indole nucleus, a fact of substantial assistance in their NMR detection.

V. Detection of α, β, and N–H Signals

As already mentioned, the 2-proton resonance undergoes such a large downfield shift in tetrahydrofuran, that it often overlaps the complex benzenoid multiplet. In these cases, the location of the 2-proton resonance was usually determined by decoupling it from the 1-proton by rapid base-catalyzed exchange of the latter with a trace of indolylsodium[38] and observing which peak in the NMR spectrum collapsed from a quartet ($J_{1,2} = 2.4$ Hz; $J_{2,3} = 3.3$ Hz)[29, 30] to a doublet. Coupling to the N-proton can also be eliminated by deuteration (Fig. 2A and B).

A severe broadening obscures the signal of the N–H proton, which is often buried in the complex aromatic absorptions. The approximate location of the N–H signal, under these circumstances, is obtained from the decrease in integrated areas for the N-deuterated compounds. N-Deuteration[23] is more rapid for the indoles containing electronegative substituents in the α or β positions. For indoles and α-substituted indoles, considerable deuteration also occurred at the β position as detected by the lower integrated area for the β-proton signal in the product.[28]

[14]N-Decoupling by double irradiation has also proved to be a very promising method for improving the resolution as well as for allowing the accurate determination of the signal of the N–H proton.[39–42]

[38] S. Gronowitz, A. B. Hornfeldt, B. Gestblom, and R. A. Hoffman, *Ark. Kemi* **18**, 133 (1961).
[39] L. H. Piette, J. D. Ray, and R. A. Ogg, *J. Mol. Spectrosc.* **2**, 66 (1958).
[40] J. D. Baldeschwieler and E. W. Randall, *Chem. Rev.* **63**, 81 (1963).
[41] H. Kamei, *Jap. J. Appl. Phys.* **4**, 212 (1965).
[42] S. Shimokawa, J. Sohma, and M. Itoh, *Bull. Chem. Soc. Jap.* **40**, 693 (1967).

Comparison with the spectra of N-substituted indoles also provides a means of detecting the N–H signals. Figure 4A and B illustrate[23] the application of this device in the case of indole and N-methylindole, wherein the triplets due to α- and β-protons of indole are reduced to doublets in N-methylindole, forming the simpler AB pattern. The N–H peak disappears in Fig. 4B and, hence, provides the location of the 1-proton signal in indole.

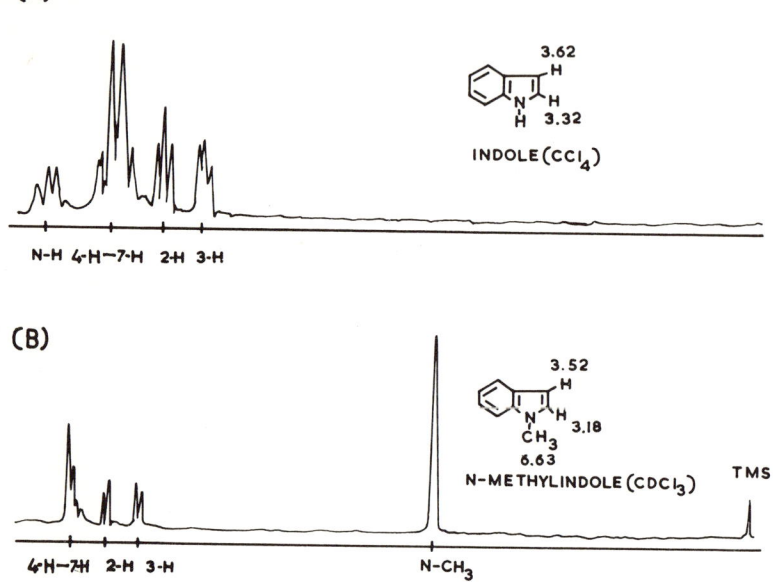

FIG. 4. Spectra of (A) indole, (B) N-methylindole, in CDCl$_3$ at 60 MHz. (From Cohen et al.,[23] by permission of the American Chemical Society.)

The use of the indole resonance[15, 43] technique simplifies the spectrum and also aids in resolving ambiguities.

VI. Benzenoid Signals

A. APPEARANCE OF BENZENOID SIGNALS

The proper assignment of benzenoid signals, especially when there is no substitution in the benzenoid ring, is often a difficult task owing to the

[43] P. J. Black and M. L. Heffernan, Aust. J. Chem. **18**, 357 (1965).

complicated patterns arising from four spin-coupled nuclei. In such cases, often only the range of signals ascertained by the calculated integrated areas can be given. A thorough investigation of the spectral parameters of indole was made[43] by examining its NMR spectrum (Fig. 5A) in acetone at 100 MHz.

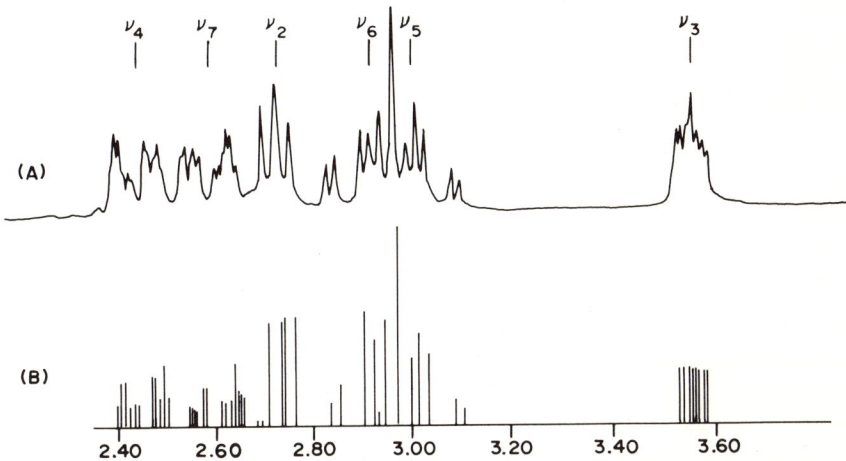

Fig. 5. (A) Experimental and (B) calculated spectrum of indole (0.051 gm/ml in acetone) at 100 MHz. (From Black and Heffernan,[43] by permission of the Commonwealth Scientific and Industrial Research Organization (CSIRO).)

Field-sweep double resonance was employed to give the chemical shift differences between the three pairs of protons—4 and 5, 7 and 6, and 3 and 7 (the second named proton being irradiated in each case). Double resonance of the N–H proton revealed a long-range coupling (0.8 Hz) to H-4. Assignment of lines to individual transitions was made by calculating a series of trial spectra with various reasonable values of the spectral parameters until qualitative agreement with the observed spectra was obtained. The analysis of the spectrum was carried out using an iterative procedure[44] (Fig. 5B). The chemical shifts and spin-spin coupling constants derived from the analyses are given[43] in Table III. The estimated precision of the chemical shifts is ± 0.01 units and that of the coupling constants is ± 0.1 to ± 0.15 Hz.

The heteroatom exerts its slight, but consistent, effect on the chemical shifts and coupling constants of the adjacent ortho (heterocyclic ring)

[44] J. D. Swalen and C. A. Reilly, *J. Chem. Phys.* **37**, 21 (1962).

TABLE III

CHEMICAL SHIFTS AND COUPLING CONSTANTS
OF INDOLE AT 100 MHz[a]

Signals	Positions of the signals in τ (ppm)	Coupling constants, J in Hz
1-H	-0.12	$J_{1,2}=2.5$; $J_{1,3}=2.0$; $J_{1,4}=0.8$; $J_{2,3}=3.1$; $J_{3,7}=$
2-H	2.73_5	0.7_0; $J_{4,5}=7.8_4$; $J_{4,6}=1.2_3$; $J_{4,7}=0.9_4$; $J_{5,6}=$
3-H	3.55_4	7.0_7; $J_{5,7}=1.2_9$; $J_{6,7}=8.0_7$
4-H	2.45_0	
5-H	3.00_5	
6-H	2.92_1	
7-H	2.60_1	

[a] Spectrum is taken with 0.051 gm/ml of indole in acetone. From Black and Heffernan,[43] by permission of the Commonwealth Scientific and Industrial Organization (CSIRO).

and peri (benzenoid ring) protons. The 2- and 7-protons are observed at lower fields than the 3- and 6-protons, the α-protons being more deshielded by the ring current than the β-protons, owing to the usual diamagnetic anisotropy effect. It is of interest to note that the coupling constant $J_{6,7}$ in the benzenoid ring is greater than $J_{4,5}$, which, in turn, is greater than $J_{5,6}$, whereas $J_{1,2}$ in the heterocyclic ring is less than $J_{2,3}$ (short-range influences on coupling constants are expected to be due predominantly to variations in the σ-electron structure and are perhaps connected with the small changes in bond angles, whereas changes in chemical shifts are presumably dominated by the π-electron structure of the molecule). This effect has been observed in indazole,[45] benzofuran,[43] and other systems.[46,47] In indolizine (2),[48] the observations made in the heterocyclic ring of indole are once again repeated, $J_{2,3}$ and $J_{5,6}$ being respectively less than $J_{1,2}$ and $J_{7,8}$.

(2)

[45] P. J. Black and M. L. Heffernan, *Aust. J. Chem.* **16**, 1051 (1963).
[46] P. J. Black and M. L. Heffernan, *Aust. J. Chem.* **17**, 558 (1964).
[47] J. P. Kokko and J. H. Goldstein, *Spectrochim. Acta* **19**, 1119 (1963).
[48] P. J. Black, M. L. Heffernan, L. M. Jackman, Q. N. Porter, and G. R. Underwood, *Aust. J. Chem.* **17**, 1132 (1964).

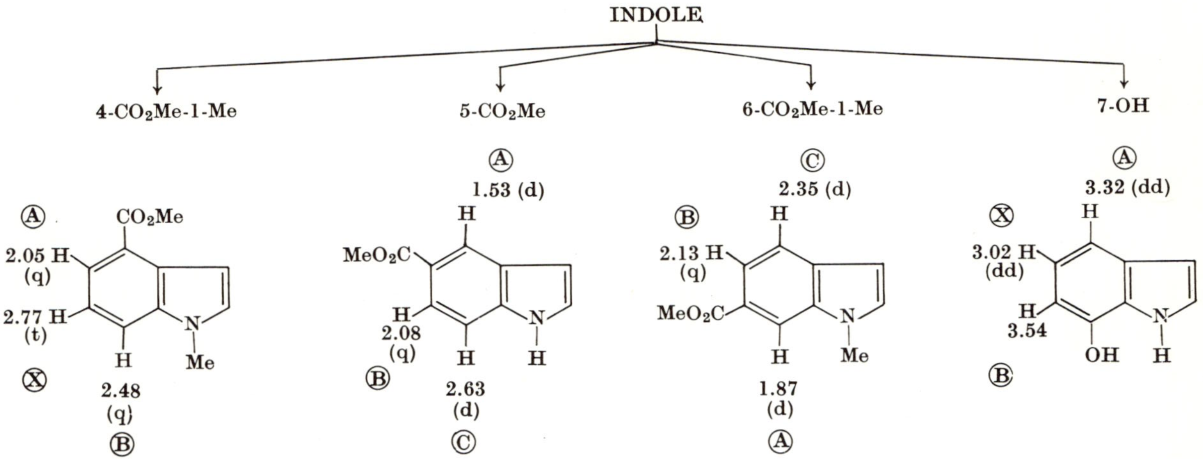

Sec. VI.A.] NMR SPECTROSCOPY OF INDOLES 289

	I	II	III	IV
INDOLE →	4-CO$_2$Me-1-Me	5-CO$_2$Me	6-CO$_2$Me-1-Me	7-OH
	J in Hz: $J_{5,6} = 7.5$; $J_{6,7} = 8$; $J_{5,7} = 1.5$	$J_{4,6} = 1.5$; $J_{6,7} = 8.5$	$J_{4,5} = 8.5$; $J_{5,7} = 1.5$	—
Pattern:	ABX	ABC	ABC	ABX
	$J_{AX} \approx J_{BX}$	$J_{BC} \gg J_{AB}$	$J_{BC} \gg J_{AB}$	$J_{AX} \neq J_{BX}$

Structure I (4-CO$_2$Me-1-Me): Ⓐ 2.05 H (q), Ⓧ 2.77 H (t), Ⓑ 2.48 H (q), N–Me

Structure II (5-CO$_2$Me): Ⓐ 1.53 H (d), Ⓑ 2.08 H (q), Ⓒ 2.63 H (d), N–H

Structure III (6-CO$_2$Me-1-Me): Ⓒ 2.35 H (d), Ⓑ 2.13 H (q), Ⓐ 1.87 H (d), N–Me

Structure IV (7-OH): Ⓐ 3.32 H (dd), Ⓧ 3.02 H (dd), Ⓑ 3.54 H, OH

SCHEME 1

Monosubstitution in the benzenoid ring gives rise to spectra of type ABX or ABC which, in many cases, are not too far from first-order. The coupling constants for ortho-hydrogens are found to range from 7 to 10 Hz, and those for meta-hydrogens, from 2 to 3 Hz. Scheme 1 lists the change in patterns of the three spin-coupled nuclei on going from 4- to 7-substitution.*

Disubstitution in the benzenoid ring provides simpler spectra in the aromatic region. Thus, 4,5- 4,6-, 6,7-, or 5,7-disubstitutions make way for the remaining 2-protons to form an easily analyzable AB pattern producing thereby the two ortho or meta-coupled doublets, depending on the substitution pattern present. 5,6-Disubstitution, however, often gives rise to singlets for the 4- and 7-protons owing to the nonobservable para coupling which is often less than 1 Hz. Finally, trisubstitution in the benzenoid ring leads to a singlet structure for the remaining proton, excluding long-range effects.

B. Assignment of Benzenoid Signals and Elucidation of Benzenoid Substitution

The appearance of benzenoid signals in analyzable patterns, the coupling constants of benzenoid protons, the comparison of spectra with those of analogous compounds, the solvent and concentration dependence of the 7-proton resonance, spin decoupling experiments, and some additional factors which are summarized below, may lead to the proper assignment of the benzenoid signals, and thus to the substitution pattern.

The additional factors are as follows:

1. 7-Halogeno[49, 50] and 7-nitro[51] substitution shifts the 1-Me resonance downfield. 7-Alkyl derivatives reveal a mutual paramagnetic shift of the 1-alkyl and 7-alkyl resonances.[52]

2. The powerful deshielding effect of an ester group at position 3 also extends to the 4- and 5- positions of the benzenoid ring.[53]

---

* The reference for 7-CO_2Me-indole could not be obtained and, therefore, a 7-hydroxyindole is accommodated in the chart.

[49] R. M. Acheson and J. M. Vernon, *J. Chem. Soc.* 1907 (1963).
[50] R. A. Heacock, O. Hutzinger, B. D. Scott, J. W. Daly, and B. Witkop, *J. Amer. Chem. Soc.* **85**, 1825 (1963).
[51] W. E. Noland and K. R. Rush, *J. Org. Chem.* **31**, 70 (1966).
[52] G. R. Allen, C. Pidacks, and M. J. Weiss, *Chem. Ind. (London)*, 2096 (1965).
[53] R. M. Acheson, *J. Chem. Soc.*, 2630 (1965).

An example of the elucidation of substitution orientation by NMR methods[52] follows.

In a Nenintzescu indole synthesis with 2-alkyl-1,4-benzoquinones (3) and 3-aminocrotonates (4) (Scheme 1), two isomeric products (5) and (6) were formed. The compound 5 was already identified as 5-hydroxy-6-alkylindole-3-carboxylic ester. For compound 6, there was ambiguity as to whether there was 4- or 7- alkyl substitution. The following solution was obtained:

1. A J value of 2.5 to 3.0 Hz indicated that meta benzenoid protons are present. Had the substituent R_1 occupied the 4-position, the coupling constant would be of the order of 7–10 Hz, arising from the ortho 6–7-coupling.

2. If the substituent R_1 had occupied position 4, decarboethoxylation would produce an upfield shift of the 4-alkyl resonance. Instead, there was a pronounced diamagnetic shift (0.5–0.6 ppm) for the 4-proton.

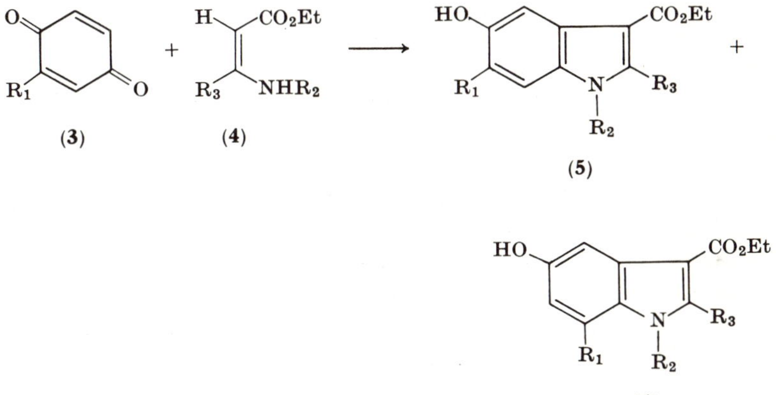

3. A comparison of the N–H and N-alkyl derivatives of 6 showed the expected paramagnetic shifts for the 7- and N-alkyl resonances.

VII. Substituted Indoles

To facilitate the systematic study of various substituted indoles, spectra were taken at 60 and 100 MHz in common solvents like CCl_4, $CDCl_3$, $(CH_3)_2SO$, and CH_3COCH_3, and the indoles categorized accordingly.

A. Indoles Substituted in the Benzenoid Ring Only

Comparison of the spectral assignments of various monomethylindoles[29] with those of simple indole (Table IV, solvent CCl_4) shows that the usual shielding effect of the methyl group on neighboring protons[54] does not extend considerably to the 2- and 3-protons. However, the powerful deshielding effect[55] of the ester group, as is evident from the spectrum of 5-carbomethoxyindole,[53] not only affects the neighboring protons in the carbocyclic ring, but also shifts the C–H signals of the heterocyclic ring downfield.

Further, the change in the spectrum of 5-methoxycarbonylindole on addition of piperidine[53] shows, as in the case of pyrroles,[38] the elimination of the N–H coupling with the other protons of the five-membered ring.

The value of $J_{3,7}$ in monomethylindoles is not constant, but varies slightly from compound to compound; the largest value observed is for 6-methylindole.

The spectra of hydroxyindoles,[56] in comparison with that of indole, show considerable deshielding of the 2-proton, especially in the case of 5- and 7-hydroxyindoles. The benzenoid protons, which usually absorb in the region between 2.9–3.1τ in indole itself (in $CDCl_3$, Table IV), are shifted to as high as 3.60τ in hydroxyindoles. That the deuteration of hydroxyindoles under controlled conditions is selective is confirmed by NMR spectroscopy.[56]

5,6-Diacetoxy- and dimethoxyindole[50] show two sharp singlets in the aromatic region, which correspond to the 4- and 7-hydrogens. The peak found at higher field is assigned to the 7-proton because it is shielded to a greater extent by the adjacent oxygen, as well as nitrogen atoms, than the 4-proton with its single ortho-oxygen atom. 7-Iodination brings about a downfield shift of all remaining protons.[50]

B. 1-Substituted Indoles with or without Substituents in the Benzenoid Ring

N-Substitution eliminates the 1,2- and 1,3-interactions, so that the triplets of the 2-and 3-proton signals of indole are reduced to doublets.

N-Substitution affects mainly the 2- and 3-proton signals. The deshielding effect of a 1-ester group is so great as to bury the 2-proton

---

[54] B. P. Daily and J. N. Shoolery, *J. Amer. Chem. Soc.* **77**, 3977 (1955).
[55] P. Diehl, *Helv. Chim. Acta* **44**, 829 (1961).
[56] J. W. Daly and B. Witkop, *J. Amer. Chem. Soc.* **89**, 1032 (1967).

signal in the complex aromatic absorption, as well as to shift the benzenoid signals downfield (to 2.34–2.83τ).[60]

It is difficult, however, to correlate the chemical shifts and the effect of substituents when there are several substituents of different character (electron-withdrawing and electron-donating) present at different sites in the same molecule. In general, from the study of the various compounds available, it can be concluded that the effect $(+I$ or $-I)^{57}$ is additive for multiple substitution of the same kind, the position closest to the substituent being changed most and the effect being attenuated with distance. Thus, the deshielding effect of the ester group on the 3-proton is found to be greater when it is at the 4-position than when at the 6-position. In the former case the signal of the 3-proton is observed at as low field as that of the 2-proton.

Iodination[50] in the 7-position seems to bring about remarkable downfield shifts of the N-alkyl protons. The effect is greatest in N-isopropylindoles (1.48 ppm) in which the tertiary hydrogen is likely to be kept in close juxtaposition to the iodo substituent because of steric factors.[50, 61] Acetoxy groups are similarly affected.

The location of the 2-proton absorption at 2.68τ in 1-phenylindole which appears with the benzenoid proton signals, was confirmed[58] by decoupling. Irradiation at 39.0 Hz to low field of the 3-proton doublet at 3.34τ caused it to collapse to a singlet.

C. 2-Substituted Indoles with Substituents in the Benzenoid Ring

The fact that the 1- and 3-proton signals are affected most by the $+I$ or $-I$ effect of the substituents on the 2-position is evident in the spectra of the various compounds investigated.

The 3-proton of 2-ethoxycarbonyl-4-nitroindole absorbs, as might be expected, at unusually low field (1.91τ). In the 7-nitro derivative the deshielding is not so pronounced (2.71τ).[57]

The protons ortho and para to the nitro groups are subject to downfield shifts attributable to mesomeric effects; ortho protons also experience a large proximity effect resulting in further deshielding.

[57] Y. Nomura and Y. Takeuchi, *Tetrahedron Lett.*, 5585 (1968).
[58] C. R. Ganellin and H. F. Ridley, *J. Chem. Soc. C*, 1537 (1969).
[59] E. Houghton and J. E. Saxton, *J. Chem. Soc. C*, 595 (1969).
[60] S. Kăspárek and R. A. Heacock, *Can. J. Chem.* **44**, 2805 (1966).
[61] W. Powell, R. A. Heacock, G. L. Mattok, and D. L. Wilson, *Can. J. Chem.* **47**, 467 (1969).

TABLE IV

1-SUBSTITUTED INDOLES WITH OR WITHOUT SUBSTITUENTS IN THE BENZENOID RING

Indole	Solvent	Positions of the proton signals in τ (ppm)							Coupling constants, J in Hz	Ref.
		Ring protons						Substituent protons		
		1-H	2-H	3-H	4-H	5-H	6-H	7-H			
1-Allyl-	—	—	3.38 (d)	3.57 (d)	←——— 2.2 to 3.1 (m) ———→				N-CH$_2$-CH=CH$_2$: -CH$_2$, 6.1 (m); -CH=, 4.2 to 4.9; (m) =CH$_2$, 5.1 to 5.6 (m)	$J_{2,3}=3.0$	59
1-Benzyl-	CCl$_4$	—	3.04, 3.09 (d)	3.55, 3.60 (d)	—	—	—	—	C$_6$H$_5$CH$_2$-: 4.83 (s)	—	58
1-(α-Bibenzyl)-	CCl$_4$	—	3.11, 3.18 (d)	3.58, 3.62 (d)	—	—	—	—	C$_6$H$_5$-C-H: 4.27, 4.39, 4.51 (t)	—	57
1-Carboethoxy-	CDCl$_3$	—	—	3.45 (dd)	←——— 2.34 to 2.83^a (m) ———→				C$_6$H$_5$, CH$_2$: 6.41, 6.53 (d)	$J_{2,3}=3.7$; $J_{3,7}=0.7$	60
1-Isopropyl-5,6-diacetoxy-	CDCl$_3$	—	2.80 (d)	3.55 (d)	2.64 (s)	—	—	2.83 (d)	1-R: CH, 5.49; (septet) CH(CH$_3$)$_2$, 8.56 (d) (6H) -OCOCH$_3$: 7.72 (s) (6H)	$J_{2,3}=3.2$; $J_{3,7}=1.0$	50, 63
1-Isopropyl-5,6-diacetoxy-7-iodo-	CDCl$_3$	—	2.67 (d)	3.57 (d)	2.64 (s)	—	—	—	1-R: CH, 4.01; CH(CH$_3$)$_2$, 8.52 (d) (6H) -OCOCH$_3$: 7.71 (s) (3H) 7.61 (s) (3H)	$J_{2,3}=3.3$	50, 61
1-(p-Methoxybenzyl)-	CDCl$_3$	—	~3.2	3.51 (d)	←——— 2.3 to 3.3 (m) ———→				1-CH$_2$-: 4.76 (s) -OCH$_3$: 6.33 (s)	—	62
1-Methyl-	CDCl$_3$	—	3.18 (d)	3.52 (d)	←——— 2.83 (m) ———→				1-CH$_3$: 6.63 (s)	—	23, 58

Compound	Solvent								Other	J (Hz)	Ref
1-Methyl-3,5,6-triacetoxy-7-iodo-	CDCl$_3$	—	2.73 (d)	—	—	2.67 (s)	—	—	—	—	61
1-Methyl-4-carbomethoxy-	CDCl$_3$	—	2.87 (s)	2.87 (s)	—	—	2.05 (q)	2.77 (t)	2.48 (q)	-COOC$\underline{H}_3$: 6.03 (s); 1-C$\underline{H}_3$: 6.28 (s)	$J_{5,7}=1.5$; $J_{5,6}=7.5$; $J_{6,7}=8.0$	53
	(CD$_3$)$_2$SO	—	2.45 (d)	3.02 (d)	—	—	2.18^a (d)	2.72 (q)	2.18^a (d)	—	$J_{2,3}=3.0$; $\Sigma J=15.5$	53
1-Methyl-5,6-diacetoxy-	CDCl$_3$	—	3.05 (d)	3.63 (d)	2.67 (s)	2.67 (s)	—	—	2.92 (d)	1-C$\underline{H}_3$: 6.91 (s); -OCOC$\underline{H}_3$: 7.72 (s) (6H)	$J_{2,3}=3.1$; $J_{3,7}=1.0$	50, 61
1-Methyl-5,6-diacetoxy-7-iodo-	CDCl$_3$	—	3.04 (d)	3.63 (d)	2.67 (s)	—	—	—	—	1-C$\underline{H}_3$: 5.93 (s); -OCOC$\underline{H}_3$: 7.71 (s), 7.62 (s) (3H)	$J_{2,3}=3.1$	50, 61
1-Methyl-5,6-dihydroxy-	CD$_3$OH	—	3.18 (d)	3.82 (d)	3.02 (s)	3.07 (s)	—	—	3.23 (s)	1-C$\underline{H}_3$: 6.47 (s)	—	50
1-Methyl-5,6-dihydroxy-7-iodo-	CD$_3$OH	—	3.14 (s)	3.83 (d)	—	—	—	—	—	1-C$\underline{H}_3$: 5.93 (s)	—	50
1-Methyl-6-carbomethoxy-	CDCl$_3$	—	2.81 (d)	3.46 (d)	2.35 (d)	—	2.13 (q)	—	1.87^b	1-C$\underline{H}_3$: 6.21 (s); -OCOC$\underline{H}_3$: 6.07 (s)	$J_{2,3}=3.0$; $J_{4,5}=8.5$; $J_{5,7}=1.5$; $J_{2,3}=3.0$	53
	(CD$_3$)$_2$SO	—	2.39 (d)	3.45 (d)	2.30^a (s)	2.30^a (s)	—	—	1.85^{c,d}	—	—	53
1,4-Dimethyl-5,6-diacetoxy-	CDCl$_3$	—	3.04 (d)	3.57 (d)	—	—	—	—	3.05 (d)	—	$J_{2,3}=3.0$; $J_{3,7}=0.8$	63, 61
1,4-Dimethyl-5,6-diacetoxy-7-iodo-	CDCl$_3$	—	3.04 (d)	3.62 (d)	—	—	—	—	—	—	$J_{2,3}=3.0$	63, 61
1,7-Dimethyl-5,6-diacetoxy-	CDCl$_3$	—	3.05 (d)	3.61 (d)	2.78 (s)	—	—	—	—	—	$J_{2,3}=3.0$	61
1,7-Dimethyl-6-methoxy-	CCl$_4$	—	3.38 (d)	3.84 (d)	2.84 (d)	2.65 (d)	3.38 (d)	—	—	—	—	63
	CDCl$_3$	—	3.19 (d)	3.67 (d)	—	3.21 (d)	—	—	—	—	63
1-Phenyl-	CDCl$_3$	—	2.68 (d)	3.34 (d)	—	—	—	—	—	—	—	58

^{a} Almost superimposed.
^{b} Broad.
^{c} With some further splitting.
^{d} A small amount of coupling between 3- and 7-hydrogen atoms occurs.
^{e} One of the peaks is due to 2-H.

[62] K. M. Biswas and A. H. Jackson, *Tetrahedron* **25**, 227 (1969).
[63] R. A. Heacock and O. Hutzinger, *Can. J. Chem.* **43**, 2535 (1965).

TABLE V
2-Substituted Indoles with Substituents in the Benzenoid Ring

Indole	Solvent	Positions of the proton signals in τ (ppm)							Coupling constants, J in Hz	Ref.
		Ring protons						Substituent protons		
		1-H	2-H	3-H	4-H	5-H	6-H	7-H			
2-Benzyl-	CCl_4	—	—	3.78, 3.81 (d)	—	—	—	—	$C_6H_5C\underline{H}_2$-: 6.02 (s)	—	57
	$CDCl_3$	—	—	3.85 (m)	←——— 2.40 to 3.10 ———→ (m)				2-CH_2-C_6H_5: $C\underline{H}_2$, 6.20 (s) $C_6\underline{H}_5$, 2.90 (s)	—	62, 66
2-Benzoyl-	$CDCl_3$	0.08	—	2.54^a (m)	—	—	—	—	—	—	28
2-Carboethoxy-	$CDCl_3$	0.44	—	2.77^a (m)	—	—	—	—	—	—	28
	CH_3COCH_3	−0.79	—	2.78^a (m)	—	—	—	—	—	—	28
	$(CH_3)_2SO$	−1.87	—	2.80^a (m)	—	—	—	—	—	—	28
2-Carboethoxy-4-nitro-	CD_3COCD_3	—	—	1.91 (s)	←——— 1.92 to 2.38 ———→ (m)				-COOR: $C\underline{H}_2$, 5.56; (q) $C\underline{H}_3$, 8.60 (t)	$J_{CH_2,CH_3}=7.0$	51
2-Carboethoxy-7-nitro-	CD_3COCD_3	−0.46	—	2.71 (s)	1.85^a (t)	2.73 (t)	1.85^a (t)	—	-COOR: $C\underline{H}_2$, 5.58; (q) $C\underline{H}_3$, 8.60 (t)	$J_{av\,6,5+4,5}=7.0$ $J_{av\,5,6+5,4}=7.0$ $J_{CH_2,CH_3}=7.0$	51
2-Carbomethoxy-	$CDCl_3$	0.80^b	—	2.78 (d) 2.80 (s)	2.26 (q) 2.32 (q)	←——— 2.5 to 2.9 ———→ (m)			-$COOC\underline{H}_3$: 6.04 (s)	$J_{1,3}=3.0; J_{4,5}=7.0;$ $J_{4,6}=2.0$	53
	$CDCl_3$ + piperidine	—	—	—	←——— 2.5 to 2.9^c ———→ (m)				—	$J_{4,5}=7.0; J_{4,6}=2.0$	53

Sec. VII.C.] NMR SPECTROSCOPY OF INDOLES

Compound	Solvent					Other	J values	Ref
2-Carboxy-	$(CH_3)_2SO$	—	2.80, 2.82 (d)	—	—	—	—	28
2-Chloro-	$CDCl_3$	—	3.56 (d)	—	—	—	$J_{1,3} = 2.0$	37
2-[3-(1,2-Dimethyl-piperidyl)methyl]-	$CDCl_3$	—	3.78 (s)	—	—	$>CH-CH_3$: 8.83 (d); $>N-CH_3$: 7.73 (s)	$J_{CH_3,CH} = 6.0$	67
2-Ethyl-	$CDCl_3$	—	3.85	2.55	2.40 to 3.30^e (m)	—	$2-C_2H_5$: CH_2, 7.60 (q); CH_3, 8.90 (t)	$J_{CH_3,CH_3} = 7.0$	66
2-Isopropyl-	$CDCl_3$	—	3.75 (d)	2.50	2.40 to 3.30^e (m)	—	$2-CH(CH_3)_2$: CH, 7.10 (septet); $(CH_3)_2$, 8.70 (d) (6H)	$J_{1,3} = 2.1$	66
2-Methyl-	CCl_4	(3.1)	4.06^a (m)	—	—	—	—	28
	$CDCl_3$	(3.0)	3.95^a (m)	—	2.92 (m)	$2-CH_3$: 7.80 (s)	$J_{CH,CH_3} = 7.0$	23
	CH_3COCH_3	0.33	3.86^a (m)	—	—	—	—	28
	$(CH_3)_2SO$	−0.82	3.88^a (m)	—	—	—	—	28
2-Methyl-4-iodo-5,6-dimethoxy-	$CDCl_3$	2.42	4.15^b	—	—	3.15 (s)	$-OCH_3$: 6.19 (s), 6.15 (s) (3H); $2-CH_3$: 7.64 (s)	—	50
2-Methyl-5-chloro-	$CDCl_3$	—	3.90	2.56	2.98	2.98 (s)	—	—	68
2-Methyl-5,6-diacetoxy-	$CDCl_3$	1.98	3.91^b	2.82 (s)	—	2.94 (s)	$2-CH_3$: 7.64 (s); $-OCOCH_3$: 7.72 (s) (6H)	—	50
2-Methyl-5,6-diacetoxy-7-iodo-	$CDCl_3$	2.10	3.94^b	2.86 (s)	—	—	$2-CH_3$: 7.65 (s); $-OCOCH_3$: 7.71 (s), 7.65 (s) (3H)	—	50
2-Methyl-5,6-dihydroxy-	CD_3OH	—	4.08^b	3.09 (s)	—	3.20 (s)	$2-CH_3$: 7.75 (s)	—	50
2-Methyl-5,6-dihydroxy-7-iodo-	CD_3OH	—	3.92^b	3.15 (s)	—	—	$2-CH_3$: 7.74 (s)	—	50
2-Methyl-5,6-dimethoxy-	$CDCl_3$	2.25	3.86^b	2.98 (s)	—	3.25 (s)	$-OCH_3$: 6.15 (s), 6.08 (s) (3H)	—	50

continued

TABLE V—continued

Indole	Solvent	Positions of the proton signals in τ (ppm)							Coupling constants, J in Hz	Ref.
		Ring protons						Substituent protons		
		1-H	2-H	3-H	4-H	5-H	6-H	7-H			
2-Methyl-5,6-dimethoxy-7-iodo-	CDCl$_3$	2.19	—	3.73[b]	3.03 (s)	—	—	—	2-C$\underline{H}_3$: 7.63 (s); -OC$\underline{H}_3$: 6.12 (s), 6.09 (s); (3H) (3H)	—	50
2-[2-(N-methyl piperidyl)	CDCl$_3$	—	—	3.67 (d)	—	—	—	—	2-C$\underline{H}_3$: 7.58 (s) >N-C$\underline{H}_3$: 7.97 (s)	$J_{1,3} = 2.0$	67
2-(2-Piperidyl)-	CDCl$_3$	0.8	—	3.68	—	—	—	—	—	—	67
2,5-Dimethyl-	CDCl$_3$	—	—	3.90	—	—	—	—	—	—	68
	CCl$_4$	—	—	4.01	2.83	—	3.16	3.16	2-C$\underline{H}_3$: 7.72 (s); 7-C$\underline{H}_3$: 7.58 (s)	—	33
2,6-Dimethyl-	CCl$_4$	—	—	4.01	2.75	3.23	—	3.48	2-C$\underline{H}_3$: 7.91 (s); 6-C$\underline{H}_3$: 7.63 (s)	—	33
2,7-Dimethyl-	THF	—	—	4.00	2.77	3.26	—	3.03	2-C$\underline{H}_3$: 7.78 (s); 7-C$\underline{H}_3$: 7.75 (s)	—	33
	CCl$_4$	—	—	3.92	—	—	—	—			33
	THF	—	—	3.87	—	—	—	—	2-C$\underline{H}_3$: 7.60 (s); 7-C$\underline{H}_3$: 7.57 (s)	—	33

[a] Poorly resolved signals.
[b] Broad.
[c] Simplified in this solvent.
[d] With some further splitting.
[e] One of them is due to 1-H.

[64] R. L. Hinman and E. R. Shull, *J. Org. Chem.* **26**, 2341 (1961).
[65] Y. Omote, N. Fukada, and N. Sugiyama, *Nippon Kagaku Zasshi* **90**, 1283 (1969); *Chem. Abstr.* **72**, 78134 (1970).
[66] M. H. Palmer and P. S. McIntyre, *J. Chem. Soc. B*, 446 (1969).
[67] L. J. Dolby and G. W. Gribble, *Tetrahedron* **24**, 6377 (1968).
[68] Varian NMR Spectra Catalogs, 1 and 2, Varian Associates, Palo Alto, California (1962, 1963).

There is evidence of coupling between the α- or β-ring protons and the hydrogens of an alkyl group at the adjacent carbon of the heterocyclic ring. Thus, the methyl resonance of 2-methylindole appears as a doublet with a coupling constant of about 1 Hz.[64]

It was observed that the 5- and 7-proton signals of 2-substituted indoles are shifted upfield on formation of the sodium salt.[65]

Even a 2-methyl resonance is affected by substituents present in the carbocyclic ring.[28, 50]

D. 3-Substituted Indoles with Substituents in the Benzenoid Ring

In dimethyl sulfoxide, 3-acetylindole and 3-ethoxycarbonylindole show a sharply defined singlet at 1.66 and 1.88τ, respectively, for the α-proton. Spin-spin coupling with the proton on the adjacent nitrogen atom was not observed.[28] The latter compound exhibits the same phenomenon in triethylamine, but in other solvents (e.g., acetone and dioxan) doublets due to spin-spin coupling were clearly shown. This phenomenon is due to the exchange of the indole NH group, the rapidity of which is dependent on the solvent.

In addition to their effects on the chemical shift of the 1- and 2- proton signals, the 3-substituents produce pronounced shifts of the 4-proton also. The indole NH group exhibits a striking dependence on the 7-substituent as confirmed by the spectrum of 3,7-dinitroindole.[51]

For compounds in which the 3-proton was replaced by either sulfur or halogen, the deshielding effect of the α-proton due to these substituents was less than that found for carbonyl compounds.[28]

The change in the spectrum of 3-methoxycarbonylindole on addition of piperidine to eliminate 1,2-coupling reveals a weak coupling between the 1-proton and a proton in the carbocyclic ring.[53]

The existence of measurable indolenine tautomer in 3-methylindole (7), is excluded,[23] as the isomeric indolenine form (8) would show a

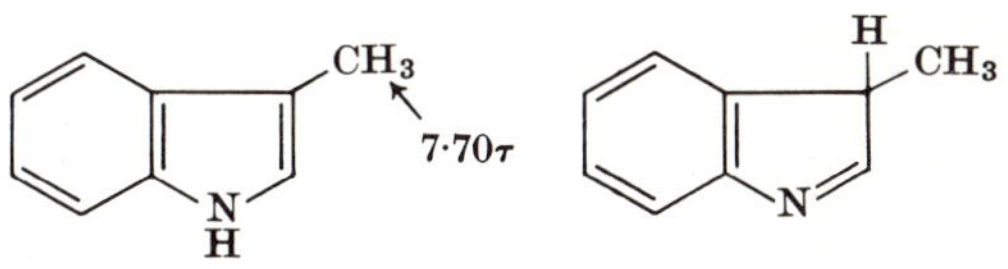

doublet in the vicinity of 8.6τ. However, no such evidence has been observed.

TABLE VI
3-Substituted Indoles with Substituents in the Benzenoid Ring

Indole	Solvent	Positions of the proton signals in τ (ppm)							Substituent protons	Coupling constants, J in Hz	Ref.
		Ring protons									
		1-H	2-H	3-H	4-H	5-H	6-H	7-H			
3-Acetyl-	$(CH_3)_2SO$	−1.92	1.66 (s)	—	—	—	—	—	—	—	28
3-(2-Acetamidoethyl)-5-methoxy- (Melatonin)	$CDCl_3$	—	3.02 (s)	—	2.98 (d)	3.30 (dd)	3.30 (dd)	2.76 (d)	—	—	56
3-Allyl-	$CDCl_3$	~2.3	3.20 (s)	—	←――― 2.3 to 3.1 (m) ―――→				$-CH_2-$: 6.55 (m); $-CH$: ~4.1; $=CH_2$ 4.9 (m) $-NH$: 8.72	—	69
3-(2-Aminoethyl)- (Tryptamine)	$CDCl_3$	1.33	3.08	—	←――― 2.78 to 2.83 (m) ―――→			—	—	—	23
3-(2-Aminoethyl)-5-hydroxy-	D_2O	—	2.98 (s)	—	3.14 (d)	—	3.34 (dd)	2.84 (d)	—	—	56
3-(2-Aminoethyl)-5-hydroxy-4-D-	D_2O	—	2.97 (s)	—	—	—	3.32 (d)	2.83 (d)	—	—	56
3-(α-Aminopropionic acid)- (Tryptophan)	D_2O	—	2.72	—	—	—	—	—	$3-CH_2 \cdot CH \cdot COOH:$ NH_2 $CH_2, 6.90, 6.67; (d)$ $CH, 6.25$	—	14
3-(α-Aminopropionic acid)-5-hydroxy- (Tryptophan-5-hydroxy-)	D_2O	—	2.92 (s)	—	3.05 (d)	—	3.30 (dd)	2.82 (d)	—	—	56
3-(α-Aminopropionic acid)-5-hydroxy-4-D- (Tryptoplan-5-hydroxy-4-D-)	D_2O	—	2.92 (s)	—	—	—	3.30 (d)	2.80 (d)	—	—	56
3-Benzoyl-	$(CH_3)_2SO$	−2.00	(2.0)	—	—	—	—	—	—	—	28
3-Benzyl-	CCl_4	—	3.29, 3.31 (d)	—	—	—	—	—	$C_6H_5CH_2-: 5.94$ (s)	—	57
	$CDCl_3$	—	3.42 (m)	—	←――― 2.3 to 3.3 (m) ―――→				$CH_2 \cdot C_6H_5: 5.98$ (s)	—	
3-Bromo-	$CDCl_3$	(2.8)	2.96, 3.00	—	—	—	—	—	—	—	62

Substituent	Solvent	1-H	2-H	4-H	5-H	6-H	7-H	Other	J (Hz)	Ref.
	CH₃COCH₃	—	—	2.21 (d) 1.97 2.01 (d) 1.88 (s)	—	—	—	—	—	28
3-Carbomethoxy-	(CH₃)₂SO	—	—	2.10 (d)	—	1.79ᶜ (q)	←——— 2.5 to 2.8 (m) ———→	-COOCH₃: 6.07 (s)	—	53
	CDCl₃ + piperidine	1.2ᵇ	2.10 (s)	—	1.82ᶜ (q)	—	←—2.5 to 2.7 (m)—→ 2.80 (q)	—	$J_{1,2}=3.5$, $\Sigma J=10.0$; $J_{6,7}=6.5$; $J_{5,7}=1.5$; $\Sigma J=10.0$	53
3-Carboxy-	(CH₃)₂SO	—	—	1.82, 1.86 (d)	—	—	—	—	—	—	28
3-(o-Carboxybenzyl)-	CH₃COCH₃	1.5ᵇ	—	—	—	—	—	-COOH: 0.1 (s)ᵇ; 3-CH₂: 5.45 (s)	—	62
3-Carboxymethyl-4,6-dimethoxy-	(CD₃)₂SO	-0.58	3.13 (d)	—	—	—	3.92 (d)	3.57 (d)	-COOH: -1.84 (s); -OCH₃: 6.26 (s); 6.28 (s) (3H); 3-CH₂: 6.33 (s) (3H)	—	58
3-Chloro-	CDCl₃	—	2.87 (d)	—	—	—	—	—	—	$J_{1,2}=3.0$	37
3-(2′-Dibenzylamino-vinyl)-	CDCl₃	—	—	—	—	—	—	—	-N(CH₂·C₆H₅)₂: 5.70 (?); -CH=CH·NR₂: =CH·NR₂, 4.52 (d); -CH=, 2.93 (d)	$J_{CH,CH}=14.0$	70
3-(2′-Dibenzylamino-vinyl)-5-benzyloxy-	CDCl₃	—	—	—	—	—	—	—	-N(CH₂·C₆H₅)₂: CH₂: 5.72 (?); -CH=CH·NR₂: =CHNR₂, 4.54 (d); -CH=, 3.08 (d); 5-OCH₂·C₆H₅·CH₂, 4.97 (?)	$J_{CH,CH}=15.0$	70
3-(2′-Diisopropylamino-vinyl)-	CDCl₃	—	—	—	—	—	—	—	-N[CH(CH₃)₂]₂: (CH₃)₂, 8.80 (d) (6H), CH, 6.28 (septet); -CH=CH·NR₂: =CHNR₂, 4.59 (d); -CH=, 3.12 (d)	$J_{CH,CH}=14.5$	70
3-[2-(Dimethylamino)-ethyl]-	CDCl₃	0.90	3.13 (s)	—	—	←——— 2.78 to 2.83 ———→	—	-N(CH₃)₂: 7.65	—	23
3-[2-(Dimethylamino)-ethyl]-4-nitro-5-benzyloxy-	CDCl₃	—	2.85 (s)	—	—	3.06 (d)	2.67 (d)	—	$J_{6,7}=9.0$	72
3-[2-(Dimethylamino)-ethyl]-5-hydroxy-	D₂O	—	3.0 (s)	—	3.10 (d)	—	3.35 (dd)	2.87 (d)	—	—	56

continued

TABLE VI—continued

Indole	Solvent	Ring protons						Substituent protons	Coupling constants, J in Hz	Ref.
		1-H	2-H	3-H	4-H	5-H	6-H	7-H			
3-[2-(Dimethylamino)-ethyl]-5-hydroxy-4-D-	D$_2$O	—	3.00 (s)	—	—	—	3.33 (d)	2.85 (d)	—	—	56
3-Dimethylaminomethyl-4,6-dimethoxy-	(CD$_3$)$_2$SO	−0.62	3.17 (d)	—	—	3.94	—	3.61 (d)	−OC$\underline{H}_3$: 6.26 (s), 6.31 (s); (3H) (3H) −N(C$\underline{H}_3$)$_2$: 7.88 (s); (6H)	—	58
3-Ethylthio-	CDCl$_3$	—	e	—	—	2.0 to 3.0 (m)		—	3-C$\underline{H}_2$: 6.43 (s) −C$_2$H$_5$: C$\underline{H}_2$, 7.3 (q); C$\underline{H}_3$, 8.80 (t)	J_{CH_2,CH_3}=7.0	72
3-Formyl-4,5,6-trimethoxy-	(CD$_3$)$_2$SO	−2.00	2.01 (s)	—	—	—	—	3.15 (s)		—	58
3-Glyoxylamide N,N-dimethyl-	CDCl$_3$	—	2.31	—	1.73	—	—	—		—	68
3-(p-Methoxybenzoyl)-	CF$_3$COOH	—	—	—	—	1.65 to 2.8 (m)			−OC$\underline{H}_3$: 5.94 (s)	—	62
3-(p-Methoxybenzyl)-	CDCl$_3$	—	3.30 (m)	—	—	2.3 to 3.3 (m)			3-C$\underline{H}_2$: 6.01 (s); −OC$\underline{H}_3$: 6.32 (s)	—	62
3-Methyl-	CCl$_4$	(3.1)	3.65^a (m)	—	—	—	—	—		—	28
	CDCl$_3$	(2.9)	3.39^a (m)	—	—	2.82 (m)			3-C$\underline{H}_3$: 7.70 (s)	—	23
	CH$_3$COCH$_3$	+0.48	≤3.0	—	—	—	—	—		—	28
	(CH$_3$)$_2$SO	−0.69	2.94^a (m)	—	—	—	—	—		—	28
3-[2-(Methylamino)-ethyl]-	CDCl$_3$	0.70	3.12	—	—	2.78 to 2.83 (m)			−NH·C$\underline{H}_3$: N$\underline{H}$, 8.73; C$\underline{H}_3$, 7.58	—	23
3-[5-(2,4-Pentadienal)]-	—	—	2.02 (s)	—	—	2.60 (m) (6H:4(Bz) + 2(vinyl))			−C$\underline{H}$O: 0.31 (d)	$J_{CHO,CH}$=8.0	73
3-Phenyl-	CDCl$_3$	—	2.97	—	—	2.83				—	23
3-Propyl-	CDCl$_3$	2.3	3.11 (s)	—	—	2.3 to 3.1 (m)			3-C$\underline{H}_2$·C$\underline{H}_2$·C$\underline{H}_3$: C$\underline{H}_2$, 7.29 (t); C$\underline{H}_2$, 8.35 (m);	—	69

	$(CH_3)_2SO$	1.17	2.45 (d)	—	2.32 (m)	—	—	—	—	$J_{1,2}=2.67$	34
	CH_3COCH_3	−0.25[d]	2.50 (d)	—	2.25 (m)	—	—	—	—	$J_{1,2}=2.4$	34
3-Vinyl-	CCl_4	2.37, 2.31, 2.24, 2.17 (q)	3.29 (s)	—	←	3.00 (m)		→	(X) –CH=C< H(B) : H(A); 5.02, 4.99; (d) 4.83, 4.80; (d) 4.60, 4.57; (d) } AB region; 4.29, 4.26 (d) }; $-C\underline{H}$: 3.50, 3.35 (d) } X region	—	75
3,4-Dinitro-	CD_3COCD_3	−2.17[b]	1.35 (s)	—	—	←	2.30 (m)	→	—	—	51
3,5,7-Trinitro-	CD_3COCD_3	< −3.3	1.18 (s)	—	0.93 (d)	—	0.63 (d)	—	—	$J_{4,6}=2.0$	51
3,7-Diformyl-4,6-dimethoxy-	$(CD_3)_2SO$	−2.00	2.26 (d)	—	—	3.44 (s)	—	—	$-C\underline{H}O$: −0.18 (s), −0.25 (s) (1H) (1H); $-OC\underline{H}_3$: 5.96 (s), 6.03 (s) (3H) (3H)	—	58
3,7-Dimethyl	$CDCl_3$	—	3.08 (m)	—	←	2.4 to 3.1 (m)	→	—	$3-C\underline{H}_3$, 7.68 (s); $7-C\underline{H}_3$, 7.58 (s)	—	76
3,7-Dinitro-	CD_3COCD_3	< −3.3	1.75 (s)	—	1.43 (t)	2.39 (t)	1.43 (t)	—	—	$J_{av\,5,6+5,4}=8.0$, $J_{av\,6,5+4,5}=9.0$	51

[a] Poorly resolved signals.
[b] Broad.
[c] With some further splitting.
[d] Very broad.
[e] Buried in the complex aromatic region.

69 A. H. Jackson and A. E. Smith, *Tetrahedron* **21**, 989 (1965).
70 J. W. Daly and B. Witkop, *J. Org. Chem.* **27**, 4104 (1962).
71 W. F. Gannon, J. D. Benigni, J. Suzuki, and J. W. Daly, *Tetrahedron Lett.*, 1531 (1967).
72 R. V. Jardine and R. K. Brown, *Can. J. Chem.* **43**, 1293 (1965).
73 J. C. Powers, *J. Org. Chem.* **30**, 2534 (1965).
74 L. B. Agenas, *Ark. Kemi* **23**, 157 (1964).
75 W. E. Noland and R. J. Sundberg, *J. Org. Chem.* **28**, 884 (1963).
76 G. S. Bajwa and R. K. Brown, *Can. J. Chem.* **47**, 785 (1969).

The spectra of N-alkylated tryptamines (9)[23] show a consistent difference in chemical shift for N(a) and N(b) substitution, the N(a) peak appearing at lower field. There is no evidence in the spectra of tryptamine and its derivatives for the existence of a cyclic tautomer (10).

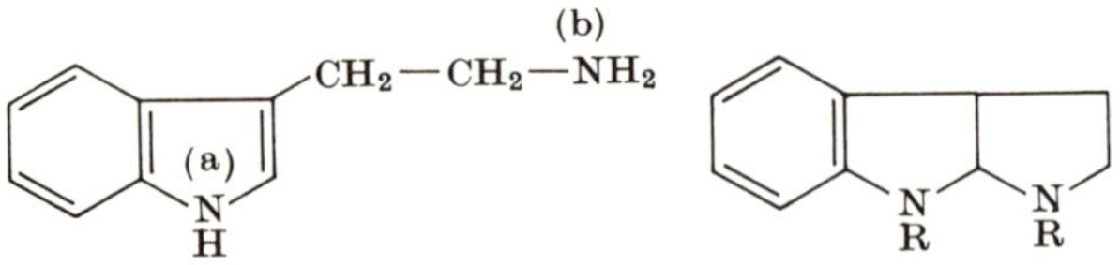

The signal position for the N–H proton of di(3-indolyl)selenide, which was readily assigned because of its disappearance in the presence of D_2O, is strongly influenced by the solvent. The N–H signal in dimethyl sulfoxide solution is much less broad than in acetone or dioxane. Supporting evidence for the structure comes from the similarity of its NMR spectrum in acetone[34] to that of 3-selenocyanoindole,[74] in which the sharp 2-H doublet occurs at lower field ($\tau = 2.31$), presumably as a result of the electronic effect of the more polar SeCN substituent.

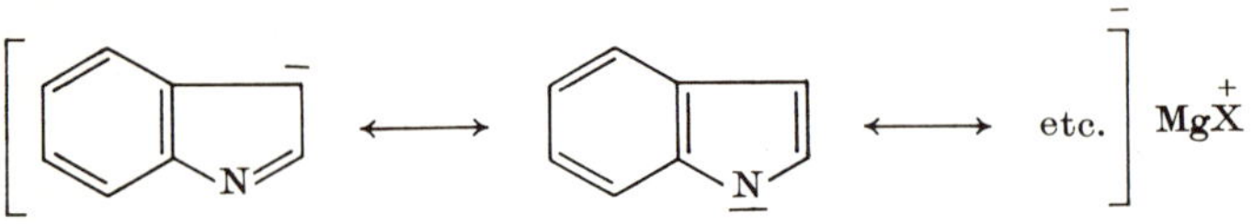

NMR studies indicate that, in tetrahydrofuran, the indole Grignard reagent is predominantly the ionic resonance hybrid (11).[77]

E. Indoles with 1,2- 1,3-, and 2,3-Substitution with Substituents also in the Benzenoid Ring

In each of the above cases, the remaining proton in the heterocyclic ring appears as a singlet (discounting long-range effects), except in the case of 2,3-disubstituted indoles wherein the N–H signal appears as a diffuse hump.

[77] M. G. Reinecke, H. W. Johnson, and J. F. Sebastian, *Tetrahedron Lett.*, 1183 (1963).

The NMR spectrum of 5-bromo-1-hydroxyindole-2-carboxylic acid,[78] in comparison with the spectra of related compounds, suggested that the compound is present as the taútomeric 3H-indole 1-oxide in $CDCl_3$ (11a).[79]

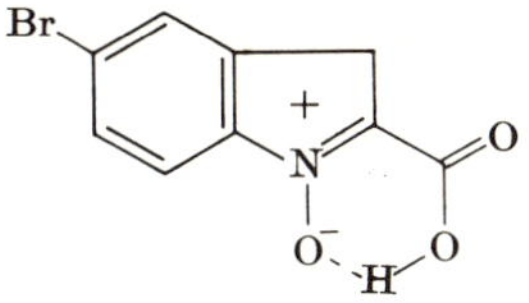

(11a)

The large downfield shift of the N-methyl resonance due to halogenation in the 7-position is shown in molecules of this group.[50] Deshielding also occurs with the 2-methyl group as a result of dichlorophosphoryl substitution at the 3-position.[37]

It is not surprising that the spectra of 2-methyl-3-propylindole and 2-propyl-3-methylindole are almost identical.[69]

F. 1,2,3-Trisubstituted Indoles with Substituents in the Benzenoid Ring

NMR confirmed the structures of 1-methyl-2,3,4,5-tetramethoxycarbonylindole (12) and 1-methyl-2,3,6,7-tetramethoxycarbonylindole (13).[53]

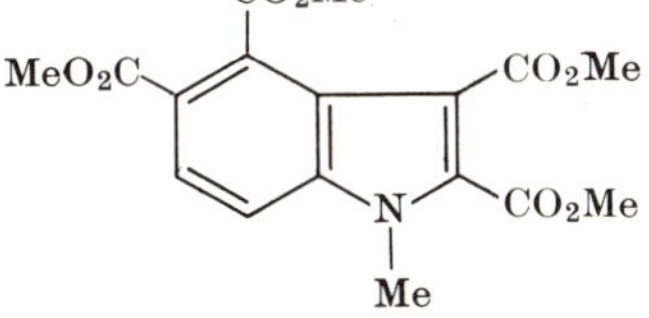

(12) 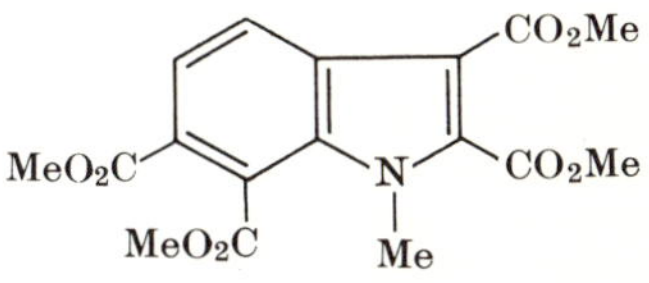   (13)

For both indoles, the large coupling constants (9.0 and 8.5 Hz) require that the hydrogen atoms are ortho to each other in the benzenoid ring. Therefore, two of the four ester groups should be present at positions 2 and 3. In conjunction with the data for indole monoesters,[53] the lowest field proton of the tetraesters can only be accommodated at

[78] R. M. Acheson, C. J. Q. Brookes, D. P. Dearnaley, and B. Quest, *J. Chem. Soc. C*, 504 (1968).
[79] M. Maisseron-Canet and J. P. Boca, *C. R. Acad. Sci.* **260**, 2851 (1965).

TABLE VII

1,2-DISUBSTITUTED INDOLES WITH SUBSTITUENTS IN THE BENZENOID RING

Indole	Solvent	Positions of the proton signals in τ (ppm)							Coupling constants, J in Hz	Ref.
		Ring protons						Substituent protons		
		1-H	2-H	3-H	4-H	5-H	6-H	7-H			
1-Benzyl-2-chloro-	CDCl$_3$	—	—	3.4 (d)	—	—	—	—	—	—	37
1-Benzyl-2-phenyl-5-hydroxy-7-methyl-	CDCl$_3$	—	—	3.60 (s)	3.18 (d)	—	3.33 (d)	—	1-C$\underline{H}_2$·C$_6$H$_5$: C$\underline{H}_2$, 4.6 (s); C$_6$H$_5$, 2.83 to 3.67 (m); 2-C$_6$H$_5$: 2.73 (s); 5-O$\underline{H}$: 4.98^c; 7-C$\underline{H}_3$: 7.63 (s)	$J_{4,6}=J_{6,4}=2.0$	80
1-Butyl-2-phenyl-5-hydroxy-7-methyl-	CDCl$_3$	—	—	3.65 (s)	3.14 (d)	—	3.46 (d)	—	1-C$\underline{H}_2$·(C$\underline{H}_2$)$_2$·C$\underline{H}_3$: C$\underline{H}_3$, 9.25 (t); (C$\underline{H}_2$)$_2$, 8.78 (m); C$\underline{H}_2$, 5.78 (t); 2-C$_6$H$_5$: 2.58 (s); 5-O$\underline{H}$: 4.93^c; 7-C$\underline{H}_3$: 7.37 (s)	$J_{4,6}=J_{6,4}=2.5$	80
1-Hydroxy-2-carbomethoxy-	CDCl$_3$	—	—	3.02 (s)	←—— 2.3 to 3.05 (m) ——→				1-O$\underline{H}$: −0.03^b; 2-COOC$\underline{H}_3$: 6.12 (s)	—	78
1-Isobutyl-2-phenyl-5-hydroxy-7-methyl-	CDCl$_3$	—	—	3.62 (s)	3.11 (d)	—	3.43 (d)	—	1-C$\underline{H}_2$·C$\underline{H}$(CH$_3$)$_2$: (CH$_3$)$_2$, 9.47 (d); C$\underline{H}$, 8.17 (septet); C$\underline{H}_2$, 5.91 (d); 2-C$_6$H$_5$: 2.58 (s); 5-O$\underline{H}$: 4.8^c; 7-C$\underline{H}_3$: 7.38 (s)	$J_{4,6}=J_{6,4}=2.0$	80

Compound	Solvent					Other	Ref.
1-Methoxy-2-carbomethoxy-	CDCl₃	—	—	2.88 (s)	2.25 to 2.95 (m)	1-OCH₃: 5.80 (s); 2-COOCH₃: 6.07 (s)	78
1-Methoxy-2-carbomethoxy-5-bromo-	CDCl₃	—	—	2.98 (s)	2.2 to 2.80 (m)	1-OCH₃: 5.80 (s); 2-COOCH₃: 6.04 (s)	78
1-Methoxy-2-carbomethoxy-6-bromo-	CDCl₃	—	2.48 (d)	2.92 (s)	2.74 (q)	1-OCH₃: 5.78 (s); 2-COOCH₃: 6.05 (s)	$J_{4,5}=8.0$, $J_{5,7}=1.7$	78
1-Methyl-2-[2-(dimethylamino)ethyl]-	CDCl₃	—	—	3.73 (s)	—	2.30^a	1-CH₃: 6.38 (s); 2-(CH₂)₂: ~7.2 (m) (4H) -N(CH₃)₂: 7.65 (s) (6H)	—	58
1-Methyl-2-phenyl-	CDCl₃	—	—	3.41 (s)	2.83 (m)	1-CH₃: 6.43 (s)	—	23
1-Phenyl-2-[2-(dimethylamino)ethyl]-	CDCl₃	—	—	3.63 (s)	—	2-(CH₂)₂-: ~7.4 (m); -N-(CH₃)₂: 7.95 (s)	—	58
1-Propyl-2-phenyl-5-hydroxy-7-methyl-	CDCl₃	—	3.15 (d)	3.67 (s)	3.47 (d)	1-CH₂·CH₂·CH₃: CH₃, 9.46 (t); CH₂, 8.62 (sextet); CH₂, 5.86 (t); 2-C₆H₅: 2.62 (s); 5-OH: 4.51^c; 7-CH₃: 7.41 (s)	$J_{4,6}=J_{6,4}=2.0$	80

^{a} Shows signals of splitting.
^{b} Vanishes on adding D₂O.
^{c} Broad.

TABLE VIII

1,3-DISUBSTITUTED INDOLES WITH SUBSTITUENTS IN THE BENZENOID RING

Indole	Solvent	Ring protons (Positions of the proton signals in τ (ppm))						Substituent protons	Coupling constants, J in Hz	Ref.
		1-H	2-H	3-H	4-H	5-H	6-H	7-H			
1-Allyl-3-methyl-	CDCl₃	—	3.21 (s)	—	←——— 2.3 to 3.1 (m) ———→				3-C$\underline{H}_3$: 7.71 (s); −C$\underline{H}_2$: ~5.5 (m); −C$\underline{H}$: ~4.2 (m); =C$\underline{H}_2$: ~5.0 (m)	—	69
1-Amino-3-methyl-	CCl₄	—	3.42 (s)	—	—	2.99 (m)	—	—	3-C$\underline{H}_3$: 7.78 (s); 1-N$\underline{H}_2$: 5.85 (s)	—	81
1-Benzyl-3-(p-methoxy-benzyl)-	CDCl₃	—	3.20 (d)	—	←——— 2.3 to 3.3 (m) ———→				1-C$\underline{H}_2$: 4.80 (s); 3-C$\underline{H}_2$: 5.97 (s); −OC$\underline{H}_3$: 6.27 (s)	—	62
1-Benzyl-3-phenyl-	CDCl₃	—	c	—	—	—	—	—	1-C$\underline{H}_2$C₆H₅: 4.78 (s)	—	58
1-Butyl-3-methyl-	CDCl₃	—	3.30 (s)	—	←——— 2.3 to 3.1 ———→				1-C$\underline{H}_2$·C$\underline{H}_2$·C$\underline{H}_2$·C$\underline{H}_3$: 6.13 (t); C$\underline{H}_2$·C$\underline{H}_2$, 8.2 to 8.9 (m) C$\underline{H}_3$, 0.90 (t)	—	69
1-(3-Dimethylamino-1-phenylpropyl)-	CDCl₃	3.01 (d)	3.44 (d)	—	—	—	—	—	1-C$\underline{H}$(C₆H₅)−: 4.37 (t); −(C$\underline{H}_2$)₂−: ~7.7 (m); (4H) −N(C$\underline{H}_3$)₂: 7.85 (s) (6H)	—	58
1-(3-Dimethylamino-1-phenylpropyl)-3-phenyl-	CDCl₃	—	c	—	—	—	—	—	1-C$\underline{H}$(C₆H₅)−: 4.35 (t); −(C$\underline{H}_2$)₂−: ~7.6 (m); (4H) −N(C$\underline{H}_3$)₂: 7.88 (s) (6H)	—	58
1-Isopentyl-3-methyl-	CDCl₃	—	3.18 (s)	—	←——— 2.3 to 3.1 (m) ———→				1-C$\underline{H}_2$·C$\underline{H}_2$·C$\underline{H}$(CH₃)₂: C$\underline{H}_2$, 6.0 (t); C$\underline{H}_2$, 8.2 to 8.6; (m) (C$\underline{H}_3$)₂, 9.07 (d) (6H)	—	69
1-Methoxy-3-carbomethoxy	—	—	3.95 (s)	—	—	2.29 (m)	—	—	1-OC$\underline{H}_3$: 5.91 (s); 3-COOC$\underline{H}_3$: 6.03 (s)	—	82
1-(p-Methoxybenzyl)-3-benzyl-	CDCl₃	—	3.21 (d)	—	←——— 2.3 to 3.3 (m) ———→				1-C$\underline{H}_2$: 4.93 (s); 3-C$\underline{H}_2$: 5.93 (s); −OC$\underline{H}_3$: 6.32 (s)	—	62
1-Methyl-3-(2-	CDCl₃	—	3.20	—	←——— 2.78 to 2.83				1-C$\underline{H}_3$: 6.36 (s)	—	23

NMR SPECTROSCOPY OF INDOLES

Compound	Solvent						Ref.
1-Methyl-3-(1-pyrrolyl)-	CCl₄	3.30 (s)	—	—	←―― 2.3 to 3.00 (m) ――→	—NH·C$\underline{H}_3$: 7.62 (d) 1-C$\underline{H}_3$: 6.56 (s); α-pyrryl-$\underline{H}$: 3.18 (t); (2H) β-pyrryl-$\underline{H}$: 3.78 (t); (2H)	$J_{\alpha,\beta}$(pyrryl-H) = 2.1	83
1-Methyl-3,4-dicarbomethoxy-	CDCl₃	2.33 (s)	—	—	←―― 2.43 to 2.73 (m) ――→	1-C$\underline{H}_3$: 6.29 (s); 3-COOC$\underline{H}_3$: 6.15 (s); 4-COOC$\underline{H}_3$: 6.04 (s)	—	53
1-Methyl-3,4-dicarbomethoxy-4,5-dihydro-	CDCl₃ + piperidine	1.81	—	2.24 (m)	—	—	—	—
	CDCl₃	2.78 (s)	5.42 (t)	6.81^a (d)	←― 2.5 to 2.70 ―→ 4.05^b (d) 4.08^b (s)	1-C$\underline{H}_3$: 6.50 (s); 3-COOC$\underline{H}_3$: 6.30 (s); 4-COOC$\underline{H}_3$: 6.30 (s)	—	53
1-Methyl-3,5,6-triacetoxy-	CDCl₃	2.78 (s)	2.68 (s)	—	3.93 (s)	1-C$\underline{H}_3$: 6.43 (s); -OCOC$\underline{H}_3$: 7.71 (s); (9H)	—	50
1-Methyl-3,5,6-triacetoxy-7-bromo-	CDCl₃	2.73 (s)	2.68 (s)	—	—	1-C$\underline{H}_3$: 5.91 (s); -OCOC$\underline{H}_3$: 7.71 (s); (3H) 7.68 (s); 7.62 (s) (3H) (3H)	—	50
1-Methoxy-3,5,6-triacetoxy-7-iodo-	CDCl₃	2.75 (s)	2.68 (s)	—	—	1-C$\underline{H}_3$: 5.90 (s); -OCOC$\underline{H}_3$: 7.71 (s); (3H) 7.68 (s); 7.61 (s) (3H) (3H)	—	50
1-Propyl-3-methyl-	CDCl₃	3.26 (s)	—	—	←― 2.3 to 3.1 (m) ―→	1-C$\underline{H}_2$·C$\underline{H}_2$·C$\underline{H}_3$: C$\underline{H}_2$, 6.13 (t); C$\underline{H}_2$, 8.1 to 8.6; (m)	—	69
						3-C$\underline{H}_3$: 7.72 (s)		
1,3-Dibenzyl-	CCl₄	3.30 (s)	—	—	—	1-C$\underline{H}_2$C₆H₅: 4.88 (s); 3-C$\underline{H}_2$·C₆H₅: 5.94 (s)	—	57
1,3,7-Trimethyl-	CDCl₃	3.32 (s)	—	—	←―― 2.5 to 3.2 (m) ――→	1-C$\underline{H}_3$: 6.08 (s); 3-C$\underline{H}_3$: 7.73 (s); 7-C$\underline{H}_3$: 7.29 (s)	—	76

^{a} —CH₂ group.
^{b} Assignments possibly inverted.
^{c} Indistinguishable from the aromatic proton signals.

80 S. N. Betkerur and S. Siddappa, unpublished results (1970).
81 D. I. Haddlesey, P. A. Mayor, and S. S. Szinai, *J. Chem. Soc.*, 5269 (1964).
82 V. Askam and R. H. L. Deeks, *J. Chem. Soc. C*, 1243 (1968).
83 C. B. Hudson and A. V. Robertson, *Aust. J. Chem.* **20**, 1699 (1967).

TABLE IX

2,3-Disubstituted Indoles with Substituents in the Benzenoid Ring

Indole	Solvent	Positions of the proton signals in τ (ppm)								Coupling constants, J in Hz	Ref.
		Ring protons									
		1-H	2-H	3-H	4-H	5-H	6-H	7-H	Substituent protons		
2-Benzyl-3-(*p*-methoxybenzyl)-	$CDCl_3$	—	—	—	← 2.3 to 3.3 (m) →				2-C$\underline{H}_2$–: 5.92 (s); 3-C$\underline{H}_2$–: 5.92 (s); –OC$\underline{H}_3$: 6.26 (s)	—	62
2-Butyl-3-methyl-	$CDCl_3$	~2.3	—	—	← 2.3 to 3.1 (m) →				2-$\underline{CH_2}\cdot\underline{\underline{(CH_2)_2}}\cdot CH_3$: C$\underline{H}_2$, 7.33 (t), (C$\underline{H}_2$)$_2$, 8.2 to 8.9 (m); C$\underline{H}_3$, 9.09 (t) 3-C$\underline{H}_3$: 7.77 (s)	—	69
2-Carboethoxy-3,4-dinitro-	CD_3COCD_3	—	—	—	← 2.00 (m) →				2-COOC$_2$H$_5$: C$\underline{H}_2$, 5.53 (q); C$\underline{H}_3$, 9.63 (t)	$J_{CH_2,CH_3}=7.0$	51
2-Carboethoxy-3,7-dinitro-	CD_3COCD_3	−2.26	—	—	1.57[a] (t)	2.42 (t)	1.57[a] (t)	—	2-COOC$_2$H$_5$: C$\underline{H}_2$, 5.49 (q); C$\underline{H}_3$, 8.58 (t)	$J_{av\,5,6+5,4}=8.0$; $J_{av\,6,5+4,5}=8.0$; $J_{CH_2,CH_3}=7.0$	51
2-Chloromethyl-3-carboethoxy-4,5,6,7-tetrabromo-	$CDCl_3$/ $(CD_3)_2SO$	−2.37	—	—	—	—	—	—	2-C$\underline{H}_3$Cl: 5.08 (s) 3-COOC$_2$H$_5$: C$\underline{H}_2$, 5.62 (q) C$\underline{H}_3$, 8.62 (t)	$J_{CH_2,CH_3}=7.0$	84
2-Chloromethyl-3-carboethoxy-4,5,6,7-tetrachloro-	$(CD_3)_2SO$	−2.62 (s)	—	—	—	—	—	—	2-C$\underline{H}_2$Cl: 4.99 (s); 3-COOC$_2\underline{H}_5$: C$\underline{H}_2$, 5.60 (q); C$\underline{H}_3$, 8.61 (t)	$J_{CH_2,CH_3}=7.0$	84
2-Diethylaminomethyl-3-carboethoxy-4,5,6,7-tetrachloro-	$CDCl_3$	1.02 (s)	—	—	—	—	—	—	2-C$\underline{H}_2$NR$_2$: 6.02 (s); –N(C$_2$H$_5$)$_2$: (C$\underline{H}_2$)$_2$, 7.35 (q); (4H) (C$\underline{H}_3$)$_2$, 8.93 (t) (6H) 3-COOC$_2$H$_5$: C$\underline{H}_2$, 5.60 (q); C$\underline{H}_3$, 8.59 (t)	$J_{CH_2,CH_3}=7.0$; $J_{CH_2,CH_3}=7.0$	84

Compound	Solvent					Assignments	Coupling constants	Ref.
carboxyethyl)-						−COOCH_3: 8.12 (s); −CH_A=CH_BHc: −CH_A, 3.5 to 4.00 (q); =CH_2, 4.63 to 5.23 (m)	J_{AB}=18.0 J_{AC}=10.0	85
2-Dithiocarboethoxy-3-ethylthio-	CDCl$_3$	—	—	—	—	−CH_2CH_3−: CH_2, 6.52 (m); CH_3, 4.63 to 5.23 (m)	J_{CH_2,CH_3}=7.5	72
2-Isopentyl-3-methyl-	CDCl$_3$	~2.3	—	2.0 to 3.0 (m)	—	−CH_2·CH_2·CH(CH_3)$_2$: CH_2, 7.39 (t); CH_2, 8.3 to 8.7 (m); (CH_3)$_2$, 9.10 (d);	—	69
2-(p-Methoxybenzoyl)-3-benzyl-	CDCl$_3$	—	—	2.3 to 3.3 (m)	—	3-CH_3: 5.81 (s); −OCH_3: 6.23 (s)	—	62
2-(p-Methoxybenzyl)-3-benzyl-	CDCl$_3$	—	—	2.3 to 3.3 (m)	—	2-CH_2−: 5.90 (s); 3-CH_2−: 6.03 (s); −OCH_3: 6.28 (s)	—	62
2-Methyl-3-acetyl-	(CH$_2$)$_2$SO	−1.8	—	—	—	—	—	28
3,3'-Azobis[2-Methyl-	(CD$_3$)$_2$SO	1.52 (2H)	1.6(q) (2H)	2.5 to 3.0 (m) (6H)	—	−CH_3: 7.21 (s) (6H)	$J_{4,5}$=$J_{5,6}$=$J_{6,7}$=6.0 $J_{4,6}$=$J_{5,7}$=3.0	86
3,3'-Azobis[5-chloro-2-methyl-	(CD$_3$)$_2$SO	−1.76 (2H)	1.6 (d) (2H)	2.85 (q) (2H)	—	−CH_3: 7.18 (s) (6H)	$J_{6,7}$=9.0 $J_{4,6}$=2.0	86
2-Methyl-3-carboethoxy-4,5,6,7-tetrachloro-	CDCl$_3$/ (CD$_3$)$_2$SO	−1.67 (s)	—	—	—	2-CH_3: 7.38 (s); 3-COOC$_2$H$_5$: CH_2, 5.62 (q); CH_3, 8.60 (t)	J_{CH_2,CH_3}=7.0	84
2-Methyl-3-carboethoxy-5-hydroxy-	CH$_3$OH	—	2.62 (d)	3.30 (d) 3.16 (d)	2.79 (d)	2-CH_3: 7.28 (s); Ester CH_3: 8.50 (t)	$J_{4,6}$=2.0; $J_{6,7}$=9.0; J_{CH_2,CH_3}=8.0	87
2-Methyl-3-carboethoxy-5-hydroxy-6-methyl-	CH$_3$OH	—	2.51 (s)	—	2.90 (s)	2-CH_3: 7.65 (s); 6-CH_3: 7.30 (s); Ester CH_3: 8.50 (t)	J_{CH_2,CH_3}=8.0	87

continued

84 W. Ried and P. Weidemann, *Chem. Ber.* **102**, 2684 (1969).
85 E. Houghton and J. E. Saxton, *J. Chem. Soc. C*, 1003 (1969).
86 A. S. Bailey and J. J. Meerer, *J. Chem. Soc. C*, 1345 (1966).
87 S. A. Monti, *J. Org. Chem.* **31**, 2669 (1966).

TABLE IX—continued

Indole	Solvent	Ring protons						Substituent protons	Coupling constants, J in Hz	Ref.
		1-H	2-H	3-H	4-H	5-H	6-H	7-H		
2-Methyl-3-carboethoxy-5,6-dimethoxy-	CDCl$_3$	1.60^e (s)	—	—	2.37 (s)	—	—	3.22 (s)	2-C$\underline{H}_3$: 7.34 (s); 3-COOC$_2$H$_5$: C$\underline{H}_2$, 5.64 (q); C$\underline{H}_3$, 8.57 (t); -OC$\underline{H}_3$: 6.18 (s), 6.08 (s) (3H) (3H)	J_{CH_2,CH_3}=6.6	88
2-Methyl-3-carboethoxy-5-hydroxy-7-methyl-	CH$_3$OH	—	—	—	2.65 (d)	—	3.44 (m)	—	2-C$\underline{H}_3$: 7.52 (s); 7-C$\underline{H}_3$: 7.27 (s); Ester C$\underline{H}_3$: 8.50 (t)	J_{CH_2,CH_3}=7.0	87
2-Methyl-3-chloro-	CDCl$_3$	—	—	—	—	—	—	—	2-C$\underline{H}_3$: 7.78 (s)	—	37
2-Methyl-3-(dichlorophosphoryl)-	CDCl$_3$	—	—	—	—	—	—	—	2-C$\underline{H}_3$: 7.62 (s)	—	37
2-Methyl-3-(p-nitrophenyl)-	CDCl$_3$	—	—	—	←——— 2.20 to 3.0^e ———→				2-C$\underline{H}_3$: 7.43 (s); 3-(p-C$_6$H$_4$NO$_2$): -H$\underline{C}$—NO$_2$, 1.73 (d); =H$\underline{C}$ (2H) -C-C$\underline{H}$—, 2.28(d); C$\underline{H}$= (2H)	J_{ortho}=9.0	66
2-Methyl-3-phenyl-	CDCl$_3$	—	—	—	←——— 2.30 to 3.00^e (m) ———→				3-C$\underline{H}_3$: 7.75 (s)	—	66
	CH$_3$COCH$_3$	—	—	—	←——— 2.30 to 3.00^e (m) ———→				3-C$_6$H$_5$: 2.60 (s)	—	66
2-Methyl-3-propyl	CDCl$_3$	~2.3	—	—	←——— 2.3 to 3.1 (m) ———→				2-C$\underline{H}_3$: 7.80 (s); 3-C$\underline{H}_2$·C$\underline{H}_2$·C$\underline{H}_3$: C$\underline{H}_2$, 7.38 (t); C$\underline{H}_2$, ~8.4 (m); C$\underline{H}_3$, 9.11 (t)	—	69
2-Methyl-3-(toluene-p-sulfonamido)-	(CD$_3$)$_2$SO	−0.84	—	—	←——— 2.3 to 3.4 (8H) (m) ———→				>N$\underline{H}$: 8.50; 4'-C$\underline{H}_3$: 7.65 (s); 2-C$\underline{H}_3$: 8.04 (s)	—	86

[Sec. VII.F.] NMR SPECTROSCOPY OF INDOLES 313

Compound	Solvent					Assignments	J	Ref.
triacetoxy-						7.72 (s); 7.69 (s) (3H) (3H) (s)		
2-(p-Nitrophenyl)-3-acetyl-t-butylamino-	CDCl$_3$	0.4 (s)	—	2.54 (m)	—	—C(CH$_3$)$_3$: 8.64 (s) (9H); —COCH$_3$: 8.02 (s) (3H); 2-(p-C$_6$H$_4$NO$_2$): —HC═NO$_2$, 1.64 (d); (2H) —HC═ , 2.02 (d) (2H)	$J_{ortho+para}=9.0$	89
2-(p-Nitrophenyl)-3-t-butylamino-	CDCl$_3$	—	—	2.66^d (m)	—	—C(CH$_3$)$_3$: 9.0 (s) (9H); 2-C$_6$H$_4$NO$_2$^{d}: 1.92 (s) (5H)	—	89
2-Phenyl-3-acetyl-t-butylamino-	CDCl$_3$	—	—	2.54 (m)	—	3-NH–: 6.48 (m) —C(CH$_3$)$_3$: 8.68 (s) (9H) —COCH$_3$: 8.04 (s)	—	89
2-Propyl-3-methyl-	CDCl$_3$	~2.3	—	2.3 to 3.1 (m)	—	2-CH$_2$·CH$_2$·CH$_3$: CH$_2$, 7.38 (t); CH$_2$, 8.45 (m); CH$_3$: 9.09 (t); 3-CH$_3$: 7.80 (s)	—	69
2-(4-Pyridyl)-3-cyano-5,6-dimethoxy-	CF$_3$COOH	—	2.76 (s)	—	2.81 (s)	—OCH$_3$: 5.91 (s); 5.88 (s) (3H) (3H) Pyridyl: 0.95 to 1.48 (broad multiplet)	—	90
2,3-Dimethyl-	CDCl$_3$	—	2.55	—	2.95^c (m)	2-CH$_3$: 7.95^b (s); 3-CH$_3$: 7.88^b (s)	—	23, 66

^{a} With some further splitting.
^{b} The order of assignments is tentative.
^{c} One of them is due to 1-H.
^{d} Values for the phenyl groups are assigned in analogy with the compounds synthesized in this laboratory.[80]
^{e} Broad.

88 T. Kametani, T. Yamanaka, and K. Ogesawara, *J. Chem. Soc. C*, 1616 (1969).
89 J. A. Deyrup, M. M. Vestling, W. V. Hagan, and H. Y. Yun, *Tetrahedron* **25**, 1467 (1969).
90 J. T. Suh and B. M. Puma, *J. Org. Chem.* **30**, 2253 (1965).

position 4, thus giving rise to a unique structure (**13**) for one of them. The resonance position of the more shielded proton (at 2.02τ) of the isomeric indole is at too high a field for it to be peri to a 3-ester group and this again leads to a unique structure (**12**).[53]

The structure of compound **14**, deduced earlier from chemical studies, was confirmed by NMR.[53] The bridgehead proton couples with the adjacent proton, giving rise to a doublet at 5.17τ. Also, the large difference between $J_{5,6}$ and $J_{6,7}$ (5.5 and 10.5 Hz) is consistent with the butadiene-like character of the carbocyclic ring.[91]

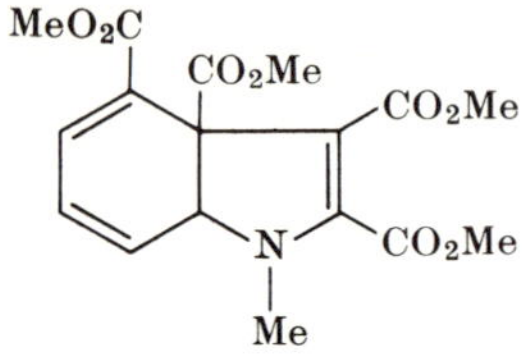

(14) 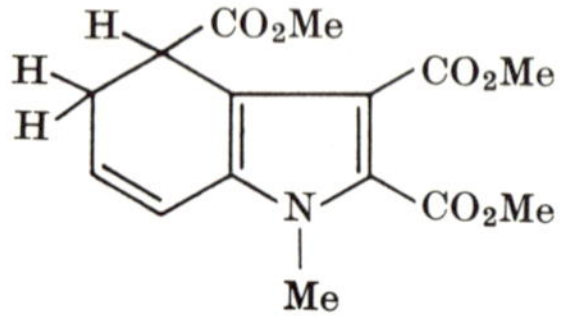   (15)

The NMR spectrum of the triester (**15**) showed the carbocyclic double bond to be at the 6,7-position. The single proton at position 4 appears as a triplet and the 5-methylene group as a doublet which does not seem to couple appreciably with the adjacent olefinic proton at position 6.[53]

On bromination trimethyl 1-methylindole-2,3,4-tricarboxylate (**16**) yielded a mixture of the 6-bromo derivative (**17**) and a second compound [reaction (2)]. This latter was revealed by NMR to be the 5 bromo-oxindole derivative (**18**).[92] The large coupling constant ($J = 8.5$ Hz) for

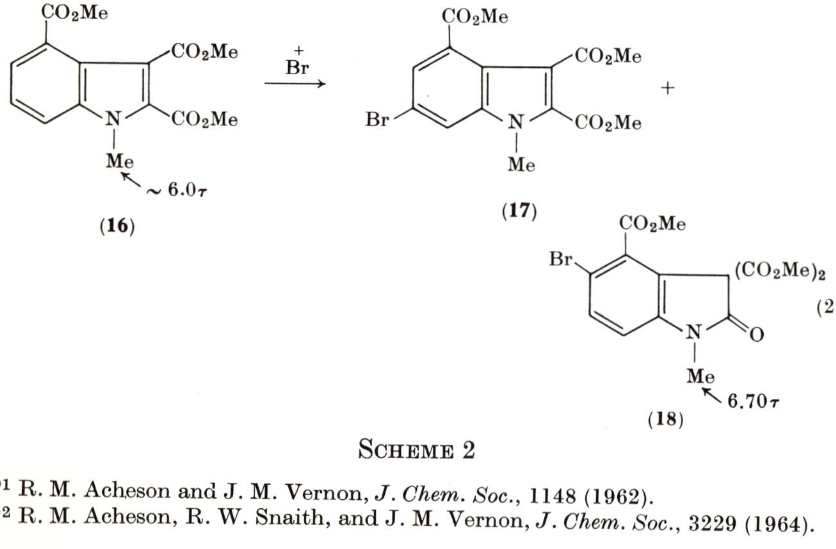

(16)   (17)   (18)   (2)

SCHEME 2

[91] R. M. Acheson and J. M. Vernon, *J. Chem. Soc.*, 1148 (1962).
[92] R. M. Acheson, R. W. Snaith, and J. M. Vernon, *J. Chem. Soc.*, 3229 (1964).

TABLE X

1,2,3-TRISUBSTITUTED INDOLES WITH SUBSTITUENTS IN THE BENZENOID RING

Indole	Solvent	Positions of the proton signals in τ (ppm)							Coupling constants, J in Hz	Ref.
		Ring protons						Substituent protons		
		1-H	2-H	3-H	4-H	5-H	6-H	7-H			
1-Amino-2,3-dimethyl-	CCl$_4$	—	—	—	←——— 2.7 to 3.4 (m) ———→				1-N$\underline{H}_2$: 6.55 (s); 2-C$\underline{H}_3$: 8.00 (s)^a 3-C$\underline{H}_3$: 7.90 (s)^a	—	81
1-(Dimethylamino)- 2-methyl-3-phenyl-	CCl$_4$	—	—	—	←——— 2.3 to 3.2 (m) ———→				1-N(C$\underline{H}_3$)$_2$: 6.94 (s); 2-C$\underline{H}_3$: 7.6 (s); 3-C$_6$$\underline{H}_5$: 2.3 to 3.2 (m)	—	93
1-Ethyl-2-methyl-3- formyl-4-chloro-5- dimethylamino-	—	—	—	—	—	—	2.98^a (d)	2.77^a (d)	1-C$\underline{H}_2$·C$\underline{H}_3$: 6.35 (q); 3-C$\underline{H}$O: −1.77 (s); 5-N(C$\underline{H}_3$)$_2$: 7.57 (s)	$J_{6,7} = 8.0$	94
1-Ethyl-2-methyl-3-phenyl-	—	—	—	—	←——— 2.3 to 3.1 (m) ———→				1-C$_2$$\underline{H}_5$: C$\underline{H}_2$, 5.9 (q); C$\underline{H}_3$, 8.7 (t); 2-C$\underline{H}_3$: 7.6 (s)	—	95
1-Hydroxy-2-carboxy-3- cyano-	CDCl$_3$	—	—	—	←——— 2.05 to 2.5 (m) ———→				1-O$\underline{H}$ and 2-COO$\underline{H}$: 3.3 to 4.2^b	—	78
1-Methoxy-2-carbo- methoxy-3,5-dibromo-	CDCl$_3$	—	—	—	2.23 (d)	—	←— 2.35 to 2.65 (m) —→		1-OC$\underline{H}_3$: 5.80 (d); 2-COOC$\underline{H}_3$: 6.00 (s)	$J_{4,6} = 1.5$	78
1-Methoxy-2-carbo- methoxy-3,6-dibromo-	CDCl$_3$	—	—	—	2.13 (d)	2.67 (q)	—	2.32	1-OC$\underline{H}_3$: 5.79 (s); 2-COOC$\underline{H}_3$: 5.98 (s)	$J_{4,5} = 7.0$; $J_{5,7} = 1.5$	78
1-Methoxy-2-carbo- methoxy-3-cyano-5- bromo-	CDCl$_3$	—	—	—	2.01 (d)	—	←— 2.25 to 2.5 (m) —→		1-OC$\underline{H}_3$: 5.71 (s); 2-COOC$\underline{H}_3$: 5.91 (s)	$J_{4,6} = 2.0$	78
1-(o-Methoxycarbonyl- phenyl)-2-(p-chloro- phenyl)-3-carbomethoxy-	CDCl$_3$	—	—	—	—	—	—	—	1-(o-C$_6$H$_4$COOC$\underline{H}_3$): 6.55 (s); 3-COOC$\underline{H}_3$: 6.22 (s)	—	96
1-(o-Methoxycarbonyl- phenyl)-2-phenyl-3- carbomethoxy-	CDCl$_3$	—	—	—	—	—	—	—	1-(o-C$_6$H$_4$COOC$\underline{H}_3$): 6.57 (s); 3-COOC$\underline{H}_3$: 6.25 (s)	—	96

continued

93 R. F. Meyer, *J. Org. Chem.* **30**, 3451 (1965).
94 W. A. Remers and M. J. Weiss, *J. Amer. Chem. Soc.* **87**, 5262 (1965).
95 R. J. Sundberg and T. Yamazaki, *J. Org. Chem.* **32**, 290 (1967).
96 J. W. Schulenberg, *J. Amer. Chem. Soc.* **90**, 7008 (1968).

TABLE X—continued

Indole	Solvent	Positions of the proton signals in τ (ppm)							Substituent protons	Coupling constants, J in Hz	Ref.
		Ring protons									
		1-H	2-H	3-H	4-H	5-H	6-H	7-H			
1-Methyl-2-chloromethyl-3-carboethoxy-4,5,6,7-tetrachloro-	(CD$_3$)$_2$SO	—	—	—	—	—	—	—	1-C$\underline{H}_3$: 5.88 (s); 2-C$\underline{H}_2$Cl: 4.93 (s); 3-COOC$_2$H$_5$: C$\underline{H}_2$, 5.60 (q); C$\underline{H}_3$, 8.63 (t)	$J_{CH_2,CH_3}=7.0$	84
1-Methyl-2-dimethyl-aminomethyl-3-carboethoxy-4,5,6,7-tetrachloro-	CDCl$_3$	—	—	—	—	—	—	—	1-C$\underline{H}_3$: 5.80 (s); 2-C$\underline{H}_2\cdot$NR$_2$: 6.25 (s); $-$N(C$_2$H$_5$)$_2$: (C$\underline{H}_2$)$_2$, 7.45 (q) (4H) (C$\underline{H}_3$)$_2$, 8.98 (t) (6H)	$J_{CH_2,CH_3}=7.0$	84
1-Methyl-2,4-dicarbomethoxy-3,6-dibromo-	CDCl$_3$	—	—	—	—	2.34^a	—	2.44^a	3-COOC$_2$H$_5$: C$\underline{H}_2$, 5.60 (q); C$\underline{H}_3$, 8.62 (t)		84
	CDCl$_3$	—	—	—	—				$-$C$\underline{H}_3$ and $-$COOC$\underline{H}_3$: 6.02 (s), 6.07 (s), (3H) (3H) 6.20^c (s) (3H)	$J_{5,7}=2.0$	92
1-Methyl-2,3,3a,4-tetracarbomethoxy-3a,7a-dihydro-	CDCl$_3$	—	—	—	—	2.9 (d)	3.68 (q)	4.01 (q) 7a: 5.17 (d)	1-C$\underline{H}_3$: 7.20 (s); 2-COOC$\underline{H}_3$: 6.21^a (s); 3-COOC$\underline{H}_3$: 6.24^a (s); 3a-COOC$\underline{H}_3$: 6.37 (s); 4-COOC$\underline{H}_3$: 6.10^a (s)	$J_{5,6}=5.5$; $J_{6,7}=10.5$; $J_{7,7a}=3.0$	53
1-Methyl-2,3,3a,4-tetracarbomethoxy-6,7,7a-trihydro-	CDCl$_3$	—	—	—	—	2.99 (t)	7.78^d (m)	8.06^d (m) 7a^c: 5.99 (t)	1-C$\underline{H}_3$: 7.20 (s); 2-COOC$\underline{H}_3$: 6.24^a (s); 3-COOC$\underline{H}_3$: 6.28^a (s); 3a-COOC$\underline{H}_3$: 6.39 (s); 4-COOC$\underline{H}_3$: 6.12^a (s)	$J_{5,6}=4.0$; $J_{7,7a}=5.0$	53
1-Methyl-2,3,4-tricarbomethoxy-	CDCl$_3$	—	—	—	—	2.18 (q)	←— 2.3 to 2.72 (m) —→		1-C$\underline{H}_3$: 6.15^a (s); 2-COOC$\underline{H}_3$: 6.10^a (s); 3-COOC$\underline{H}_3$: 6.12^a (s); 4-COOC$\underline{H}_3$: 6.05^a (s)	$J_{5,6}=6.5$; $J_{5,7}=1.5$	53
	CHCl$_3$	—	—	—	—	Complex	—	—	$-$C$\underline{H}_3$ and COOC$\underline{H}_3$: 6.03^c (s), 5.98 (s), (3H) (3H) 6.03 (s), 6.05 (s) (3H) (3H)	—	92

Sec. VII.F.]   NMR SPECTROSCOPY OF INDOLES   317

Compound	Solvent								Chemical shifts	J (Hz)	Ref.
1-Methyl-2,3,4-tricarbomethoxy-6-bromo-	CHCl₃	—	—	—	—	2.08ª	—	2.28ª	–CH₃ and COOC$\underline{H}_3$: 6.09ᶜ (s) 6.01 (s), (3H) (3H) 6.05 (s), 6.08 (s) (3H) (3H)	$J_{5,7}=2.0$	92
1-Methyl-2,3,4-tricarbomethoxy-4,5-dihydro-	CDCl₃	—	—	5.49 (t)	6.80ᵈ	4.06ᵍ (d)	4.09ᵍ (d)		1-C$\underline{H}_3$: 6.35 (s); 2-COOC$\underline{H}_3$: 6.17 (s); 3-COOC$\underline{H}_3$: 6.26 (s); 4-COOC$\underline{H}_3$: 6.35 (s)	$J_{4,5}=6.5$; $J_{67}=11.0$	53
1-Methyl-2,3,4-tricarbomethoxy-4,5,6,7-tetrahydro-	CDCl₃	—	—	6.16ᶠ	8.05ᵈ (m)	8.05ᵈ (m)	7.46ᵈ		1-C$\underline{H}_3$: 6.33ª; 2-COOC$\underline{H}_3$: 6.16ª; 3-COOC$\underline{H}_3$: 6.22ª; 4-COOC$\underline{H}_3$: 6.33ª	—	53
1-Methyl-2,3,4-tricarbomethoxy-6-bromo-	CCl₄	—	—	—	2.08	—	2.28ª		–CH₃ and COOC$\underline{H}_3$: 6.01, 6.05, 6.08, 6.09ᶜ	$J_{5,7}=2.0$	92
1-Methyl-2,3,4,5-tetracarbomethoxy-	CDCl₃	—	—	—	—	2.02 (d)	2.55 (d)		1-C$\underline{H}_3$: 6.01ª (s); 2-COOC$\underline{H}_3$: 5.94ª (s); 3-COOC$\underline{H}_3$: 5.98ª (s); 4-COOC$\underline{H}_3$: 5.94ª (s); 5-COOC$\underline{H}_3$: 5.92ª (s)	$J_{6,7}=9.0$	53
1-Methyl-2,3,6,7-tetracarbomethoxy-	CDCl₃	—	—	1.69 (d)	2.05 (d)	—	—		1-C$\underline{H}_3$: 6.20ª (s); -COOC$\underline{H}_3$: 6.05 (s); 6.05 (s) (3H) (3H) 5.93 (s), 5.93 (s) (3H) (3H)	$J_{4,5}=8.5$	53
1,2-Dimethyl-3-carboethoxy-4,5,6,7-tetrachloro-	CDCl₃	—	—	—	—	—	—		1-C$\underline{H}_3$: 6.08 (s); 2-C$\underline{H}_3$: 7.26 (s); 3-COOC₂H₅: C$\underline{H}_2$: 5.60 (q); C$\underline{H}_3$, 8.61 (t)	$J_{CH_2,CH_3}=7.0$	84
1,2-Dimethyl-3-(toluene-p-sulfonamido)-	(CD₃)₂SO	—	—	—	← 2.3 to 3.4 (m) (8H) →				>NH: 0.75; 1-C$\underline{H}_3$: 6.41 (s); 2-C$\underline{H}_3$: 7.98 (s); 4-C$\underline{H}_3$: 7.65 (s)	—	86

ª Tentative assignments.
ᵇ Vanishes on adding D₂O.
ᶜ One of these is an N–CH₃ group.
ᵈ CH₂ group.
ᵉ Partially obscured.
ᶠ Part of an unresolved 4-proton peak.
ᵍ Assignments possibly inverted.

the aromatic hydrogen atoms of **18** indicated that these are ortho to each other and therefore the bromine atom cannot occupy position 6. The ambiguity between positions 5 and 7 was solved,[92] first on the basis that the *N*-methyl resonance would show a considerable downfield shift (of about 0.50τ) from that of simple 1-methyloxindole (for which the N–Me signal was observed at 6.79τ), if the bromine atom had entered position 7, and second, introduction of the bromine atom at position 5 causes downfield shifts of the ortho-hydrogens (6-H) and upfield shifts of the meta-hydrogens (7-H). The observed changes were consistent with the 5-bromo structure.

VIII. Other Applications

In addition to its use in structural establishment of indoles, NMR has been widely applied to the study of (*a*) the protonation of indoles, (*b*) the stereochemistry of the side-chain substituents, and (*c*) the mechanism of the Fischer indole synthesis.

A. Protonation of Indoles

Direct spectroscopic evidence showed that indole and methylindoles are protonated at the 3-position in strongly acidic solutions.[97, 98] In the spectrum of skatole (3-methylindole) dissolved in 12 *M* sulfuric acid, the 3-methyl peak is shifted to higher field and is split into a doublet ($J = 7.5$ Hz). A quartet appears due to the added proton in the 3-position, coupling with the methyl group, and the 2-hydrogen peak is split into a doublet by coupling with the N–H proton. 1,2-Dimethylindole in sulfuric acid gives a singlet (area 2) at the same field as the skatole quartet. Similar results were found for 2,3-dimethyl- and 1,2,3-trimethylindoles.

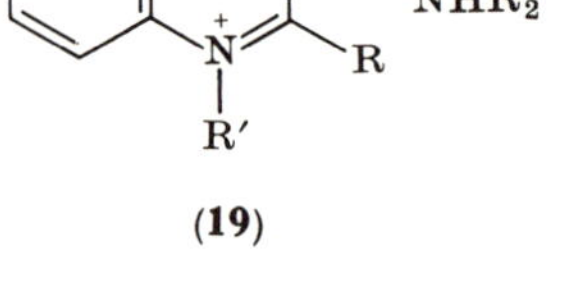

(**19**)

[97] R. L. Hinman and J. Lang, *Tetrahedron Lett.*, 12 (1960).
[98] R. L. Hinman and E. B. Whipple, *J. Amer. Chem. Soc.* **84**, 2534 (1962).

NMR spectrum also shows that tryptamine and several of its N- and C-methyl derivatives are protonated at the 3-position of the indole nucleus in strongly acidic media, to give the corresponding 3H-indolium salts (19).[99]

B. STEREOCHEMISTRY OF THE SIDE-CHAIN SUBSTITUENTS

NMR is also used in stereochemical assignments,[70] e.g., the geometrical structure of indole-3-acrylic acid[100] was assigned by comparing its NMR spectrum with that of several cis and trans isomers of acids of general type, R·CH=CH·COOH. From the observation that the coupling constant was 17.5 Hz, a trans structure was assigned.

C. MECHANISM OF THE FISCHER INDOLE SYNTHESIS

The formation of the new C–C bond (stable b) in the Robinson mechanism[101] of the Fischer indole synthesis, could be justified only after the isolation of the dienone imine intermediate during the indolization of the phenylhydrazone. Many attempts were made[102-106] to isolate such an intermediate. The reported α-ketimino-β-(o-aminophenyl)-γ-butyrolactone hydrochloride,[104, 107] and 3-acetylamino-2-(o-aminophenyl)-2-butene[106] were proved by NMR not to be the dienone imines, but were found to have structures 20[103] and 21,[108] respectively. Recently, the

---

[99] A. H. Jackson and A. E. Smith, *J. Chem. Soc.*, 5510 (1964).
[100] C. Rappe, *Acta Chem. Scand.* **18**, 818 (1964).
[101] G. M. Robinson and R. Robinson, *J. Chem. Soc.* **113**, 639 (1918).
[102] R. B. Carlin, W. O. Henly, and D. P. Carlson, *J. Amer. Chem. Soc.* **79**, 5712 (1957).
[103] R. J. Owellen, J. A. Fitzgerald, B. M. Fitzgerald, D. A. Walsh, D. M. Walker, and P. L. Southwick, *Tetrahedron Lett.*, 1741 (1967).
[104] H. Plieninger, *Ber.* **83**, 273 (1950).
[105] F. P. Robinson and R. K. Brown, *Can. J. Chem.* **42**, 1940 (1964).
[106] N. N. Suvorov, N. P. Sorokina, and Yu. N. Sheinker, *J. Gen. Chem. USSR.* **28**, 1058 (1958).
[107] H. Plieninger and I. Nogradi, *Ber.* **88**, 1964 (1955).
[108] R. H. C. Elgersma and E. Havinga, *Tetrahedron Lett.*, 1735 (1969).

successful isolation of such a dienone imine intermediate (**23**) has been reported.[109] But there exists an ambiguity between structures **22** and **23**,

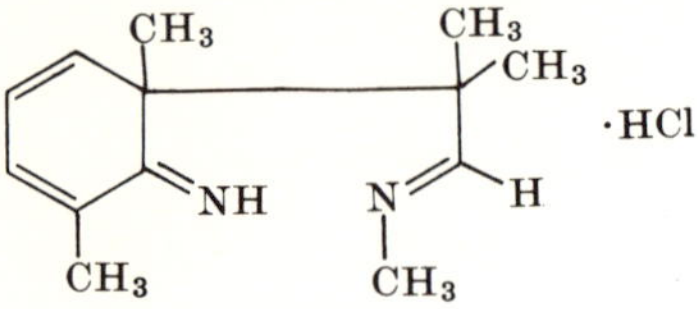

(**22**)

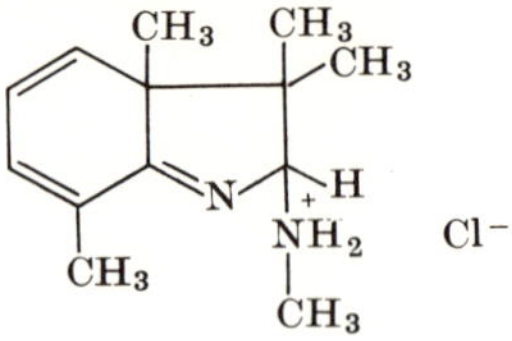

(**23**)

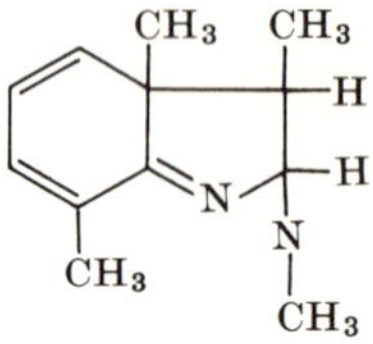

(**23a**)

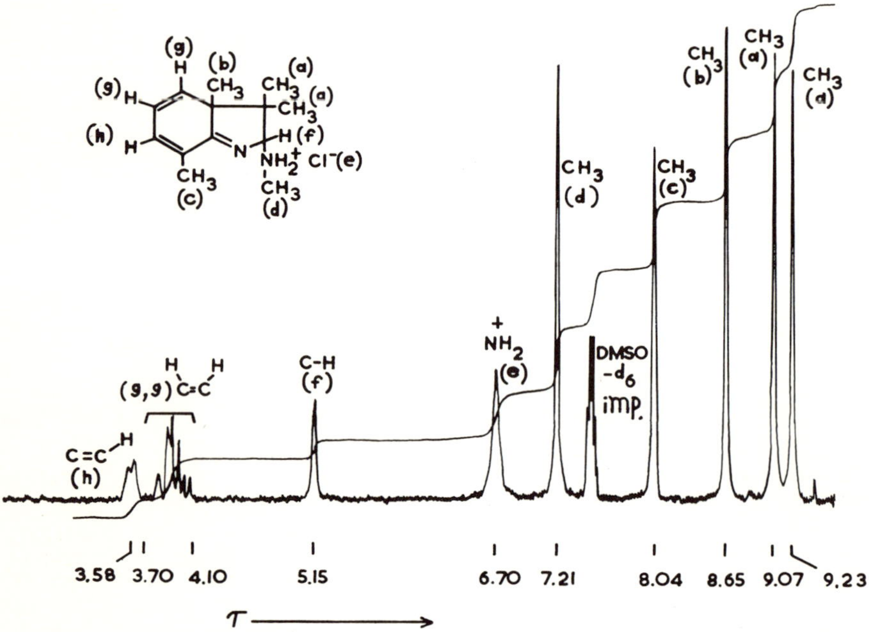

FIG. 6. 100 MHz spectrum of **23** (hydrochloride) in $(CD_3)_2SO$. (From Bajwa and Brown,[109] by permission of the National Research Council of Canada.)

[109] G. S. Bajwa and R. K. Brown, *Can. J. Chem.* **46**, 3105 (1968).

as the elemental analysis, UV and IR data were the same for both. However the NMR spectra of the salt (**23**) and the free base derived from it (**23a**) (Figs. 6 and 7) in deuterated dimethyl sulfoxide clearly showed

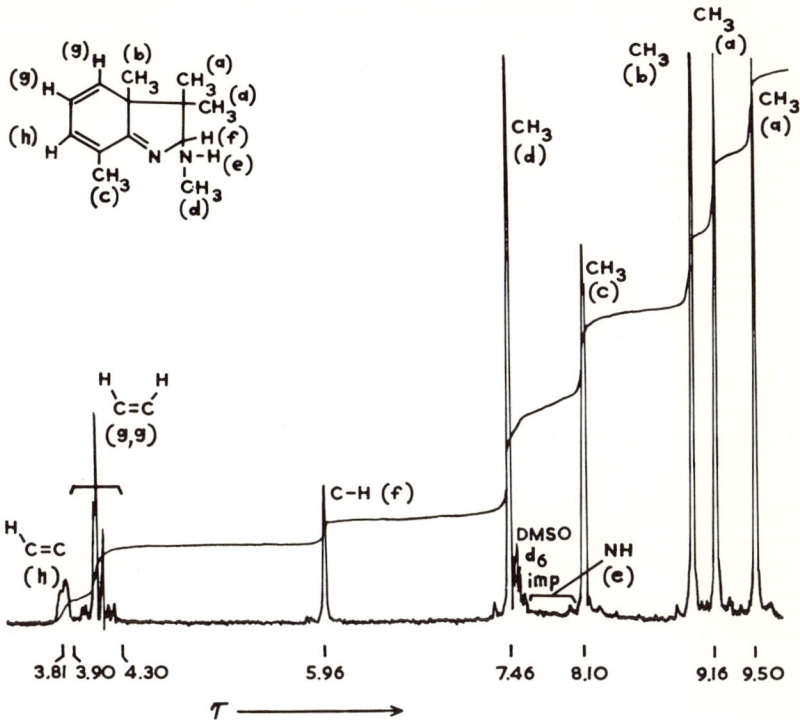

FIG. 7. 100 MHz spectrum of **23a** (free amine) in $(CD_3)_2SO$. (From Bajwa and Brown,[109] by permission of the National Research Council of Canada.)

that **23** was the correct structure. The structure was resolved[109] as follows.

1. The signals (each 1 H) at 5.15 (Fig. 6) and 5.96τ (Fig. 7) were assigned to the former aldehydic proton f in structures **23** and **23a**, since this proton is very much like the anomeric proton in glycosides or the acetals (O–CH–O). If **22** were the correct structure of the dienone imine, the signal for this proton would appear well downfield in the region 0–1.0τ.[109]

2. A small, long-range coupling (0.7 Hz) between the protons f and h was observed. Such a coupling could occur only if the compound had structure **23**.

More recently, a solid mixture has been obtained by the condensation of N'-methyl-2,6-dimethylphenylhydrazine hydrochloride with propionaldehyde, whose NMR spectrum showed that it was a mixture containing chiefly 1,3,7-trimethylindole, a substantial amount of 3,7-dimethylindole and a small amount of the dienone imine (**24**).[76]

(**24**)

IX. Conclusions

To date, NMR has been used to study (*a*) the structures of indoles, substituted in both heterocyclic as well as the benzenoid ring (recently, there have been a number of publications wherein NMR is used to elucidate the complicated structures of biologically important antibiotics[95, 110-112] and alkaloids[113-117] containing the indole skeleton); (*b*) tautomeric structures and equilibria;[96, 118-120] (*c*) the stereochemistry of side-chain substituents; and (*d*) the protonation of indoles. It is hoped that in the future many significant applications of NMR in the indole field, such as a study of intermolecular and intramolecular reactions and the rates of fast reactions, will be found.

[110] G. R. Allen and J. M. Weiss, *J. Org. Chem.* **30**, 2904 (1965).
[111] G. R. Allen, J. F. Poletto, and M. J. Weiss, *J. Org. Chem.* **30**, 2897 (1965).
[112] W. A. Remers, R. H. Roth, and M. J. Weiss, *J. Org. Chem.* **30**, 2910 (1965).
[113] S. R. Johns, J. A. Lamberton, and J. L. Occolowitz, *Aust. J. Chem.* **19**, 1951 (1966).
[114] J. Pecher, R. H. Martin, N. Defay, M. Kaisin, J. Pectors, and G. van Binst, *Tetrahedron Lett.*, 270 (1961).
[115] R. Romanet, A. Chemizart, S. Duhoux, and S. David, *Bull. Soc. Chim. Fr.* 1048 (1963).
[116] F. Walls and X. Arevalo, *Bol. Inst. Quim. Univ. Nac. Auton. Mex.* **15**, 3 (1963); *Chem. Abstr.* **61**, 8260 (1964).
[117] M. R. Yagudaev, V. M. Malikov, and S. Yu. Yunosov, *Khim. Prir. Soedin.* **6**, 89 (1970); *Chem. Abstr.* **73**, 35564 (1970).
[118] S. Hünig, H. Buysch, H. Hoch, and W. Lendle, *Chem. Ber.* **100**, 3996 (1967).
[119] G. Klose and E. Uhlemann, *Tetrahedron* **22**, 1373 (1966).
[120] L. W. Reeves, *Can. J. Chem.* **35**, 1351 (1957).

Four main difficulties are encountered at present. First, it is difficult to demonstrate empirically the theoretically expected linear relation between chemical shifts and electron densities. The more subtle chemical shift variations still await satisfactory explanation. The chemical shifts for the aromatic ions are influenced appreciably by polar solvent effects and by ring-current shifts. When these corrections can be made, a rough linear relation between chemical shift and electron density can be established. It is, however, now possible to calculate the shielding contributions associated with the diamagnetic anisotropy of π-electron systems (ring currents)[121] and unequal π-electron charge densities[122] and to relate these contributions to experimental data. LCAO-MO[123-128] calculation of electron density maps[129] is a powerful tool. Recently, the variable electronegativity self-consistent field molecular orbital method (VESCF),[130, 131] which is based on the conventional SCF[132] procedure, has been of great help in calculating the π-electron distribution. Second, there have been few and scattered reports of the systematic study of the spectra of indoles. Third, much work is to be done on the elucidation of benzenoid substitution, e.g., by examining various orderly substituted indoles (several of these indoles have now been synthesized in this laboratory[133]). Finally, little research has been carried out on nuclei other than hydrogen. A report has recently appeared on the ^{13}C NMR (CMR) spectra of indoles.[128] Using single-frequency and off-resonance proton-decoupling techniques, the chemical shifts of the carbon atoms of indole and its seven monomethyl derivatives have been measured and assigned.[134] There has been little general interest in the ^{14}N magnetic

---

[121] G. G. Hall, A. Hardisson and L. M. Jackman, *Tetrahedron* **19** (Suppl. 2), 101 (1963).
[122] B. P. Dailey, A. Gawer, and W. C. Neikam, *Discuss. Faraday Soc.* **34**, 18 (1962).
[123] K. Fukui, T. Yonezawa, C. Nagata, and H. Shingu, *J. Chem. Phys.* **22**, 1433 (1954).
[124] R. McWeeny, *Mol. Phys.* **1**, 311 (1958).
[125] J. A. Pople, *Mol. Phys.* **1**, 175 (1958).
[126] C. Aussems, S. Jaspers, G. Leroy, and F. Van Remoortere, *Bull. Soc. Chim. Belg.* **78**, 479 (1969); *Chem. Abstr.* **72**, 66225 (1970).
[127] R. E. Moore and H. Rapoport, *J. Org. Chem.* **32**, 3335 (1967).
[128] R. G. Parker and J. D. Roberts, *J. Org. Chem.* **35**, 996 (1970).
[129] P. S. Song and W. E. Kurtin, *Photochem. Photobiol.* **9**, 175 (1969); *Chem. Abstr.* **70**, 81031 (1969).
[130] R. D. Brown and M. L. Heffernan, *Aust. J. Chem.* **12**, 319, 330, 543, 554 (1959).
[131] R. D. Brown and M. L. Heffernan, *Aust. J. Chem.* **13**, 38, 49 (1960).
[132] G. G. Hall and A. Hardisson, *Proc. Roy. Soc. Ser. A* **268**, 328 (1962).
[133] G. A. Bhat and S. Siddappa, unpublished results (1971).
[134] J. D. Roberts, F. J. Weigert, J. I. Kroschwitz, and H. J. Reich, *J. Amer. Chem. Soc.* **92**, 1338 (1970).

resonance studies, owing to rapid quadrupole relaxation, even though quadrupole line broadening can be minimized by the use of low viscosity solvents.[135]

To overcome these problems would be most rewarding.

Acknowledgments

We wish to thank Professor Wayland E. Noland and Dr. Akram Sandhu, University of Minnesota, Professor C. R. Kanekar and Dr. G. Govil, Tata Institute of Fundamental Research, Bombay, and Dr. N. S. Narasimhan, Poona University, for having read the manuscript and making invaluable comments and suggestions which have helped us to improve this article. We also wish to thank Professor S. Siddappa, Karnatak University, for his keen interest and encouragement. One of the authors (R.S.H.) acknowledges the University Grants Commission for a scholarship.

[135] D. Herbison-Evans and R. E. Richards, *Mol. Phys.* **8**, 19 (1964).

Author Index

Numbers in parentheses are reference numbers and indicate that an author's work is referred to although his name is not cited in the text.

A

Aamodt, L. C., 88, 93(72)
Abe, K., 238, 246(23), 247, 248(120), 256(23, 120, 124), 257(23, 120, 195, 196), 259(196), 262(23, 120), 264
Abe, T., 41
Abraham, R. J., 279
Abramovitch, R. A., 182, 254, 273
Ache, H. J., 144, 147
Acheson, R. M., 14, 24(51), 25(51), 31(51), 290, 292(53), 295(53), 296(53), 299(53), 301(53), 305(53), 306(78), 307(78), 309(53), 314(53), 315(78), 316(53, 92), 317(53, 92), 318(92)
Achmatowicz, B., 116
Acree, T. E., 243
Acton, E. M., 246
Adamcik, J. A., 214, 224(96), 225(96)
Adams, A., 241
Adsetts, J. R., 183
Agenas, L. B., 302(74), 303, 304(74)
Agui, H., 114, 129(66), 130(66)
Ahmad, A., 18, 33(72a)
Ahrens, R. W., 147
Ainsworth, C., 204
Ajello, T., 89, 90(78), 91
Akiyoshi, S., 248
Akkerman, A. M., 259
Albert, A., 189, 212(20), 5, 6(13), 14(12), 17, 23, 24, 32, 33(98), 34, 36, 38, 43(143), 49, 53(225), 54(225), 55
Albertson, N. F., 224
Alessandri, L., 80, 81(38), 95(38)
Aliprandi, B., 178
Alkaitis, A., 190, 211(30a)
Allcock, H. R., 3
Allen, C. W., 103, 108(21), 110(21), 116(21), 126(21), 126(21)
Allen, D. W., 19
Allen, G. R., 290, 219(52), 322

Allen, M. J., 200, 229(50), 244
Alt, G. H., 8
Ammon, H. L., 188, 190(16), 191, 192(17), 194(15, 16), 195(15, 16)
Anderson, A. G., 188, 189, 190(11, 16), 191(11), 194(15, 16), 195(15, 16), 196(41a), 198(41a)
Anderson, D. J., 29
Anderson, D. W., 240, 244(41)
Anderson, E. L., 116, 131(72)
Anderson, R. G., 188, 189, 190(11, 16), 191(11)
Andrews, E. G., 14
Angeli, A., 72, 80, 81, 92, 95(39, 99)
Angyal, S. J., 140, 146(17)
Apelgot, S., 141
Arai, H., 250
Arakawa, H., 202
Arcamore, F., 38
Archer, S., 250, 264(159), 265
Arevalo, X., 322
Ariyan, Z. S., 17
Armit, V. W., 189
Arnall, P., 188, 189(7), 202(7), 204(66), 204(66)
Arndt, F., 247
Arnold, W. H., 245
Arnone, A., 58
Aronova, N. I., 224, 226, 228, 229(136)
Artusi, G. C., 96
Ashby, R. A., 154, 171(90)
Asher, J. D. M., 39
Ashworth, G., 242, 245(78), 246(78)
Askam, V., 308(82), 309
Aston, J. G., 34
Atkinson, J. H., 69, 82(9), 83(9)
Atkinson, R. S., 69, 82(9), 83(9)
Atlani, P. M., 42
Auerbach, J., 75
Aussems, C., 323
Avinur, P., 178

326   AUTHOR INDEX

Ayca, E., 247
Aylward, G. H., 137

B

Backeberg, O. G., 242
Badger, G. M., 36, 48
Baer, E., 108
Bär, F., 244
Baer, H. H., 116
Baeyer, A., 40
Bailey, A. S., 311, 312(86), 317(86)
Bailey, D. T., 113, 127(59a), 131(59a), 135(59a)
Bailey, J. R., 188, 189(3), 202(3)
Bailey, P. S., 94
Bailey, W. C., 247
Bajwa, G. S., 303, 309(76), 320, 321, 322(76)
Baker, W., 9
Balaban, A. T., 178, 180(135)
Baldeschwieler, J. D., 285
Balikian, J. M., 257
Ball, J. S., 45
Ballantine, J. A., 57
Bambas, L. L., 234, 239(4), 240(4), 241, 242(4)
Bamberger, E., 5, 6
Banwell, C. N., 281
Barakat, M. Z., 11
Baron, M., 5
Barrett, J. H., 59
Bart, J. C. J., 235, 236(9)
Bartlett, R. G., 241
Bartok, W., 71
Barton, D. H. R., 39, 123
Baruffini, A., 241, 248(52)
Basangoudar, L. D., 278
Bassler, G. C., 278, 279(19)
Basu, U., 202, 205(67a)
Battersby, A. R., 106, 119, 136(36)
Bauer, H., 71, 89(17)
Bauer, K., 252
Baum, T., 258
Baumgarten, R. J., 265
Baumgartner, H., 72
Baumler, R., 91
Baxter, J. N., 237, 238(16)
Bayer, M., 24
Beak, P., 182

Bean, G. P., 180
Bebie, J., 240
Becke, F., 241
Becke-Goehring, M., 12
Beeck, O., 154
Beer, R. J. S., 57
Beger, J., 192
Belegrates, K., 213
Bell, J. F., 251, 260(160), 261(160), 262(160), 263(160), 266, 268(160), 268(250), 273(250)
Bellingham, P., 179
Bencze, W. L., 200, 229(50, 51)
Benedetto, L., 270
Benigni, J. D., 303
Berchtold, G. A., 15
Berends, W., 61
Bergmann, E. D., 48
Bergson, G., 192
Beringer, F. M., 247, 255(134)
Berkowitz, D. M., 50
Bernal, I., 17
Bernheim, F., 72
Bernstein, H. J., 278, 282(21)
Berry, K. H., 54
Berson, J. A., 190, 192, 194(31c)
Betkerur, S. N., 278, 306(80), 307(80), 309
Beukers, R., 61
Bevington, J. C., 4
Bevis, M. J., 100, 124(3)
Bhat, G. A., 323
Biava, C. G., 244
Bickel, W. D., 72
Biddiscombe, D. P., 167(113), 168
Biedermann, A., 19
Biellmann, J. F., 42
Bieron, J. F., 275
Biyloo, J. D., 51
Bisset, N. G., 38
Biswas, K. M., 294(62), 295, 296(62), 300(62), 301(62), 302(62), 308(62), 310(62), 311(62)
Bitter, K., 219
Bizioli, G., 38
Black, P. J., 286, 287(43), 288
Blackburn, G. M., 61
Blackwood, J. E., 237
Blackwood, R. K., 243
Bláha, K., 118
Blanchard, W. M., 245

Blau, W., 48
Blicke, F. F., 104
Blinder, S. M., 88, 93(72)
Bloch, K., 152
Bloodworth, A. J., 12
Bloomfield, J. J., 10
Blunck, F. H., 274
Boaz, H. E., 266, 269(250), 273(250)
Bobbitt, J. M., 101, 103(9), 104(24, 26), 105(9, 25), 106(25), 108(21), 110(9, 19, 20, 21, 22, 23, 24, 25, 26), 116(9, 19, 21, 22, 24, 25), 117(9), 118(20), 119(9, 23, 26), 120(26), 121(22), 126(21), 130(22, 23), 133(9, 19, 21, 24, 25), 134(9), 135(26), 136(26)
Boca, J. P., 305
Bocchi, V., 80, 81, 82(50), 83(46, 50), 84(50), 96
Bock, H., 2
Böhme, H., 238, 248, 251(25), 253, 256(140), 262(25), 273(25)
Boekelheide, V., 190
Böshagen, H., 237, 238(18), 267(18)
Bond, G. C., 155
Bonham, J., 182
Bonnett, R., 50
Borgel, C., 102, 106(13), 107(13), 109(13), 129(13)
Borgna, P., 241
Borne, R. F., 111, 112(51)
Borsdorf, R., 192, 193, 194(32)
Bosshard, H. H., 253
Bott, R. W., 152
Bowie, W. T., 58
Boxhill, G. C., 229
Boykin, D. W., Jr., 85, 88(63), 90, 91(63), 92(63), 93(63)
Boyland, E., 244
Bradenberg, C. F., 188, 189(8), 202(8)
Bradley, R. M., 243
Bradshaw, H., 266, 267(247)
Brand, J. C. D., 278
Brannock, K. C., 219
Brassel, J., 35
Braun, J., 246, 257(119)
Braunton, P. N., 19
Bredereck, H., 26
Brennan, J. A., 72
Brierley, A., 247, 255(134)
Brodie, B., 5

Brookes, C. J. Q., 305, 306(78), 307(78), 315(78)
Brossi, A., 111, 112, 113, 114(54), 122(59), 123, 124, 125(82), 127(56), 129(54), 131(54, 56, 59, 59a), 132(54, 56, 59), 133(54, 87), 135(59a)
Brown, B. R., 25
Brown, C. A., 151, 152(66)
Brown, D. J., 32, 49
Brown, D. W., 101(49), 106, 107(35), 109(39), 110(35, 45), 117(48), 118(45), 119(49), 120(49), 130(35, 45, 49), 131(49), 134(49), 136(49, 76)
Brown, G. B., 146
Brown, H. L., 151, 152(66)
Brown, J. P., 23, 30
Brown, R. D., 323
Brown, R. E., 229, 230
Brown, R. K., 281, 282(28), 285(28), 296(28), 297(28), 299(28), 300(28), 301(28, 72), 302(28, 72), 303, 311(28, 72), 319, 320, 321, 322(76)
Brown, S. D., 122
Brown, W. G., 139, 144(10)
Brunetti, H., 24
Bryan, G. T., 244
Buchardt, O., 61, 190
Buchwald, A., 246, 247(121), 251(121), 252(121), 262(121)
Buck, K. T., 126
Buckley, D., 8
Budzikiewicz, H., 137, 238, 239(20), 262(20), 264(20), 265(20), 271(20), 274(20)
Buechel, K. H., 50
Bullitt, O. H., 190
Bungard, G., 244
Burckhardt, C. A., 200, 229(51)
Burdon, M. G., 274
Burk, E. H., 30
Burpitt, R. D., 63
Burr, G. O., 36
Burstein, S., 114, 130(63)
Busch, D. H., 7
Buysch, H., 322

C

Cacace, F., 144, 147, 178
Caignault, C., 114, 130(64), 131(64)

Calf, G. E., 151, 158, 159, 160(63, 70, 72), 161(70, 72), 164(72), 166(105), 173(105)
Cals, S., 230
Calvin, M., 190, 211(30a)
Campaigne, E., 25
Campbell, H. A., 39
Canevazzi, G., 38
Carelli, V., 58
Carlin, R. B., 319
Carlos, D. D., 30
Carlson, D. P., 319
Caro, A. N., 259
Carpenter, D. C., 5
Carstensen-Oeser, E., 53
Casini, A., 58
Casinovi, C. G., 38
Casnati, G., 80
Caton, M. P. L., 49
Cattadori, F., 90
Catterall, G., 55
Cava, M. P., 117, 126, 134(73)
Cavagnol, J. C., 247, 256(125), 260(125), 261(125), 262(125), 263(125), 273(231)
Cerutti, P., 45, 51
Červinka, O., 118
Cescon, L. A., 47
Cetina, R., 282
Chaddock, T. E., 243
Challenger, F., 20, 51(82)
Chandra, P., 61
Chang, A., 258, 266(208)
Chang, T.-C. L., 202
Chapman, A. W., 265
Chapman, D., 278, 300(14)
Chapman, O. L., 15
Chargaff, E., 39
Chattaway, F. D., 253
Chatterjee, A., 8
Chaudhuri, N. K., 224
Cheeseman, G. W. H., 24
Chemizart, A., 322
Chenicek, J. A., 72
Chien-Pen Lo, 249
Chierici, L., 80, 83(46), 96
Childress, S. J., 258
Choufoer, H. J., 258
Chovanec, G. F., 190, 211(28b)
Chow, B. F., 71
Chriss, R. J., 71, 86(14), 87(14)

Christensen, J. E., 246
Christman, D. R., 139
Chu, C. C., 68
Chukhadzhyan, G. A., 229
Ciamician, G., 68, 89, 91
Cignarella, G., 269
Ciranni, E., 147
Cirrani, G., 147
Clark, F. E., 245, 270(99)
Clark, J., 34
Clark, V. W., 50
Clark-Lewis, J. W., 56, 64
Clauss, K., 243
Clemo, G. R., 21
Coates, C. E., 245
Cohen, G., 115, 122
Cohen, L. A., 222, 225(116), 279, 285(23), 286(23), 294(23), 297(23), 299(23), 300(23), 301(23), 302(23), 304(23), 307(23), 308(23), 309(23), 313(23)
Cohen, R., 190
Coke, J. L., 39
Collin, P. J., 164
Collins, M., 115, 122
Colman, J., 55
Colomb, H. O., 94
Colonge, J., 203, 211(68a)
Comes, R. A., 259
Commerford, J. D., 257
Conant, J. B., 71
Cook, G. L., 45
Cooksey, A. R., 91
Cookson, R. C., 13
Coraor, G. R., 47
Cordier, P., 216, 224(99)
Corrodi, H., 115
Corrsin, L., 164
Coufoer, H. J., 258
Cramer, F., 39
Crawford, B. R., 144, 145(36), 146
Crawford, E., 150, 153(61), 158(61), 160(61), 170(61)
Crelling, J. K., 107, 108(38), 110(38), 130(38)
Crutchley, D. J., 57
Curtis, N. F., 18
Cusmano, G., 92, 95(99)
Cutter, H. B., 245, 246(105)
Cuvigny, Th., 246
Cymerman-Craig, J., 237, 238(16)

D

D'Adamo, A., 139
Daglish, A. F., 21
Daily, B. P., 292, 323
Daisley, R. W., 35
D'Alelio, G. F., 242, 258(58), 266(208)
Dall'Olio, A., 96
Dalton, C. K., 107, 108(38), 110(38), 130(38)
Dalton, D. R., 107, 108(38), 110(38), 130(38)
Daly, J. W., 279, 285(23), 286(23), 290, 292(50), 293(50), 294(23, 50), 295(50), 297(23, 50), 298(50), 299(23, 50), 300(23, 56), 301(23, 56, 70), 302(23, 56), 303, 304(23), 305(50), 307(23), 308(23), 309(23, 50), 313(23, 50), 319(70)
Darlak, R. S., 190, 211(28b)
Dascola, G., 96
David, S., 322
Davidson, J. N., 39
Davies, A. G., 12
Davies, J. V., 45
Davies, R. J. H., 61
Davies, W., 60, 242
Davy, M., 114, 130(64), 131(64)
Dawn, H.-S., 235, 242(11), 253, 254(174)
Dawson, R. F., 139
Deady, L. W., 124
Dearnaley, D. P., 305, 306(78), 307(78), 315(78)
De Christopher, P. L., 265
Deeks, R. H. L., 114, 120(65), 129(65), 308(82), 309
Defay, N., 322
De Garmo, O., 242, 245(78), 246(78)
Dehn, W. M., 252
Delpierre, G. R., 27
de Mandirola, O. B., 5
de Martilis, F., 249
DeMatte, M. L., 190, 211(28b)
de Mayo, P., 62, 73
Dennstedt, M., 89
de Roode, R., 242
de Schryver, F. C., 28
Dessauer, R., 47
Detar, R. L., 229
Dev, S., 225

De Varda, G., 89
Dewar, M. J. S., 230
Deyrup, J. A., 313
Diaz, E., 282
Dick, W., 6
Diehl, P., 292
Dietrich, R., 198
Dinan, F. J., 275
Dinneen, G. U., 72
Djerassi, C., 137
Dmitrieva, N. D., 224
Dobbie, J. J., 33
Dodd, G. M., 57
Dohme, A. R. L., 261
Dolby, L. J., 93, 94, 297(67), 298
Dolci, M., 249
Donahoe, H. B., 257
Dorfman, L. M., 147
Dornow, A., 207
Doudoroff, M., 38
Dougherty, G., 17
Dovinola, V., 81
Draber, W., 56
Dreux, J., 203, 211(68a), 214, 224(93)
Drexler, M., 247, 255(134)
Druge, L. W., 72
Drushel, H. V., 202
Dufraisse, C., 76, 78
Duhamel, L., 108
Duhamel, P., 108
Duhoux, S., 322
Dukes, C. E., 244
Dunathan, H., 58
Dunbar, R. E., 252
Duquesne, M., 141
Dutt, S. B., 68
Dutta, C. P., 103, 104(24), 110(24), 116(24), 133(24)
Dutta, S., 103, 105(25), 106(25), 110(25), 116(25), 133(25)
Dyer, J. R., 278, 286(15)
Dyke, S. F., 101(49), 106, 107(35), 109(35), 110(35, 45), 115, 117(48), 118(45, 71), 119(71a), 120(49, 71), 127(71), 128, 130(35, 45, 49), 131(49), 134(49), 136(49, 71, 76)
Dzantiev, B. G., 139
Dziomko, V. M., 265

E

Eaborn, C., 152
Eaker, C., 242, 245(78), 246(78)
Eberhardt, H., 198
Ebermann, R., 101, 103(9), 105(9), 110(9), 116(9), 117(9), 119(9), 133(9), 134(9)
Eccleston, B. H., 141, 142(26), 148(26)
Eckenroth, H., 245, 250(103), 251(103, 153), 257(103, 107, 153), 264(103)
Eder, M., 213
Effenberger, F., 26
Egelhaaf, A., 40
Eglite, D., 98
Eguchi, T., 188, 189(2), 202(2), 218(2)
Eiden, F., 248, 256(140)
Eidinoff, M. L., 152, 174(79)
Eisenbrand, J., 240
Eisner, U., 61
Ekstrom, A., 138, 141, 148(8)
Elad, D., 47, 61(216)
Elderfield, R. C., 19, 187, 188(1), 189(1)
Eley, D. D., 154
Elgersma, R. H. C., 319
Elliott, I. W., 58
Ellis, A. C., 115, 119(71a)
Elvidge, J. A., 278, 279(17), 281, 285(29), 292(29)
El-Zimaity, T., 219, 222(112)
Emerson, W. S., 105, 126(30)
Emsley, J. W., 278
Enander, B., 147
Epsztajn, J., 188, 192(18), 194(18), 210, 211, 212
Erdtman, H., 40
Ernst, I. T., 164
Errede, L. A., 164
Ertürk, E., 244
Étienne, A., 28, 76
Evans, D. F., 68
Evans, E. A., 138, 141(5), 142(5), 146(5), 174(5), 178(5)
Evleth, E. M., 192, 194(31c)

F

Fahlberg, C., 234, 239(6, 7), 240(6, 7), 245, 255, 256(7)
Fahr, E., 1, 20(1), 61
Falk, J. E., 38
Fanta, P. E., 49
Farkas, A., 154
Farkas, L., 154
Farquhar, D. K., 205, 206(70), 207(70)
Farray, N., 242, 256(66)
Favini, G., 192, 194(31a)
Fawcett, F. S., 27
Fayez, M. B. E., 106, 110(34), 122(34), 126(34), 130(34), 133(34)
Feeley, A., 61
Feeney, J., 278
Feigl, D. M., 242, 258(58), 266(208)
Feitelson, J., 192
Feller, H., 61
Feng, P. Y., 141, 143(21)
Fernandez, C. M., 140, 146(17)
Fessler, W. A., 242, 258(58), 266(208)
Fierz-David, H. E., 35
Fieser, L. F., 7, 38
Fieser, M., 7, 38
Fila-Hromadko, S., 63
Filer, L. J., 240, 244(41)
Filler, R., 28
Fischer, E., 11, 100
Fischer, H., 69, 70(7), 72, 81(7), 82, 83(53) 86(7), 90, 91
Fischer, H. O. L., 108
Fischer, R., 242
Fischer, B. D., 151, 158
Fitzgerald, B. M., 319
Fitzhugh, O. G., 244
Fitzgerald, J. A., 319
Fitzpatrick, M., 227, 228(141), 229(141)
Fletcher, G., 141, 148(23)
Flippen, J. L., 47
Flores, R. J., 214, 224(96), 225(96)
Foerst, W., 239, 240(29), 244(29)
Fondovila, M. E., 253
Fong, D., 183
Fono, A., 254
Forbes, E. J., 100, 124(3)
Forino, C., 86, 90(65)
Forkey, D. M., 189
Forneau, J. P., 114, 130(64), 131(64)
Forrester, A. R., 30
Forsyth, R., 100
Foster, R. G., 281, 285(29), 292(29)
Foucaud, A., 229
Fowkes, F. S., 241, 250(42), 251(42)

AUTHOR INDEX 331

Fox, B. J., 164
Fox, M., 182
Franchi, G., 249
Franck, R. W., 13
Frank, A. W., 100, 101, 108, 119(11), 136(11)
Frank, F., 86
Frank, R. L., 42, 54(193)
Frank, R. W., 75
Franke, J., 189
Frankel, J. J., 13
Fraser, R. R., 158, 171(101)
Frawley, J. P., 244
Freeman, F., 196, 197(43), 205(44), 206 (44, 70), 207(70), 225(44)
Freese, K. J., 50
Freimanis, J., 217
Freri, M., 94, 95(107)
Fresenius, P., 253
Freytag, J., 202
Friedman, O. M., 114, 130(63)
Fritchie, C. V., 201
Fritsch, P., 273
Fritz, H., 44, 105, 106, 111(31), 127(31), 129(31)
Froelich, R., 243
Frydman, B. J., 247
Fuchs, H., 4, 35
Fuchs, R., 246, 247(121), 251(121), 252 (121), 262(121)
Fürst, G., 61
Fuhlhage, D. W., 44
Fukada, N., 298, 299(65)
Fukui, H., 279
Fukui, K., 12, 323
Fukumoto, K., 112, 113(54), 114, 127(55), 129(66), 130(66), 132(55)
Fukushima, D. K., 152, 174
Funasaka, W., 202
Furey, R. L., 13
Furukawa, H., 123

G

Gabriel, S., 55
Gänicke, K., 68
Gait, A. J., 241, 260(50)
Gallagher, T. F., 152, 174
Gamble, N. W., 60

Ganellin, C. R., 293, 294(58), 295(58), 301(58), 302(58), 303(58), 307(58), 308(58)
Ganguly, M., 8
Gannon, W. F., 303
Gant, P. L., 140
Garcia-Muñoz, G., 215, 225(97)
Gardent, J., 107, 110, 127(40)
Gardini, G. P., 80, 81, 82(50), 83(46, 50), 84(50)
Garnett, J. L., 137, 138, 139, 140, 141, 142(20), 144(20), 145(20, 36), 146 (17, 20), 147, 148(8, 23), 150, 151, 152, 153, 154, 155(58, 82), 157(82), 157(58, 80, 82), 159(57, 82), 160(63, 70, 72), 161(70, 72), 162(82), 164(72), 166(82, 105), 170(98), 171(82, 90), 173(105), 175(56, 68), 176(124, 125), 177(127), 178(125, 126), 183, 185(64)
Gattermann, L., 19
Gaudiano, A., 270
Gavrilov, V. E., 88
Gawer, A., 323
Gay, R. L., 241, 248(56), 254(56)
Geiger, W., 237, 238(18), 267(18)
Gell, R. J., 21
Gensler, W. J., 99, 108(1), 110, 114, 126(62), 130(47), 133(47), 134(47)
Genunche, A., 178, 180(135)
George, H., 28
Gerding, H., 11
Gerver, J. H., 242
Gestblom, B., 285, 292(38)
Ghanem, N. A., 147
Gialdi, F., 241, 248(52)
Giambrone, S., 90, 91
Giddey, A., 50
Gilardi, R. D., 47
Gilchrist, T. L., 29
Giletti, B. J., 139
Gindler, E. M., 247, 255(134)
Giori, C., 96
Giza, Y., 258, 266(208)
Gladys, C. L., 237
Glos, M., 198
Glover, E. E., 30
Godar, E., 204, 205(68b), 211(68b)
Godefroi, E. F., 222, 225(115a)
Godfrey, J. J., 214, 225(92)
Göth, H., 45

Göttlicher, S., 59
Gofen, G. I., 43
Gold, V., 183
Goldman, I. M., 14
Goldstein, J. H., 288
Gompper, R., 243, 244(80), 246(80), 247
Goodman, L., 246
Gorbatschow, S. W., 247, 251(122), 254(122)
Gordon, M. P., 146
Gorin, E., 178, 180(132)
Gortner, R. A., 36
Gott, P. G., 63
Govier, W. M., 229
Govindachari, T. R., 114, 124(61)
Gowenlock, B. G., 12
Gradsten, M. A., 16
Graf, R., 35
Graymore, J., 17
Greenlee, T. W., 141, 143(21)
Greenwald, R. B., 200, 201(52), 229(52)
Gregory, H., 31
Grein, A., 38
Grethe, G., 111, 112, 113, 114(54), 122(59), 125, 127(56, 59a), 129(54), 131(54, 56, 59, 59a), 132(54, 56, 59), 133(54, 87), 135(59a)
Gribble, G. W., 297(67), 298
Grimm, D., 97
Grinsteins, V., 217, 221(102), 222(102), 224(102), 229(102), 230(102)
Gripenberg, J., 37
Gritter, R. J., 71, 86(14), 87(14)
Grogan, C. H., 245, 255(113), 257(113), 258(112), 262, 266(232), 273(232)
Grohe, K., 214
Gronowitz, S., 285, 292(38)
Grumbach, F., 229, 230
Grundkoffer, M., 198
Grundmann, C., 9
Grunwell, J. R., 15
Guljé, A. R., 94
Gundermann, K. D., 256, 257(197)
Gurevoch, A. I., 231
Gusten, H., 54
Guthrie, D. A., 101, 119(11), 136(11)

H

Habermehl, G., 59
Hackerman, N., 171
Haddlesey, D. I., 308(81), 309, 315(81)
Hafner, K., 59
Hagan, W. V., 313
Hagen, H., 241
Hahn, W. E., 188, 192(18), 194(18), 210, 211, 212
Haines, W. E., 45
Hall, G. G., 323
Halpern, B., 147
Halvarson, K., 176, 178(129), 180(129)
Hamer, F. M., 34
Hammond, P. D., 17
Hamor, G. H., 242, 243, 245, 246(106), 250, 256(64, 66, 106), 257(106), 267(155), 272(155)
Hanker, I., 189
Hanngren, A., 140
Hanske, W., 57
Hantzsch, A., 34, 42
Harbridge, J. B., 247
Harcourt, D. N., 111, 112(52), 113(52), 121(52), 127(52), 132(52), 134(52)
Hardegger, E., 246
Hardisson, A., 323
Hardy, G., 101(49), 110, 117(48), 119(49), 120(49), 130(49), 131(49), 134(49), 136(49, 76)
Hare, L. E., 111, 112(51)
Harley-Mason, J., 64
Harper, R. J., 159
Harpp, D. N., 62
Harris, M., 63
Harris, R. L. N., 38
Harris, T. M., 182
Harrison, W. F., 188, 189, 190(11), 191(11)
Hart, H., 259
Hart, L. E., 241, 250(42), 251(42)
Hartley, W. N., 33
Hartmann, P., 82, 83(53)
Hartough, H. D., 41, 55
Haskelberg, L., 31
Hassid, W. Z., 38
Hata, N., 61, 190
Hatcher, W. H., 5
Hauck, F. P., 42, 224
Hauser, C. R., 241, 248(56), 254(56), 269(135), 270(135)
Havinga, E., 319
Haworth, R. D., 242
Hay, R. W., 18

Hayasaka, T., 114, 129(66), 130(66)
Hayes, N. F., 174
Heacock, R. A., 36, 290, 292(50), 293(50), 294(50, 60, 61, 63), 295(50, 61), 297(50), 298(50), 299(50), 305(50), 309(50), 313(50)
Hecht, G., 15
Hedaya, E., 50
Heffernan, M. L., 286, 287(43), 288, 323
Heibronner, E., 2
Heider, K., 246, 257(119)
Heine, H. W., 15
Heifner, C., 62
Heller, G., 246, 247(121), 251(121), 252(121), 262(121)
Hellmann, H., 249
Helm, R. V., 45
Helmick, L. S., 180, 182(141)
Helmy, E., 87
Henbest, H. B., 8
Henderson, L. J., 137, 170
Henderson, W. E., 245, 270(100)
Hendrickson, Y. G., 72
Henecka, H., 15, 224
Henly, W. O., 319
Henry, R. A., 252
Henseleit, E., 26
Henshall, T., 222, 225(114)
Henze, M., 28
Herbison-Evans, D., 324
Herington, E. F. G., 167(113), 168
Herout, V., 37
Herr, W., 144, 147
Herzog, W., 242
Hesse, H., 250
Hettler, H., 235, 236(12), 237(12), 238(13, 14), 239(20), 241, 244, 246(17), 251, 252(17, 167), 262(17, 20, 96, 167), 263(13, 17, 96), 264(17, 20, 96), 265(17, 20, 28, 96), 266(13), 267(13, 17, 28), 268 (13, 17), 271(20, 28, 43, 96), 272(43), 273(28, 43), 274(20, 28, 167), 275(28, 96, 253), 276(14, 43)
Hideg, K., 18
Higashida, S., 27
Higuchi, T., 235, 242(11), 253, 254(174)
Hiiragi, M., 114, 115, 129(66), 130(66, 67, 67a)
Hill, C. A., 256
Hill, J. H. M., 50

Hillers, S., 173
Hills, J., 31
Hilmy, M. K., 254
Hine, J., 247
Hinman, R. L., 50, 298, 299(64), 318
Hinton, I. G., 111, 113(53), 127(53), 131(53), 132(53)
Hiremath, S. P., 278
Hirota, K., 151, 158
Hitchings, G. H., 33, 40(136)
Ho, T. C., 68
Hoch, H., 322
Hoch, J., 102, 106(13), 107(13), 109(12, 13), 129(13)
Hodgkins, J. E., 122
Hodson, H. F., 52
Hodges, R. J., 150, 175(56), 176(124, 125), 178(125, 126)
Hodgman, C. D., 262
Höft, E., 70
Hoffman, H., 198
Hoffman, R. A., 281, 285, 292(38)
Hoffmann, H. M. R., 274
Hoffmann, R. Z., 57(2a)
Holland, V. R., 30
Hollemann, A. F., 258
Holmes, J. L., 51
Hopff, H., 248
Horak, V., 189
Horie, M., 34
Horie, T., 23
Horii, Z., 226
Horiuti, J., 154
Hornfeldt, A. B., 285, 285(38)
Horning, E. S., 244
Horrex, C., 158
Hoshino, O., 112, 115(57), 121, 127(57, 58), 128(77, 78), 129(57, 58), 131(57)
Hoskinson, R. M., 146
Hosmane, R. S., 278
Hough, E., 123
Houghton, E., 293, 294(59), 311
Houk, K. N., 182
Hradetzky, F., 213
Hudec, J., 13
Hudson, C. B., 309
Huebner, C. F., 35
Hüni, A., 86
Hünig, S., 35, 322
Huffman, J. W., 107, 133(39)

Hughes, W. L., 174
Humi, H., 242
Hummel, G., 90
Hunebelle, G., 174
Hunt, R. R., 49
Hunter, J. H., 269, 270(257)
Hurd, C. D., 274
Hurst, D. T., 49
Hussain, A. A., 235, 242(11), 254
Hutzinger, O., 290, 292(50), 293(50), 294(50, 63), 295(50), 297(50), 298(50), 299(50), 305(50), 309(50), 313(50)
Huyser, E. S., 159

I

Iddon, B., 45
Iffland, D. C., 243
Igeta, H., 43, 59
Ikehara, M., 39
Ikushima, K., 206
Illari, G., 96
Imoto, E., 45
Intrieri, O. M., 146
Isen, Zh. I., 226, 229(137)
Ishiguro, T., 206
Ishihara, H., 61
Ishimaru, H., 114, 129(66), 130(66, 67, 67a)
Isono, C., 241, 245(55), 247(55)
Ito, K., 123
Ito, M., 115, 130(70)
Ito, T. I., 197, 205(44), 206(44), 225(44)
Itoh, M., 285
Ivanchukov, N. S., 88, 93(72)
Iwata, C., 226

J

Jackman, L. M., 278, 284(16), 288, 323
Jackson, A. H., 44, 117, 125, 134(74), 294(62), 295, 296(62), 300(62, 69), 301(62), 302(62, 69), 303, 305(69), 308(62, 69), 309(69), 310(62, 69), 311(62, 69), 312(69), 313(69), 319
Jacobson, P., 19
Jacquier, R., 114, 130(64), 131(64)
James, F. C., 60
Jamrozik, J., 64

Janfaza, M., 256
Jaques, B., 114, 120(65), 129(65)
Jardine, R. V., 281, 282(28), 285(28), 296(28), 297(28), 299(28), 300(28), 301(28, 72), 302(28, 72), 303, 311(28, 72)
Jaspers, S., 323
Jastorff, B., 263
Jefcoate, C. R., 84
Jenkins, G. I., 154
Jenkins, P. N., 123
Jensen, B., 190
Jensen, H., 243
Jensen, H. B., 70, 71(12), 81(12)
Jensen, L. H., 188, 192(17)
Jerchel, D., 42
Jesurun, J. A., 234, 251(3), 252(3), 256(3), 261(3), 262(3), 263(3), 266(3), 269(3), 271(3), 273(3)
Jo, Y., 257
Johns, H. E., 61
Johns, S. R., 322
Johnson, A. L., 59
Johnson, A. W., 38, 69, 82(9), 83(9)
Johnson, C. D., 175, 178(122), 179(122), 181(122)
Johnson, E. M., 15
Johnson, H. W., 282, 283(33), 284(33), 285(33), 298(33), 304
Jones, R. A., 36
Jonsson, G., 115
Jorgenson, M. J., 61
Josephson, K., 246
Jucker, O., 246
Junek, H., 195, 196, 197(42), 198(42)
Jungel, H. J., 234, 239(5)
Jurd, L., 21

K

Kadunce, R. E., 245, 258(108), 264(108)
Kadyrov, C. S., 43
Kainer, H., 87, 88(69), 92, 93(71)
Kaisin, M., 322
Kalb, M., 34
Kamal, A., 18, 33(72a)
Kamei, H., 285
Kametani, T., 112, 113(54), 114, 127(55), 129(66), 130(66, 67, 67a), 132(55), 312(88), 313

Kan, R. O., 1, 13, 20(1), 62
Kaneko, C., 59
Kapadia, G. J., 106, 110(34), 122(34), 126(34), 130(34), 133(34)
Karklins, A., 217, 220(104a, 104b)
Karle, I. L., 47, 51
Karle, J., 51
Karmil'chik, A. Ya., 173
Karminski, W., 49
Karr, C., 202
Karrer, P., 33, 46, 51(209)
Kašpárek, S., 36, 293, 294(60)
Katritzky, A. R., 70, 83, 175, 178(122), 179(122), 180, 181(122), 190
Kato, K., 248
Kauffman, G. M., 182
Kauffmann, T., 49
Kawazoe, Y., 182
Kay, I. T., 38
Kazuka, S., 190, 211
Keller-Schierlein, W., 40
Kelly, C. I., 100
Kelly, F. W., 274
Kemball, C., 150, 153(61), 155(59), 158(61), 159, 160(61), 170(61), 173
Kempter, G., 195, 196(41b), 198(41b)
Kenyon, W. G., 15
Kessar, S. V., 224
Khanna, K. L., 101, 103(9), 105(9), 106(32), 110(9, 19, 32), 111(32), 116(9, 19), 117(9), 119(9), 126(32), 130(32), 131(32), 132(9, 19, 32), 134(9)
Kharasch, M. S., 254
Khattak, M. N., 46
Kiely, J. M., 101, 103(9), 105(9), 110(9, 19, 20), 116(9, 19), 117(9), 118(20), 119(9), 132(9, 19), 134(9)
Kieran, P., 173
Kiesslich, G., 35
Kigasawa, K., 114, 129(66), 130(66, 67, 67a)
Kikukawa, O., 48
Kinsman, R. G., 106, 107(35), 109(35), 110(35), 130(35)
Kirby, G. W., 39
Kirksey, C. H., 58
Kirzner, N. A., 50, 57(234)
Kitano, H., 12
Kittleson, A. R., 250

Klages, A., 266, 273(252)
Klason, P., 12
Klein, K., 245, 257(107)
Kleineberg, G., 213
Kleinicke, W., 246, 247(121), 251(121), 252(121), 262(121)
Kleopfer, R., 61
Klötzer, W., 247
Klose, G., 322
Klose, W., 90
Kloss, J., 246, 247(121), 251(121), 252(121), 262(121)
Klubes, P., 148
Knoll, J. E., 152, 174(79)
Knowles, W. S., 8
Kny, H., 279, 285(23), 286(23), 294(23), 297(23), 299(23), 300(23), 301(23), 302(23), 304(23), 307(23), 308(23), 309(23), 313(23)
Kobayashi, S., 8, 16(22), 21
Koerppen, G., 245, 250(103), 251(103, 153), 257(103, 153), 264(103)
Kogan, G. A., 108
Kogan, I. M., 265
Kohler, E. P., 245
Kojima, T., 202
Kokko, J. P., 288
Kokoreva, I. Yu, 226
Kolbe, J., 202
Kolosov, M. N., 231
Kolthoff, I. M., 242
Kondo, T., 114, 130(60)
Kormendy, M., 242, 256
Kornblum, N., 243
Korobko, V. G., 231
Kortge, P., 141, 142(27, 29)
Kosuge, T., 44
Kosyakovskaya, M. N., 43
Kovačević, K., 63
Kováts, E. sz., 18
Kozuka, S., 190, 211(28a)
Kraaijeveld, A., 259
Krämer, D. M., 61
Krauch, C. H., 61
Kreile, D., 98
Kremer, C. B., 4, 14(5)
Krizek, H., 139, 144(10)
Kroschwitz, J. I., 323
Krüger, U., 237, 238(14), 276(14)
Kucherov, V. F., 108, 226, 228, 229(136)

Kühn, A., 40
Kugita, H., 37
Kuhn, R., 48, 87, 88(69), 92, 93(71)
Kulicki, Z., 49
Kumar, A., 224
Kumler, P. L., 190
Kupchan, S. M., 106, 119(37), 126(37), 136(37)
Kurtin, W. E., 323

L

Lagercrantz, C., 73
Lagowski, J. M., 83, 190
Lahav, M., 28
Lakshmikantham, M. V., 117, 134(73)
Lamberton, A. H., 48
Lamberton, J. A., 322
Lamchen, M., 27
Lander, G. D., 264
Landesberg, J. M., 182
Landquist, J. K., 7, 8(19)
Lang, J., 318
Lapworth, A., 206, 242
Larson, J. K., 14
Laslett, R. L., 36
Latham, D. R., 188, 189(8), 202(8)
Laubach, G. D., 266
Lauer, W. M., 158, 164
Lautsch, W., 83
Law, S. W., 140, 142(20), 144(20), 145(20), 146(20), 147
Lawrenson, I. J., 167(113), 168
Lechner, A., 211
Lee, H. L., 111, 112, 114(54), 125, 127(56), 129(54), 131(54, 56), 132(54, 56), 133(54)
Leese, C. L., 24
Legrand, L., 260
Lemberger, A. P., 239
Lemke, L., 245
Lemmon, R. M., 147
Lendle, W., 322
Leonard, N. J., 42
Leroy, G., 323
Leser, E. G., 13
Letcher, R., 123

Levey, S., 245, 246(105)
Levina, R. Y., 224
Levitt, G., 259
Ley, H. L., 244
Liberatore, F., 58
Liberles, A., 76
Libermann, D., 229, 230
Libman, D. D., 35
Lightner, D. A., 73, 97
Likforman, J., 107, 110, 127(40)
Linabergs, J., 217, 220(100), 224(100)
Lindroth, H., 147
Lindsay Smith, J. R., 84
Linhart, F., 35
Link, K. P., 35, 39
Linn, W. J., 190
Linnell, R. H., 68
Linstead, R. P., 22
List, A., 245, 255
Litvan, F., 199
Lloyd, D., 18
Lochte, H. L., 188, 189(5, 6), 202(5, 6), 204(5, 6), 205(5, 6), 206(5)
Loening, K. L., 237
Loev, B., 242, 256
Logothetis, A., 51
Lombardino, L., 241, 242(57)
Long, M. A., 150, 183
Longeraz, R., 224
Lord, N. W., 88, 93(72)
Lord, R. C., 164
Lorenc, L., 14
Lorenz, M., 251, 252(162, 163, 164, 165), 273(162, 163, 164, 165)
Los, M., 192
Losin, E. T., 187, 188(1), 189(1)
Low, L. K., 97
Lowe, A. R., 22
Lozac'h, N., 260
Lucas, K., 33
Ludwig, A., 245, 254
Lüttke, W., 55
Lumpkin, C. C., 247, 255(134)
Lustgarten, D. M., 229, 230
Lutri, C., 80, 95(39)
Lutz, R. E., 85, 88(63), 90, 91(63), 92(63), 93(63)
Lynch, T. R., 19
Lyssi, H., 248

M

McCarty, C. G., 264
McCaskill, E. S., 58
McClelland, E. W., 241, 250(42), 251(42), 260(50)
Macdonald, C. G., 151, 158(62), 160(62), 161(162), 163(62), 164, 170(62)
McEwen, W. E., 100
McGeachin, S. G., 6
McIntosh, C. L., 15
McIntyre, P. S., 296(66), 297(66), 298, 312(66), 313(66)
McKibben, M., 241
McKillop, A., 10, 43(33)
McLafferty, F. W., 137
McLamore, W. M., 266
McOrmie, J. F. W., 49
McPhail, A. T., 39
McWeeny, R., 323
Mader, F., 45
Madonia, P., 84, 90
Magee, M., 174
Magidson, O. J., 247, 251(122), 254(122)
Magidson, O. Y., 214
Magnus, P. D., 278, 300(14)
Maisseron-Canet, M., 305
Makrides, A. C., 171
Malikov, V. M., 322
Mameli, E., 265, 266(244), 267(244)
Mamuzić, R. I., 14
Manatt, S. L., 192, 194(31c)
Maneck, G., 36
Mann, F. G., 111, 113(53), 127(53), 131(53), 132(53)
Mannessier, A., 260, 261(219), 262(19)
Mannessier-Mameli, A., 255, 261(189), 265, 266(189, 225, 229, 244), 267(244), 268(225, 228, 246), 270(227)
Manns, E., 241
Manske, R. H. F., 206
Mansouri, M., 102, 106(13), 107(13), 109(13), 129(13)
Mantescu, C., 178, 180(135)
Mao, C.-L., 248, 269(135), 270(135)
Mapstone, G. E., 202
Marais, J. L. C., 242
Marburg, M., 110, 130(47), 133(47), 134(47)
March, J., 10, 11(32), 35(32)

Marchand, A., 103, 108(21), 110(21), 116(21), 126(21), 133(21)
Marco, G. J., 250
Mariella, R. P., 204, 205(68b), 211(68b)
Marini-Bettolo, G. B., 38
Marino, J. P., 274
Mark, H. F., 239, 240(30), 243(30), 244(30)
Marondel, G., 54
Marschall, H., 33
Martel, J., 76, 249
Martello, R. F., 190
Martin, D., 274
Martin, J. C., 63, 219
Martin, J. F., 167(113), 168
Martin, R. H., 322
Martin, R. L., 17
Martin-Smith, M., 26
Marullo, N. P., 265
Marumoto, R., 8, 16(22)
Marvel, C. S., 2
Marvel, J. R., 239
Marzec, A., 180
Maselli, C., 248, 262, 273
Mason, S. F., 192, 194(31b)
Massingill, J. L., 122
Mateos, J. L., 282
Matsen, F. A., 171
Matskanova, M. A., 32
Matsui, K., 257, 264(198)
Matsumara, K., 198
Matthews, B. W., 61
Mattok, G. L., 293, 294(61), 295(61)
Mayer, H., 27
Mayer, R., 189
Maynard, J. T., 190
Mayor, P. A., 308(81), 309, 315(81)
Meadow, J. R., 238, 247, 256(125), 260(125), 261(125), 262(24, 125), 263(125), 273(24)
Medenwald, H., 237, 238(18), 267(18)
Medzhidov, A. A., 88
Meen, R. H., 63, 219
Meerer, J. J., 311, 312(86), 317(86)
Mehta, S. J., 250, 267(155), 272(155)
Meisel, S. L., 55
Melander, L., 176, 179(129), 180(129)
Melson, G. A., 7
Meltzer, R. I., 229, 230
Meriwether, H. T., 144

Merritt, L. L., 245, 246 (105)
Mertes, M. F., 111, 112 (51)
Meth-Cohn, O., 30
Metlesics, W., 8
Metzger, W., 69, 70 (7), 81 (7), 82, 86 (7)
Meyer, R. F., 315
Meyers, A. I., 215, 225 (97), 227, 228 (141), 229 (141), 230
Micheel, F., 251, 252 (162, 163, 164, 165), 273 (162, 163, 164, 165)
Michelman, J. S., 182
Midorikawa, H., 212
Mihailović, M. L., 14, 40
Mildner, P., 61
Millar, I. T., 19
Miller, A. H., 79
Miller, F. M., 55
Miller, R. C., 72
Miller, S. I., 107, 108 (38), 110 (38), 130 (38)
Mills, W. H., 43
Mine, S., 250
Mingoia, Q., 254
Mistretta, C. M., 243
Mistryukov, E. A., 222, 223 (119), 224, 225 (117), 226, 228, 229 (135, 136, 146)
Mizuguchi, J., 240
Mochel, W. E., 27
Möhlau, R., 95
Moffat, J. G., 274
Moffitt, W., 192
Moggi, A., 80
Moller, K., 198
Mollier, Y., 260
Mondelli, R., 58, 80, 83 (46)
Mondt, J., 59
Montefinale, G., 147
Monti, L., 249
Monti, S. A., 311, 312 (87)
Moore, D. R., 19
Moore, D. W., 54
Moore, R. E., 323
Moore, T. E., 103, 110 (23), 119 (23), 130 (23)
Morgan, J. F., 72
Morgan, K. J., 91, 92 (97)
Moriarty, R. M., 51
Morimura, S., 27
Morita, K., 8, 16 (22), 21
Morita, Y., 206
Morozov, E. N., 98

Morrey, D. F., 91, 92 (97)
Morrison, H., 61
Mory, R., 253
Moser, F. H., 22
Mosher, H. S., 215
Mowry, D. T., 249
Moyes, R. B., 158
Moyeux, M., 229
Moynahan, E. B., 110, 111 (50), 117 (50), 119 (50), 136 (50)
Müller, A., 52
Müller, E., 246, 257 (119)
Müller, H., 218
Müller, J., 108
Muha, G. M., 54
Mukharji, P. C., 224
Muller, G. J., 11
Mulliken, R. S., 156, 194
Mumm, O., 250
Munch, R. J., 242, 245 (78), 246 (78)
Murata, K., 250
Murata, N., 250
Murayama, K., 27
Murphy, P. J., 98
Murray, A., 174
Mustafa, A., 45, 254
Muth, C. W., 190, 211 (28b)
Mychaylyszya, V., 230

N

Nabih, I., 87
Nace, H. R., 275
Nagasawa, A., 12
Nagata, C., 323
Naik, N. N. 100, 124 (3)
Nakayima, M., 43
Nakamura, M., 230
Nakao, H., 115, 130 (70)
Nazarov, I. N., 214, 222 (95), 224, 225 (95), 227, 228, 229 (140), 230
Negoro, T., 190, 211 (28a)
Neikam, W. C., 323
Neiland, O., 217, 220 (100), 224 (100)
Neilands, L., 217, 218 (101, 108), 219 (103), 220 (103, 104a), 221 (102), 222 (102), 224 (101, 102, 103, 105), 229 (102), 230 (102, 108)

Neilson, D. G., 264
Neiman, M. B., 88
Neiss, E. S., 25
Nelson, A. A., 244
Nerdel, F., 33
Nesbit, M. R., 70
Nesmeyanov, A. N., 139
Neuse, E., 207
Neygenfind, H., 241, 267, 271(43, 254), 273(254), 276(43, 254)
Nicholaus, B. J. R., 33
Nicolaus, R. A., 86, 90, 95(67)
Nicolella, V., 38
Nielsen, A. T., 54
Nierenstein, M., 21
Nijveld, W. J., 11
Nilles, G. P., 25, 31(103), 37(103)
Nir, A., 178
Nishiwaki, T., 42, 56(191)
Nitta, Y., 241, 245(55), 247(55)
Nixon, A. C., 202
Noel, F., 72
Nogradi, I., 319
Noland, W. E., 290, 296(51), 299(51), 303(51), 310(51)
Nomura, Y., 293, 294(57), 296(57), 300(57), 309(57)
Norman, R. O. C., 84
Normant, H., 246
Novello, F. C., 242
Novotony, L., 37
Numerof, P., 139

O

Oae, S., 148, 190, 211(28a, 29b)
Occolowitz, J. L., 322
Ochiai, M., 8, 16(22)
O'Connor, G. L., 275
Oddo, B., 89, 254
Odell, A. L., 150
Oei, A. T. T., 151
Offenhauer, R. D., 72
Ogesawara, K., 312(88), 313
Ogg, R. A., 285
Ogino, K., 190, 211(28a)
Ohnishi, M., 182
Okamura, K., 37
Okaya, Y., 235

O'Keefe, J. H., 147
Okuda, C., 43
Okutsu, E., 190
Olejniczak, B., 211
Ollis, W. D., 9
Olofson, R. A., 182
Olsson, S., 176, 180(130)
Omote, Y., 298, 299(65)
Onishchenko, P. P., 37
Onoprienko, V. V., 231
Opfer, H., 238, 251(25), 262(25), 273(25)
Opliger, C. E., 107, 133(39)
Ordish, H. G., 43
Orelup, J. W., 240
Orezzi, P., 38
Oriente, G., 90
Osborne, A. G., 188, 190(11), 191(11)
Osborne, M. W., 229
Oser, B. L., 244
Ostman, B., 176, 180(130)
O'Sullivan, D. G., 238
Oswald, A. A., 72
Otsuji, Y., 45
Ouannes, C., 75
Owellen, R. J., 319

P

Pain, D. L., 35
Palmer, D. R., 57
Palmer, M. H., 296(66), 297(66), 298, 312(66), 313(66)
Pandler, W. A., 180, 182(141)
Pany, J., 141, 142(31)
Paquette, L. A., 59
Parameswaran, K. N., 114, 130(63)
Parcell, R. F., 224
Parham, W. E., 5
Parker, R. G., 323
Parmeggiani, G., 248
Parmentier, J. H., 146
Parnell, E. W., 222, 225(114)
Paul, A. G., 105, 106(32), 110(32), 111(32), 126(32), 130(32), 131(32), 133(32)
Paul, A. P., 200, 215
Pauling, L., 5
Pazdro, K. M., 54
Pearson, R. G., 247, 262(127)
Pecher, J., 322
Pectors, J., 322

Peeling, E. R. A., 152
Peets, E. A., 139
Pelcere, J., 217
Peller, M. L., 88, 93(72)
Penco, S., 38
Perkin, A. G., 21
Perrier, L., H., 261
Person, H., 229
Peterson, D. F., 174
Petrarca, A. E., 237
Pettit, G. R., 245, 246(111), 258(108), 264
Pfaender, P., 44
Pfitzner, K. E., 274
Pfleiderer, W., 35, 40, 55
Pfundt, G., 20
Phillips, G. O., 45
Piasek, E. J., 28
Pickles, V. A., 151, 160(72), 161(72), 164(72)
Pictet, A., 52
Pidacks, C., 290, 291(52)
Pieroni, A., 72, 80, 81, 82, 83(51), 84(51), 86(51), 95(51)
Piette, L. H., 285
Piloty, O., 90, 91
Pineau, R., 102, 106(13), 107(13), 109(13), 129(13)
Pinhey, J. T., 21
Pirelli, A., 38
Pirzada, N., 124
Pitkin, R. M., 240, 244(41)
Pitman, I. H., 235, 242(11), 253, 254(174)
Pittman, A. G., 188, 189(5, 6), 202(5, 6), 204(5, 6), 205(5, 6), 206(5)
Plancher, G., 90
Plieninger, H., 83, 319
Plötz, E., 72
Plummer, A. J., 139
Polaczkowa, W., 54
Polanyi, M., 154
Poletto, J. F., 322
Pomeranzew, Yu., 238
Ponci, R., 241, 248(52)
Pople, J. A., 278, 281, 282(21), 323
Popp, F. D., 100, 110, 111(50), 117(50), 119(50), 125, 136(50)
Porai-Koshits, E. A., 50, 54(234)
Porter, Q. N., 242, 288
Pouchnot, O., 71, 76(16), 88(16)
Powell, W., 293, 294(61), 295(61)

Powers, J. C., 284, 297(37), 301(37), 302(73), 303, 305(37), 306(37), 312(37)
Pozdeev, V. V., 139
Pratesi, P., 93, 95(101, 102)
Pratt, T. H., 140
Prelog, V., 40, 188, 189(10), 207
Price, J. M., 244
Probst, F., 35
Prochazka, M., 59
Profft, E., 239, 240(31), 245
Protiva, M., 230
Puma, B. M., 313
Pummerer, R., 54
Purrmann, R., 33, 40(136)
Purtshert, B., 254
Purves, C. B., 100, 101, 108, 119(11), 136(11)
Pyman, F. L., 100

Q

Quelet, R., 102, 105(17), 106(13), 107(13), 109(12, 13, 14, 16, 17), 110(17), 129(13, 14, 16, 17), 134(14)
Quest, B., 305, 306(78), 307(78), 315(78)
Quilico, A., 95
Quistad, G. B., 73, 97
Qureshi, A. A., 18, 33(72a)

R

Radda, G. K., 84
Raimondi, M., 192, 194(31a)
Ralhan, N. K., 230
Ramage, G. R., 7, 8(19)
Ramasseul, R., 88, 89(75), 97
Rambaud, R., 234, 235(2), 240(2)
Ramirez, F., 200, 215
Ramp, F. L., 5
Rampal, A. L., 224
Randall, E. W., 285
Randall, J. C., 182
Randall, W. W., 242
Ranjon, A., 71, 76(16), 78, 88(16)
Rao, K. B., 227, 228(141), 229(141)
Rapoport, H., 323
Rapoport, L., 9, 12(27), 17(27)
Rappe, C., 319
Rasmussen, C., 259

Rassat, A., 88, 89(75), 97
Ratts, K. W., 250
Ray, A. C., 68
Ray, J. D., 285
Reavlin, B. L., 242, 256(64)
Redlich, A., 95
Redvanly, C. S., 148
Rees, C. W., 29
Reese, C. B., 188, 190, 191(12), 194(12), 195(12)
Reeves, L. W., 322
Reich, F., 55
Reich, H. J., 323
Reichert, B., 211
Reid, E. E., 238, 245, 255(113), 257(113), 257(112), 262(24), 266(232), 273(24, 232)
Reid, K. F., 235, 261
Reid, S. T., 1, 20(1), 26, 73
Reilly, C. A., 287
Reine, A. H., 227, 228(141), 229(141)
Reinecke, H., 83
Reinecke, M. G., 282, 283(33), 284(33), 285(33), 298(33), 304
Reissert, A., 241
Rembarz, G., 243, 248(83), 249(83), 258(83)
Remers, W. A., 315, 322
Remsen, I., 234, 239(6, 7), 240(6, 7), 245, 256(7), 261, 269, 270(99, 257)
Renaud, R. N., 158, 171(101)
Replogle, L. L., 195, 196(41a), 198(41a)
Reynolds, W. A., 240, 244(41)
Rhoades, D. F., 46, 47
Ricci, J. S., 17
Rice, H. L., 245, 246(111)
Rice, L. M., 245, 255(113), 257(113), 258(112), 262, 266(232), 273(232)
Rice, W. Y., Jr., 39
Richards, R. E., 324
Richmond, H. D., 256
Richter, R., 20, 56(83), 63(83)
Rideal, E. K., 154
Ridgewell, B. J., 179
Ridley, H. F., 293, 294(58), 295(58), 301(58), 302(58), 303(58), 307(58), 308(58)
Ried, W., 310(84), 311, 316(84), 317(84)
Rigby, G. W., 207, 211
Rinkes, I. J., 92

Rio, G., 71, 76, 78, 85, 88(16), 89
Rist, N., 229, 230
Ritchie, E., 21
Rittenberg, D., 152
Robbins, K. E., 45
Roberts, J. D., 278, 279, 282(18), 323
Robertson, A. V., 309
Robertson, J. M., 39
Robertson, M. S., 58
Robinson, B., 52
Robinson, F. P., 319
Robinson, G. M., 319
Robinson, R., 100, 105(8), 107, 110(8), 129(8), 131(8), 189, 319
Robinson, T., 38, 39, 52
Robison, B. L., 188
Robison, M. M., 188, 189(9), 190, 207, 225
Roch, J., 51
Rochen, L. K., 4, 14(5)
Rodia, R. M., 93, 94
Roe, A. M., 247
Roedeg, A., 214
Roger, R., 264
Rogers, D., 123
Rokos, H., 55
Romain, C. R., 51
Romanet, R., 322
Rooney, J. J., 155, 160(95)
Rose, N. C., 242
Rosenblum, C., 141, 142(28), 144
Rosenfeld, D. D., 71
Rosenthal, I., 47, 61(216)
Rosevear, P. E., 49
Ross, S. T., 131(90, 99)
Roth, H. J., 28
Roth, R. H., 322
Rothemund, P., 36
Rotschy, A., 52
Rouaix, A., 229
Rowland, B. I., 271
Rowland, F. S., 138, 139
Roy, D. N., 103, 108(21), 110(21), 116(21), 126(21), 133(21)
Rozantsev, E. G., 88
Rubessa, F., 257
Rudenko, V. A., 214, 222(95), 224, 225(95), 227, 228, 229(140)
Rüdiger, W., 90
Rügheimer, L., 100
Rumyantsev, Yu. M., 139

Runge, F., 202
Runge, J., 250, 252(157)
Rush, J. E., 237
Rush, K. R., 290, 296(51), 299(51), 303(51), 310(51)
Russell, P. B., 33, 40(136)
Rydberg, J., 140
Rydon, H. L., 24

S

Sachs, F., 245, 254
Saenger, W., 235, 236(12), 237(12)
Saha, M., 258, 266(208)
Sainsbury, M., 101(49), 106, 107(35), 109(35), 110(35, 45), 117(48), 118(45), 119(49), 120(49), 130(35, 45, 49), 131(49), 134(49), 136(49, 76)
Saito, S., 37, 114, 130(67a), 230
Sakurai, A., 212
Salt, M. L., 139
Santacroce, C., 80, 86(44)
Sanz, M. C., 46, 51(209)
Sasse, W. H. F., 48, 59
Sasson, S., 47, 61(216)
Sates, V., 98
Sauer, M. C., 147
Saunders, B. C., 30
Savige, W. E., 60
Sawanishi, H., 44
Saxton, J. E., 293, 294(58), 311
Scarpati, R., 80, 81, 82(50), 86(44), 90(65)
Schaaf, E., 253, 254(174)
Schäfer, H., 198
Schaefer, J. P., 10
Schaur, R. J., 195, 196, 197(42), 198(42)
Schenck, G. O., 20
Schenk, H.-U., 36
Schenker, F., 124
Schiebel, H. M., 238, 239(20), 262(20), 264(20), 265(20), 271(20), 274(20)
Schlaf, H., 35
Schlittler, E., 108
Schlobach, H., 40
Schmid, H., 45, 51
Schmid, M., 253
Schmid, P., 45
Schmidt, C.-H., 31
Schmidt, G. M. J., 28

Schmidt, R. A., 124
Schmidt, R. R., 16
Schmitt, W., 33, 40(136)
Schneider, J. P., 216, 224(99)
Schneider, W. G., 278, 282(21)
Schneller, J., 230
Schön, P., 100
Schönberg, A., 11, 45
Schofield, K., 222, 225(115b)
Scholl, M., 76, 85, 89
Schrader, E., 266, 268(251), 270(251), 273(251)
Schreiner, H., 176, 180(128)
Schriesheim, A., 71
Schroeder, H. E., 206
Schröder, P. M. R., 253
Schubert, A. R., 139
Schuetz, R. D., 25, 31(103), 37(103)
Schulenberg, J. W., 315, 322
Schulenberg, W., 250, 264(159), 265
Schultz, L., 224
Schultze, M. O., 148
Schultz, L., 224
Schultze, M. O., 148
Schulze, P. E., 141, 142(22, 30), 145(30)
Schumann, W., 253, 254(175)
Schutte, E., 141, 142(29)
Schwartz, F. G., 141, 142(26), 148(26)
Schweizer, E. H., 26
Schwyzer, R., 33
Scott, B. D., 290, 292(50), 292(50), 294(50), 295(50), 297(50), 298(50), 299(50), 305(50), 309(50), 313(50)
Searle, N. E., 245
Sease, J. W., 51
Sebastian, J. F., 282, 283(33), 284(33), 285(33), 298(33)
Seebach, D., 69, 82, 83
Seide, S., 9
Seidel, F., 6
Senn, O. F., 240
Serjeant, E. P., 32
Sethi, M. L., 106, 110(34), 122(34), 126(34), 130(34), 133(34)
Shah, P. K. J., 114, 120(65), 129(65)
Shallenberger, R. S., 243
Shamasundar, K. T., 110, 130(47), 133(47), 134(47)
Shannon, J. S., 151, 158(62), 160(62), 161(62), 163(62), 164, 170(62)

Sharp, H., 137
Shatenshtein, A. I., 182
Shavel, J., 259
Sheinker, Yu, N., 238, 319
Shen, T. V., 200, 201(52), 229(52)
Sheppard, H., 139
Sheppard, N., 281
Shibuya, S., 103, 104(26), 110(26), 119(26), 120(26), 135(26), 136(26)
Shima, S., 250
Shimanskaya, M. V., 173
Shimazu, H., 8, 16(22)
Shimizu, F., 41
Shimokawa, S., 279, 285
Shindo, M., 241, 245(55), 247(55)
Shingu, H., 323
Shioyama, I., 250
Shoolery, J. N., 292
Shuekhgeimer, G. A., 224
Shull, E. R., 298, 299(64)
Shusherina, N. P., 224
Shvekhgeimer, G. A., 214, 222(95), 225 (95)
Shvetsov, N. I., 228
Siddappa, S., 278, 306(80), 307(80), 309, 323
Siedel, W., 87, 88(70)
Siegel, S., 159
Sigillo, G., 89
Sih, J. C., 103, 110(22), 116(22), 121(22), 130(22)
Silber, P., 68
Silins, E., 217
Silversmith, E. F., 47
Silverstein, R. M., 278, 279(19)
Silverton, J. V., 39
Sim, G. A., 39
Simanyi, L. H., 222, 225(115a)
Simmons, J. E., 59
Singer, G. M., 182, 273
Singh, R. P., 266
Singh, S., 227, 228(141), 229(141)
Sircar, J. C., 227, 228(141), 229(141)
Sirsi, M., 278
Skita, A., 19
Skorianetz, W., 18
Slack, R., 35
Slater, G., 198, 199(47)
Slavinskaya, V. A., 98
Smiley, R. A., 243

Smith, A. E., 44, 300(69), 302(69), 303, 305(69), 308(69), 309(69), 310(69), 311(69), 312(69), 313(69), 319
Smith, C. D., 265
Smith, C. L., 180, 182(140)
Smith, E., 273
Smith, E. B., 70, 71(12), 81(12)
Smith, E. M., 254
Smith, G. F., 41
Smith, G. G., 274
Smith, K. M., 38
Smith, P. V., 42, 54(193)
Smolin, E. M., 9, 12(27), 17(27)
Smolinski, S., 64
Snaith, R. W., 314, 316(92), 317(92), 318(92)
Sohma, J., 279, 285
Soine, T. O., 245, 246(106), 256(106), 257(106)
Sokolov, D. V., 226, 229(137)
Soldan, F., 253
Sollich, W. A., 137, 151, 152, 153, 154, 155, 158, 170(98), 175(68)
Sollich-Baumgartner, W. A., 151, 152, 153, 154, 155(82), 157(82), 158(80, 82), 159(82), 162(82), 164, 166(82), 171 (82), 175, 178(126)
Somerville, A. W., 198, 199(47)
Sommers, A. L., 202
Song, P. S., 323
Šorm, F., 37
Sorokina, N. P., 319
Southwick, P. L., 319
Späth, A., 253, 254(175)
Speziale, A. J., 250
Sperber, H., 4
Sprio, V., 84, 90
Squire, R. C., 158
Srinivasan, P. C., 254
Srivastava, R. C., 254
Stafford, W. H., 192
Stahmann, M. A., 35
Stanaback, R. J., 229, 230
Starkova, O. I., 98
Stasiak, S., 211
Steadman, T. R., 214, 225(92)
Stecher, P. G., 10, 11(29), 12(29), 15(29), 16(29), 17(29) 19(58), 22(29), 31(29), 39
Stedman, G., 158

Stefanovic, G., 14
Steinfeld, A. S., 103, 105(25), 106(25), 110(25), 116(25), 133(25)
Steinfeld, J., 244
Steinkopf, W., 51, 57
Stell, W. G., 199
Stepanova, R. N., 108
Stephen, E., 246, 250(117), 251(117), 252(166), 266(166), 267(166), 273(166)
Stephen, H., 246, 250(117), 251(117), 252(166), 266(166), 267(166), 273(166)
Sternbach, L. H., 8
Sternhell, S., 281, 285(30)
Stewart, G. W., 117, 125, 134(74)
Stjernström, N. E., 40
Stobbe, H., 209
Stock, J., 90
Stone, H., 246
Stoven, O., 114, 130(64), 131(64)
Striegler, C., 209
Streitweiser, A., 194
Strohmeier, W., 147
Stumpf, W., 4, 14(5)
Sturm, E., 266, 273(252)
Süs, O., 198
Sugahara, H., 114, 129(66), 130(66)
Sugasawa, S., 230
Sugiyama, N., 298, 299(65)
Sugimoto, N., 37
Sugimoto, T., 48
Suh, J. T., 313
Sullivan, W. R., 35
Sumpter, W. C., 55
Sundberg, R. J., 303, 315, 322(95)
Sunin, A. N., 88
Sutcliffe, L. H., 278
Suvorov, N. N., 319
Suzuki, J., 303
Suzumura, H., 202, 205(56, 57)
Swalen, J. D., 287
Swallow, A. J., 147
Swenson, W. J., 252
Sydykov, B. T., 226, 229(137)
Symon, T., 72
Szabo, L., 242
Szinai, S. S., 308(81), 309, 315(81)
Szmant, H. H., 237
Szmuszkovicz, J., 24
Szpilfogel, S., 188, 189(10), 207

T

Taddei, F., 182
Takasu, T., 241, 245(55), 247(55)
Takeuchi, Y., 293, 294(57), 296(57), 300(57), 309(57)
Takido, M., 105, 106(32), 110(32), 111(32), 126(32), 130(32), 131(32), 133(32)
Talbert, B. M., 147
Tale, I., 217, 220(104b)
Tamagaki, S., 190, 211(28a, 29b)
Tamura, Y., 26, 276
Tanaka, F., 190
Tanaka, H., 123
Tanaka, I., 61
Tanaka, S., 114, 130(60), 202
Tanaka, T., 37
Tanaseichuk, B. S., 98
Taure, L., 217
Taylor, E. C., 10, 43(33), 62, 201
Taylor, J. H., 174
Taylor, T. I., 153, 156(85)
Taylor, W. I., 225
Tchiroukine, E., 102, 106(13), 107(13), 109(13), 129(13)
Tebby, J. C., 19
Teeters, W. O., 16
Teitel, S., 113, 127(59a), 131(59a), 135(59a)
Teotino, U., 269
Terapane, J. F., 265
Terayama, Y., 112, 115(57), 127(57), 129(57), 131(57)
Tertov, B. A., 37
Thanaseichuk, B. S., 88
Theobald, C. W., 27
Theodoropulos, S., 50
Thesing, J., 27, 36, 52
Thiemann, A., 144, 147
Thien, T. V., 28
Thiers, M., 203, 211(68a)
Thomas, A. L., 22
Thomas, K., 42
Thomas, K. E., 12
Thompson, M. J., 56
Thompson, R. B., 72
Thompson, W. C., 188, 189(3, 4), 202(3)
Thomson, R. H., 30
Thorp, J. M., 150
Thorpe, J. E., 48

AUTHOR INDEX

Thorpe, R. F., 202
Tidd, B. K., 23
Tiers, G. V. D., 137
Till, A. R., 144
Todd, A., 50
Todesco, P. E., 182
Toffoli, F., 249, 270
Tomita, K., 47, 93
Topsom, R. D., 124
Tosik, B. K., 210
Trager, L., 142, 143(32)
Traverso, J. T., 251, 255(161), 260(160), 261(160), 262(160), 263(160), 266(161), 267(161), 268(160), 269(250), 273(250)
Traynelis, V. J., 190
Treibs, A., 90, 97
Treibs, W., 192, 195, 196(41b), 198(41b), 250, 252(157)
Tretter, J. R., 14
Tröger, J., 15
Troparesky, A., 247
Tsien, W. H., 139
Tsuchiya, T., 43, 59
Tsuge, O., 34
Tsuyimoto, N., 26
Tuck, B., 24
Turnbull, K., 147
Turovskii, I. V., 217, 220(100), 224(100)
Turton, C. N., 138

U

Uchimura, M., 26
Ueda, T., 151
Uff, B. C., 100, 124(3)
Uhlemann, E., 322
Uhlenbroek, J. H., 51
Ukhova, L. I., 227, 228, 229(140)
Ullyot, G. E., 116, 131(72)
Ulrich, H., 20, 56(83), 68(83), 273
Umar, S., 68
Umezawa, B., 112, 115(57), 121, 127(57, 58), 128(77, 78), 129(57, 58), 131(57)
Underwood, G. R., 288
Urban, E. J., 47
Uskoković, M., 111, 112, 113, 114(54), 122(59), 125, 127(56), 129(54), 131(54, 56, 59), 132(54, 56, 59), 133(54, 87)

V

Valnot, J.-Y., 108
Vanags, G., 32, 217, 218(101), 219(103), 220(104a), 224(101, 103, 105)
Varacca, V., 96
van Binst, G., 322
Van der Werf, C. A., 44
Van Remoortere, F., 323
van Tamelen, E. E., 258
van Thielen, J., 28
Varghese, A. J., 29, 46, 47
Vartanyan, S. A., 229
Veiss, A., 217, 220(104a, 104b)
Verbiscar, A. J., 139, 144(10)
Veremeenco, P., 81, 82, 83(51), 84(51), 86(51), 95(51)
Verly, W. G., 174
Verma, S. S., 254
Vernon, J. M., 290, 314, 316(92), 317(92), 318(92)
Vestling, M. M., 313
Veveris, A., 217, 221(102), 222(102), 224(102), 229(102), 230(102)
Vigier, A., 214, 224(93)
Vilek, J. O., 230
Vill, J. J., 214, 225(92)
Vinot, N., 102, 105(17), 106(13, 15), 107(13, 18), 108(18), 109(12, 13, 14, 15, 16, 17, 18), 110(17), 117(18), 127(18), 129(13, 14, 15, 16, 17, 18), 134(14, 15, 18)
Vinutha, A. R., 182
Viscontini, M., 40
Vlasova, S. L., 88, 98
Vogin, E. E., 244
Volter, J. J., 155
v. Eller, H., 54
v. Reibnitz, B., 18
von Wolff, E., 245

W

Wacker, A., 61, 143, 143(32)
Wacker, O., 105, 106, 111(31), 127(31), 129(31)
Wada, S., 248
Waigh, R. D., 111, 112(52), 113(52), 121(52), 127(52), 132(52), 134(52)

Wake, S., 45
Walker, A. J., 273
Walker, J., 35
Walker, D. M., 319
Walker, R. L., 205, 206(70), 207(70)
Wang, S. Y., 46, 47
Walls, F., 322
Walsh, D. A., 319
Warren, E., 242
Wasserman, H. H., 76, 79
Watanabe, H., 241, 248(56), 254(56), 269(135), 270(135)
Watson, J. G., 244
Webster, D. E., 152
Weidermann, P., 310(84), 311, 316(84), 317(84)
Weidler, A. M., 192
Weigert, F. J., 323
Weinblum, D., 61
Weise, A., 274
Weisgraber, K. H., 103, 105(25), 106(25), 110(25), 116(25), 133(25)
Weiss, M. J., 290, 291(52), 322
Weisse, G., 9
Wells, J. L., 201
Wells, R. J., 222, 225(115b)
Welstead, W. J., 93
Weniger, J., 266, 273(252)
Wenzel, M., 141, 142(22, 27, 30), 145
West, J. C., 176, 177(127)
Westerkamp, J. F., 5
Westermark, T., 147
Weyerstahl, P., 33
Whaley, W. M., 114, 124(61)
Whalley, W. B., 21
Whipple, E. B., 318
Whisman, M. L., 141, 142(26), 148(26)
Whistler, R. L., 38
White, A. M., 179
White, R. F. M., 278
Whitehead, C. W., 251, 255(161), 260(160), 261(160), 262(160), 263(160), 266(161), 267(161), 268(160), 269(250), 273(250)
Whittemore, I. M., 147
Wibaut, J. P., 94
Wiberg, K. B., 271
Widdowson, D. A., 123
Widmann, H., 227, 228(141), 229(141)
Wiechers, A., 39

Wienhöfer, E., 49
Wieser, H., 61
Wiggins, L. F., 31
Wildman, W. C., 113, 127(59a), 131(59a), 135(59a)
Wiles, L. A., 17
Wiley, P., 56
Wiley, R. H., 56
Willard, J. E., 147
Willard, P. W., 251, 260(160), 261(160), 262(160), 263(160), 268(160), 260(250), 273(250)
Wille, E., 55
Willett, J. D., 15
Williams, D. H., 137
Williams, J. R., 61
Williams, T., 113, 122(59), 124, 131(59), 132(59)
Williams, T. L., 98
Willis, J. B., 237, 238(16)
Wilshire, J. F. K., 282, 303(34), 304(34)
Wilson, D. L., 293, 294(61), 295(61)
Wilson, J. W., III., 116, 131(72)
Wilson, T., 75
Wilzbach, K. E., 138, 139, 145, 146, 147, 148
Winberg, H. E., 27
Winbury, M. M., 229
Winkelmann, E., 253, 254(175)
Winkler, F., 87, 88(70)
Winter, D. P., 103, 110(20), 118(20)
Wischin, R., 258
Witkop, B., 222, 225(116), 279, 285(23), 286(23), 290, 292(50), 293(50), 294(23, 50), 295(50), 297(23, 50), 298(50), 299(23, 50), 300(23, 56), 301(23, 56, 70), 302(23, 56), 303, 304(23), 305(50), 307(23), 308(23), 309(23, 50), 313(23, 50), 319(70)
Wohler, F., 12
Wolf, A. P., 139, 148
Wolff, I., 35
Wolfgang, R. L., 138, 139, 140
Wollenberg, H., 141, 142(29, 30), 145(30)
Woltermann, A., 49
Wong, D., 58
Woods, P. S., 174
Woodward, R. B., 2, 57(2a)
Worrall, D. E., 62
Wright, J. B., 255

Wu, H. C., 68
Wynberg, H., 5, 51
Wyss, U., 248

Y

Yagi, H., 114, 129(66), 130(66)
Yagudaev, M. R., 322
Yamabe, T., 12
Yamamoto, H., 5, 6(13), 8(13), 14(12), 23, 24, 33(98)
Yamamoto, K., 10, 22(31)
Yamamoto, S., 257, 264(198)
Yamamuro, T., 61
Yamanaka, T., 312(88), 313
Yamanashi, Y., 112, 121(58), 127(58), 128(77, 78), 129(58)
Yamazaki, T., 315, 322(95)
Yanagi, K., 34, 248
Yang, K., 140
Yanovskaya, L., A., 108
Yates, P., 19, 58, 61
Yavorsky, P. M., 178, 180(132)
Yhland, M., 73
Yip, R. W., 62
Yoewell, D. A., 106, 119, 136(36)
Yokogawa, H., 43
Yoneda, S., 49
Yonezawa, T., 323
Yorke, M., 30
Yoshida, N., 47, 93
Yoshida, O., 244
Yoshida, Z., 48
Yoshioka, T., 27, 82, 83(53)
Yoshioka, Y., 182
Yoshitake, A., 106, 119(37), 126(37), 136(37)
Yost, W. L., 200, 229(51)
Young, P. C., 100, 105(8), 107, 110(8), 129(8), 131(8)
Young, S., 253, 254(174)
Yui, F. B., 50, 57(234)
Yun, H. V., 313
Yunosov, S. Yu., 322
Yura, Y., 115, 130(70)

Z

Zahradnik, R., 189
Zanker, V., 45
Zealley, T. S., 9
Zechmeister, L., 51
Zelck, U., 243, 248(83), 249(83), 258(83)
Zenda, H., 44
Ziegler, E., 213
Ziegler, H., 24
Ziegler, K., 68, 253, 254(175)
Ziffer, H., 61
Zigeuner, G., 24
Zincke, T., 218
Zinner, H., 243, 248(83), 249(83), 258(83)
Zinnes, H., 259
Zollinger, H., 253
Zoltewicz, J. A., 180, 182(140)
Zuleski, F. R., 259

Cumulative Index of Titles

A

Acetylenecarboxylic acids, reactions with heterocyclic compounds, **1**, 125
t-Amino effect, **14**, 211
Aminochromes, **5**, 205
Anthranils, **8**, 277
Aromatic quinolizines, **5**, 291
Aza analogs, of pyrimidine and purine bases, **1**, 189
Azines, reactivity with nucleophiles, **4**, 145
Azines, theoretical studies of, physicochemical properties and reactivity of, **5**, 69
Azinoazines, reactivity with nucleophiles, **4**, 145
1-Azirines, synthesis and reactions of, **13**, 45

B

Benzisothiazoles, **14**, 43
Benzisoxazoles, **8**, 277
Benzoazines, reactivity with nucleophiles, **4**, 145
Benzofuroxans, **10**, 1
Benzo[*b*]thiophene chemistry, recent advances in, **11**, 178
Benzo[*c*]thiophenes, **14**, 331

C

Carbenes, reaction with heterocyclic compounds, **3**, 57
Carbolines, **3**, 79
Chemistry
  of benzo[*b*]thiophenes, **11**, 178
  of diazepines, **8**, 21
  of furans, **7**, 377
  of lactim ethers, **12**, 185
  of mononuclear isothiazoles, **14**, 1
  of 4-oxy- and 4-keto-1,2,3,4-tetrahydroisoquinolines, **15**, 99
  of phenanthridines, **13**, 315
  of phenothiazines, **9**, 321
  of 1-pyridines, **15**, 197
  of 1,3,4-thiadiazoles, **9**, 165
  of thiophenes, **1**, 1
Claisen rearrangements, in nitrogen heterocyclic systems, **8**, 143
Complex metal hydrides, reduction of nitrogen heterocycles with, **6**, 45
Covalent hydration, in heteroaromatic compounds, **4**, 1, 43
Cyclic enamines and imines, **6**, 147
Cyclic hydroxamic acids, **10**, 199
Cyclic peroxides, **8**, 165

D

Development of the chemistry of furans, 1952–1963, **7**, 377
2,4-Dialkoxypyrimidines, Hilbert-Johnson reaction of, **8**, 115
Diazepines, chemistry of, **8**, 21
Diazomethane, reactions with heterocyclic compounds, **2**, 245
1,2-Dihydroisoquinolines, **14**, 279
Diquinolylmethane, and its analogs, **7**, 153
1,2- and 1,3-Dithiolium ions, **7**, 39

E

Electronic aspects of purine tautomerism, **13**, 77
Electronic structure of heterocyclic sulfur compounds, **5**, 1
Electrophilic substitutions of five-membered rings, **13**, 235

F

Ferrocenes, heterocyclic, **13**, 1
Five-membered rings, electrophilic substitutions of, **13**, 235
Furan chemistry, development of the chemistry of (1952–1963) **7**, 377

H

Halogenation of heterocyclic compounds, **7**, 1

Hammett equation, applications to heterocyclic compounds, **3**, 209
Hetarynes, **4**, 121
Heteroaromatic compounds, free-radical substitutions of, **2**, 131
   nitrogen, covalent hydration in, **4**, 1, 43
   prototropic tautomerism of, **1**, 311, 339; **2**, 1, 27
Heteroaromatic substitution, nucleophilic, **3**, 285
Heterocycles
   photochemistry of, **11**, 1
   by ring closure of ortho-substituted t-anilines, **14**, 211
Heterocyclic chemistry
   application of NMR spectroscopy to, **15**, 277
   literature of, **7**, 225
Heterocyclic compounds
   application of Hammett equation to, **3**, 209
   halogenation of, **7**, 1
   isotopic hydrogen labeling of, **15**, 137
   mass spectroscopy of, **7**, 301
   quaternization of, **3**, 1
   reactions of, with carbenes, **3**, 57
   reaction of acetylenecarboxylic acids with, **1**, 125
   reactions of diazomethane with, **2**, 245
   sulfur, electronic structure of, **5**, 1
N-Heterocyclic compounds, electrolysis of, **12**, 213
Heterocyclic diazo compounds, **8**, 1
Heterocyclic ferrocenes, **13**, 1
Heterocyclic oligomers, **15**, 1
Heterocyclic pseudo bases, **1**, 167
Heterocyclic syntheses, from nitrilium salts under acidic conditions, **6**, 95
Hilbert–Johnson reaction of 2,4-dialkoxypyrimidines, **8**, 115
Hydroxamic acids, cyclic, **10**, 199

I

Imidazole chemistry, advances in, **12**, 103
Indole Grignard reagents, **10**, 43
Indoles
   acid-catalyzed polymerization **2**, 287
   and derivatives, application of NMR spectroscopy to, **15**, 277
Indoxazenes, **8**, 277
Isoindoles, **10**, 113
Isothiazoles, **4**, 107
Isotopic labeling of heterocyclic compounds, one-step methods, **15**, 137
Isoxazole chemistry, recent developments in, **2**, 365

L

Lactim ethers, chemistry of, **12**, 185
Literature of heterocyclic chemistry, **7**, 225

M

Mass spectrometry of heterocyclic compounds, **7**, 301
Metal catalysts, action on pyridines, **2**, 179
Monoazaindoles, **9**, 27
Mononuclear thiazoles, recent advances in chemistry of, **14**, 1
Monocyclic pyrroles, oxidation of, **15**, 67
Monocyclic sulfur-containing pyrones, **8**, 219

N

Naphthyridines, **11**, 124
Nitrilium salts, heterocyclic syntheses involving, **6**, 95
Nitrogen heterocycles, reduction of, with complex metal hydrides, **6**, 45
Nitrogen heterocyclic systems, Claisen rearrangements in, **8**, 143
Nuclear magnetic resonance spectroscopy, applications to heterocyclic chemistry, **15**, 277
Nucleophiles, reactivity of azine derivatives with, **4**, 145
Nucleophilic heteroaromatic substitution, **3**, 285

O

1,3,4-Oxadiazole chemistry, recent advances in, **7**, 183
1,3-Oxazine derivatives, **2**, 311

Oxazolone chemistry, recent advances in, **4**, 75
3-Oxo-2,3-dihydrobenz[*d*]isothiazole-1,1-dioxide (saccharin) and derivatives, **15**, 233
4-Oxy- and 4-keto-1,2,3,4-tetrahydroisoquinolines, chemistry of, **15**, 99

P

Pentazoles, **3**, 373
Peroxides, cyclic, **8**, 165
Phenanthridine chemistry, recent developments in, **13**, 315
Phenothiazines, chemistry of, **9**, 321
Phenoxazines, **8**, 83
Photochemistry of heterocycles, **11**, 1
Physicochemical aspects of purines, **6**, 1
Physicochemical properties
    of azines, **5**, 69
    of pyrroles, **11**, 383
3-Piperideines, **12**, 43
Prototropic tautomerism of heteroaromatic compounds, **1**, 311, 339; **2**, 1, 27
Pseudo bases, heterocyclic, **1**, 167
Purine bases, aza analogs, **1**, 189
Purines
    physicochemical aspects of, **6**, 1
    tautomerism, electronic aspects of, **13**, 77
Pyrazine chemistry, recent advances in, **14**, 99
Pyrazole chemistry, progress in, **6**, 347
Pyridazines, **9**, 211
Pyridine(s)
    action of metal catalysts on, **2**, 179
    effect of substituents in, **6**, 229
1-Pyrindines, chemistry of, **15**, 197
Pyridopyrimidines, **10**, 149
Pyrimidine bases, aza analogs of, **1**, 189
Pyrones, monocyclic sulfur-containing, **8**, 219
Pyrroles
    acid-catalyzed polymerization, of, **2**, 287
    physicochemical properties of, **11**, 383
Pyrrolizidine chemistry, **5**, 315

Pyrrolopyridines, **9**, 27
Pyrylium salts, preparations, **10**, 241

Q

Quaternization of heterocyclic compounds, **3**, 1
Quinazolines, **1**, 253
Quinolizines, aromatic, **5**, 291
Quinoxaline chemistry, recent advances in, **2**, 203
Quinuclidine chemistry, **11**, 473

R

Reduction of nitrogen heterocycles with complex metal hydrides, **6**, 45
Reissert compounds, **9**, 1
Ring closure of ortho-substituted *t*-anilines, for heterocycles, **14**, 211

S

Saccharin and derivatives, **15**, 233
Selenazole chemistry, present state of, **2**, 343
Selenophene chemistry, advances in, **12**, 1
Synthesis and reactions of 1-azirines, **13**, 45

T

Tautomerism, prototropic, of heteroaromatic compounds, **1**, 311, 339; **2**, 1, 27
1,2,3,6-Tetrahydropyridines, **12**, 43
Theoretical studies of physicochemical properties and reactivity of azines **5**, 69
1,2,4-Thiadiazoles, **5**, 119
1,2,5-Thiadiazoles, **9**, 107
1,3,4-Thiadiazoles, chemistry of, **9**, 165
1,2,3,4-Thiatriazoles, **3**, 263
Thiophenes, chemistry of, recent advances in, **1**, 1
Three-membered rings, with two hetero atoms, **2**, 83
1,6,6aS^{IV}-Trithiapentalenes, **13**, 161

DATE DUE

DEMCO 38-297